Handbook of Mobile Systems Applications and Services

Handbook of Mobile Systems Applications and Services

Edited by

Anup Kumar and Bin Xie

CRC Press is an imprint of the
Taylor & Francis Group, an **informa** business
AN AUERBACH BOOK

CRC Press
Taylor & Francis Group
6000 Broken Sound Parkway NW, Suite 300
Boca Raton, FL 33487-2742

Printed in the United States of America on acid-free paper
Version Date: 20120320

International Standard Book Number: 978-1-4398-0152-9 (Hardback)

Visit the Taylor & Francis Web site at
http://www.taylorandfrancis.com

and the CRC Press Web site at
http://www.crcpress.com

To my parents, wife, and twin daughters

Anup Kumar

To my wife Pei Zhang and my boy Derock and baby girl Daisy

Bin Xie

Contents

Preface

The service-oriented mobile computing paradigm delivers integrated functionalities to support the availability of a wide variety of applications in the mobile environment. The technical advances of mobile services over the last few years have successfully fostered a variety of new mobile applications that are available on an as-needed basis. It is expected that these applications will offer more exciting and efficient services for users anytime and anywhere, across heterogeneous access and networking technologies. The success of the next generation of mobile communication will depend highly on the seamless mobility of these services and applications. These mobile services have shown a great commercial potential; for example, the mobile services for entertainment such as mobile TV, music, and games. The mobile market is expected to be valued at about $64 billion by 2012 according to a recent market prediction.

Due to diversified mobile environments, the design issues of mobile services are extremely complicated and a number of technical challenges need to be explored, such as mobile system protocols, distributed algorithms, and computational methods for service-oriented computing (SOC). The challenges also include the design, implementation, usage, and evaluation of mobile computing and wireless systems, applications, and services. For example, it is critical to define the functionalities and behaviors of mobile services such that they can be effectively described, advertised, discovered, and composed by others. Therefore, a well-designed service platform allows innovative services to be created, deployed, and managed, addressing the needs of both the customer and the provider. In addition, semantic technologies help to structure contextual knowledge about the user environment. This book is designed to provide a complete understanding of these technologies, and investigates the evolution of these technologies from the basic concepts to implementation protocols, from the fundamentals to practical applications. It stands as a comprehensive reference for students, instructors, researchers, engineers, and other professionals, building their understanding of mobile service computing.

The book addresses various fundamental design issues for mobile services. The book is organized in three major sections. The first section of the book provides background and building blocks for mobile services architecture, the second section gives details of middleware support for mobile services, and the third section discusses security and applications of mobile services.

The first section includes Chapters 1 through 6. Chapter 1 illustrates the service-oriented architecture (SOA) and also discusses why SOA/SOC is important for the development of mobile services today and in the future. It discusses SOA-based architectural styles and protocols, such as SOAP/REST Web services, interaction models, and service management, in mobile services. Chapter 2 covers current service discovery frameworks developed in industry and research communities and provides a way to classify them based on the taxonomy presented. The strengths and weaknesses of different approaches are discussed. The second part of the chapter focuses on considerations and challenges that are specific to service discovery for mobile computing. Chapter 3 covers the basics of mobile services and applications that are deployed in various contexts and surroundings. This chapter illustrates how these services can be context aware, having the contextually relevant information of the users. The main focuses are service adaptation, context management, and context evaluation. Chapter 4 presents an Event-based Location Aware Query (ELAQ) model that continuously aggregates data in specific areas around mobile sensors of interest to provide mobile data services with location awareness and event detection for the users. Chapter 5 discusses the complex network environment of mobile collaboration, identifies the challenging design issues, and explores a variety of service collaboration protocols and architectures in mobile systems. Chapter 6 provides an insightful study on mobile agent technology and its importance in the design of mobile services.

The second section of the book includes Chapters 7 through 12. Chapter 7 discusses the middleware for mobile and pervasive services. It provides core classification approaches and future trends in this area. Chapter 8 presents a layered architecture of context-aware middleware and describes the functional components of each individual layer as system services, followed by a survey of recent middleware systems and their classification according to a taxonomy derived from the layered architectural model. The chapter also discusses future research challenges in context-aware middleware technologies. Chapter 9 discusses architecture for supporting multimode terminals in integrated heterogeneous wireless networks. It provides an overview of virtual wireless services to terminals with multiple interfaces. Chapter 10 discusses quality-of-service (QoS) protocols and their enhancements in supporting user mobility. The unique issues for enabling QoS in the mobile wireless network are investigated from the mobility protocols to advanced QoS adaptation schemes. The chapter further discusses the QoS-aware middleware that implements the QoS standards in such a way that the QoS reservation and adaptation are effectively hidden from the applications, which facilitates the implementation of QoS-enabled applications in a mobile environment. Chapter 11 provides an architecture called Web service coordination management middleware. It is a simple but powerful enhancement that enables Web services to locally manage their workflow dependencies and to handle messages resulting from multiple workflows.

The third section of the book includes Chapters 13 through 16. Chapter 13 presents and analyzes the security vulnerabilities to which mobile services are subject. It further provides taxonomy for attacks against mobile services and outlines the solutions that have been proposed for mitigating them. It also introduces best practices for security design that mobile service developers can follow to improve security of the mobile systems. Chapter 14 addresses the problem of location-based access control in the context of privacy protection. It discusses the impact of mobility on two current privacy laws—the E.U. Data Privacy Directive and the Health Insurance Portability and Accountability Act in the United States. It also proposes a new location-based access control model GEO-PRIVACY. Chapter 15 addresses the network availability constraint (*insufficient radio resources*) to serve all the mobile services originating from a single-user terminal. The key thrust behind this chapter is that a mobile user terminal may obtain a service by attaching itself to multiple attachment points, that is, base station/access points. The main goal of Chapter 16 is to discuss the development of assistive technology solutions that will facilitate day-to-day activities for people who are blind or visually impaired.

Editors

Dr. Anup Kumar completed his PhD from NCSU and is currently a professor of CECS Department at the University of Louisville and the director of Mobile Information Network and Distributed Systems (MINDS) Lab. His research interests include wireless and mobile systems, routing in ad hoc and sensor networks, distributed algorithm implementation, QoS-based Web services, seamless mobile computing environment, and wireless content delivery over the Internet. He has been PI and Co-PI of several federal grants funded by National Science Foundation, Department of Treasury, Kentucky Science Foundation, Department of Hometown Security, etc. He is the associate editor of *IEEE Transactions on Services Computing*. He is also the associate editor of the *Internal Journal of Web Services Research and International Society of Computers* and their *Application Journal*. He was a member of IEEE Distinguished Visitor Program (2006–2008). In GLOBECOM-2010, he had organized the Cloud Computing Forum and has given tutorials on Cloud Computing at CyberC-2010, ICCNT-2010, and SCC-2009. He is currently serving on the organizing committees of many international conferences. He was the Chair of IEEE Computer Society Technical Committee on Simulation (TCSIM) (2004–2007). He has published and presented over 175 papers. Some of his papers have appeared in *ACM Multimedia Systems Journal*, several *IEEE Transactions, Wireless Communication and Mobile Computing, Journal of Parallel and Distributed Computing, IEEE Journal on Selected Areas in Communications*, etc. He was the associate editor of the *International Journal of Engineering Design and Automation* (1995–1998). He has served on many conference programs and organizing committees such as CyberC-2009, MASS-2008, SCC-2008, ICWS-2008, IEEE ISCC 2007, IEEE ICSW-2006, IEEE MASS-2005, IEEE SCC-2005, IEEE ICWS-2005, CIT-2005, IEEE MASCOTS, and ADCOM '97 and '98. He has also edited special issues in *IEEE Internet Magazine* and *International Journal on Computers and Operations Research*. He is a senior member of IEEE.

Dr. Bin Xie received his MSc and PhD (with honors) in computer science and computer engineering from the University of Louisville, Kentucky, in 2003 and 2006, respectively. He was a post doc at the University of Cincinnati from 2006 to 2008. He thereafter was a visiting scholar at the NEC C&C Innovation Research Laboratories and Carnegie Mellon University CyLab Japan. Dr. Xie is the founder and currently the president of InfoBeyond Technology LLC. InfoBeyond researches, simulates, prototypes, develops, and delivers useful networking products for wireless mobile communications and information processes. The research works at InfoBeyond have been broadly supported by the army, navy, air force, and the State of Kentucky. Dr. Xie is the author of *Handbook/Encyclopedia of Ad Hoc and Ubiquitous Computing* (World Scientific: ISBN-10: 981283348X) and *Heterogeneous Wireless Networks—Networking Protocol to Security* (VDM Publishing House: ISBN: 3836419270). He has published over 60 papers in the IEEE conferences and journals. Some of his papers have appeared in the most-cited journals in the areas of telecommunication and computer architecture. His research interests focus on mobile computing and applications, such as wireless sensor network, ad hoc networks, mesh networks, 4G, and network security. In these areas, he has carried out substantial research on the fundamental issues of network deployment, network coverage, network connectivity, performance evaluation, and Internet/wireless infrastructure security. His recent research encompasses image process supported by the Navy Research Laboratory. He has successfully established many mathematical models (graph theory, information/code theory, game theory, linear programming, and optimization) for communication networks and their security. He is an editor member of the *Journal of International Journal of Information Technology, Communications and Convergence* (*IJITCC*), and has edited a special issue on the cyber-enabled information discovery. He has recently edited a special issue on clustering and clouding computing for *Elsevier Future Generation Computer Systems* (*FGCS*). Dr. Xie has served as the program chair or TPC member for over 30 conferences and workshops. He is a senior member of IEEE.

Contributors

Sanjuli Agarwal
InfoBeyond Technology, LLC
Louisville, Kentucky

Berthold Agreiter
Institute of Computer Science
University of Innsbruck
Innsbruck, Austria

M. Iyad Alkhayat
Department of Computer Engineering and Computer Science
University of Louisville
Louisville, Kentucky

Antoine B. Bagula
Department of Computer Science
University of Cape Town
Cape Town, South Africa

Janaka Balasooriya
School of Computing, Informatics, and Decision Systems Engineering
Arizona State University
Tempe, Arizona

Mieso K. Denko
Department of Computing and Information Science
University of Guelph
Guelph, Canada

Schahram Dustdar
Distributed Systems Group
Vienna University of Technology
Vienna, Austria

Matthias Farwick
Institute of Computer Science
University of Innsbruck
Innsbruck, Austria

Abraham George
Kyocera Wireless India Ltd.
Bangalore, India

Ratan K. Ghosh
Department of Computer Science and Engineering
Indian Institute of Technology
Kanpur, India

Bing He
Department of Computer Science
University of Cincinnati
Cincinnati, Ohio

Liang Hong
School of Computer Science
Wuhan University
Wuhan, People's Republic of China

Patrick C. K. Hung
Business and Information Technology
University of Ontario Institute of Technology
Oshawa, Canada

Basel Katt
Institute of Computer Science
University of Innsbruck
Innsbruck, Austria

Robert Kelley
Department of Computer Engineering and Computer Science
University of Louisville
Louisville, Kentucky

Anup Kumar
Department of Computer Engineering and Computer Science
University of Louisville
Louisville, Kentucky

Jingli Li
TopWorx, Inc. and InfoBeyond Technology, LLC
Louisville, Kentucky

Shamkant B. Navathe
College of Computing
Georgia Institute of Technology
Atlanta, Georgia

Mohammad Oliya
Department of Computer Science
National University of Singapore
Singapore

Anala Aniruddha Pandit
Department of Computer Engineering and Computer Science
University of Louisville
Louisville, Kentucky

Sushil K. Prasad
Department of Computer Science
Georgia State University
Atlanta, Georgia

Hung Keng Pung
Department of Computer Science
National University of Singapore
Singapore

A. S. Shaik
Department of Electrical and Computer Engineering
The University of Memphis
Memphis, Tennessee

Sang H. Son
Department of Computer Science
University of Virginia
Charlottesville, Virginia

Thomas Trojer
Institute of Computer Science
University of Innsbruck
Innsbruck, Austria

Hong-Linh Truong
Distributed Systems Group
Vienna University of Technology
Vienna, Austria

Yafeng Wu
Department of Computer Science
University of Virginia
Charlottesville, Virginia

Bin Xie
InfoBeyond Technology, LLC
Louisville, Kentucky

Wenwei Xue
Department of Computer Science
National University of Singapore
Singapore
and
Nokia Research Center
Beijing, People's Republic of China

M. Yeasin
Department of Electrical and Computer Engineering
University of Memphis
Memphis, Tennessee

Yingbing Yu
Department of Computer Science and Information Technology
Austin Peay State University
Clarksville, Tennessee

M. Zennaro
Department of Mathematics
Università degli Studi di Trieste
Trieste, Italy

Jian Zhu
Department of Computer Science
National University of Singapore
Singapore

I
BUILDING BLOCKS FOR MOBILE SERVICES ARCHITECTURE

Chapter 1

Service-Oriented Architecture for Mobile Services

Hong-Linh Truong and Schahram Dustdar

Contents

1.1 Introduction

Over the last few years, Service-Oriented Computing (SOC) and Service-Oriented Architecture (SOA) have demonstrated their capabilities in tackling integration and interoperability challenges raised by complex, heterogeneous data, services, and applications spanning different organizations. SOC is a computing paradigm in which services are considered as the fundamental elements for building distributed applications [1]. In SOC, SOA plays a major role to define how service-based systems are designed [1]. Various standards and technologies have been developed to support SOC/SOA concepts, such as SOAP (Simple Object Access Protocol [http://www.w3.org/TR/soap12-part1/]), WSDL (Web Services Description Language [http://www.w3.org/TR/wsdl]), and BPEL (Business Process Execution Language [http://www.oasis-open.org/committees/wsbpel/]). Among enabling technologies for SOA, Web services technologies, mostly based on XML, HTTP, WSDL, and SOAP, are the most widely implemented choice for SOA-based solutions. Web services technologies have been widely applied in small- and large-scale distributed systems built atop different platforms, such as high-end machines, mid-range servers, and workstations, for real-world applications.

As SOC/SOA offers many advantages for integrating heterogeneous data, services, and applications, SOC/SOA is a powerful enabling model for the development of mobile services. Originally, mobile services were considered as part of telecommunication service providers but today mobile services have evolved dramatically in many aspects [2]. The term "mobile services" is typically used to indicate services for mobile users. Although mobile services are usually considered to provide Web content to mobile users from anywhere, at anytime, and with any device, as indicated by Häyrynen in *Enabling Technologies for Mobile Services: The MobiLife Book* [3], the concept of mobility is not limited to the access of Web content and services from mobile devices. In our study, mobile services offer functionality for mobile applications to *access and provide* distributed data, content, and services. Used by mobile users, mobile applications typically rely on mobile devices and mobile networks, which can be dedicated (e.g., Internet and telecommunication network) or nondedicated (e.g., a mobile *ad hoc* network—MANET) established in specific situations. Thus, it is clear that different levels of movements of users, devices and services exist, and these movements must be taken into account and supported by mobile services [4]. To date, mobile services for mobile applications hosted in dedicated infrastructures of telecommunication and Internet service providers have been widely and well developed. However, with the recent development and deployment of powerful mobile devices (e.g., personal digital assistants (PDAs) and smart phones) and mobile network infrastructures and techniques (e.g., (free) abundant WiFi hotspots and MANET), mobile services are increasingly developed and deployed on mobile devices to offer on-demand or personal services to users.

In this chapter, we study state-of-the art SOC/SOA for mobile services. We consider software services, not business services, and concentrate on mobile services based on Web services technologies* and on mobile devices owing to several reasons. First, various mobile applications on mobile devices have been developed based on Web services technologies because these technologies enable applications on mobile devices to access data, information, and services via standard protocols (e.g., see many StrikeIron Web services at http://www.strikeiron.com/StrikeIronServices.aspx). Although Web services technologies are largely used for developing mobile applications at the client side, supporting Web services for mobile devices on the client side is not as strong as that in other platforms. Furthermore, recently, the ways of developing and using mobile applications have been changed dramatically to meet the requirements of user participation† and mass customization.‡ Therefore, it makes sense to study existing solutions to distill specific versus generic, and inflexible versus reusable, protocols and models for mobile Web services.

Second, many research efforts have focused on providing programming tools and software engineering methodologies for developing applications for mobile devices (e.g., PDAs, smart phones, subnotebooks, and laptops). In most cases, existing tools

* Hence, we use the term "mobile Web services" to refer to mobile services implemented with Web services technologies.

† http://en.wikipedia.org/wiki/Web_2.0

‡ http://en.wikipedia.org/wiki/Mass_customization

and methodologies aim at supporting the development of mobile applications acting as a client to access data and services from dedicated, high-end systems. This is partially due to constrained resources of mobile devices on which mobile applications function, and the usage mode in which mobile devices tend to access services rather than to provide them. However, many mobile applications and scenarios have shown the need to have Web services on mobile devices [5–7]. The lack of studies of how we can reuse/integrate current mobile applications for/into existing, diverse Web services available in today's pervasive environments has also shaped our focus in this study.

Finally, mobile services that are not based on Web services or SOA-based technologies on telecommunication or Internet service providers are well developed and discussed in the literature. Some parts of mobile services systems are built on the basis of typical, dedicated distributed systems (e.g., the service side of mobile services) that are addressed well by utilizing existing SOA techniques. Therefore, we will not concentrate on nonmobile Web services and on the service side of mobile services developed in conventional distributed systems.

The rest of this chapter is organized as follows: Section 1.2 discusses why SOC/SOA for mobile services is important. Section 1.3 discusses current SOA techniques for mobile services, with a focus on fundamental architectural styles and protocols. We present mobile Web services programming supports in Section 1.4. Then, we discuss a real-world application in Section 1.5. Different SOA techniques are discussed in Section 1.6. Section 1.7 outlines research challenges. We conclude the chapter in Section 1.8. This chapter also suggests a further reading list in Section 1.9 and finally, a list of possible exercises.

1.2 Why SOC/SOA for Mobile Services?

According to the GSMA (GSM Association [http://www.gsmworld.com/]), at the time of writing the mobile world has reached 4 billion connections and by 2013 it is forecasted to have 6 billion connections [8]. With such a large number of mobile devices (and mobile users), obviously there is a strong need for accessing and sharing data, information, content, and services by using mobile devices.

Before explaining the role of SOC/SOA for mobile services, it makes sense to discuss typical mobile services. Various authors have presented different classifications of mobile services. For example, mobile services have been classified according to type of consumptions, context, social setting, and relationship between consumer and service providers in Reference [9]. As SOC/SOA is a computing model used to address the integration and interoperability challenges, it would be better to classify mobile services in a way that highlights the main benefit of SOC/SOA. Within this view, we divide mobile services into two main classes: *individual-oriented* and *group-oriented* mobile services. The first class, depicted in Figure 1.1, indicates mobile

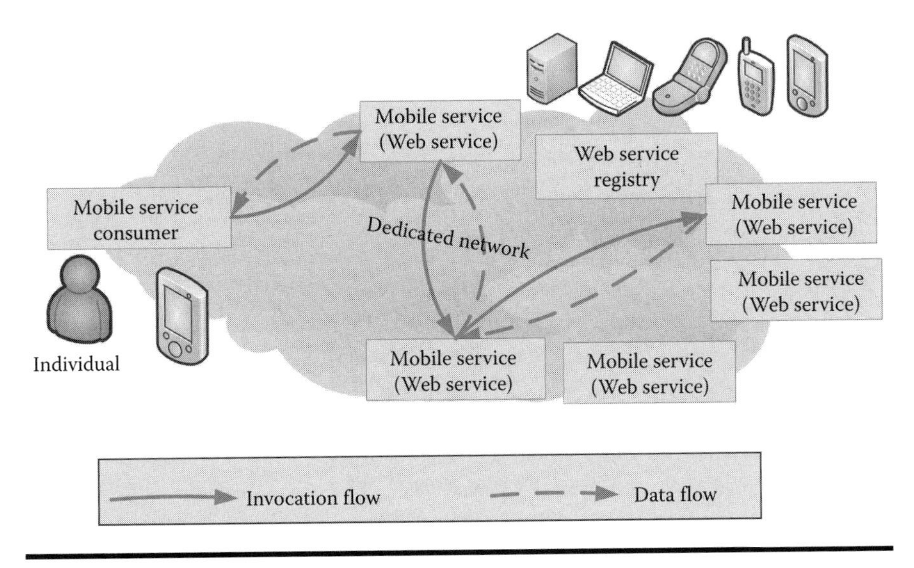

Figure 1.1 Individual-oriented mobile services.

services designed and developed for individuals* who (occasionally) utilize the services for different purposes and have few interactions with others in close time and/or space dimensions. Here, individual-oriented mobile services are tailored for individual use. In the second class, mobile services are designed for a group of people working together in close time and/or space dimensions, shown in Figure 1.2. In this case, there are a high number of interactions among people through mobile services in a close/distributed space in a short/long period of time. This kind of mobile service might be deployed on a nondedicated infrastructure. Although this simple classification provides nonorthogonal classes, it helps us to identify some major issues that SOC/SOA can help. Table 1.1 presents some typical properties of the two classes.

For the first class, we can utilize SOC/SOA to achieve the interoperability and integration of different services to provide added-value services for individuals. Let us consider diverse and rich sets of available services in the market for individual use. SOC/SOA will be useful for developing and integrating mobile applications utilizing these services due to various factors. First, there are new business models for individual-oriented mobile services which are targeted to normal people. Such models can rely on a vast source of data, contents, and services provided by different vendors through mobile networks or the Internet, to create converged services. On the one hand in today's market, data, information, and service providers want to maximize the number of their customers. Therefore, they provide their services with well-defined interfaces based on SOC/SOA models to simplify the access of

* We do not mean that the service is not by a single person but rather that users' usages are separated, sometimes unrelated, from each other.

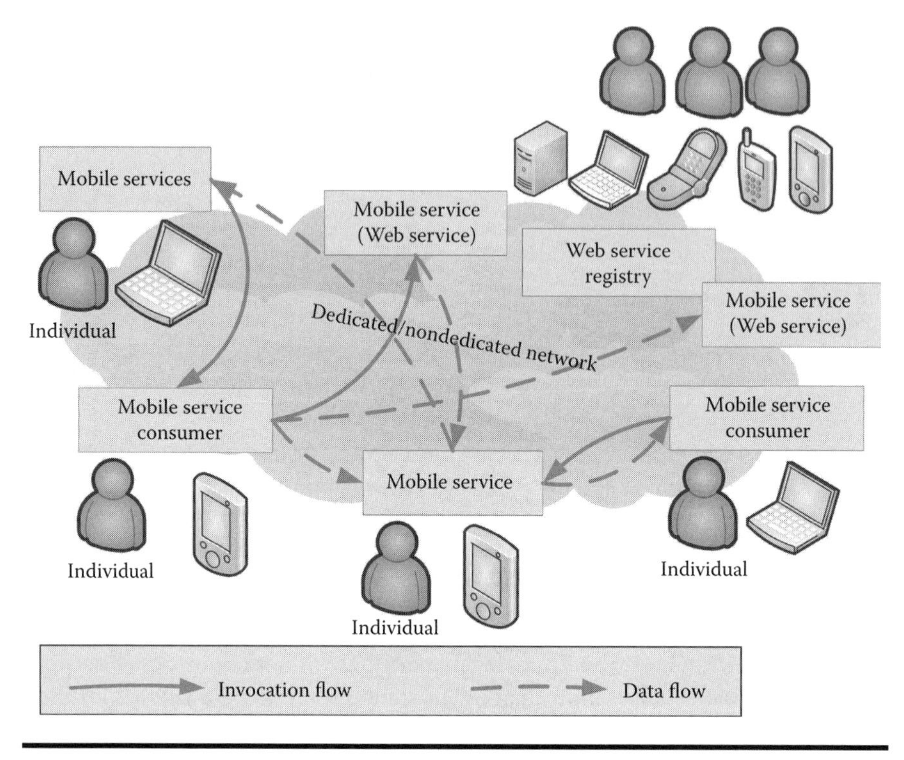

Figure 1.2 Group-oriented mobile services.

Table 1.1 Properties of Individual-Oriented and Group-Oriented Mobile Services

Properties	*Individual-Oriented*	*Group-Oriented*
Infrastructure	Dedicated infrastructure	Dedicated and nondedicated infrastructure
User involvement	Single user	Multiple users
Interaction among users	Very loosely	Tightly, concurrent, near-real-time interactions
Typical applications	E-commerce, booking, payment, travel information, e-government	Multiplayer game, collaborative work, mobile field assistant, mobile asset management, CRM (customer relationship management) field applications

their services and to foster the service composition and the integration among services to provide value-added services. On the other hand, there is a strong need to access these services by normal people during their movement, for example, booking flights, searching information, making payments, and accessing government documents. The end user also wants to participate in the Web, for example, to perform mash-ups of data and contents from different (Web) services and to provide their own mobile services. Therefore, SOC/SOA-based techniques could help mobile applications utilize SOA-based services in a standard way in order to fulfill the need of the user. Non-SOA solutions would limit mobile applications to access diverse and powerful Web services.

In the group-oriented class, there are group-oriented tasks where mobile services on mobile devices play a critical role. Examples of these tasks are collaborative work performed by a team using mobile devices [5], distributed healthcare [10,11], and disaster responses [12]. These tasks not only require access to diverse data and services hosted in distributed organizations but also need to access and share information in environments with nondedicated infrastructures. Thus, interoperable solutions for mobile services must be supported. In particular, mobile devices have been considered to be very useful in *ad hoc* team collaborations where dedicated infrastructures are not available. Such collaborations normally require flexible and interoperable applications to access as well as offer services. This raises the question of how to provide pervasive and mobile devices with middleware and applications so that the devices can provide collaboration services accessible through standard interfaces and protocols. SOC/SOA, which has introduced means to foster the interoperability, flexibility, and reusability of software, can be used to develop mobile Web services for these scenarios.

1.3 Architectural Styles and Protocols for Mobile Web Services

As well documented in the literature, in SOC/SOA, there are three fundamental, conceptual entities:

- *Service*: a service offers a concrete functionality that can be loosely coupled through the network based on a well-defined interface [13]. A service is expected to be autonomous, platform-independent and its functionality can be published and discovered [1].
- *Service consumer* (also called service client/requester): a service consumer is to consume a service. It makes requests to services and receives responses from the services.
- *Service registry*: a service registry provides facilities for services to publish information about their functionality, interfaces, and locations so that service consumers can search and select relevant services.

These roles are conceptual because an application might function as both a service and a service consumer; a service might act as a service consumer of another service; and a service registry might not be a separate element when its registry function is embedded into an application. Often, we distinguish the *client side* and *service side* when referring to service consumer components and service components, respectively.

There are various ways to design and implement SOC/SOA solutions for mobile services. As other techniques have been well discussed in different places (see a further reading list in Section 1.9), we will focus on Web services technologies. Generally, Web services and Web service consumers can be implemented using different (de facto) standards and protocols. These standards and protocols are used to define, for example, exchange message, communication protocol, service description, and security [13]. Main standards and protocols are:

- XML: is used to describe data exchanged and information about services in Web services.
- WSDL: is used to describe the interface of Web services. Through the interface description of a Web service, Web service consumers can invoke corresponding functions provided by the service.
- SOAP: defines the structure of messages exchanged between Web service consumers and Web services. SOAP is based on XML.
- HTTP: is one of the most popular communication protocols for sending and receiving messages between Web service consumers and Web services.

Although interactions between mobile Web services and their consumers are implementation specific, we can conceptually simplify these interactions into two main types, shown in Figure 1.3. In the first type, a mobile Web service consumer (depicted by the block in Figure 1.3a), through a mobile/*ad hoc* network, will access a mobile application gateway (also known as mobile Web service/application proxy in the literature) which, in turn, will process the consumer's requests and pass the requests to corresponding mobile Web services. We call this model *indirect interaction*. In the second type, a mobile Web service consumer (depicted by the block in Figure 1.3b) will utilize mobile Web services by invoking them directly. We call this case *direct interaction*. Note that direct interaction means that the requests and responses between services and their consumers may be relayed through intermediate services that perform the request/response routing but do not change the request/response content. In both cases, the mobile Web service consumer will be executed on mobile devices, whereas mobile services might or might not be deployed in mobile devices. The mobile devices hosting mobile Web services and mobile Web service consumers can be mobile phones, smart phones, PDAs, and laptops. Networks among mobile Web service consumers and mobile Web services can include the Internet, WiFi *ad hoc* networks, UMTS, EDGE, and GPRS/GMS, as well as a mix of them. Network and device capacities strongly impact on the

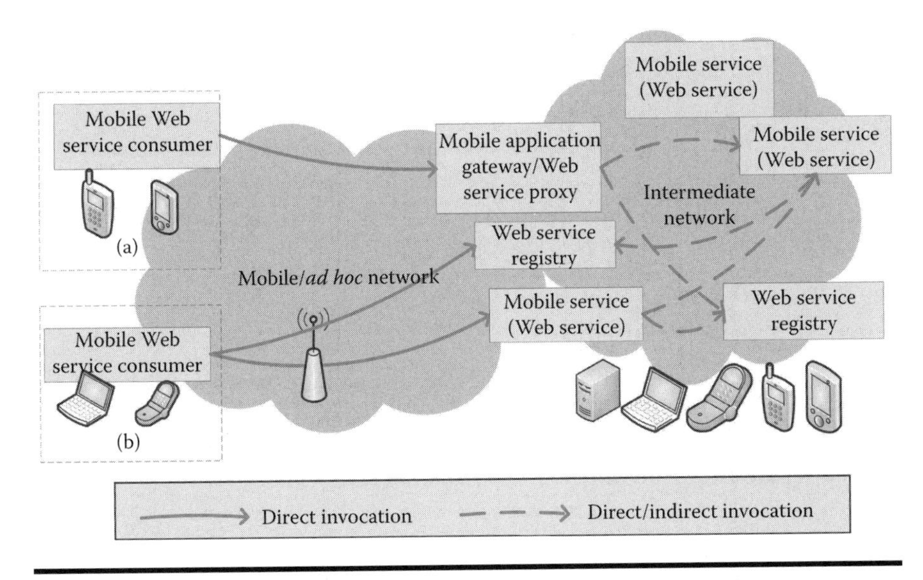

Figure 1.3 Generic interaction models between mobile Web service consumer and mobile services: (a) indirect interaction and (b) direct interaction.

selection of the architectural style for the design and implementation of mobile Web services.

The first type is very popular when mobile Web service consumers are executed in mobile devices that do not have enough capabilities to handle complex data types and to present results in rich, interactive graphical user interfaces. This type is also well suited for bandwidth-constrained networks. However, it does not support complex interactions among mobile users. Therefore, it is suitable for individual-oriented mobile services. In many cases, the mobile application gateway is not a Web service but is based on Web application technologies (referred to as mobile Web applications, discussed in Section 1.6.4). In the second type, the mobile Web service consumer can invoke mobile services directly as these services might provide data in a simple way or the mobile device hosting consumers have the capability to handle complex data. Furthermore, in this type, more complex interactions can be implemented, and thus it is suitable for both individual- and group-oriented mobile services.

In both types, mobile Web service consumers can invoke the service registry to find relevant services or mobile services might find relevant services on behalf of the consumer. Furthermore, services can be found from querying a dedicated registry or from an overlay network in which a dedicated registry does not exist, such as in Vimoware [5]. While mobile Web services are normally not executed on resource-constrained devices, many real-world applications have demonstrated why and how mobile services should be run in mobile devices. For example, in Vimoware [5], device sensors and context management services for disaster responses are developed as mobile Web services. Sliver [14] is another example that demonstrates how a BPEL engine can be implemented as a mobile Web service. In some cases, when a service is hosted in a mobile device, there is an intermediate to act as a proxy for other clients to access the service. For example, in the mobile service platform [15], the service consumer executes a surrogate host that acts a proxy for the service running on mobile devices. These examples indicate an increasing use of mobile Web services on mobile devices to provide service features.

1.3.1 Architectural Styles

From architectural styles, SOC/SOA Web service solutions for mobile services can be built on the basis of (i) SOAP-based Web services, (ii) REST-based Web services, and (iii) mixed SOAP/REST-based Web services with other technologies. As a specific solution can use different techniques, we will discuss only fundamental architectural styles.

1.3.1.1 SOAP-Based Web Services

In SOAP-based Web services, the interface of a Web service is described by WSDL. The messages exchanged between a Web service and its consumers are based on SOAP. SOAP-based messages are designed to be transferred by using different protocols, such as RPC, HTTP, and SMTP. However, HTTP is the most popular protocol used in SOAP-based Web services. These services achieve the integration and interoperability through agreed interfaces and, additionally, agreed service contracts. They support complex interaction models based on different styles of message-oriented and RPC communications, including point-to-point, broker, and P2P. Another advantage of this style is that interfaces can be published and searched, thus simplifying the service discovery process. Therefore, two different services provided by different organizations can easily be integrated with each other. From an architecture point of view, SOAP-based Web services are suitable for integrating mobile Web services on the service side and for group-oriented mobile services.

To date, SOAP-based Web services are strongly supported in mobile application development. Standard protocols and various programming toolkits have been developed for writing SOAP-based mobile services in mobile devices, such as PDAs and smart phones, using different programming languages, for example, C/C++, C#, and Java. However, most toolkits focus on the development of client-side components. The development and deployment of service-side components on mobile devices are not well supported, only in a few specific toolkits that offer limited features.

1.3.1.2 RESTful Web Services

REST-based Web services are built on the basis of the concept of REST (REpresentation State Transfer) [16]. REST is a design principle for Web services that supports stateless services. Services offer their capabilities through resources, each identified by a unique URI and accessed and manipulated by using HTTP methods such as GET, POST, PUT, and DELETE. By using these four basic methods, any Web service consumer can read, update, create, and delete resources offered by a REST-based Web services. REST techniques offer more simple mechanisms for developing Web services by not relying on a large set of standards and by aiming at supporting request–reply interactions. Therefore, the REST architectural style fits very well to many mobile applications, in particular to individual-oriented ones that are typically request–reply-based client/service and stateless. Furthermore, as REST-based Web services mainly utilize HTTP, they are quite suitable for mobile devices because they use less processing power. This is proved through a wide range of real mobile applications based on REST such as commerce [17], data mash-up [18], and GIS for enterprise field workers [19]. However, in many cases, if we need to support more complex interaction models, such as P2P interactions, then REST is not suitable. Further discussion on the choice of REST solutions over SOAP solutions can be found in Reference [20].

1.3.1.3 Mixed Architectural Styles

In many cases, mobile services solutions are not completely based on either SOAP or REST Web services but a mix of them, and they also use other technologies. Figure 1.4 shows some possible mixed architectural styles used for mobile Web services.

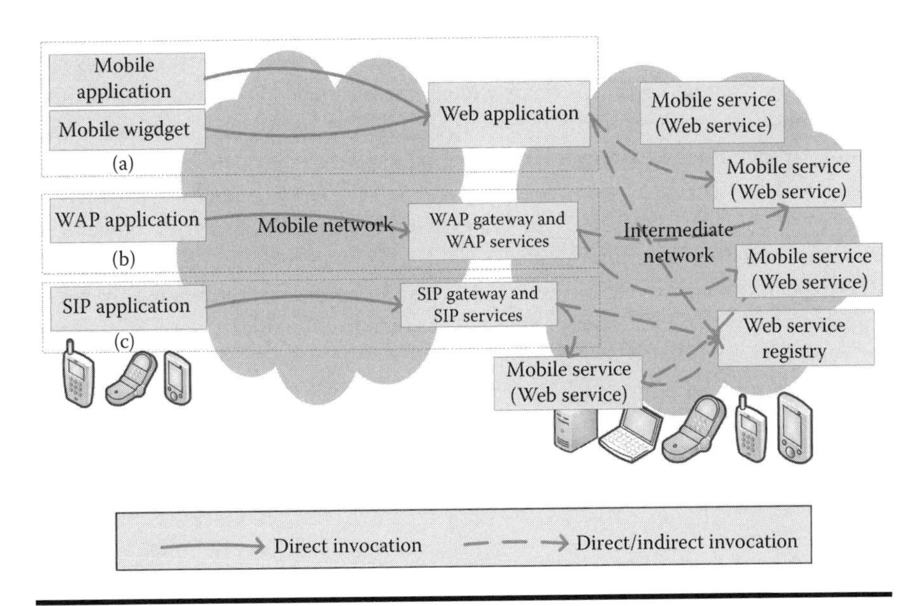

Figure 1.4 Mixed architectural styles for mobile service-oriented architecture services.

- Using a REST-based model for connecting a client to a gateway and using SOAP-based services for service integration at the service side: First, this architecture aims at addressing the simplicity and performance in the interaction between mobile service consumers and mobile services. Second, by using SOAP-based styles at the back end of the service side, the service provider can integrate different types of services from different providers to offer many benefits to consumers.
- Using SIP (Session Initiation Protocol), Web applications, and WAP (Wireless Application Protocol) between consumers and appropriate gateways, and SOAP-based solutions for integrating gateways with other services: This approach is able to deal with limited capability and infrastructure available at the client site to provide converged mobile services [7], such as voice communication integrated with Web content and IP multimedia subsystem application servers [21].

1.3.2 Communication Patterns

Communication models for mobile services naturally follow existing models such as request–reply, broadcast/multicast, and P2P. The design of SOAP allows different one-way or two-way, asynchronous/synchronous, one-to-one, one-to-many, broadcast, P2P communication patterns [22]. The most common communication patterns implemented in SOAP-based mobile Web services are synchronous request–reply, multicast, and P2P using RPC, message-oriented communication, and intermediate services. However, similar patterns have not been observed in REST-based Web services that mostly support only request–reply patterns.

Figure 1.5 illustrates some patterns on service invocation. In the request–reply pattern, a mobile service consumer sends a request to a mobile Web service and obtains the response. This communication can be synchronous or asynchronous, but a synchronous request–reply pattern is the most popular one implemented in contemporary mobile Web services applications, both in SOAP and REST styles, such as in References [23–26]. This pattern is suitable for individual-oriented mobile services; however, it does not work well in group-oriented mobile services of which mobile applications need to be informed with new information instantly, such as in a situational change in disaster responses. Asynchronous request–reply, broadcast/multicast, and P2P patterns are typically supported through specific SOAP/REST-based call-back and polling techniques, asynchronous HTTP, or WS-Notification implementations, such as in References [5,27]. They can be implemented well with SOAP-based mobile services. However, most programming toolkits require the developer to do this without any support. The one-way invocation pattern is also supported by SOAP standards.

1.3.3 Service Discovery and Composition

Service discovery and composition are fundamental processes of SOA. To support these processes, services have to be well defined and described. In general, with

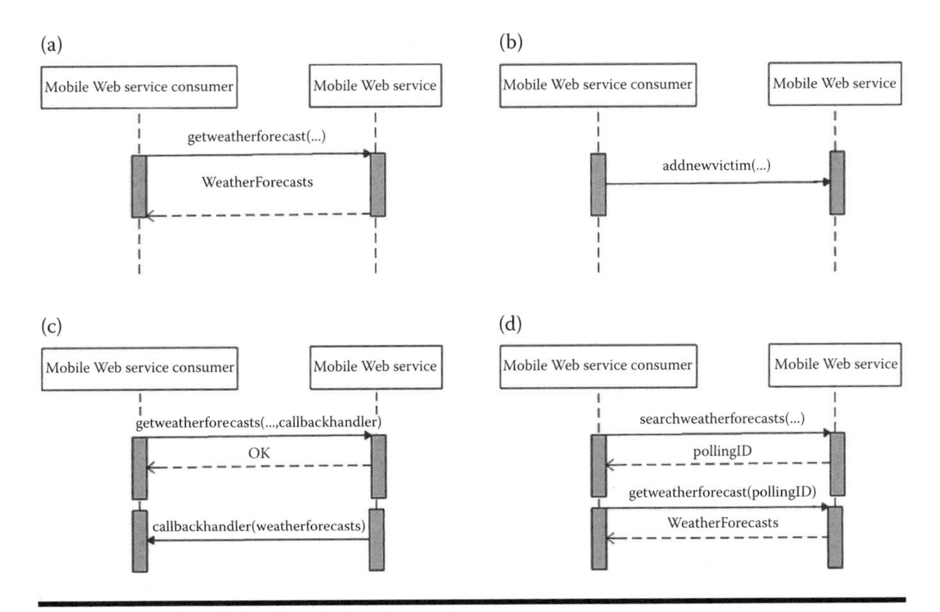

Figure 1.5 Examples of invocation patterns. (a) Synchronous request-replay invocation. (b) One-way invocation. (c) Asynchronous invocation using callbacks. (d) Asynchronous invocation using polling.

SOAP-based services, the use of WSDL and other metadata has facilitated the discovery process. Semantic representations of mobile Web services are also used increasingly. In principle, mobile Web service consumers find SOAP-based Web services from service registries, such as Universal Description, Discovery, and Integration (UDDI). In practice, this model does not work well [28]. However, the REST-based Web services discovery process is largely negligible at the moment. In fact, there is a lack of mechanisms to describe REST-based Web services that can facilitate the discovery process. One of few efforts is the proposal of Web Application Description Language (WADL) [29] for describing REST resources. However, it has not been widely adopted.

Figure 1.6 shows basic models of Web services discovery that work with mobile Web services. For multicast discovery, these models can be supported by SLP (Service Location Protocol [http://tools.ietf.org/html/rfc2608]), UPnP (http://www.upnp.org), WS-Discovery (http://specs.xmlsoap.org/ws/2005/04/discovery/ws-discovery.pdf), and tool-specific implementation, and can be implemented using UDP/HTTP multicast, SOAP, and WS-Notification. As mobile devices have some limitations, in many cases, mobile Web service consumers have not actually implemented the discovery process, especially in individual-oriented mobile services. The user just provides his/her known services (based on other sources, e.g., search engines) to mobile applications, for example, as in Reference [30]. For group-oriented mobile services, various attempts

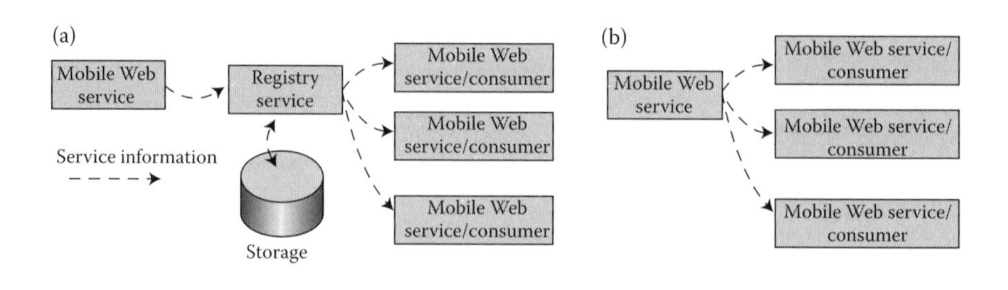

Figure 1.6 Examples of discovery models. (a) Centralized and multicast discovery with dedicated registry. (b) Multicast discovery without dedicated registry.

have been made to support broadcast/multicast and P2P service discovery. A JXTA-based P2P discovery for mobile Web services is presented in References [31,32]. In another effort, SLP-based/specific discovery techniques have been developed in RESCUE [33,34]. The WSAMI middleware [24] also supports an SLP-based discovery. Broadcast/multicast and SLP-based discovery methods are useful for mobile services in a network; however, they do not scale well in a large-scale system of mobile services. Therefore, they are more suitable for a small group of services.

The composition of mobile Web services occurs mostly on the service side in nonmobile platforms by using well-known techniques. However, service aggregation techniques on mobile devices are very limited. Some efforts have been devoted to service aggregation by utilizing workflows [35], in particular using BPEL, in the mobile devices, for example, Sliver [14]. This type of service aggregation is an attempt to support the user to coordinate tasks and to perform business processes using mobile devices. To support the user to compose his/her own services, some research efforts have been investigated [36]. However, currently research issues in this direction are very open.

1.3.4 Data Handling

Data handling in mobile applications is strongly dependent on specific services and device capabilities. Typical data handled by mobile applications are Web contents and multimedia. Various standards have been proposed for handling these types of data, such as Mobile Media application programming interface (API) (JSR 135), Mobile 3D Graphics API (JSR 184), and 2D Scalable Vector Graphics API (JSR 226). Many types of mobile Web services do not transfer rich data or large amounts of data. However, recent advanced mobile Web services, such as location-based services and GIS-based collaborative tools, require the transfer of rich data. In particular, many mobile Web services support user participation and mass customization by offering a mash-up of contents from different sources; these mash-ups possibly include text, graphics, audio, and videos.

As handling complex, voluminous data transferred through Web services require strong processing capabilities, mobile Web services have to utilize different techniques to improve the data handling. One technique is not to use XML but a different data format exchanged between mobile Web service consumers and mobile Web services. A popular technique that works well with REST services is to utilize JSON (JavaScript Object Notation [http://www.json.org/]) which requires less processing capability and has smaller data size. Listing 1.1 presents an example of JSON-based data returned by the REST-based ImageSearch Service from Yahoo! (http://search.yahooapis.com/ImageSearchService/V1/imageSearch) and Listing 1.2 presents a sample of code to process the JSON-based data on mobile devices using the JSONObject library (http://www.json.org/java/index.html). Another technique is to compress the SOAP body transferred using compression techniques [37]. This technique, however, requires the control of both mobile Web service consumers and mobile Web services. Therefore, it is not interoperable. Another way is to utilize

```
{"ResultSet": {
  "Result": [
          {
                "ClickUrl": "http://www.residenzjoop.com/files/
                image/lage/hundertwasser.jpg",
                "FileFormat": "jpeg",
                "FileSize": 104448,
                "Height": "429",
                "RefererUrl": "http://www.residenzjoop.com/
                location",
                "Summary": "Hundertwasserhaus",
                "Thumbnail": {
                     "Height": "97",
                     "Url": "http://thm-a01.yimg.com/
                     image/0acfb0befb605e0e",
                     "Width": "145"
                },
                "Title": "hundertwasser jpg",
                "Url": "http://www.residenzjoop.com/files/
                image/lage/hundertwasser.jpg",
                "Width": "640"
          },
     //....
     ],
     "firstResultPosition": 1,
     "totalResultsAvailable": "556",
     "totalResultsReturned": 10
}}
```

Listing 1.1 Simplified example of JSON-based data.

```
//...
ImageSearch imagesearch = new ImageSearch("YahooDemo",
"Hunderwasserhaus", "json");
String result = imagesearch.submit();

JSONObject resultJSON = new JSONObject(result);
JSONObject resultSetObject = resultJSON.
getJSONObject("ResultSet");
int totalResultsReturned = resultSetObject.
getInt("totalResultsReturned");
if (totalResultsReturned < 1) {
       return ;
}
JSONArray resultObject = resultSetObject.
getJSONArray("Result");
for (int i = 0; i < resultObject.length(); i++) {
  JSONObject objectItem = resultObject.getJSONObject(i);
  Iterator iterator = objectItem.keys();
  while (iterator.hasNext()) {
      String key = (String) iterator.next();
      String data = objectItem.getString(key);
      if (!key.equals("Thumbnail")) {
          System.out.println(key + ":\t" + data);
      }
  }
}
//...
```

Listing 1.2 Example of processing JSON-based data.

binary formats of XML (http://en.wikipedia.org/wiki/Binary_XML) to reduce the size of transferred data.

1.3.5 Service Management

The management of mobile Web services is naturally similar to that of general Web services such as accessing service information, monitoring service status, and supporting runtime deployment. Various specifications have been developed for managing Web services, in particular SOAP-based Web services, such as WS-Management (http://www.dmtf.org/standards/wbem/wsman) and Web Services Distributed Management (WSDM) (http://www.oasis-open.org/committees/tc_home.php?wg_abbrev=wsdm#overview). These specifications focus on the access and exchange of management information of services. However, to date these standards have not been well supported in existing mobile Web services.

Reference [23] is one of the few works that provide service information using rules and notifications.

Different from other platforms, service management for mobile Web services has to deal with two major issues: service continuity and service runtime deployment and execution. Service continuity is of paramount importance in mobile services because the high level of mobility increases the high level of service disruption and mobility. Service runtime deployment and execution are other important issues because mobile devices are typically not assumed to support preinstalled services or always-on, preprovisioning services. This is particularly true when services are executed on a nondedicated infrastructure and mobile devices.

The importance of service continuity for mobile services has been identified and a modeling technique has been proposed in Reference [38]: a service continuity layer is proposed to handle tasks to ensure the continuity of services such as monitoring and handover management. In the Suspend–Relocate–Resume for Web Services (SRR-WS) framework [39], service continuity is ensured by suspending services and relocating client's data and resuming services with the client data. SRR-WS achieved this mechanism by utilizing the proxy model. To work with SRR-WS, mobile Web service consumers need to include a specific SRR-WS client library that handles caching and session relocation. The SRR-WS framework includes a proxy with suspend/resume and session relocation modules. By using a proxy mechanism, SRR-WS does not require a change in mobile services. However, consumers have to be reprogrammed to work with SRR-WS. SRR-WS has been implemented with SOAP over HTTP for mobile devices. Another technique is to migrate mobile Web services when the hosting environment is unable to host the services, for example, due to low battery or the mobility of the service provider [26]. In this way, migration requests are explicitly made by the service provider (which is also hosted on a mobile device). In principle, service code and description can be migrated. However, this solution cannot work alone without ensuring correct information about services.

Service runtime deployment and execution for mobile applications have been researched intensively. As mobile Web service consumers are part of mobile applications, it is expected that existing runtime deployment and execution techniques could be applied to mobile Web services.

1.3.6 Security and Identity Support

Security and identity support are of paramount importance for mobile users and are strongly dependent on the architectural styles used for mobile Web services. With SOAP-based solutions, various security standards have been proposed, such as the OASIS WS-Security and the Liberty ID-WSF/ID-SIS. The REST-based solutions typically rely on HTTP and SSL security techniques. Furthermore, depending on the integration of mobile Web services, different identity management techniques can be used.

For SOAP-based Web services, the OASIS WS-Security (http://www.oasis-open.org/committees/tc_home.php?wg_abbrev=wss) specifies protocols for securing message exchanges in Web services. WS-Security utilizes XML Digital Signature, XML Encryption, and X.509 certificates. Its key feature is to protect SOAP messages exchanged. The support of WS-Security standards on mobile Web service development toolkits is strong. Most toolkits support XML signatures and encryption. For REST-based Web services, HTTP Basic/Digest and SSL are popular techniques for implementing authentication, authorization, and encryption in mobile Web services because the communication of REST-based Web services is based on HTTP. Because these techniques are well supported in mobile device operating systems and programming libraries, they can be easily used by REST-based mobile Web services and clients. Furthermore, one practical method applied to the authentication of the access to REST-based Web services is to use tokens embedded in service interfaces (e.g., user ID, email, and development key), such as in StrikeIron and Amazon Web services.

The OAuth (http://oauth.net) is also increasingly used for authorizing resource access in mobile Web services. The OAuth is an authorization delegation protocol that defines how a user can grant service consumers access to a user's private resources hosted in another service. The OAuth is built atop HTTP and works on the basis of the exchange of two types of tokens asking for authorization access and for accessing resources on behalf of users. The protocol includes three main steps (see details in http://oauth.net/core/1.0/) to allow the consumer to obtain an unauthorized request token that is authorized by the user, to use the request token to obtain an access token, and finally to access resources. All requests are signed and parameters of requests can be described in the HTTP Authorization header, HTTP POST body, or URLs of HTTP GET. At the time of writing, the OAuth has been implemented in many programming languages, such as C#, Java, JavaScript, Python, and Ruby. As OAuth is designed for HTTP and Web resource access, it is very suitable for REST-based Web services.

Identity management techniques are critical when a mobile Web service consumer utilizes different mobile Web services, potentially provided by different providers. Being able to associate user identity with his/her selected mobile service, these services can improve the authentication and authorization, reduce user intervention, such as support single sign-on, and improve service provisioning. Two popular identity management frameworks are the ID-WSF/ID-SIS and the OpenID. The IdentifyWeb Services Framework (ID-WSF) and the Identity Service Interface Specifications (ID-SIS) (http://www.projectliberty.org/liberty/resource_center/specifications) define mechanisms enabling identity-based services. By associating the identity of individual users with Web services, ID-WSF/ID-SIS-enabled systems can establish a federated identity. By doing so, the user does not need to provide his/her identity many times when using different services and the service provider can optimize the composition of services. The ID-WSF version 2.0 supports SAML, authentication, single sign-on, and identity mapping by utilizing XML-Signature, XML-Encryption, and WS-Security.

The OpenID (http://openid.net) is another protocol to support single sign-on. OpenID is strongly supported by the industry (see the list of providers at http://openiddirectory.com/openid-providers-c-1.html). The idea of OpenID is to provide a single identity management protocol for accessing resources from different service providers. In the OpenID version 2.0 (http://openid.net/specs/openid-authentication-2_0.html) a user is identified by an OpenID identifier described in an URL or XRI. An OpenID identifier is managed by an identity provider or OpenID provider that carries out the user authentication. When a user accesses a service that offers OpenID authentication, the service will request the user to provide the user OpenID. The service will communicate with the identity provider of the user to verify the user. In case an authentication is required, the service will redirect the user to the identity provider for authentication (e.g., entering user name and password). OpenID utilizes HTTP, SSL/TSL, and URL; thus it is quite suitable for REST-based services. As OpenID focuses on authentication and single sign-on, it is suitable for service integration on the service side.

Although ID-WSF/ID-SIS and OpenID are widely supported on the service side of mobile Web services in many platforms, their availability on mobile devices where applications act on behalf of the user is limited. OpenID has been supported in many libraries but it has not been well integrated with SOAP/REST. For example, it is relatively easy to send an OpenID request to a Web service; however, authentication requiring user intervention is not quite clear (see a discussion on OpenID binding to SOAP/REST at http://wiki2008.openid.net/REST/SOAP/HTTP_Bindings).

1.4 Mobile Web Services Programming Support

1.4.1 Standards and Specifications

Various standards and specifications have been developed for Web services. However, not all of them are fully supported in mobile Web services. Fundamental standards and specifications such as SOAP, XML, and HTTP are well supported in most mobile platforms. Other standards such as OASIS WS-Security and Liberty ID-WSF are supported in some platforms.

Apart from the above-mentioned generic standards and specifications for Web services, some have been specifically designed for mobile Web services. The J2ME Web services specification (JSR-172) (http://jcp.org/en/jsr/detail?id=172) defines optional packages for the development of Web service applications on mobile devices. In particular, it focuses on XML processing capabilities and on APIs for J2MS Web service clients based on RPC communication. JSR-172 also defines the mapping from a subset of the WSDL to Java code, suitable for J2ME. This specification has been supported in many programming toolkits such as Nokia Web services platform, Sun J2ME Wireless Toolkit (http://java.sun.com/products/sjwtoolkit/), and Netbeans (http://www.netbeans.org).

The mobile service architecture (MSA) specification (JSR 248) has been introduced recently to create an environment for Java-enabled mobile devices. MSA includes several existing standards for handling security and commerce, graphics, communications, personal information, and application connectivity.

1.4.2 Mobile Web Service Toolkits

1.4.2.1 General-Purpose Web Service Toolkits

These toolkits aim at addressing broad mobile Web services applications. The main toolkits that have been developed are discussed in this section.

1.4.2.1.1 kSOAP2

kSOAP2 [40] is one of the very first and popular SOAP Web service libraries for developing Web services on constrained devices. kSOAP provides various facilities that have been utilized by other SOC/SOA work on mobile devices, such as Sliver [14], a BPEL engine, and RESCUE [33,34]. It can be used to develop both Web service clients and services. However, kSOAP does not provide advanced features for handling security, service discovery, and service management. In principle, all of these features have to be implemented by the developer.

1.4.2.1.2 gSOAP

gSOAP is an open-source C/C++ Web service development toolkit [41]. gSOAP supports XML data-binding solutions through autocoding techniques. It also provides tools for generating C/C++ code from WSDL/XSD files and supports different platforms including mobile devices. gSOAP is well studied in Reference [42]. Although gSOAP provides security support, similar to kSOAP it focuses mainly on the development of Web services and their consumers. Therefore, it does not support service discovery and aggregation.

1.4.2.1.3 Google Android

Android (http://code.google.com/android) is not a particular mobile Web service toolkit but a software platform including an operating system, middleware, and applications for mobile devices. Android provides various facilities for developing mobile applications but it does not provide a toolkit for writing SOAP-based Web services and clients. However, it provides various classes for writing mobile Web clients by providing HTTP, XML, JSON, and OAuth classes for implementing mobile Web service clients. The clients can use HTTP actions to invoke REST-based Web services. Both synchronous and asynchronous HTTP invocations are supported. The returning results can be in XML, JSON, RSS, Atom, and so on.

1.4.2.1.4 .NET

The Microsoft .NET Compact Framework (CF) (http://msdn.microsoft.com/en-us/netframework/aa497273.aspx) supports the development of applications on smart/mobile devices. The .NET CF provides various facilities for developing mobile applications, including mobile Web service applications on the client side based on HTTP, SOAP, and XML. It supports both asynchronous and synchronous invocations of Web services, HTTP Basic/Digest authentication, SSL, and SOAP extension. However, the development of Web services on mobile devices is not supported. The .NET CF also provides useful tools to support the developer to write mobile Web service clients. For example, the .NET CF Service Model Metadata Tool (netcfsvcutil.exe) can be used to generate client proxy codes to simplify the development of clients consuming Web services on the device.

1.4.2.1.5 Java FX Mobile

The Sun Java FX (http://www.javafx.com) is a platform for developing rich Internet applications and Java FX Mobile is the version of Java FX for mobile devices. The key technologies provided by Java FX are in a rich set of libraries for handling graphics, media data, and Web services. By utilizing Web service libraries, we can develop mobile service applications on mobile devices. Java FX Mobile provides only HTTP-based libraries for the development of Web service applications on the client side. Therefore, only REST-based Web service clients can be built. This library includes APIs for handling HTTP requests and XML/JSON. Java FX Mobile does not include facilities for writing REST-based Web services. Java FX is well integrated into different programming development environments such as NetBeans and Eclipse IDE.

1.4.2.1.6 Nokia Web Service Platform

The Nokia Web Services Platform [43] includes various development facilities for mobile service applications on Nokia's S60 or Series 80 platform. It supports both Java technology and Symbian OS C++. Apart from other facilities, for Java-based applications, the Nokia WSP supports the J2MEWeb Services Specification (JSR-172). At the moment, only client-side functionality is supported. The C+ version provides APIs for handling requests to and responses from Web services. These APIs are built on the basis of a Web Services Framework (WSF) API or the Liberty

Identity-Based Web Services Framework (ID-WSF). OASIS WS-Security and Liberty ID-WSF are supported.

1.4.2.1.7 Apache Muse

The Apache Muse Project (http://ws.apache.org/muse/) implements the Web Services Resource Framework (WSRF), Web Services BaseNotification (WSN), and Web Services Distributed Management (WSDM) specifications. Therefore, it supports SOAP-based Web services based on WSRF. By using WSN, Web services based on Muse can also implement eventing based on Web services. The WSDM allows the developer to include management features in Web services, for example, service capabilities and status. Muse can be deployed in different platforms including Java SE/Java EE and Open Services Gateway initiative (OSGi). When deployed in OSGi, Muse can be used to develop mobile Web services.

Table 1.2 summarizes some properties of the above-mentioned general-purpose mobile Web service toolkits.

1.4.2.2 Specific Toolkits

1.4.2.2.1 Vimoware/RESCUE

Vimoware [5] is a Web service-based toolkit that can be used to develop Web services on mobile devices and to conduct *ad hoc* team collaborations by executing predefined or situational flows of tasks. One of the main components is a *light-weight Web services middleware* that supports the SOAP-based Web services. This middleware was later developed into a separate middleware named RESCUE [33,34]. In Vimoware/RESCUE, Web services are developed by applying the POJO (Plain Old Java Object) principle. The developer creates a Web service by extending an abstract Java class. This class requires the specification of a service description and provides basic methods for extracting metadata about the service. Service operations are implemented as normal Java methods. On the basis of the description provided by the service developer and on the metadata, WSDL files can also be created.

The main specific characteristic of Vimoware/RESCUE is that it is specifically designed for Web services in the *ad hoc* network of mobile devices. Thus, it also provides runtime and reconfigurable service provisioning and service discovery facilities. Vimoware/RESCUE uses kSOAP2 and reuses parts of Sliver [14] to deploy Web services. The transport communication can be configured either by HTTP, which is realized by a light-weight version of the Jetty (http://jetty.mortbay.org) engine, or by direct TCP socket communication. In Vimoware/RESCUE, three interaction patterns exist: (a) one-way interactions, (b) synchronous request–response interactions, and (c) real asynchronous request–response. Services can be deployed into Vimoware/RESCUE at runtime. In Vimoware/RESCUE, a

Table 1.2 Summary of General-Purpose Mobile Web Service Toolkits

Toolkit	*Architectural Style*		*Client/Service*		*Language*			*Security/Identity*			*Management*
	SOAP	*REST*	*Client*	*Service*	*C/C++*	*Java*	*C#*	*WS-Based*	*HTTP-Based*	*Identity*	
kSOAP/ kXML	x		x	x		x					
gSOAP	x		x	x	x			x			
Android		x	x			x			x	x	
.NET	x	x	x		x		x	x	x		
Java FX mobile		x	x			x			x		
Nokia WSP	x		x	x	x	x		x	x	x	
Muse	x		x	x		x		x			x

P2P-based subscription/notification mechanism is supported for service advertisement and discovery. This mechanism is implemented on the basis of UDP multicasts.

1.4.3 Software Development Process for Mobile Web Services

Unlike Web services on other platforms, the design and development of mobile Web services lack supporting tools. Model-driven development (MDD) techniques that seem very suitable for modeling diverse mobile platforms have been investigated for mobile applications [44,45]. However, unlike the support of MDD for other types of applications, MDD for SOA-based mobile Web services has not been well studied. SPATEL [46], developed in the SPICE project, is a UML-based language for modeling composite telecommunication services. This language has been demonstrated with different GUI frameworks like S60 Nokia smart phones. In Reference [45], from a meta-model, DSL (domain-specific language) is generated and can be used for domain-specific modeling (DSM). In DSM, mobile services are modeled and XForm code can be generated for use in mobile applications. However, this MDA approach is implemented for a specific platform that is not Web services.

In Reference [38], a composition model for mobile services (not necessary mobile Web services) has been proposed to include four main components, namely, service logic (describing service functions), service data (describing data used in services), service content (describing data product of services), and service profile (describing user/device properties). From the composition model, components of a mobile service can be mapped into different distribution models, representing the concrete deployment of components of mobile services, such as monolithic, client–server, peer-to-peer, or multiple distributed components [38]. This conceptual model can be applied for modeling mobile Web services. However, we are not aware of any existing tool to support this model.

1.4.4 Writing Mobile Web Services and Consumers

The writing code for mobile Web service consumers to invoke mobile Web services is straightforward and similar to that for Web services in other environments. Most programming toolkits support very primitive code generation for SOAP-based service consumers. A typical way of writing a SOAP-based service consumer code is to import the WSDL file of the service and generate a stub code from that WSDL. Figure 1.7 shows a simple example of how Netbeans supports the writing of a mobile Web service consumer code for the `WeatherForecast` service (http://www.webservicex.net/WeatherForecast.asmx?wsdl). By indicating the URL of the conversion service, a code can be generated, for example, as shown in Figure 1.8.

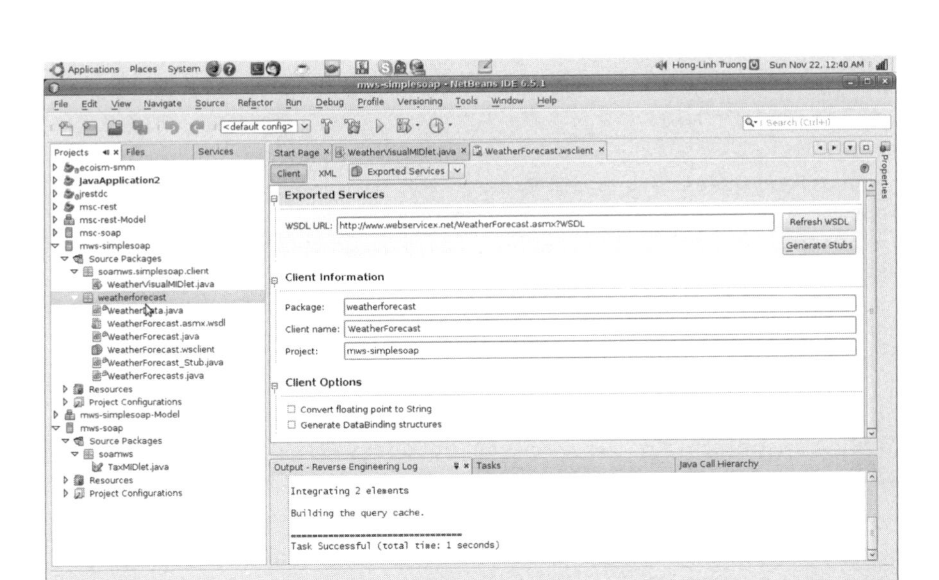

Figure 1.7 Example of generating code for a mobile Web service consumer in Netbeans.

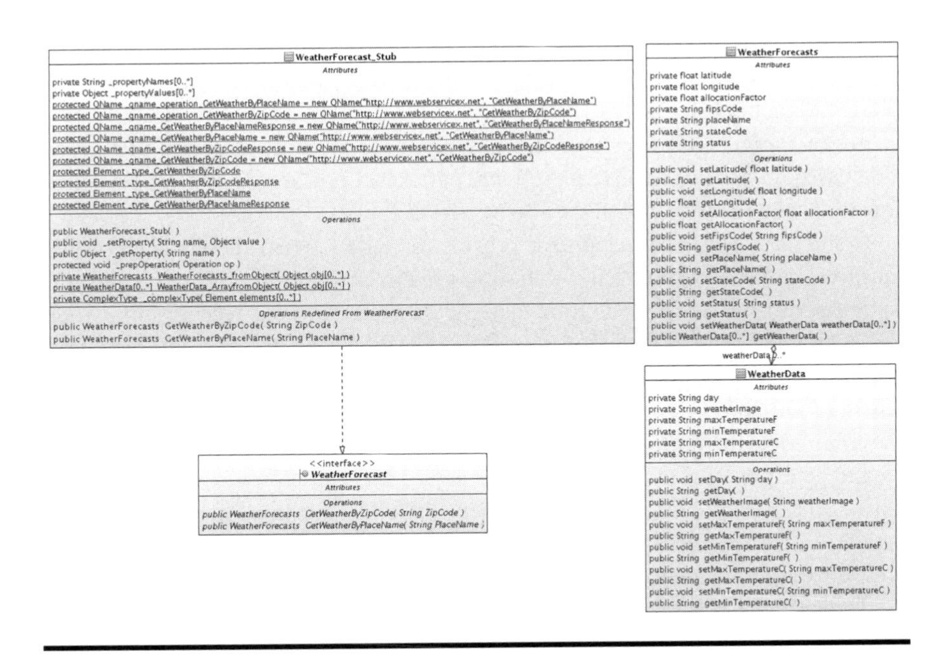

Figure 1.8 Example of the class diagram for generated code produced by Netbeans.

```
 //...
String inputZIP;
//...
try {
   WeatherForecast serviceConnection = new WeatherForecast_
Stub();
   WeatherForecasts forecasts = serviceConnection.
GetWeatherByZipCode(inputZIP);
   System.out.println(forecasts.getFipsCode());
   System.out.println(forecasts.getLatitude());
   System.out.println(forecasts.getLongitude());
   System.out.println(forecasts.getPlaceName());
   System.out.println(forecasts.getStateCode());
   System.out.println(forecasts.getStatus());
   WeatherData[] data = forecasts.getWeatherData();
   for (int i = 0; i < data.length; i++) {
          System.out.println(data[i].getDay() + ": max =" +
data[i].getMaxTemperatureC() + "min=" + data[i].
getMinTemperatureC() + "," + data[i].getWeatherImage());
   }
} catch (RemoteException ex) {
      ex.printStackTrace();
}
```

Listing 1.3 Code excerpt for invoking the `WeatherForecast` service.

A similar process can be found in other toolkits like Microsoft Visual Studio (for .NET/C#) or Eclipse.

Given the generated code, it is straightforward to write a service invocation code based on synchronous request–reply patterns; for example, Listing 1.3 shows how to invoke the `WeatherForecast` service.

A similar software development process for REST-based mobile Web service consumer code generation is not well supported in current tools. In most cases, the developer just writes the code to invoke the service. However, recently some tools also support writing WADL-based information for REST services and code generation from WADL files. These tools should be incorporated into/combined with existing IDE for code development. For example, Figure 1.9 shows the REST Describe tool (http://tomayac.de/rest-describe/latest/RestDescribe.html) that is used to generate WADL for the `Yahoo! Geocoding Service`. On the basis of the generated WADL, this tool can generate codes for invoking the service, as shown in Figure 1.10. Figures 1.11 and 1.12 illustrate the resulting service invocations for the above-mentioned examples in different platforms.

To develop a Web service on mobile devices is much more challenging because of the lack of tools. Recently, the Apache Muse supports the developer to develop Web services in mobile devices in a way similar to normal environments. With the

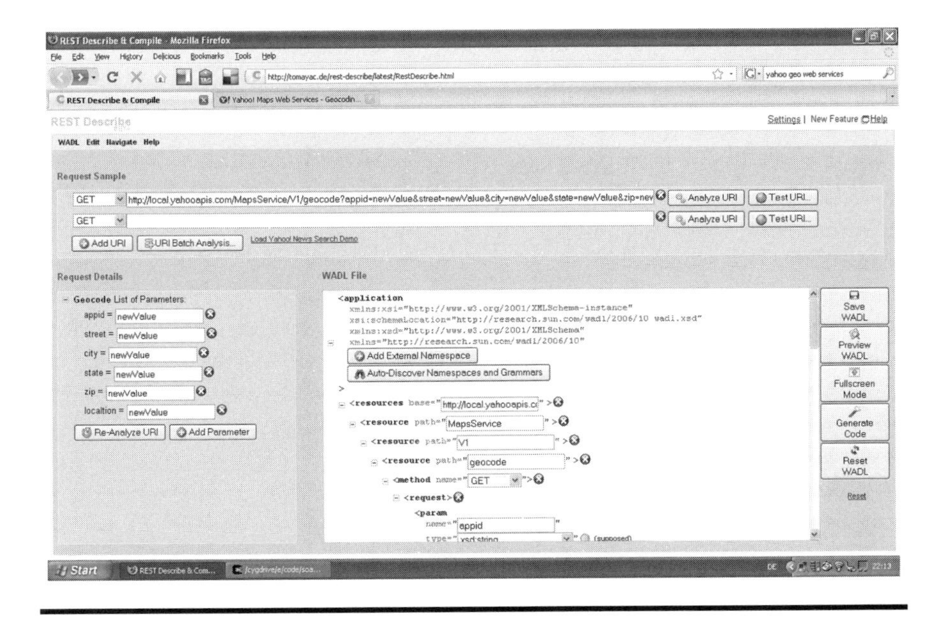

Figure 1.9 Example of using the REST Describe tool to create Web Application Description Language.

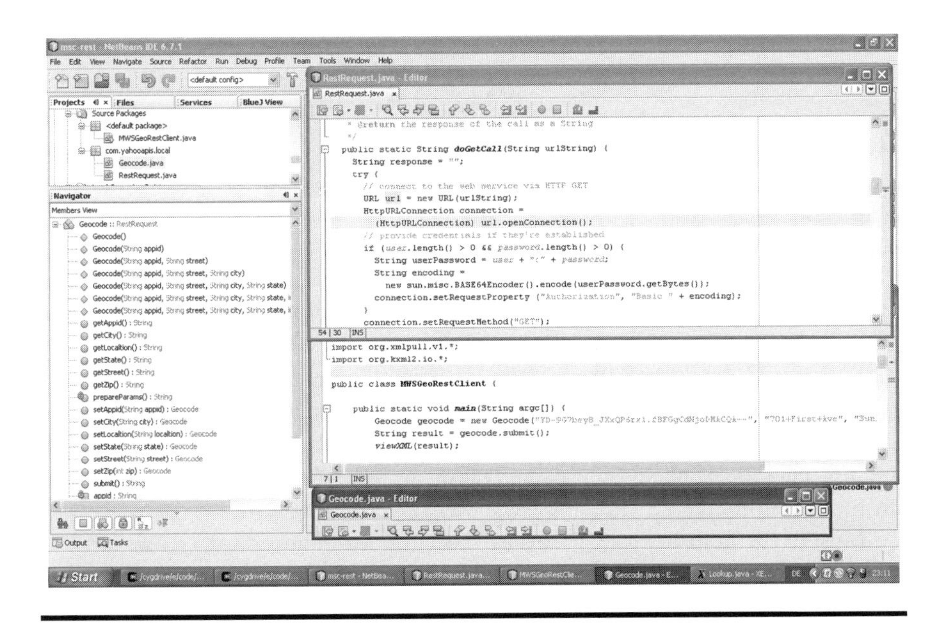

Figure 1.10 Example of code generated from Web Application Description Language using the REST Describe (visualized in Netbeans).

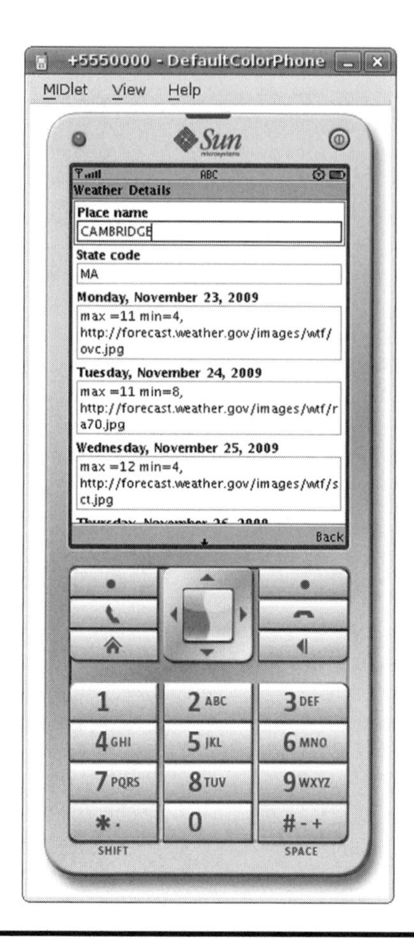

Figure 1.11 A mobile Web service consumer for the `WeatherForecast` service in an emulator.

Apache Muse, the developer can generate service skeletons from WSDL files and write his/her code based on OSGi containers. Specific tools like RESCUE allow Web services to be written like a plain Java object. However, with RESCUE, the service code has to be executed within a particular hosting environment. Listing 1.4 shows a simplified example of a Web service based on RESCUE for a medical staff that accepts reports about victims (operation `addNewVictim`) and that returns the profile of the staff (operation `getProfile`).

Most of the tasks involved to service discovery, call-back, security, and identity management have to be implemented by the developer. Currently, there is a lack of supporting code generation for these complex tasks. Thus, high-level libraries and tool assistance are needed to help the developer simplify these tasks. Listings 1.5 and 1.6 give an example of using the jSLP tool (http://jslp.sourceforge.net), a library

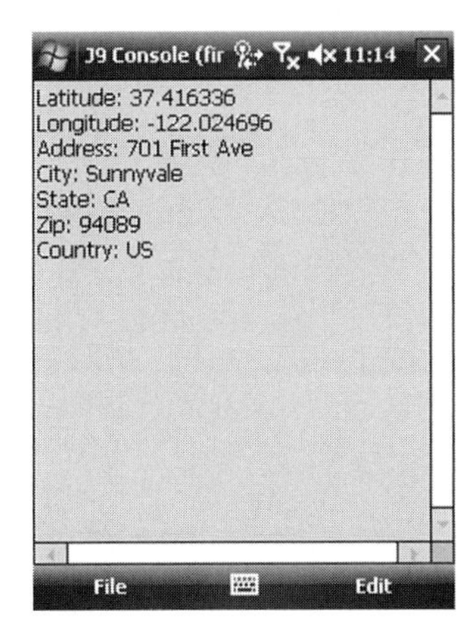

Figure 1.12 A mobile Web service consumer for the Yahoo! Geocoding service in an HTC Tytn II.

```
//....
import at.ac.tuwien.vitalab.middleware.AWebService;
public class MedicalStaffService extends AWebService {
  public MedicalStaffService() {
  super("MedicalStaffService","http://www.infosys.tuwien.ac.
at/medicalstaffservice");
  }
   public void onDeploy() {
     //...
  }
  public void onUndeploy() {
  //..
  }
  public int addNewVictim(String location, String
victimStatus) {
    //..
    return myCurrentLoad ;
  }
  public String getProfile(String typeOfStaff) {
    //...
    return myProfile ;
  }
//...
}
```

Listing 1.4 Simplified example of a Web service in RESCUE.

```
//...
Advertiser advertiser = ServiceLocationManager.
getAdvertiser(new Locale("en"));
ServiceURL myService = new
ServiceURL("service:disastersupport:
http://192.168.218.101:8080/services", 3600);
Hashtable attributes = new Hashtable();
attributes.put("ServiceID","MedicalStaffService")
attributes.put("Team","MedicalTeam");
attributes.put("TypeOfStaff","Nurse");
//...
advertiser.register(myService, attributes);
```

Listing 1.5 Simplified example of publishing mobile services with jSLP.

```
Locator locator = ServiceLocationManager.getLocator(new
Locale("en"));
// find all disaster support services belonging to medical
staff
ServiceLocationEnumeration sle = locator.findServices(new
ServiceType("service:disastersupport"), null,
"(Team=MedicalTeam)");
while (sle.hasMoreElements()) {
   ServiceURL serviceURL = (ServiceURL)sle.nextElement();
   //...
}
//..
```

Listing 1.6 Simplified example of discovering mobile services using jSLP.

implementing SLP, to publish and discover the `MedicalStaffService` above. In this example, we assume that there are several medical staff available in a large-scale disaster scenario and each staff has the `MedicalStaffService` in his/her PDA. The service publishes its hosting environment information and other metadata while the consumer would like to search only services on PDAs of nurses. Of course, the assumption is that the consumer knows how to invoke the service.

1.5 Real-World SOA Mobile Services

To demonstrate the benefit of SOC/SOA, this section presents a real-world example of mobile Web services. This example is built on our experiences from our research projects in collaborative working environments (CWEs). Modern CWEs introduce the need of various real-world SOA-based mobile Web services because today's teamwork is performed by dynamic teams that are established on demand, use various mobile devices, and need to access a vast source of data and services in order to fulfill their tasks.

Our example discusses SOAP-based mobile services and consumers for disaster responses performed in the EU WORKPAD project (http://www.workpad-project.eu). The WORKPAD system [12] supports disaster management on the basis of two main ideas: (i) multiple supporting teams working on the field using mobile devices to collect information about the disaster, and (ii) teams accessing vast sources of information from different organizations to optimize their tasks and store the collected disaster information to organizational information systems. In doing so, each member uses a mobile device, and members need to interact with each other.

SOAP-based mobile Web services and consumers have been widely used in WORKPAD tools, as shown in Figure 1.13. Mobile devices are used in *ad hoc* and team collaborations where dedicated infrastructures are not available. Such collaborations normally require flexible and interoperable services while running on mobile devices and being integrated with various other services. Therefore, WORKPAD supports mobile Web services for managing context information and

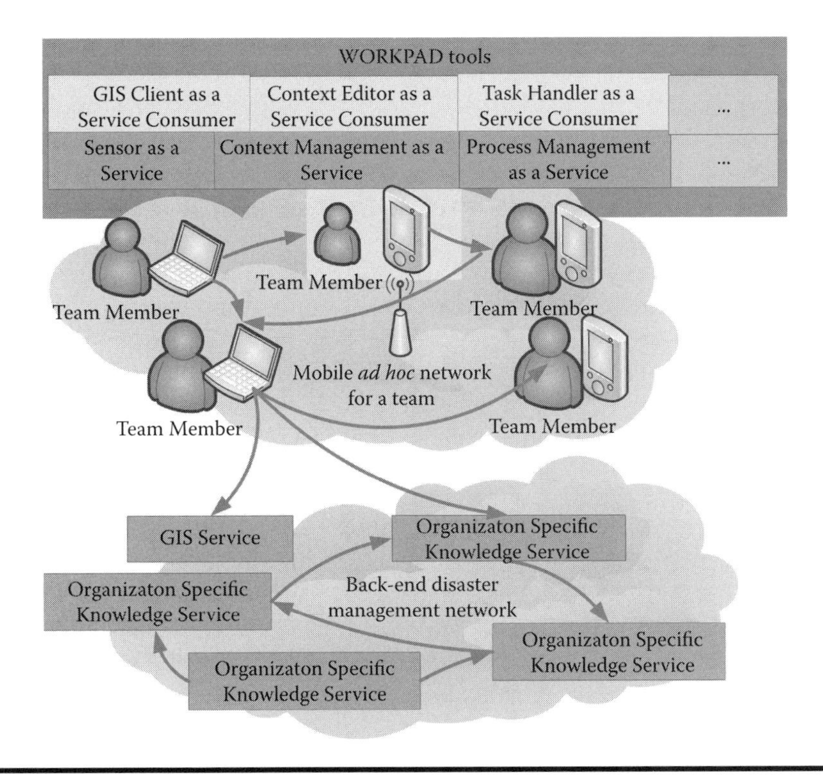

Figure 1.13 (Simplified and abstract) Simple Object Access Protocol mobile services in the WORKPAD project.

for coordinating collaborative tasks. The context management system of WORKPAD [5,27] provides various types of context information to different clients such as GIS-based mobile applications and task management. Context sensors and context management services are mobile Web services and context information is described in XML, thus facilitating the integration between the context management system and other applications. Task management provides a SOAP-based BPEL engine for coordinating tasks [47]. Furthermore, GIS-based mobile applications utilize Web service technologies to access GIS and other information from the back-end services that are also Web services.

Without using Web services, WORKPAD would face a great challenge in the integration of different sources of data. For example, GIS data are now provided through Web services (e.g., see Reference [48]). Thus, by developing SOAP-based GIS clients, WORKPAD can ensure that its GIS software is interoperable and reusable. Various information sources from different government agencies are hard to access through mobile applications, if these sources are not wrapped into Web services. If the context management system was not based on Web services solution, it would be hard to not only integrate the context information to different clients but also to share the context information among different teams. Furthermore, without using Web services, WORKPAD components would be a very specific implementation that would not be highly reused in other works. For example, the Web service-based context management system can be used in SOA-based smart-home environments [49].

1.6 SOA/Web Services and Other Technologies/Styles for Mobile Services

1.6.1 OMA/Mobile Web Services

The Open Mobile Alliance (OMA) (http://www.openmobilealliance.org) has defined the OMA mobile Web service (MWS) to provide guidelines and specifications for the integration and interoperability of Web services with the OMA architecture. With respect to architectural styles, MWS supports SOAP-based models. MWS considers cases when Web service consumers support and do not support full Web service stacks. Thus, both cases in Figure 1.3 are included: the interaction between mobile Web service consumers and mobile Web services can be direct or through proxy or intermediaries. With respect to service discovery, MWS supports centralized discovery based on UDDI and provides security guidelines for both transport and message security based on SSL/TSL and WS-Security.

1.6.2 CDC/OSGi Model

Another trend to support SOA-based mobile services is to use the OSGi technology defined by the OSGi Alliance (http://www.osgi.org). The OSGi technology specifies a dynamic component system for Java. In this system, components interact with each other within a single system. OSGi components can be published, discovered, and composed dynamically at runtime. They can also be deployed and undeployed on demand and be exposed as Web services. Various OSGi implementations have been provided for different platforms, including mobile devices.

CDC/OSGi is a capable platform for developing mobile services based on SOA. Various platforms are supported by CDC/OSGi. Examples are Sprint Titan (http://developer.sprint.com/site/global/develop/technologies/sprint_titan/p_sprint_titan.jsp) and Apache MUSE (http://ws.apache.org/muse/) that provides development of Web services atop OSGi. R-OSGi [50] is a middleware built atop OSGi that transparently connects, deploys, and executes OSGi services spanning multiple OSGi containers; thus, it can be used for developing P2P and collaborative mobile Web services.

1.6.3 Event-Driven Architecture and SOA

The event-driven architecture (EDA) [51] is a computing paradigm in which changes are sensed and captured in events and corresponding actions are performed according to events. The basic tenet of EDA is that an event reflects a significant change of something by gathering meaningful information about the change. The fundamental conceptual entities of an EDA system are event sources (event emitters) to generate events and event sinks (event consumers) to perform actions based on events (e.g., filtering, relay, or processing events). Events are sensed, transferred, and processed through a system of event channels loosely coupling distributed event sources and sinks.

EDA systems do not necessarily include SOA-based services. However, the need to integrate events from disparate sources has fostered the integration between EDA and SOA [52,53]. This integration has been proposed through standard bodies as well as architecture solutions. WSEventing (http://www.w3.org/Submission/WS-Eventing/) and WS-ECA [54] have been proposed for supporting event subscription, propagation, and event–condition–action rules.

1.6.4 Web Applications, WAP, and SIP

Web applications, WAP, and SIP models are other technologies/architecture styles that are related to SOA and mobile services (see also Section 1.3.1.3 and Figure 1.4). Although they are different technologies/styles, all of them require proxy in order for the mobile applications on mobile devices to interact with other services at the back end.

The Web application model for mobile services is widely used in practice. Conceptually, the Web application model follows a three-tier architecture: the mobile service consumer is responsible for the presentation layer; the Web application, the application tier, acts as a gateway to provide dynamic Web content to the mobile service consumer; and the mobile services provide data and content to the gateway. The mobile service consumer is normally developed based on Web forms, XForm, JavaScript, and HTML/XHTML, whereas the Web application gateway uses dynamic Web content technologies such as ASP and JSP. Typically, the Web application model widely supports synchronous interactions in which the requester pushes the requests and receives the content. Recently, advanced AJAX techniques for mobile devices also support the requester to asynchronously poll results. Some toolkits are available for mobile AJAX, such as MobileWeb (http://mymobileweb.morfeo-project.org/). A single-purpose mobile Web application can also be packaged for download and installation on mobile devices; this is called a mobile widget [55]. This model is quite suitable for applications that require Web contents. Examples of Web application model toolkits are the Yahoo! Blueprint Platform, Microsoft ASP.NET mobile, and Microsoft Mobile Internet Toolkit.

Other architectural styles for mobile services are based on WAP (http://www.wapforum.org/what/technical_1_2_1.htm) and SIP (http://www.ietf.org/html.charters/sip-charter.html). WAP defines how an application on mobile phones or PDAs can access the Internet through HTTP. WAP defines a protocol stack atop different wireless data networks, such as SMS, GPRS, and UMTS. This protocol stack includes protocols dealing with datagram, transport, transaction, session control, and environments. A WAP application accesses the Internet through a WAP gateway that acts as a proxy. WAP has been widely used in services, such as mobile mails for Mobile CRM [56], transportation information systems (http://www.tfl.gov.uk/tfl/livetravelnews/mobileservices/wap/default.asp), and mobile banking. SIP is a signaling protocol widely used in multimedia mobile services, such as streaming multimedia delivery, instant messaging, and presence service. As it focuses on multimedia aspect, using SIP alone will support only a few types of mobile services. However, SIP can be combined with SOA-based services to provide integrated communication services and enterprising business services. This type of application is particularly important for the mobile user in networked enterprises and collaborative environments. Examples of SIP together SOA are IMS [21] and Akogrimo [11,57]. A SIP for mobile devices is also developed in Java ME (JSR 180—Session Initiation Protocol for J2ME 1.0.1).

1.7 Challenges for Future Research

As partially discussed in previous sections, we need to develop further techniques to bring the full advantage of SOA-based solutions to mobile services.

1.7.1 Supporting Mobile Web Services on Mobile Devices

Supporting tools should not only focus on the client side but also on mobile Web services in mobile devices. This will require great research effort for the development of service management systems that have to take into account service mobility, run-time deployment, and service continuity on mobile devices. Here, ensuring correct information about services and service continuity is a great challenge. The issue related to service registration and discovery when Web services are hosted in mobile devices, as discussed in Reference [6], is still valid. We need to combine service migration and adaptation with service monitoring and management techniques. This issue is also related to service disruption due to failure of devices and networks.

1.7.2 Supporting REST Mobile Services

Important supports of REST-based mobile services, such as service discovery and resource management, should be provided. In particular, for publishing and discovering REST-based Web services a new set of protocols has to be developed.

1.7.3 Programming Supporting Tools

Although various programming environments are provided, the developer lacks useful tools to debug, profile, and analyze their mobile services. In particular, given the strong connection between performance and energy consumption in mobile devices, tools for analyzing energy consumption and performance are particularly interesting. Currently, only a few works have been devoted for this research issue.

1.7.4 Context Sharing in Mobile Services

When mobile services are integrated from different providers, sharing context information among these services is challenging. Context sharing is a key to reduce user intervention and improve service provisioning and adaptation. Existing standards have proposed sharing user identity but it is not enough as the context associated with a usage is much more than identity. Thus, standard protocols for specifying, propagating, and managing context in a federation of services are desired. This research is also involved with context-aware computing domains.

1.7.5 User-Defined Opportunistic Composition and Creation of Mobile Service

When a lot of mobile services are available, the user will have many opportunities to compose mobile services on their own. SOA solutions have helped to simplify service discovery processes and many service composition techniques have been proposed, yet they are complex and assume advanced knowledge. How can we support a novice user to compose mobile services in a place where, by chance, his/her mobile device detects many interesting mobile services provided by other users and onsite service providers? In addition, how can we support the user to define his/her own mobile services? Although some initial ideas have been proposed [36,58], these are premature.

1.8 Conclusion

Over the last few years, the increasing mobility of the user and the advanced, powerful mobile networks and devices have fostered emerging mobile services for different scenarios. Mobile services are no longer just for accessing Web contents and for personal use but also for enterprise use and collaborative work. Furthermore, there is no longer a clear boundary between mobile services hosted in mobile networks and on the Internet. This requires mobile applications to access vast information and services from anywhere with any device. As a result, SOC/SOA technologies, in particular, Web service technologies, have been utilized in mobile services, solving many integration problems.

In this chapter, we analyze state-of-the-art SOC/SOA for mobile services with a focus on mobile Web services. Both SOAP-based and REST-based architectural styles have been discussed. On the one hand, many Web service solutions, based on SOAP and REST technologies, have been introduced for mobile services, enhancing substantially how mobile services should provide and interact with the mobile user. On the other hand, Web service solutions for collaborative works, distributed health care, and enterprises have been developed, extending the use of mobile services for individual need to team work and enterprise business. Both REST and SOAP mobile Web services have different capabilities and this difference needs to be analyzed to select the right architectural style. Other chapters in this book provide further state-of-the-art research on particular topics that are sketched in this chapter, such as security, service discovery, monitoring, and performance analysis.

1.9 Further Reading

Several books have presented many fundamental concepts of SOC/SOA, such as *Web Services: Concepts, Architecture and Applications* [59] and *SOA Principles of Service Design* [60]. With respect to the development of mobile services, there are valuable References for further reading. *Mobile Web Services: Architecture and Implementation* [61] presents main concepts in designing Web services for mobile systems. Generic concepts such as addressing, service discovery, identity management, and security in SOA-based environments are covered. This book also presents implementation techniques and examples of mobile Web services. However, it focuses on SOA based on Nokia service development APIs for S60 platform.

Mobile Web Services [62] presents main technologies for implementing wireless mobile services with a focus on mobile Web services providing Web contents.

Among others, it covers very well WAP, wireless content representations, location management, privacy, and mobile user context. However, it does not focus on SOA solutions for mobile Web services.

Enabling Technologies for Mobile Services: The MobiLife Book [3] presents a compelling landscape about mobile services based on the user-centered design process. It analyzes requirements of the mobile world and discusses the mobile services architecture accordingly. Key enabling technologies such as context management framework, multimodal and personalization technologies, and trust and privacy are well presented. This book also presents Reference applications, best practices, and marker analysis of mobile services. It is an excellent reading source for understanding mobile applications and services.

Mobile service-oriented architectures in general are discussed in Reference [63]. This discussion does not indicate any specific Web services platform. Some mobile programming toolkits were evaluated in Reference [64]. This evaluation discussed the performance of gSOAP and kSOAP Web service consumers in embedded devices.

This chapter gives an overview of the main techniques for mobile Web services with a focus on programming support for service development, service invocation, message handling, and discovery. Therefore, it does not present these techniques in detail in terms of qualitative and behavioral analysis, such as performance, scalability, energy consumption, and detailed protocol structures. Such details can be found in external References, other chapters of this book, and the above-mentioned further readings.

EXERCISES

1. Survey existing mobile services, such as Google and StrikeIron services, and map them into a table of direct/indirect interaction, SOAP, REST, and other architectural styles; and compare their strengths and weaknesses.
2. Not all published WSDL files of Web services are compatible with Java ME. Analyze existing WSDL files, find common incompatible properties, and present your suggestion.
3. Assume that some services are hosted in a mobile device and services are described by WSDL and UDDI. Analyze possible issues that arise when the service provider migrates the services to another hosting environment. What should be done to ensure that the service discovery process is not corrupted?
4. Analyze possible service discovery protocols for REST-based Web services. Implement a P2P-based service discovery for REST Web services.
5. Compare data transfer using JSON and XML format. Which format is suitable for which application types? Perform a performance comparison for similar Web services that offer both JSON and XML (e.g., Google and Yahoo! mobile services).

6. Currently, many REST-based Web services return XML and/or JSON, but programming toolkits assume the developer will utilize specific libraries (e.g., kXML or JSONObject) to parse the returned data. Develop a small project to integrate tools for generating code from WADL and tools for parsing XML/JSON data for REST clients.
7. Study service information associated with mobile Web services on mobile devices. Compare the importance of the service information of mobile Web services with that of Web services in other platforms in terms of service discovery and service management.
8. Analyze the dependency between mobile Web services and the contemporary operating systems for mobile devices (PDAs and smart phones) in terms of concurrency processing and service disruption. Analyze the impact of issues on the selection of architectural styles, service invocation models, and fault management.

Acknowledgments

The effort spent in writing this chapter is partially supported by the European Union through the FP6 projects inContext and WORKPAD, and the FP7 project SM4All. The research in this chapter is partially a result of our experiences gained from the development of SOA-based solutions for teamwork, disaster management, and smart environments in the inContext (http://www.in-context.eu), the WORKPAD (http://www.workpad-project.eu), and the SM4All (http://www.sm4all-project.eu) projects. We are grateful to all our colleagues in inContext, WORKPAD, and SM4All for fruitful discussion and sharing ideas on the development of mobile Web services. Our discussions on Vimoware/RESCUE and the WORKPAD example are based on research results conducted within the WORKPAD project.

References

1. Mike P. Papazoglou, Paolo Traverso, Schahram Dustdar, and Frank Leymann. Service-oriented computing: State of the art and research challenges. *IEEE Computer*, 40(11):38–45, 2007.
2. Ivar Jørstad, Schahram Dustdar, and Do Van Thanh. An analysis of current mobile services and enabling technologies. *IJAHUC*, 1(1/2):92–102, 2005.
3. Mika Klementtinen, editor. *Enabling Technologies for Mobile Services: The MobiLife Book*. West Sussex, England: John Wiley & Sons, October 2007.
4. Ivar Jørstad, Schahram Dustdar, and Do Van Thanh. Service-oriented architectures and mobile services. In Jaelson Castro and Ernest Teniente, editors, *CAiSE Workshops (2)*, pp. 617–631. Porto: FEUP Edicoes, 2005.

5. Hong Linh Truong, Lukasz Juszczyk, Shariq Bashir, Atif Manzoor, and Schahram Dustdar. Vimoware—a toolkit for mobile web services and collaborative computing. In *SEAA '08: Proceedings of the 34th Euromicro Conference on Software Engineering and Advanced Applications*, pp. 366–373, Parma, Italy, September 3–5, 2008.
6. Stefan Berger, Scott McFaddin, Chandra Narayanaswami, and Mandayam Raghunath. Web services on mobile devices—implementation and experience. *wmcsa*, 0:100, 2003.
7. Ari Shapiro and Andreas Frank. Mobile SOA: End-to-end Java™ technology-based framework for network services. In *Proceedings of JavaOne 2008 Conference*, San Francisco, May 5–9, 2008.
8. Mobile world celebrates four billion connections, http://www.gsmworld.com/newsroom/press-releases/2009/2521.htm. Last accessed: February 26, 2009.
9. Gerd Andersson, Adrian Bullock, Jarmo Laaksolahti, Stina Nylander, Fredrik Olsson, LinderMarie Sjö, Annika Waern, and Magnus Boman. Classifying mobile services. *SICS Technical Report T2004:04*. Swedish Institute of Computer Science ISSN 1100-3154, 2004. http://eprints.sics.se/2349/01/SICS-T–2004-04–SE.pdf
10. Marco Savini, Andreea Ionas, Andreas Meier, Ciprian Pop, and Henrik Stormer. The eSana framework: Mobile services in eHealth using SOA. In *Proceedings of the Second European Conference on Mobile Government*, Brighton, August 30–31 and September 1, 2006, ISBN: 0-9763341-1-9.
11. The Akigrimo project. http://www.mobilegrids.org/
12. Tiziana Catarci, Massimilano de Leoni, Andrea Marrella, Massimo Mecella, Berardino Salvatore, Guido Vetere, Schahram Dustdar, Lukasz Juszczyk, Atif Manzoor, and Hong-Linh Truong. Pervasive software environments for supporting disaster responses. *IEEE Internet Computing*, 12(1):26–37, 2008.
13. David Booth, Hugo Haas, Francis McCabe, Eric Newcomer, Michael Champion, Chris Ferris, and David Orchard. Web services architecture (W3C Working Group note, February 11, 2004), 2004. http://www.w3.org/TR/ws-arch/. Last accessed: February 10, 2009.
14. Gregory Hackmann, Mart Haitjema, Christopher D. Gill, and Gruia-Catalin Roman. Sliver: A BPEL workflow process execution engine for mobile devices. In Asit Dan and Winfried Lamersdorf, editors, *ICSOC*, volume 4294 of *Lecture Notes in Computer Science*, pp. 503–508. The Netherlands: Springer, 2006.
15. A. van Halteren and P. Pawar. Mobile service platform: A middleware for nomadic mobile service provisioning. In *IEEE International Conference on Wireless and Mobile Computing, Networking and Communications, 2006. (WiMob'2006)*, pp. 292–299, Montreal, Quebec, June 19–21, 2006.
16. Roy Thomas Fielding. Architectural styles and the design of network-based software architectures. Ph.D. thesis, University of California, Irvine, 2000.
17. S. McFaddin, D. Coffman, J.H. Han, H.K. Jang, J.H. Kim, J.K. Lee, M.C. Lee et al., Modeling and managing mobile commerce spaces using restful data services. In *9th International Conference on Mobile Data Management, 2008, MDM '08*, pp.81–89, Berlin, April 27–30, 2008.
18. Sami Mäkeläinen and Timo Alakoski. Fixed-mobile hybrid mashups: Applying the rest principles to mobile-specific resources. In *WISE '08: Proceedings of the 2008 International Workshops on Web Information Systems Engineering*, pp. 172–182. Berlin/Heidelberg: Springer-Verlag, 2008.

19. ArcGIS mobile blog. http://blogs.esri.com/Dev/blogs/mobilecentral/archive/2008/10/20/The-Mobile-Web.aspx. Last accessed: March 13, 2009.
20. Cesare Pautasso, Olaf Zimmermann, and Frank Leymann. Restful web services vs. "big"' web services: Making the right architectural decision. In Jinpeng Huai, Robin Chen, Hsiao-Wuen Hon, Yunhao Liu, Wei-Ying Ma, Andrew Tomkins, and Xiaodong Zhang, editors, *17th International World Wide Web Conference (WWW2008)*, pp. 805–814. Beijing, China: ACM, April 21–25, 2008.
21. Hechmi Khlifi and Jean-Charles Grégoire. IMS application servers: Roles, requirements, and implementation technologies. *IEEE Internet Computing*, 12(3):40–51, 2008.
22. XMLP scenarios. http://www.w3.org/TR/xmlp-scenarios/. Last accessed: March 12, 2009.
23. Guido Gehlen and Georgios Mavromatis. Mobile web service based middleware for context-aware applications. In *Proceedings of the 11th European Wireless Conference 2005*, Vol. 2, pp. 784–790. Nicosia, Cyprus: VDE Verlag, April 2005.
24. Valerie Issarny, Daniele Sacchetti, Ferda Tartanoglu, Françoise Sailhan, Rafik Chibout, Nicole Levy, and Angel Talamona. Developing ambient intelligence systems: A solution based on web services. *Automated Software Engineering*, 12(1): 101–137, 2005.
25. Fahad Aijaz, Bilal Hameed, and Bernhard Walke. Asynchronous mobile web services: Concept and architecture. In *Proceedings of 2008 IEEE 8th International Conference on Computer and Information Technology*, p. 6, Sydney, Australia, July 2008.
26. Yeon-Seok Kim and Kyong-Ho Lee. A light-weight framework for hosting web services on mobile devices. In *ECOWS '07: Proceedings of the Fifth European Conference on Web Services*, pp. 255–263. Washington, DC, USA: IEEE Computer Society, 2007.
27. Hong Linh Truong, Lukasz Juszczyk, Atif Manzoor, and Schahram Dustdar. Escape—An adaptive framework for managing and providing context information in emergency situations. In Gerd Kortuem, Joe Finney, Rodger Lea, and Vasughi Sundramoorthy, editors, *EuroSSC*, volume 4793 of *Lecture Notes in Computer Science*, pp. 207–222. The Netherlands: Springer, 2007.
28. Martin Treiber and Schahram Dustdar. Active web service registries. *IEEE Internet Computing*, 11(5):66–71, 2007.
29. Marc Hadley. Web application description language (WADL). *Technical Report TR-2006-153*, Sun Microsystems, April 2006.
30. Robert Steele, Khaled Khankan, and Tharam Dillon. Mobile web services discovery and invocation through auto-generation of abstract multimodal interface. In *ITCC '05: Proceedings of the International Conference on Information Technology: Coding and Computing (ITCC'05)—Volume II*, pp. 35–41. Washington, DC, USA: IEEE Computer Society, 2005.
31. Satish Narayana Srirama, Matthias Jarke, and Wolfgang Prinz. Mobile web services mediation framework. In *MW4SOC '07: Proceedings of the 2nd Workshop on Middleware for Service Oriented Computing*, pp. 6–11. New York, NY, USA: ACM, 2007.
32. Satish Narayana Srirama, Matthias Jarke, Hongyan Zhu, and Wolfgang Prinz. Scalable mobile web service discovery in peer to peer networks. In *Third International Conference on Internet and Web Applications and Services, 2008, ICIW '08*, pp. 668–674. Athens, Greece: IEEE Computer Society, June 8–13, 2008.

33. Lukasz Juszczyk and Schahram Dustdar. A middleware for service-oriented communication in mobile disaster response environments. In Sotirios Terzis, editor, *MPAC*, pp. 37–42. New York, NY, USA: ACM, 2008.
34. RESCUE—Service oriented middleware for mobile devices, http://www.infosys.tuwien.ac.at/prototyp/Rescue/Rescue_index.html. Last accessed: February 26, 2009.
35. Lasse Pajunen and Suresh Chande. Developing workflow engine for mobile devices. In *EDOC '07: Proceedings of the 11th IEEE International Enterprise Distributed Object Computing Conference*, p. 279. Washington, DC, USA: IEEE Computer Society, 2007.
36. Marco Pistore, Paolo Traverso, Massimo Paolucci, and Matthias Wagner. From software services to a future Internet of services. *Future Internet Assembly* 183–192, 2009.
37. Mia Tian, Thiemo Voigt, Tomasz Naumowicz, Hartmut Ritter, and Jochen Schiller. Performance considerations for mobile web services. *Computer Communications Journal*, 27(11):1097–1105, 2004.
38. Ivan Jorstad, Do van Thanh, and Schahram Dustdar. A service continuity layer for mobile services. *2005 IEEE Wireless Communications and Networking Conference*, 4:2300–2305, 2005.
39. Christoph Dorn and Schahram Dustdar. Achieving web service continuity in ubiquitous mobile networks: The SRR-WS framework. In Moira C. Norrie, Schahram Dustdar, and Harald Gall, editors, *UMICS*, volume 242 of *CEUR Workshop Proceedings*, CEUR-WS.org, 2006.
40. kSOAP2, http://ksoap2.sourceforge.net/. Last accessed: February 26, 2009.
41. Robert van Engelen and Kyle Gallivan. The gSOAP toolkit for web services and peer-to-peer computing networks. In *CCGRID*, pp. 128–135. Washington, DC, USA: IEEE Computer Society, 2002.
42. Vittorio Miori, Luca Tarrini, and Rolando Bianchi Bandinelli. Deliverable d2.2: Requirements analysis for footprint and power constrained devices; light—xml-innovative generation for home networking technologies. *Technical Report*, CNR, Pisa, Italy, May 2005. http://dienst.isti.cnr.it/Dienst/Repository/2.0/Body/ercim.cnr.isti/2005-PR-02/pdf?tiposearch=ercim&langver=
43. Nokia web services platform. http://www.forum.nokia.com/Resources_and_Information/Explore/Other/Web_Services/. Last accessed: March 13, 2009.
44. Peter Braun and Ronny Eckhaus. Experiences on model-driven software development for mobile applications. In *ECBS '08: Proceedings of the 15th Annual IEEE International Conference and Workshop on the Engineering of Computer Based Systems*, pp. 490–493. Washington, DC, USA: IEEE Computer Society, 2008.
45. Juergen Dunkel and Ralf Bruns. Model-driven architecture for mobile applications. In Witold Abramowicz, editor, *Business Information Systems, 10th International Conference, BIS 2007*, Poznan, Poland, April, pp. 470–483. Berlin/Heidelberg: Springer-Verlag, 2007.
46. Mariano Belaunde and Paolo Falcarin. Realizing an MDA and SOA marriage for the development of mobile services. In *ECMDA-FA '08: Proceedings of the 4th European Conference on Model Driven Architecture*, pp. 393–405. Berlin/Heidelberg: Springer-Verlag, 2008.
47. Daniele Battista, Massimiliano de Leoni, Alessio De Gaetanis, Massimo Mecella, Alessandro Pezzullo, Alessandro Russo, and Costantino Saponaro. Rome4eu: A web service-based process-aware system for smart devices. In Athman Bouguettaya, Ingolf

Krüger, and Tiziana Margaria, editors, *ICSOC*, volume 5364 of *Lecture Notes in Computer Science*, pp. 726–727. The Netherlands: Springer, 2008.
48. Shengru Tu and Mahdi Abdelguerfi. Web services for geographic information systems. *IEEE Internet Computing*, 10(5):13–15, 2006.
49. Marco Aiello and Schahram Dustdar. Are our homes ready for services? A domotic infrastructure based on the web service stack. *Pervasive and Mobile Computing*, 4(4):506–525, 2008.
50. Jan S. Rellermeyer, Gustavo Alonso, and Timothy Roscoe. R-OSGi: Distributed applications through software modularization. In Renato Cerqueira and Roy H. Campbell, editors, *Middleware*, volume 4834 of *Lecture Notes in Computer Science*, pp. 1–20. The Netherlands: Springer, 2007.
51. Brenda M. Michelson. Event-driven architecture overview. *Patricia Seybold Group*, February 2006. http://soa.omg.org/Uploaded%20Docs/EDA/bda2-2-06cc.pdf
52. Jack van Hoof. How EDA extends SOA and why it is important, September 2006. http://soa-eda.blogspot.com/2006/11/how-eda-extends-soa-and-why-it-is.html
53. Jeff Hanson. Event-driven services in SOA, January 2005. http://www.javaworld.com/javaworld/jw-01-2005/jw-0131-soa.html
54. Jae-Yoon Jung, Jonghun Park, Seung-Kyun Han, and Kangchan Lee. An ECA-based framework for decentralized coordination of ubiquitous web services. *Information & Software Technology*, 49(11–12):1141–1161, 2007.
55. Marcos Caceres. Widgets 1.0: Packaging and configuration—W3C working draft, December 22, 2008. http://www.w3.org/TR/2008/WD-widgets-20081222/
56. Kevin H.W. Shen and Daniel C.H. Lee. WAP mail service and short message service for mobile CRM. In *MSE '00: Proceedings of the 2000 International Conference on Microelectronic Systems Education*, p. 201. Washington, DC, USA: IEEE Computer Society, 2000.
57. Juergen M. Jaehnert, Stefan Wesner, and Victor A. Villagra. The Akogrimo mobile grid Ref. architecture—Overview. http://www.mobilegrids.org/modules.php?name=UpDownload&req=getit&lid=108, November 2006, whitepage.
58. Daniel Schall, Hong Linh Truong, and Schahram Dustdar. Unifying human and software services in web-scale collaborations. *IEEE Internet Computing*, 12(3):62–68, 2008.
59. Gustavo Alonso, Fabio Casati, Harumi Kuno, and Vijay Machiraju. *Web Services: Concepts, Architecture and Applications*. Berlin/Heidelberg: Springer-Verlag, 2004.
60. Thomas Erl. *SOA Principles of Service Design (The Prentice Hall Service-Oriented Computing Series from Thomas Erl)*. Upper Saddle River, NJ, USA: Prentice Hall PTR, 2007.
61. Frederick Hirsch, John Kemp, and Jani Ilkaka. *Mobile Web Services: Architecture and Implementation*. New York: John Wiley & Sons, 2006.
62. Ariel Pashtan. *Mobile Web Services*. New York, NY, USA: Cambridge University Press, 2005.
63. Jilles van Gurp, Anssi Karhinen, and Jan Bosch. Mobile service oriented architectures (MOSOA). In Frank Eliassen and Alberto Montresor, editors, *DAIS*, volume 4025 of *Lecture Notes in Computer Science*, pp. 1–15. The Netherlands: Springer, 2006.
64. Daniel Schall, Marco Aiello, and Schahram Dustdar. Web services on embedded devices. *IJWIS*, 2(1):45–50, 2006.

Chapter 2

Service Discovery for Mobile Computing

Classifications, Considerations, and Challenges

Jian Zhu, Mohammad Oliya, and Hung Keng Pung

Contents

2.1 Introduction

The rapid growth of the Internet and network systems has led to an ever-increasing number of businesses participating in e-commerce worldwide. Today, the Web has evolved from solely a repository of data to a collection of complex and heterogeneous services. Furthermore, the permeation of mobile and wireless computing further eases the process of sharing and accessing services. Mobile users may share their application services (e.g., video streaming) anytime, anywhere through their personal devices, such as hand phones and personal digital assistants (PDAs). Service discovery in this mobile and dynamic environment is thus a challenging task, as the mobility and dynamism of services may greatly affect their relevance to the user.

In network systems, a service may refer to any software or hardware facility that is available to the user, whereas in a business environment, a service is a set of singular and perishable benefits generated by functions of technical systems or by distinct activities of individuals [1]. Service discovery in the computing environment can be defined as *network protocols that allow automatic detection of hardware devices and services as well as those generated by activities of humans on a computer network*. Once the demanded service is located, the user may invoke and access it if it is a Web service or go to the actual location of its provider to get served.

Each service discovery protocol (SDP) consists of two basic participating entities: the *service provider* and the *service user*. The service provider is the entity that offers the service, whereas the service user is the one that is interested in finding and using a service. In conventional frameworks, the role of an entity is fixed for both the service provider and the user; however, with the advance of recent peer-to-peer (P2P) techniques, the emphasis on the role becomes vague. In certain cases, the service entity can be both the provider and the user at the same time. For service discovery, a considerable number of frameworks already exist, but the core mechanism for service discovery remains the same—*advertise*, *discover*, and *access*. The advertisement or the process of service publication is usually carried out by deploying a third participating entity known as the *service directory*. This yellow page is responsible for hosting partially or entirely all the service information for service matching in the discovery phase. However, as we will see later, a service directory is not a compulsory entity in a service discovery framework and has its own advantages and disadvantages. Service discovery is the process of finding services, where the mechanism used is strongly related to the service organization scheme in the framework as well as to the underlying network types. As an example, the organization can be centralized or distributed, and the network can be wired or wireless. Finally, service access refers to the method used to invoke and utilize the service if it is a Web service.

This chapter covers current service discovery frameworks developed in industry and research communities and provides a way to classify them on the basis of the taxonomy presented here. The strengths and weaknesses of different approaches are discussed. The second part of the chapter focuses on considerations and challenges that are specific to service discovery for mobile computing. Potential research areas are also indicated at the end of the chapter. We also note that there are already several review papers on service discovery in recent years, such as those by Marin-Perianu et al. [2], Zhu et al. [3], Hosseini-Seno et al. [4], Ververidis et al. [5], and Meshkova et al. [6]. Whereas Hosseini-Seno et al. [4] and Ververidis et al. [5] focus on service discovery frameworks in mobile *ad hoc* networks (MANETs), Meshkova et al. [6] have done a broader survey for general resource discovery mechanisms including P2P frameworks. However, taxonomies adopted in References [4–6] are more from a network perspective rather than a service-related point of view. For instance, they address packet propagation and routing methods, and issues such as service matching and invocation are not emphasized. On the basis of this judgment, surveys elaborated in References [2,3] are closer to our approach, but our review is more detailed and has a clear taxonomy based on the core mechanism of service discovery (i.e., advertise, discover, access). Besides, considerations of service discovery for mobile computing are emphasized in particular.

The rest of this chapter is organized as follows. In the rest of the Introduction, we provide a taxonomy of major components used in service discovery frameworks and present existing frameworks deployed in industry. In Section 2.2, detailed service discovery approaches are described and classified on the basis of the taxonomy. In addition, their strengths and weaknesses are discussed. Section 2.3 presents considerations of service discovery, especially for mobile computing. Section 2.4 discusses challenges and issues identified as future research directions. Finally, we conclude the chapter in Section 2.5.

2.1.1 Service Discovery Components

Different terminologies are used in existing service discovery frameworks to reflect their unique properties. Nevertheless, a common taxonomy is required to analyze the strengths and weaknesses of the frameworks. Furthermore, this common base could help identify potential areas in service discovery for mobile computing that need more elaboration. In the study by Hosseini-Seno et al. [4], a taxonomy is defined as consisting of eight components, most of which are based on network considerations, such as network type and packet propagation method. Their paper does not emphasize the differences between service and resource discovery as stated by the authors. Zhu et al. [3] consider SDP components on the service application level in a flat view and classify existing frameworks into 10 choices, including service attribute naming and service invocation. We argue that it is better to design the taxonomy on the basis of the core mechanism for service discovery, as it clearly presents the components needed by service discovery and the design choices for each component. Thus, in this chapter, the taxonomy of the components of a service discovery framework in Figure 2.1 is adopted. In addition, in Section 2.2, we will further extend the taxonomy for some components to give a more detailed classification.

2.1.2 Industry Standards

In this section, we would like to briefly introduce existing frameworks adopted by industry to give readers an overview of service discovery in real-world scenarios. The industry standards presented in this chapter include Universal Description, Discovery, and Integration (UDDI) [7], Jini [8], Universal Plug and Play (UPnP) [9],

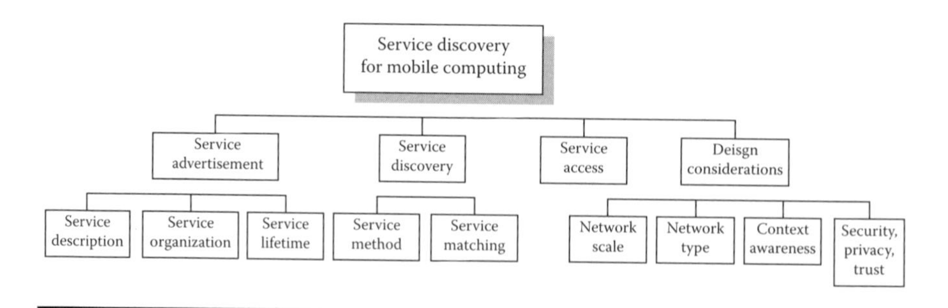

Figure 2.1 A taxonomy of components of a service discovery framework.

Bluetooth [10], Service Location Protocol (SLP) [11], Salutation [12], Bonjour [13], and DEAPspace [14].

2.1.2.1 Universal Description, Discovery, and Integration

The UDDI project is an industry initiative to define publishing and discovering Web services and to advance the B2B integration on the Internet [7]. The most important component in the framework is the Universal Business Registry (UBR), which is a master directory for all publicly available Web services. Four types of information are stored in the UBR, namely, business information, service information, binding information, and information about specifications for services. Moreover, to better understand services such as the format of exchanged data and the protocol used, the technical model (tModel) is linked with each service template. For service registration and discovery, the UDDI provides a set of application programming interfaces (APIs) for service providers and consumers to interact with the registry. For example, service providers can use *save_xx()* and *delete_xx()* to update and delete their entities from the registry. Similarly, service consumers can use *find_xx()* and *get_xx()* to find and get detailed information about the entities in which they are interested.

UDDI has contributed to the ease of administration control and service management, which is also superior to service discovery in local and static environments. The disadvantage, however, is the lack of scalability in the design of even the latest version—UDDIv3 supports affiliations of registries. Besides, it is designed to support Web service discovery only. From this perspective, users of UDDI must have prior knowledge regarding Web service protocols, for example, Web Services Description Language (WSDL), and Simple Object Access Protocol (SOAP). Another major problem of UDDI is that it does not guarantee any quality (e.g., availability) of the registered services, an issue investigated by Hagemann et al. [15]. There are proposals to address this problem by extending UDDI, as found in References [16,17].

2.1.2.2 Jini

The Jini framework [8], developed by Sun Microsystems and now transferred to the Apache River project [18], specifies a way for clients and services to find each other on the network and to work together to accomplish a task. The whole architecture is built on top of the Java Virtual Machine (JVM), and thus is platform-independent. It organizes services into service groups called *djinn*, and each group has a DNS-style name associated with it. Within *djinn*, there is a central registry called the Jini Lookup Server (JLS) that maintains a flat collection of service items. Each service item represents an instance of a service, and is described by a universally unique identifier (UUID) as well as a collection of attributes. When looking for a service, clients must specify a *serviceTemplate* including a UUID (if known), service type, service attributes, and so on. A match is made between the service description (i.e., *serviceItem*) and *serviceTemplate* by using a set of rules. Note Jini also specifies how to discover one or more JLSs by using multicast and unicast network communication protocols.

A special feature of Jini is the mobile Java code, which can be moved among requestors, services, and registries. The requestors may look up registries in other networks by using a multicast protocol. In this way, the Jini framework achieves system scalability. However, the problem of Jini is that it requires all entities involved in service discovery to have a Java runtime environment. Moreover, not all routers in the Internet support multicast protocols.

2.1.2.3 Universal Plug and Play

The goals of the UPnP protocol [9] are to allow devices to connect to networks with zero configuration and to simplify the process of using devices from a wide range of vendors. There are three basic components in the protocol: *devices*, *services*, and *control points*. A UPnP device is a container of services and nested devices; a service is the smallest unit of control; and a control point is a controller capable of discovering and controlling other devices. Whenever a service wants to join the network, it needs to send out a message to notify others about its presence. The advertisement is done using a multicast protocol. However, before the device advertises its services, it needs to obtain an Internet protocol (IP) address for device identification. UPnP supports the AutoIP protocol; that is, the device can get its IP address even without a dynamic host configuration protocol server. Once the device advertises its service information including the service URL, other devices in the network can observe the advertisement and control points can record it. When a client wants to discover a service, the simple SDP over IP networks is adopted. The client can do a multicast of its query to other devices in the network. The query can be answered either by the control point, which has the requested service information, or by the service directly.

A trivial problem of UPnP is its extensive usage of the multicast mechanism that results in scalability issues; however, it is assumed to be applicable to the home or office only.

2.1.2.4 Bluetooth

Bluetooth [10], as a recent industrial standard, is a cable-replacement technology designed to wirelessly connect peripherals, which is unlike UPnP's wired constraint. The technology allows a Bluetooth-enabled device to communicate via short-ranged radio links with low power and low cost. For service discovery, the SDP is defined at the application layer. Services in the SDP are represented by a *service record* form in the same way as Jini, and each service is an instance of a service class that provides an abstraction for attributes associated with each service record; however, users may create their own attributes for the service provided.

Service matching is done on the basis of the values of service attributes contained in those service records. The SDP also allows clients to browse for services that are currently available.

Similar to UPnP, the Bluetooth protocol is designed for personal area network and with low energy cost in mind; therefore, it only supports short-range communication, for example, 1–100 m.

2.1.2.5 Service Location Protocol

The SLP [11] is an Internet Engineering Task Force (IETF) standard that supports automatic resource discovery. Similar to UPnP, this protocol is only for IP-based networks, though it scales better than UPnP. The SLP defines three types of agents in the service discovery framework: *user agent* (UA), *service agent* (SA), and *directory agent* (DA). The UAs acquire service handles for end-user applications that request the services. The SAs are responsible for advertising service handles to the DAs. The DA maintains a list of the advertised services in the network. Services are described using unique URLs and a set of attribute–value pairs. Like the Jini framework, the SLP uses a multicast protocol for service registration. Once SAs join the network they multicast a registration message. Also, if there is a new DA available, it will send an unsolicited advertisement infrequently to inform the UAs and SAs of its presence. Service discovery can be done either by multicasting a service request message so as to be directly answered by existing SAs or by unicasting the request to a known DA to get relevant information.

Services in the SLP are grouped together using *scope*, which could indicate a location, administrative grouping, and proximity in a network topology or some other category. SAs and DAs are always assigned a scope string. A UA is normally assigned a scope string, in which case the UA will only be able to discover that particular group of services, allowing a network administrator to provide services to users. Note that due to the IP multicast constraint, the SLP does not scale well in the Internet, and, as claimed, it is intended to function within networks under cooperative administrative control and enterprise networks with shared services.

2.1.2.6 Salutation

The Salutation protocol [12] is an open standard service discovery and session management protocol, although currently dissolved. The architecture defines an entity called the Salutation Lookup Manager (SLM), which, similar to JLS in Jini, acts as a service broker for services in the network. Services are described by using their different functions called Function Units that represent essential features of services and are kept by the local SLM. Service discovery can be performed across multiple SLMs either by unicasting to a central directory server (e.g., DA in SLP) or by broadcasting. The matching is done by sending the required service type from the local SLM to the remote SLM through the Remote Procedure Call (RPC) for comparisons.

The Salutation architecture defines a Transport Manager (TM) component that implements the detailed transport functionality to operate on different network technologies such as Infrared and Bluetooth. An SLM may have multiple TMs for the underlying communication methods. The Salutation protocol also has a light version, called Salutation-Lite, which is a smaller footprint version and can be run on the Windows CE operating system for the mobile communications market. Therefore, Salutation makes the least assumption of the underlying protocol stack and computing resources, and it can be easily ported to low-power handheld devices.

2.1.2.7 Bonjour

Bonjour [13], previously named Rendezvous, is an SDP developed by Apple for its implementation of zero-configuration networking. It locates devices such as printers, as well as the services provided by them in a local network (e.g., iTunes music), on the basis of standard IPs. The core component of service discovery in Bonjour is the Multicast Domain Name System (MDNS) that is used for service name-to-address translation. More specifically, service discovery in Bonjour is accomplished by *browsing*. The DNS-format queries (defined in the DNS-based service discovery [19]) are sent over the local network using IP multicast for a given service type and domain, and any matching services reply with their names. The result is a list of available services to choose from. Because these DNS queries are sent to a multicast address, no single DNS server with global knowledge is required to answer the queries.

The MDNS in Bonjour allows name-to-address translation, so queries can be made according to the type of services needed rather than the specific hosts providing them. In this case, even when host addresses change, Bonjour can still make the application connect to the service by using the naming mechanism. Besides, Bonjour provides four mechanisms to reduce overheads due to multicasting, including caching, suppression of duplicate responses, exponential back-off, and service announcement [13].

2.1.2.8 DEAPspace

The DEAPspace project [14], started by IBM Research Zurich Laboratory, tries to address P2P networking of pervasive devices instead of client–server networking. All service information is stored on service providers. The DEAPspace algorithm uses a push-based approach, in which all devices hold a list of all known services called the *world view*. The world view of each device is further broadcast to its neighbors in short range to make all devices in the network know services on other devices. To indicate service expiry time, the time-to-live is broadcast along with the service information. DEAPspace can be viewed as a representative for P2P-based service discovery in a wireless medium. Each device can be either the service provider or the client or both at the same time.

2.2 Classifications of Service Discovery Frameworks

2.2.1 Taxonomy of Service Advertisement

Components for service advertisement include the service description mechanism, the service organization scheme, and the service lifetime maintenance method.

2.2.1.1 Service Description

Services must be described before they can be discovered, and, as discussed in Section 2.2.2.2, the quality of service (QoS) matching is strongly correlated to the richness of service description. The simplest description format is the *attribute–value pair* structure that is deployed in Jini/Bluetooth; they describe all the service information using a list of attributes in the service template/record. To further reflect the hierarchical relationships among attributes and to allow flexible design, an *XML-style* representation is adopted. For instance, the UDDI describes service entity with the outermost XML element defining its type (e.g., *businessEntity*, *businessService*, and *tModel*) and the inner element defining its detailed attributes (e.g., *Identifier*). The UPnP, SDS (Service Discovery Service) [20], and INS/TWINE [21] also adopt a similar approach. Note that service attributes can be represented in a richer format rather than in text format only. As a case, in Jini, Java objects are allowed to be used as attributes that associate extra descriptive information with a service. Besides, for Web services, a specific markup language is used to describe their operations and messages involved, that is, WSDL [22]. It is a W3C standard for describing network services as a set of endpoints operating on messages containing either document-oriented or procedure-oriented information in XML format.

With the effort of a semantic Web, a richer and precise description is used to define a service, such as that based on ontology language. The purpose of ontology is to make a description expressive enough for a computer to determine the meaning automatically. The commonly used ontology languages are those from W3C standards, such as OWL-S [23], WSMO [24], and WSDL-S [25]. OWL-S (formerly known as DAML-S) facilitates the description of services in terms of four different ontologies: *Service* ontology, *ServiceProfile* ontology, *ServiceModel* ontology, and *ServiceGrounding* ontology. The *Service* ontology links to the other three ontologies. The *ServiceProfile* ontology defines some nonfunctional properties of the service and exposes some of the functionality by referencing inputs, preconditions, and results from the *ServiceModel* ontology. The *ServiceModel* ontology describes the behavior of the service in terms of processes and control constructs. The *ServiceGrounding* ontology binds processes, inputs, and outputs in the process model to some transport protocol described in a WSDL document. Compared to OWL-S, WSMO provides a more complete and detailed model for service description. For instance, it separates what the user wants (i.e., *GoalModel* ontology) from what the service provides (i.e., *ServiceModel* ontology) and allows the definition of multiple interfaces for a single service. The WSDL-S, on the other hand, gives a simple and lightweight solution for the existing WSDL standard. It annotates operations and messages in the original WSDL file with semantic meanings. The annotation is independent of the ontology languages used and thus can support interoperability among multiple ontologies.

2.2.1.2 Service Organization

To support service discovery, service information must be stored in one or more places. The designed network scale and type are very important to affect the storage design or service organization scheme as we will discuss in Sections 2.3.1 and 2.3.2. In the following sections, we discuss three basic architectures to support service management, as well as their considerations, strengths, and weaknesses. Meanwhile, the representative approaches are presented.

2.2.1.2.1 Centralized Architecture

In this type of architecture, there is a clear *directory server* to store the service information. Service providers are requested to upload their service information to the server for registration. Service discovery is done by directly sending requests to the server and getting candidate services. This architecture, as mentioned for UDDI, is good for administration control and resource management, but it lacks scalability and may cause a single point of failure. Besides, the reliability of the retrieved service information is usually not guaranteed, which may degrade user experience for service discovery in highly changing network topologies such as MANETs. Depending on the scope hosted by the server, that is, global or local, the centralized architecture can be further divided into two types: *pure centralized* and *distributed centralized*.

2.2.1.2.1.1 Pure Centralized — In this case, a single directory server hosts all the service information, where global service discovery can be supported easily. To minimize the issue of a single point of failure, *replication* is usually required for a directory server. The UDDI supports data replication among two or more registry nodes. In addition, to take into consideration data consistency and freshness, UDDI deploys a logical ring structure for all sites involved in the replication process. Then a propagation process is started periodically propagating all updates since the last propagation [7]. Of course, the time interval for propagation is important for system overhead consideration. Sun et al. [26] have compared the UDDI's replication strategy, that is, lazy replication with eager replication, in LAN and WAN environments, concluding that the two strategies perform similarly in LANs and the lazy approach is faster in WANs; nevertheless, the eager approach favors the lazy strategy with regard to overhead and staleness of data.

2.2.1.2.1.2 Distributed Centralized — To minimize the chance of a single point of failure and to relieve the workload of the global directory server in the pure centralized approach, distributed centralized architecture is proposed. In this approach, service information is shared among several directory servers and each server is only in charge of a partial scope. The advantage of such frameworks is that they can achieve certain levels of scalability; they can also take into consideration practical service discovery concerns such as service administration and privacy protection. Local service discovery can be achieved in the same manner as that in pure centralized architecture by approaching the local directory server. However, global discovery is hard to support unless there is a backbone of these local directory servers. Existing distributed centralized architectures can be mainly divided into two types according to the backbone deployed: *topology-specific* and *nontopology-specific* architecture.

In topology-specific architectures, the backbone of directory servers is predefined by using structures such as hierarchy, ring, or grid. Useful criteria to define such a topology include those based on administrative domains (company divisions), network topology (network hops), network metrics (bandwidth or delay), and geographic location (physical distance). The four major topologies—*hierarchy*, *ring*, *grid*, and *hybrid*—are presented in the following.

Hierarchy Topology. In reality, many Internet wide-deployed efficient systems use a hierarchical approach, as it not only reduces the number of resources to a manageable size but also allows autonomy for different parts of the system. The hierarchical structure usually follows the DNS or LDAP [27] model, where directory servers are organized in a tree structure. Service queries are propagated up or down through the hierarchy. The representatives are Bonjour [13], NOMAD [28], Ad-UDDI [29], and SDS [20]. Apple's Bonjour, as presented before, is based on the DNS model to store service information. The European project NOMAD combines UDDI and LDAP information models to build a distributed service discovery framework in mobile networks. The Ad-UDDI framework extends the UDDI by updating service information in different registries in an active manner. It organizes all the UDDI registries according to the hierarchy defined by the Global Industry Classification Standard (GICS) [30]. Registries in the same industry classification are established with a neighboring relationship. Service queries are routed to the registry that is in charge of the classification. If they cannot be resolved locally, they will be routed to the neighboring and upper-level registries. The secure SDS from the Berkeley Ninja Project allows directory servers to be organized into multiple hierarchies. To guarantee a query can reach all SDS servers, one particular hierarchy (primary hierarchy) must be supported by all servers, for example, a hierarchy based on administrative domains.

Ring Topology. This kind of approach links directory servers in a ring structure, and most rely on existing P2P techniques to help efficient query routing. In the frameworks presented in Reference [31] and SPiDeR [32], the authors build the backbone of service directories (called superpeers in SPiDeR) on top of the P2P protocol Chord [33]. Service registration is carried out implicitly by first embedding semantic information into the peer identifiers, then grouping peers by service categories, and lastly forming islands on the ring topology. The routing mechanism in Chord is adopted to achieve *O(logN)* scale for efficient service discovery.

Grid Topology. These approaches are used more in MANETs and the backbone of service directories is formed in a grid network. Tchakarov and Vaidya [34] propose a Geography-based Content Location Protocol (GCLP) for service discovery. Service registration occurs along a crisscross trajectory with four directions: (i) east, (ii) west, (iii) north, and (iv) south. The provider selects four relay nodes that are the farthest in each direction and are responsible for forwarding the register packets. The register packet is transmitted in all four directions until no relay node is available. The register is broadcast along the crisscross trajectory, and, therefore, all the nodes near the crisscross trajectory are the directories. However, this approach requires heavy storage and communication consumption. The recent SGrid framework [35] addresses the issue by splitting the public area into a hierarchical grid. The proposed scheme registers the information of available services to a specific location along a predefined trajectory to avoid the sparse node network topology as for GCLP. The service description is registered to the center line of the network (i.e., the maximum grid level). When a requestor wants to access a service, he or she discovers the service toward the maximum grid level. The discovery is thus confined to a quarter area of the network. The experiments show SGrid outperforms GCLP in terms of a higher discovery success ratio and lower control overheads.

Hybrid Topology. The last topology introduced is that of hybrid topologies, such as a hierarchical ring structure. In the study by Klein et al. [36], the service rings framework establishes a hierarchical ring architecture. Rings are groups of devices that are physically close to each other and offer similar services. Each ring has a designated Service Access Point (SAP) for storing local service information. SAPs are also linked to form high-level rings, which also have SAPs that store summaries of services they provide.

In nontopology-specific architectures, there is no predefined topology structure for directory servers. As a result, the mechanism for discovering one or more directories should be defined. The examples of such an approach are Jini and SLP, which rely on multicast protocols to discover other servers. After discovering each other, an unstructured P2P network may be further formed by these directory servers. For instance, in the Meteor-S system [37], it builds a scalable P2P infrastructure of registries for semantic publication and discovery of Web services. The system, developed in a JXTA environment [38], is based on an unstructured P2P network and is mainly meant to organize service publications by identifying the most suitable registry to host a service description. However, there are exceptions that directory servers are not required to discover each other and form links, such as that in Reference [39]. The framework treats mobile users as messengers, and they can submit cached Web service information to the UDDI registry in range. In this way, other mobile users that are in the vicinity of the UDDI registry can discover services in other places. However, precisely speaking, such a mechanism does not support global discovery and data reliability is an issue.

2.2.1.2.2 Distributed Architecture

In this type of architecture, there is no dedicated directory server to maintain all the service information. P2P overlay networks, such as Gnutella [40], Chord [33], CAN [41], Pastry [42], and Tapestry [43], are deployed as platforms for service discovery. For detailed discussions on these P2P systems, readers may refer to the survey by Meshkova et al. [6]. Generally, P2P systems can be broadly divided into two categories: *unstructured* and *structured*. In this section, we show how service discovery frameworks are based on P2P systems.

2.2.1.2.2.1 Unstructured Distributed — These types of frameworks inherit the flexibility of unstructured P2P networks such as Gnutella. Each service provider joins the framework in a peer form and maintains minimum information about the network topology and information regarding services provided by other peers. Schemes for network conductivity, that is, the selection of neighbors, must be designed to ensure node reachability. For instance, Acosta and Chandra [44] propose four schemes to evaluate which peers would be better neighbors: (i) *high-degree biased*, in which nodes are biased toward selecting nodes with high degree (many neighbors); (ii) *proximity biased*, in which a proximity-biased evaluation function measures the network latency between nodes and nodes that are closer in terms of network distance are selected; (iii) *connectivity biased*, wherein an evaluation function is biased toward nodes that give access to nodes in the network that were not previously accessible through the current connections; and (iv) *proximity–connectivity hybrid*, in which the algorithm expands the evaluation function of the connectivity algorithm to also take into account proximity. The interesting points about this type of architecture include implementation simplicity, support for partial match queries, and relative resilience to peer leave/failures. Besides, when discussed in the service environment, they allow rich service matching mechanisms to be implemented and provide better privacy protection. However, owing to the lack of network knowledge, the discovery of services is blind in that it is independent of the query and usually is based on flooding, resembling broadcasting on the network layer; therefore, they tend not to be scaled for wide-area networks and queries may not always be resolved if techniques for overhead reduction are applied (e.g., setting TTL in the query).

The representative for unstructured distributed approaches is the Web Services Peer-to-Peer Discovery Service (WSPDS) [45], which utilizes Gnutella as the underlying protocol and defines a set of application protocols for neighbor selection and service matching. In the initial version, each service peer or *servent* maintains a list of the most recently active *servents* of the network, denoted as *servent* cache. Each time a *servent* is activated, it probes the *servents* listed in the *servent* cache to find *k* nodes that are still active and designates them as its neighbors. The *servent* cache also contains access points of a few WSPDS *servents* that are almost always active in case it is the first time for the *servent* to activate. The routing of queries is probabilistic, as discussed in detail in Section 2.2.2.1. An improved version adopts a content-based P2P network from QDN [46], assigning linkages between *servents* on the basis of data contents. The neighbors are selected on the basis of the similarity between the inputs/outputs of a service and those of others; besides, the selected nodes must be geographically close to the current one to reduce query latency.

To improve the performance on successful queries and response time, the replication mechanism can also be applied in unstructured distributed approaches. In Reference [47], a simple model is formulated that allows evaluation and comparison of different replication strategies used in unstructured P2P networks and the two most obvious replication strategies are: *uniform* (i.e., replicating everything equally) and *proportional* (i.e., more popular items are more replicated). These two strategies yielded identical performances on soluble queries and are not optimal solutions. The authors then propose that the *square-root replication* (i.e., the square-root of the popularity of the item) is optimal when restricted to soluble queries.

2.2.1.2.2.2 Structured Distributed — Unstructured P2P approaches have two weaknesses. First, they may result in queries that are not always resolved, and, second, they do not guarantee anything about performance while scaling. On the other hand, frameworks relying on structured P2P overlays solve these issues. Most of them rely on the distributed hash table (DHT) concept and assign key ownerships in a predetermined manner: each node in the overlay is responsible for maintaining a part of the hash range that represents an index to available resources; queries are input to a hash function and then are looked up in the nodes responsible for the resulting keys. For these approaches, query resolution path lengths usually increase logarithmically with the number of nodes in the network, that is, in $O(logN)$ scale. The early representative is the INS/TWINE protocol from MIT [21], which is based on the Chord structure and supports partial matching by hashing partitions of the resource description (i.e., attribute–value pairs). Klein et al. [36] not only propose service rings but also propose lanes [48], which adopts the overlay structure based on CAN. The overlay has two dimensions: one for propagation of service advertisements through a lane in a top-bottom manner on the y axis, and one for distribution of service requests between lanes on the x axis. Nodes in the same lane share the same anycast address that allows anycast routing for sending messages from one lane to another. Other similar DHT-based service discovery approaches can be found in Reference [49].

However, conventional DHT-based service discovery frameworks have some limitations. First, most hashing techniques adopted do not consider data semantics; thus, partial or range search is hard to support. Second, DHT-based overlays require high maintenance cost for nodes join/leave/failure, which usually requires $O(logN)$ messages. This may be an issue for large-scale dynamic networks. Finally, the constructed overlay does not consider practical QoS parameters, for example, physical distance and query response time. Therefore, further approaches are proposed to cope with these issues. The *semantic clustering* approaches are to cluster semantic-close services to support semantic search and reduce routing efforts. Schmidt and Parashar [50] map the sequence of service keywords in the *d*-dimensional CAN space to the one-dimensional linear space by using the Hilbert SFC technique. Points that are close on the curve are mapped from close points in the *k*-dimensional space. However, such mapping destroys the properties of consistent hashing, and, thus, does not ensure load balance. An *ad hoc* load balancing technique is further devised by Schmidt and Parashar [50]. ERGOT [51] solves the load balancing issue by combining DHTs and semantic overlay networks to enable distributed and semantic-based service discovery on the grid. The system involves a module to compute similarity among services on the basis of their category ontology. Semantic links are built among those semantic close service peers in addition to the conventional DHT links in Chord. Pirró et al. [51] argue that these two models can benefit from each other in the sense that semantic overlay networks can be constructed by exploiting DHT mechanisms, thus enlightening the way to semantics-free content publishing and retrieval mechanisms of the latter. However, there is no experiment to demonstrate the argument. Other semantic P2P-based approaches have also been proposed in References [52,53]. For maintenance, the JXTA protocol [54] combines a DHT approach with a limited range walker to check for nonsynchronized indices. It does not require strong-consistency DHT maintenance because a certain amount of *inconsistency is tolerated* and a limited range walker is used to resolve inconsistency of the DHT within the dynamic rendezvous network. In this way, we face less traffic overheads and the method is well adapted to *ad hoc* P2P networks with a high peer churn rate. Another commonly adopted strategy is to impose structures on the DHT-based P2P system, which is similar to the idea in the distributed centralized approach to *narrow down the maintenance scope*. Hierarchical P2P discovery frameworks such as GloServ [55] and VIRGO [56] are proposed. Both of them differentiate services on the basis of their application classes, and the hierarchy structure is constructed accordingly. Services within each class on each layer form a virtual group that is managed separately. Lastly, to take into consideration practical QoS, *QoS-related clustering* techniques may be used. In Reference [57], service peers are organized into clusters, in which every node may reach others within a given time frame. This is similar to the mechanism used in WSPDS [46]; thus, a certain QoS can be provided. In addition, in Reference [57], algorithms for identifying such clusters are discussed and demonstrated through simulations.

2.2.1.2.3 Hybrid Architecture

In this type of framework, both organization architectures discussed before are supported. The representatives are SLP and the framework in Reference [58]. If a directory server is deployed in the network and is within the vicinity of the user, the user may directly access it for efficiency; otherwise, the distributed discovery mechanisms are triggered to look for the desired services, for example, through flooding or DHT-based routing.

2.2.1.3 Service Lifetime

Services can be added or removed from the framework at any time. Thus, service availability or service lifetime maintenance is necessary. The maintenance is especially important for mobile computing, as the mobility of service providers may greatly affect service availability. Generally, there are two main strategies to maintain service lifetime. One is a simple approach that requires service providers to explicitly register or deregister their services to the framework. Service information in the framework (i.e., cache) is assumed to be persistent until the respective provider explicitly requests removal of them. This approach is implemented in the UDDI; however, this approach does not guarantee the quality of registered services, as illustrated in the survey by Bachlechner et al. [59]. Some optimization techniques may be applied, such as periodically sending ping messages to check the service availability. The other maintenance approach is based on leasing; that is, rather than granting services or resources until that grant has been explicitly canceled by the service provider, a leased resource or service grant is time based. The provider may renew or cancel the lease before the lease time expires, but in the case of no action, the lease simply expires and the corresponding service (i.e., cache information) is removed from the framework. Typical examples are Jini and SLP frameworks.

2.2.2 Taxonomy of Service Discovery

Service discovery is the most important element in the whole framework. It consists of two main components: *discovery method* and *matching strategy*. The discovery process deals with the routing of service queries to the target service provider or the node where the requested service information is stored. The search method is strongly related to the organization scheme used for service management as well as the network type. For instance, in directory-based frameworks, queries are directly sent to the directory in order to be resolved. In DHT-based P2P topology, queries are directed to the nodes according to the keys hashed from each search keyword. In MANETs, broadcast or flooding can be used as the simplest mechanism for service discovery. However, no matter what routing schemes are used, routing efficiency is the fundamental concern in the routing of queries. Once candidate services are found, the next step is to determine the services that better address the user's request. A *keyword-based matching* strategy is commonly applied for service description and user's request, but because of its inadequate expressing power, *semantic matching*, which relies on the ontology description as mentioned before, is adopted in some frameworks. Moreover, to gain more matching accuracy, different scopes of service matching are deployed, such as description level, Web service I/O level, and service process level. Besides, to enable intelligent matching, some matching enhancements such as context awareness are proposed. For example, the location of the service provider may greatly affect the user's interest in using the service, and device-aware matching capability is most desired for the user. We describe the context-aware service discovery in more detail in Section 2.3.3 as it distinguishes mobile service discovery and ordinary service discovery frameworks.

2.2.2.1 Discovery Method

There are two fundamental models for service discovery: (i) *Push model*—service providers or directory servers actively push their service information into the network so that the requestors can be aware of their existence; however, periodical announcements would waste the bandwidth and cause large network overheads. As a result, most approaches only push service information when the service is newly registered or updated and rely on the pull model for service discovery. (ii) *Pull model*—query the network for the desired service when needed. This model is client initiated, and thus is based on demand in order to reduce unnecessary announcements as in the push model. However, the detailed search method is strongly related to the service organization scheme discussed in Section 2.2.1.2. As discussed in the study by Sakaryan et al. [60], the detailed search methods for the pull model can be classified into two main categories—blind search and informed search—depending on how search query forwarding decisions are made.

2.2.2.1.1 Blind Search

It is assumed that in this type of search service peers have no information of the network topology, and, thus, have no idea where to send a query that is to reach the target service. The advantage of such a search scheme is that it is simple and does not impose the need for peers to maintain the network information. Of course, the disadvantage is that it is not efficient and causes large amounts of overhead. The first well-known strategy is to do *flooding* searches, in which when a peer receives a query, it simply forwards the query to its entire neighborhood; meanwhile, it checks whether the peer itself provides the requested service. If so, the peer will send back the reply message along the reverse direction as the query reaches the peer. Bluetooth and DEAPSpace use this kind of service discovery. To control overheads, the TTL is always specified in the query to restrict the flooding range; furthermore, propagating with a probability value as in probabilistic flooding in WSPDS [45] may be used. However, as simulated in Reference [61], the flooding-styled search is not more effective than probing randomly chosen peers because the unstructured topology is not designed with the search method in mind and the peer does not have any information about which other peers may best be able to resolve the query. The *random walk* search method is thus adopted. As the name implies, it randomly walks the network querying each peer it visits for the target service. More specifically, each peer chooses a neighbor randomly and sends the query to it, and other peers repeat the process until the service is found (see LANES [48] as an example). Cohen and Shenker [47] also propose to use a *proactive replication* algorithm to further reduce the response time. Note that for some industry standards, such as UPnP and SLP, the multicasting strategy used for service discovery (i.e., searching without the help of a directory server) can also be considered as a kind of flooding, except that it is based on a fixed IP address.

2.2.2.1.2 Informed Search

In some systems, peers may have extra knowledge of the network topology either learnt by previous discovery or predefined at the organization phase. The simplest way is to approach the directory server to retrieve relevant service information. In distributed architecture, one kind of such strategy is *content-oriented* search. To avoid flooding, each peer stores content summaries of its neighbors by either explicit information exchange or learning from past queries. Therefore, when the next query comes to the peer, the peer will check which neighbor can handle the same or similar content (e.g., service category) and then forwards the query to it. A special kind of content-oriented search is to take peer content into the organization design, as shown in WSPDS [46]. Peers that have similar content are locally set up with neighbor relationships. The neighbors can be learnt through previous queries by caching the peers visited in the query log list. Compared to the purely content-oriented search, this scheme significantly reduces the routing latency inside a specific content group, for example, a service category. The next type of informed search, *key-based* search, is actually deployed in most wide area service discovery methods for structured P2P networks. The DHT-based organization scheme maps keys onto peers in the network in a deterministic manner; thus, the search is also deterministic. The information of the service provided by each peer is hashed to a set of keys and stored on the node that is in charge of the key space. In this way, later when a query comes, the content of the query can be hashed using the same function and the problem is transformed to finding the node that stores the key. Different structures are used to enhance the searching efficiency, for example, ring, hierarchy, and so on. Nonetheless, as the search is too deterministic, it cannot support vague search or range search well.

2.2.2.2 Matching Strategy

Service matching is also an important component in service discovery, as it determines the relevance of the service on the user's request and then concludes whether it should be returned as a result. As mentioned in service description, service matching is strongly related to service description. The richer the information provided by service description, the more complex the matching strategy tends to be. On the whole, there are two main factors to consider when developing such a strategy: *matching content* and *matching technique*. Matching content refers to the information considered for service matching whereas the matching technique refers to the detailed method used for matching on the basis of content. Conventional approaches, like most industry standards, rely on service registries and search for a relevant service based on categories or keywords. That is, users specify a set of keywords in the query, and these keywords are looked up in the description documents of available services. In the UDDI framework, the API allows developers to specify keywords of particular interests and it then returns a list of Web services whose service description contains those keywords. To speed up the search process, the Vector Space Model (VSM) [62] in information retrieval is adopted in Reference [63]. Each service document or query is represented as a t-dimensional vector, where t is the total number of distinct words in all the service documents. A binary value is used to represent the existence of a particular word in the document or query. The cosine similarity function is then used to rank all the document vectors with the query vector. Moreover, to reduce the size of the vector, techniques such as removing stop words, unifying word forms by applying Porter's stemming algorithm [64] and picking words with high information content according to their term frequency and document frequency can be used. Nevertheless, the keyword-based approach usually results in unsatisfactory precision and recall as discussed in Reference [65]. The major reason is that keywords do not capture the underlying semantics of services. Relevant services may be missed due to lack of keywords specified in the query. Although the problem of missing keywords can be relaxed by using a synonym dictionary; the matching is still done at the syntactical level.

To improve matching quality, semantic service matching strategies are used. Generally, they can be distinguished by whether they rely on *external schema/ontology*. The first type of semantic matching uses external schema/ontology such as in References [66–69]. These approaches provide a vocabulary of terms or concepts whose meanings are constrained to describe a domain of interest. The relationships among the terms or concepts (e.g., containment and referential relationships) are clearly defined. In the matching process, service schema/ontology is represented by means of directed rooted graphs, where each node represents an attribute element, connected by directed links of different types. The matching takes two schemas/ontologies as input and determines logical mapping among those elements and then derives the similarity value. The similarity value can be measured on two levels: the element level and the structure level. On the element level, strings are first tokenized by recognizing punctuation, and then lemmatized to their base form. Subsequently, the Levenshtein edit distance [70] is used to compute the distance between them. In a more rigid manner, linguistic resources such as WordNet [71] are used to compute the similarity at the concept level based on lexical relationships. In Reference [72], the concept hierarchy in the WordNet is explored to compute the similarity between two terms to support partial matching. For instance, if *term1* and *term2* have hierarchical relations, their matching degree is computed as *0.6/N*, where *N* is the number of hierarchical links in ordNet. There are also other similar metrics in evaluating concept similarity. For instance, in Reference [73], they calculate the semantic relevance based on Resnik's method [74]. On the structure level, graph matching techniques such as graph–subgraph isomorphism and edit distance technique are usually adopted. Note that as the schema/ontology is a kind of directed rooted graph, the tree-based edit distance can be applied as illustrated in Reference [66]. In Reference [75], a tool for ontology matching that lists different techniques is provided so that the user may choose one or several techniques for comparison. Although schema/ontology matching allows service matching at a conceptual level and achieves high accuracy as studied in Kotis and Lanzenberger's work [76], there are issues, such as matching scalability, requirement of knowledge and effort to define the ontology, and cross-ontology mapping. For approaches without external schema/ontology, information retrieval techniques are used. The frequently used technique is the Latent Semantic Analysis (LSA) [77], which applies singular value decomposition in linear algebra on the service matrix composed of different service document vectors. The underlying idea is that the aggregate of all the word contexts in which a given word does and does not appear provides a set of mutual constraints that largely determines the similarity of meanings of words and sets of words to each other, and therefore, services can be classified and grouped under similar semantic meanings. Related services can also be retrieved when there is no keyword matching the user's query as in References [65,78]; also, in Reference [78] hierarchical clustering of services is supported. Woogle [79] classifies Web services by clustering parameter names of Web service operations into semantically meaningful concepts. The association rule mining is then used to reflect the correlation of two terms, and it is based on the heuristic that parameters tend to express the same concept if they occur together often. Nevertheless, it relies strongly on parameter names and does not deal with other textual descriptions. Note that techniques in machine learning can also be applied for service classification, such as Bayesian classifier used in References [80,81], and Support Vector Machine used in Reference [82]. However, all these approaches adopt a supervised learning approach and require manual work to label the training set.

By using semantic attributes of a service including its inputs/outputs, the matching can be done at the conceptual level to achieve better accuracy. However, to allow business developers to better describe the service they want and further increase the retrieval precision, recent works underline the need for the matching of service behaviors by using its process model [67,68,83]. A survey study of service process matching can be found in Reference [84]. In References [83,85,86], the process dissimilarity is computed by deriving the edit distance from deletion/insertion of a node/edge. However, with the recent study on graph distance measure, high degrees of matching precision are underlined. In Reference [87], the authors argue that existing graph distance measures have a low degree of precision because only node and edge information of the graphs are considered. The richness of substructure information should also contribute to the evaluation of graph distance. In the area of process matching, less work has been done on the matching of process structure level information. In Reference [88], the authors extend the work in Reference [85] by first converting the process graph into a block tree, where structure information is captured by each block in the tree, and then a binary tree vector is used to represent the tree. The distance measure is derived from binary tree vectors using the algorithm proposed in Reference [89]. However, such a distance metric is not considered as the true similarity measure and zero distance does not imply that the two trees are identical [89]. Besides, this approach requires the tree to be ordered, which is true for business processes.

2.2.3 Taxonomy of Service Access

After retrieving the desired service information, for example, either the location (can be URL) of the service or WSDL in the case of Web services, the next step is to access them. For conventional business services, such as shop or restaurant services, the user may approach the location of the service to get served, and this explains the importance of context awareness in service discovery. For Web services, communications can be done through the RPC, HTTP/SOAP protocols, and messages are usually in XML format. Note that Jini provides different mechanisms for service invocation, which utilizes Java RMI protocol. A user may also subscribe to a service he is interested in to receive information concerning changes. The subscription can be either between the user and the directory server or between the user and the service directly.

2.3 Considerations of Service Discovery for Mobile Computing

In Section 2.2, we demonstrated that many solutions are available when designing components of a service discovery framework. However, there is no standard benchmark to show which one is optimal, as the considerations introduced in this section should be taken into account when designing a service discovery framework. Note that although the considerations presented in the following are for service discovery in mobile computing, they can be applied to any discovery framework. However presence of concepts that are specific to mobile computing, such as context awareness, may not be considered in some environments.

2.3.1 Network Scale

The first consideration is the network size or discovery scale, which may vary from a couple of devices to the Internet scale. We classify them into two major categories: *local scale* and *global scale*. Where local scale refers to those personal, home, office or enterprise networks, global scale refers to wide area or the Internet scale networks.

The design and performance of service discovery frameworks are largely determined by the characteristics of the underlying network. For instance, in a local-scale network, like the LAN in a home, office or even a company, while the routing efficiency is not the primary concern, the searching effectiveness is rather highly demanded. Designers may choose to apply complex service matching strategy, such as semantic matching over Web service processes, to achieve high accuracy. On the other hand, for global-scale service discovery which may involve world-wide services, the scalability of the framework is always the major concern. Managing services or their information in an effective way to support efficient service discovery is challenging. Most existing frameworks deploy either a DNS-style distributed centralized approach or P2P-style organization scheme. Global-scale service discovery also impose high demands on the security, privacy, and trust of service information.

Industry standards such as UPnP, Bluetooth and Bonjour are applicable for personal networks, which contains nodes/devices/services with a personal relation to the user [90]. Khatib et al. [91] design a personal agent for the user mobile devices, such as PDAs. While their work is focusing on QoS, it leverages the existing SDPs (e.g., Bluetooth) to learn about all services available in the vicinity of the user to select the best quality service for communication. Ghader et al. [58] propose a multilevel organization of service information, where they differentiate services for different tiers of personal networks. The higher the tier, the larger the search scale, leading to the global scale in extreme. The Service Discovery Module (SDM) specifies the process of service discovery at each tier. In the lower tiers, the centralized approach is adopted; while in the higher tiers, the P2P approach is applied. Another recent proposal can be found in Reference [92] which defines a Virtual Personal Space (VPS) for each user so as to consider personalization. As mentioned before, papers in service discovery for personal area networks are mainly concerned about context awareness and personalization. Scalability is not an issue as the number of devices or services involved is usually limited. Broadcasting or multicasting is frequently used in these frameworks. The contexts of users and services (e.g., location) are important to keep the information discovered up-to-date, and personalization ensures better QoS.

Unlike personal networks, service discovery in enterprise/company scale focuses more on administration control and security. Most of these frameworks, such as UDDI, Jini, and SLP that are industry standards, deploy a client–server infrastructure. In each of these frameworks, they have respective security mechanisms that will be discussed in Section 2.3.4.

Indeed, for most proposals in research communities discussed in Section 2.2.1.2, they target service discovery on the global scale or the Internet scale. The major concern of these proposals is the scalability of the framework. As a result, most such systems deploy either a distributed centralized approach or a purely P2P approach or both. As a result, different overlay topologies, such as tree and grid topologies, are built to enhance searching efficiency.

2.3.2 Network Type

Most industry standards do not address the underlying network type, as they focus on application-level service organization and discovery. Examples are UDDI, Jini, and SLP. However, these examples were developed in the early days when wireless technology was not matured. Although the concepts in wired infrastructure can still be applied in wireless topology, some topology-related issues need to be addressed separately.

For MANETs, two practical issues are *limited topology knowledge* due to wireless communication range and *changing topology* due to node mobility. The distributed centralized approach [39,93,94] and the distributed approach (DEAPSpace [14], PDP [95], and HESED [96]) are deployed for service discovery in MANETs. However, in fully distributed approaches, such as DEAPSpace, broadcasting and multicasting techniques are applied for service discovery, imposing large system overheads. Different optimization techniques are thus applied. Ververidis et al. [5] have summarized four major approaches: *advertisement range bounding/scoping*; *selective, probabilistic, and intelligent (advertisement/request) forwarding*; *P2P information caching*; and *intermediate node responding to service requests*. They have also done a survey study on the respective approaches, which we will not repeat here. To further reduce the routing overheads in MANETs and improve efficiency of energy consumption, some approaches integrate the service discovery process with the routing process. This type of cross-layer design piggybacks service information onto routing messages, and service discovery is done on the network layer so as to reduce unnecessary message exchange at the application level. Examples can be found in AODV-SD [97], ODMRP [98], and MZRP [99]. To summarize, all these cross-layer approaches have shown their efficiency in terms of network throughput, service acquisition time, and energy consumption (demonstrated in References [97,100]), but this kind of approach destroys the protocol stack in the Internet OSI model and tends to be protocol specific, which may limit the interoperability of the discovery framework, especially when deployed for large-scale and heterogeneous networks.

A special type of wireless network is the wireless sensor network (WSN) that consists of spatially distributed autonomous devices using sensors to cooperatively monitor physical or environmental conditions, such as temperature, sound, vibration, pressure, motion, or pollutants, at different locations. WSNs, as compared to MANETs, impose more *resource-constrained* requirements for service discovery. The resource constraints can be in terms of a low data communication rate, a low computational power, limited node memory, and inadequate energy. For instance, the typical packet size in such networks does not exceed 128 bytes, and the maximum raw data rate is 250 kbps. The memory size for a typical mote varies from 128 kb to 1 Mb. Such constraints do not allow the direct application of well-known service discovery frameworks such as UPnP or SLP, as most of them need to consume large network resources due to multicast protocols. Therefore, service discovery in WSNs focuses more on *bandwidth* or *energy efficiency*. A service in WSNs essentially refers to data being offered by a particular sensor node; thus, a cross-layer design adopted from MANETs can be a good choice during data exchange among sensor nodes at the network layer. Examples include X-lisa [101–103] and the recent CLAWS project [104]. The survey by Meshkova et al. [6] summarizes three other optimization schemes for service discovery in WSNs: (i) *Simplify the needed protocol* by sacrificing some of its features to decrease the resource demands, as is done in NanoIP [105]; (ii) *use compressions for (service) data exchanged*, for example, WBXML representation [106]; and (iii) *aggregate data on the intermediate nodes*, for example, TAG [107]. Besides, cluster-based organization models also provide a way of saving energy for sensor nodes. For instance, in SANDMAN [108], the framework arranges nodes with similar mobility patterns in clusters. In each cluster, there is a *clusterhead* node that is always active and answers discovery requests. The rest of the nodes may periodically wake up to announce their services and sleep to save energy when idle. Simulations show the approach achieves energy savings of up to 66% compared to other approaches where all nodes are permanently active, while the discovery latencies are not increased. Further improvements include re-election of *clusterhead* regularly to ensure fairness, as discussed in Reference [108].

2.3.3 Context Awareness

Besides the network scale and type, another important consideration for service discovery in mobile computing is context awareness. The implicit context information may not be that crucial in the conventional P2P search, as the discovery is more based on stable resources, such as files or images. In fact, a service in mobile computing is a dynamic entity: its provider may be on the move, like for a taxi service. The current context of the user and that of the service may be quite different, which results in unavailability or irrelevance of the service, and thus influences the satisfaction of the user. In this chapter, we are not focusing on introducing context-aware systems. Interested readers are referred to the survey by Baldauf et al. [109]. Rather we focus on existing frameworks that use contexts of users and services to better fulfill the discovery task. In the following, we will give definitions for context and context-aware service discovery and discuss different techniques for context modeling. Finally, the applications of contextual information in service discovery are represented.

Probably the most common and systematic definition of context comes from the study by Abowd et al. [110], which is cited by many papers [111–113] on context awareness. Their definition of context is "any information that can be used to characterize the situation of an entity. An entity is a person, place, or object that is considered relevant to the interaction between a user and an application, including the user and applications themselves" [110:3–4]. They also judge a system as being context aware "if it uses context to provide relevant information and/or services to the user, where relevancy depends on the user's task" [110:6]. Here, we extend their definition by utilizing context in the application of service discovery, and define context-aware service discovery as *the process of service discovery that relies on contexts of users and services to enhance the searching experience.* Experience enhancements can be in terms of less query response time, higher result relevance, and better application adaptation.

Context can come from various sources, from lower to higher levels. Low-level contextual information includes raw data, such as location information from a GPS system, time information and those from sensors, for example, temperature and light level. High-level contextual information refers to contexts that cannot be directly retrieved and may require a reasoning process. A typical example for high-level contextual information can be a user's activity, which requires derivation on the basis of low-level contextual information, such as a user's location and objects touched. Context can also be classified on the basis of its source type, for example, environmental versus individual. A user's preferences can be categorized as an individual context. To define and store context data in a machine-readable format, a well-designed context model is needed in any context-aware system. Strang and Linnhoff-Popien [114] summarize some typical context modeling and integration requirements for ubiquitous computing, and classify existing context modeling approaches into six categories based on data structures used to exchange contextual information in the context-aware system. The various models include: *key-value models*, *markup scheme models*, *graphical models*, *object-oriented models*, *logic-based models*, and *ontology-based models*. For the detailed introduction on modeling and representative approaches, readers are referred to Reference [114], which also provides evaluations on different modeling approaches concerning the requirements proposed by the authors.

In this chapter, we classify frameworks for context-aware service discovery into three major categories, according to different usages of contextual information of users and services.

2.3.3.1 Context as a Means to Help Service Organization

These frameworks consider service contexts when organizing services. The services are organized according to some specific context elements. The most commonly used element is the service location, and the discovery framework is *location based*. Most such frameworks consider the physical location of services and cluster them on the basis of physical proximity. This approach is effective when the user is searching services nearby, and the response time can be reduced owing to the fact that routing of queries is restricted to a certain area/network. Two pioneer frameworks are the Cooltown project [115] and Splendor [116]. In both frameworks, they group physically related things into places embodied in Web servers, and address them using HTTP URLs. The beacon of that location advertises the Web-based URL periodically so that interested users can connect to the URL that describes the service (i.e., Web server pages) and can invoke the service. The major difference between the two frameworks is that, rather than simply using the centralized approach, Splendor adopts a *client–service–directory–proxy* discovery model, in which a proxy is used to offload the computational work of mobile services; meanwhile, it also enables security and privacy protection of mobile services, which is further discussed in Section 2.3.4. Lee and Helal [117] define a three-tier discovery architecture. The goal is to provide the most appropriate service to mobile users by exploiting any meaningful context information. Services are classified according to their coverage. The first tier is a proximity service for local area scale, for example, nearby printer. The second tier is a domain service, for example, area guide services, and the last one is a global service, for example, storage service. The mobile user may expand his or her search scope if he or she cannot find satisfied service in the current proximity. The recent Context-Sensitive Service Discovery (CSSD) [118] framework adopts a similar idea by allowing searches for services to be delegated among different network entities called user nodes, super nodes, and root nodes. A user node is located in the subnetwork and is an entity from which the user interacts with the CSSD system and performs searches for services. User nodes themselves can provide services. The super node is part of both a subnetwork and the interconnection network. It forwards service requests from user nodes to all other super nodes in the same interconnection network and to the local root node. The root node is responsible for forwarding service requests from super nodes to other root nodes on the Internet. The experimental analysis of service discovery times based on different scenarios is used to optimize parameter settings of the service discovery framework (e.g., time-out value) to achieve short response times.

2.3.3.2 Context as a Means to Help Service Matching

The majority of context-aware service discovery frameworks focus on context-aware service matching; that is, matching contexts of users (i.e., location, device capabilities) and services (i.e., service profile) to filter and select relevant services. In Reference [119], contexts of services are matched against users' preferences. Concept lattices [120] are used to study how services can be hierarchically grouped together according to their common attributes. The idea is that the higher the services are positioned in the lattice line diagram, the more preferences they miss and the lower in the resulting list they are ordered. The work in Reference [121] first differentiates dynamic service attributes from the static ones, and then allows weight assignment for each dynamic attribute by different clients. Services are ranked according to their weighted summarization of different attribute values. Note that clients may also specify those attributes that are not desired. This is reflected in the formula by times −1 to a specific weighted attribute value. The framework also allows administrators to specify policies when selecting services. The policy is a rule that is generally of the form of an *if–then* statement. In Reference [122], context is represented by a set of dimensions that take one or more discrete values or range over a specific domain. Combining different dimensions of context that specify the conditions under which a service exists results in a graph that represents a specific service category within the service directory. The graph is much like a decision tree: by specifying a context query that includes the current context of the user, satisfied service instances can be retrieved from the tree. Other context-aware service matching approaches mostly rely on the ontology-based context modeling and rule-based reasoning for service filtering and selection. The representatives are the RSD platform [123–125]. All these approaches utilize predefined rules for context reasoning. A special approach using machine learning techniques is proposed in Reference [126]. The SmartCon system proposed uses artificial neural networks to select the best available mobile service. The features extracted from contextual information are used for training and classification. Three types of features are considered: (i) *device-specific features* such as the Composite Capabilities/Preferences Profile (CC/PP), (ii) *user-specific features* such as preferences, and (iii) *service-specific features* such as service location. Preliminary experiments show an 87% success rate in the discovery and selection of the best or most relevant mobile service. In addition, SmartCon also supports iterative learning by using users, feedback.

2.3.3.3 Context as a Means to Help Application Adaptation

Contextual information can be used to help adaptation in application scenarios such as device interface design, QoS selection, recommendation, and reminding. However, here we focus on the *adaptation for context-aware service discovery*. The purpose of such an adaptation is to enable interoperability among different service discovery frameworks in heterogeneous networks. One representative is from the Multi-Protocol Service Discovery and Access (MSDA) framework in Reference [127]. The MSDA middleware platforms integrate existing service discovery and access protocols. They rely on a previous multiprotocol service discovery platform [128] and extend with multiprotocol service access. MSDA includes three components: (i) the *MSDA Manager* that provides service discovery and access to clients within the network; (ii) *Service Discovery and Access Plugins and Transformers* that interact with specific discovery and access domains; and (iii) *MSDA Bridges* that assist MSDA Managers in extending the service discovery and access to services present in other networks of the whole pervasive environment. MSDA provides an application-level adaptation to different network protocols. By defining services in MSDA-specific description language and storing network contexts (i.e., protocols) in the local MSDA manager, service discovery and access is done through routing among MSDA managers and transcoding of messages by MSDA plugins for a specific domain. The detailed implementation of MSDA middleware can be found in Reference [129].

Finally, most contextual information tends to be dynamic; thus, context-aware systems should handle context changes. Techniques for sensing context changes include periodic polling, advertising, and selective polling, but discussing context data management is out of the scope of this chapter and interested readers may refer to the survey by Baldauf et al. [109].

2.3.4 Security, Privacy, and Trust

2.3.4.1 Security

Although service discovery has been thoroughly studied in the past, its security has been vastly ignored. Security is extremely important for mobile computing, as the environment is changing and there are large numbers of uncontrolled data exchanges. From a service provider's perspective, the authentication and authorization of the user's identity need to be checked and service information should be stored at a trusted directory server if there is one; from a user's perspective, services retrieved must be reliable and personal information such as location and personal preferences should be kept confidential.

There are several *security requirements model* or *threat models* proposed specifically for service discovery. The earlier representative is from the study by Zhu et al. [130], which outlines a brief threat model, including the threats of disclosure, integrity, and denial of service. A detailed threat model is described in the study by Trabelsi et al. [131], including authentication, confidentiality, access control, trust, privacy, and nonrepudiation. Cotroneo et al. [132] clearly define the security requirements for service-oriented architecture in mobile computing. The requirements follow the core mechanism used for service discovery as presented in this chapter, namely, requirements for service registration and deregistration, secure discovery, secure delivery, and service availability. It also provides a detailed security analysis for existing industry standards which we present in the next section. Leung and Mitchell [133] consider security requirements from various types of environment constraints in MANETs, such as node failures, path failures, and routing failures. The detailed security requirements are close to those presented in Reference [131], but same as those in Reference [132]; it does not describe a security architecture to overcome the lack in security.

In the next section, we briefly describe the features of existing approaches to handle security concerns. Two parities are discussed here, industry standards and research proposals. The major distinction between the two parties is that they have different focuses on security concerns. In industry standards, as we introduced in Section 2.2.2.1, service discovery is usually carried out in a private (e.g., enterprise) or managed environment. The trust between service providers and directory servers are assumed, as central administration is possible. Thus, in terms of security protection, these frameworks focus on the interaction between service users and service providers or directory servers. In research proposals, the above assumptions cannot be made, as the environment is changing, and entities involved in the discovery process are usually unfamiliar. Another fact is that most research proposals introduced earlier do not address security concerns as they are not used in practice. Exceptions are Splendor [116] and SDS [20], to name a few.

The security model for a UDDI registry can be characterized by the collection of registry and node policies and the implementation of these policies by a UDDI node. The principal areas of security policies and mechanisms are related to data management, user identification, user authentication, user authorization, confidentiality of messages, and integrity of data. How these policies should be modeled is described in Reference [7]. To authorize or restrict access to data in a UDDI registry, an implementation of a UDDI node may be integrated with one or more identification systems and it must also identify the systems used. Integration of UDDI API and data with an identification system is implemented through the authentication and authorization API to provide access control. Other authentication and authorization mechanisms and policies are represented in UDDI through the use of tModels to describe the Web services of a UDDI node. UDDI also supports XML digital signatures on UDDI data to enable inquirers to verify the integrity of the data with respect to the publisher.

Jini security framework provides mechanisms and abstractions for managing security, especially in the presence of dynamically downloaded code. It deploys three major mechanisms to provide authentication, integrity, and confidentiality. These include a mechanism for clients to verify trustworthiness of proxies, a mechanism for verifying code integrity, and a mechanism for clients to dynamically grant permissions to proxies. Both the client and the service provider can impose constraints on the service object (or proxy).

The UPnP protocol, as default, does not implement any authentication, as it runs on local systems and their users are assumed to be completely trustworthy. UPnP device implementations must implement their own authentication mechanisms, or implement the Device Security Service. In Reference [134], a new entity, the Security Console for UPnP, along with the Device Security Service, is introduced which is a component running on a UPnP device or control point to provide the user interface for the human owner of a device. In the Device Security Service, security concerns are addressed with the SOAP control messages exchanged between control points and devices. The message security consists of identification, integrity, authentication, freshness, authorization, and secrecy, which is enforced by using the public key of each Security Console.

The Bluetooth Generic Access Profile, described in Reference [135], specifies three security modes to achieve authentication, integration, and confidentiality: Security mode 1 is for applications without security concerns; in security mode 2, a Bluetooth device shall not initiate any security procedure before a channel establishment request has been received or a channel establishment procedure has been initiated by itself; in security mode 3, the Bluetooth device must initiate security procedures before the channel link is set up. In reality, the user must use an eight-digit personal identification number (PIN) to secure devices, and Bluetooth devices will not communicate until they have "paired"—a one-off process in which both devices must enter the same PIN. This procedure might be modified if one of the devices has a fixed PIN, for example, for headsets or similar devices with a restricted user interface.

SLPv2 [136] supports authentication and integrity of service URLs and service attributes. It implements the public and private key PKI security mechanism that allows signing of service announcements. The keys are distributed to agents by the scope administrator. However, confidentiality is not provided by default, as the protocol is assumed to be used in nonsensitive environments.

Other industry standards, such as Salutaion and Bonjour, apply simple password-based authentication and access mechanisms for security.

Zhu et al. [130] present a proxy-based approach that uses other existing network channels to set up a secure and trustworthy relationship between unfamiliar parties to facilitate *ad hoc* wireless communications. Through Splendor [116] they further propose a proxy-based service discovery in infrastructure environments to offload tedious work from mobile services and provide security as well as privacy for them. A proxy represents a service and facilitates the processes of authentication and authorization between clients and services. Various public and symmetric key technologies are used to enable mutual authentication, service authorization, confidentiality, integrity, and nonrepudiation among clients, directories, proxies, and services. Splendor assumes there is a public key infrastructure available and the Certificate Authority (CA) signs the public key for all four components of the architecture. Similarly, there is a CA in SDS [20]. Each component in an SDS has a principal name and public key certificate that is signed by the CA. Services can be signed by such principals and clients can specify the principals that they trust and have access to by retrieving certificates from the CA, and when they pose queries, an SDS server will return only those services that are run by the specified, trusted principals. To ensure message integrity (i.e., between clients and SDS servers and between services and SDS servers), a traditional hybrid of asymmetric and symmetric key cryptography is used. A recent proposal can be found in the study by Trabelsi et al. [131]: their study not only defines the threat model as introduced earlier but also proposes a trusted infrastructure that relies on a trusted registry to establish trustworthy relationships among entities within the framework. Security policies such as those for service registration and discovery are enforced to solve the problems of client and service authentication, privacy protection, service access control, and availability check. Trabelsi et al. also extend the work in another study [137] by providing security mechanisms in decentralizing architecture such as MANETs. Additionally, public and private key cryptographic mechanisms, such as attribute-based encryption [131] and policy-based encryption [138], are used to encrypt messages.

2.3.4.2 Privacy and Trust

Privacy is defined as the right of an individual to control information disclosure. As such, an invasion of privacy is considered to occur when information regarding an entity is disclosed without the entity's explicit consent [139]. It depends on the willingness of the entity to share its sensitive information, such as service content and location information. The service provider may not want its service information to be stored outside of its current domain, and service users may not want to reveal their current positions or leave personal data to directory servers or services being used. Hashing and encryption techniques can be applied, like in SDS [20,140,141]. In SDS, bloom filters [142] are used to protect the service and personal information of the user, such as identity, certificates, and attributes. Basically, the bloom filter allows an easy test of domain membership. Once the user authenticates with the directory that represents a domain, he or she can send a service query. Furthermore, in References [140,141], directories are trusted entities for authentication and authorization. Splendor [113] deploys a passive location sensing system, so that the location information of users is kept private until they want to reveal their positions. Also, the identity of a service provider is hidden by providing the mapping in the proxy so that only a generic name such as "hospital service provider" is registered in the directory server. Cardoso et al. [139] enhance the MUlti-protocol Service Discovery and ACcess (MUSDAC) middleware by providing privacy protection. It aims at dissociating the source of a request from the request message and reducing the access of untrusted entities to service request contents. It provides the following detailed mechanisms to provide privacy awareness. (i) *Mechanism to enable clients to define geographic and domain restrictions for service request propagation*: This is achieved by including, in service requests, a list of geographical entities that adopt privacy laws trusted by the client and a list of companies that have privacy agreements accepted by the client. (ii) *Mechanism to increase sender anonymity degree during hop-by-hop routing*: This is done by creating a new query ID for next hop routing. The old ID is recorded so as to forward reply messages. (iii) *Mechanism to enable end-to-end anonymous encryption*: This relies on the mix protocol [143] to protect the relationship between the sender and receiver of a given message, providing relationship anonymity. EVEY [144] is another privacy-aware SDP that protects private information related to service discovery by introducing ambiguity in both service advertisements and service requests, so that they can possibly represent a group of services instead of a single specific service instance. However, this protocol requires further verification after services are matched. The authors have also done a detailed survey on architecting pervasive computing systems for privacy [145].

Trust is yet another important factor for service discovery, as it yields the belief of whether an entity is trustworthy to perform a particular operation; for instance, whether the service interacted is reliable and does what the user wants. From another perspective, interacting with a trusted entity provides security and privacy protection as well as a guarantee of QoS. The belief of an entity can be either derived from trust authorities, such as those entities indicated in the privacy law or agreement [139], or learnt from user experience—quality of experience and service reputation. For a detailed survey for trust system design, the interested reader may refer to the surveys by Wang and Vassileva [146] and Jøsang et al. [147].

2.4 Challenges for Future Research

In this section, we consider open issues and challenges of service discovery for mobile computing and present potential areas for researchers to further explore.

2.4.1 Interoperability

Considering the fact that there are a large number of service discovery frameworks either in industry or in research communities, the interoperability among them is an open issue. It is hard to define a "universal" protocol to unify them, as different frameworks may have different focuses, such as that for dynamic MANETs and resource-constrained sensor networks. Currently, there are two major approaches to address service discovery interoperability. One is to provide explicit translation from one protocol to another, that is, developing a plugin module used for bridging two discovery frameworks such as Jini and UPnP. The representative bridges can be found in Jini–UPnP [148], SLP–Jini [149], and Salutation–Bluetooth [150]. However, a trivial problem of this approach is the scalability. The developer is required to explicitly define bridges for all pairs of protocols used, and if later there is a new protocol to be defined, the translations for all the existing ones should be added, which could be a tedious task. The second approach to address interoperability is to introduce a new service discovery framework that acts as the middleware for all existing ones. This approach is mentioned in Reference [127] in Section 2.3.3.3. The middleware itself has a defined protocol for service discovery including definitions for service information and query format; meanwhile, it also provides adaptation for SDPs in heterogeneous networks or software platforms. The adaptation is still achieved by defining plugins for different protocols. The middleware provides a common language for communication, and the detailed network type or message format is transparent to the user. This approach provides a scalable way for service discovery interoperability, but it also requires work on designing the common protocol. The detailed components of service discovery should be well identified in order to support precise translations. For mobile service discovery, maybe the second approach is more suitable, because the environment of the user is changing frequently. It is better to use a common protocol to communicate with the framework and let the framework identify the detailed discovery protocol to be used rather than holding all the bridge plugins at the user side and letting the user decide which one to choose. However, less work has been done to evaluate service interoperability in mobile computing; thus, further investigation is still needed.

2.4.2 Context Adaptation

As discussed in the study by Ververidis et al. [5], most service discovery frameworks do not adapt their mechanisms on the basis of context. Context awareness has been primarily used for augmenting service selection rather than for self-tuning the discovery protocol. In Section 2.3.3, we have already introduced several means for context to help service discovery, but the usage of context is mainly to enhance the effectiveness of the discovery framework. For instance, context-aware service matching is to provide more relevant services and device-aware service discovery is to find those compatible services that can be either shown or invoked on the user device. The efficiency of service discovery based on context is only discussed for location-based service discovery. Indeed, most existing service discovery frameworks apply static operational parameters (e.g., connectivity number) and techniques (e.g., designated neighbors) that are designed at build-time. Less work has been done to dynamically adapt to context (e.g., user needs) to improve the routing efficiency of the framework at run-time. For instance, in a decentralized service management framework, if a domain of services catches the attention of a large number of users at a certain time, the service peers inside may increase their neighbor counts and possibly create neighboring relationships with peers in other domains, so that the response time can be decreased for further queries. After some time, if users lose interest in the domain, the service peers can then remove those extra links to reduce maintenance overheads. In mobile computing, as service providers may change their locations/networks, the adaptation based on context is extremely important to enhance the efficiency and effectiveness of the discovery framework. Of course, there are issues making dynamic adaption difficult. For instance, in the illustrated example, how to determine which domain of services attracts users and how many extra neighbors are needed for each peer will be problems. Evaluation metrics such as routing efficiency and connectivity maintenance overheads should be considered. Simulations or practical investigation will be useful in this case.

2.4.3 Other Issues

As for service discovery in mobile computing, security and privacy protection are always concerns. Most service discovery frameworks leverage on existing techniques, such as the PKI system. However, as argued by Marin-Perianu et al. [2], security has to be included from the very beginning in the design of any SDP. Besides security and privacy, connection maintenance is also another issue. Both service providers and users can be on the move. In this case, both service providers and users may change their IP address and the connection between them may get interrupted. Proper mechanisms should be provided to handle this problem in the discovery framework, for example, by using the mobile IP protocol [151]. Other issues such as the lack of a benchmark for evaluating service discovery frameworks and protocols are discussed in Reference [5] as well.

2.5 Conclusion

In this chapter, we have presented a large number of service discovery frameworks from industry standards and research communities. We have defined a set of taxonomies used to classify them in Section 2.2. The taxonomies are from the core mechanisms of a service discovery perspective: *service advertisement, service discovery,* and *service access.* For each process, the components involved are further discussed. In Section 2.3, different considerations are proposed for service discovery in mobile computing, which include network scale, network type, context awareness, security, privacy, and trust. The representatives are illustrated accordingly. Finally, we have identified several open issues and challenges for service discovery.

We conclude that despite the amount of work already done, service discovery is still an open field, where optimal solutions for large-scale wireless and hybrid networks are yet to be developed. The provisioned framework should be more adaptive and intelligent to satisfy different user needs and to provide security and privacy protection.

SAMPLE QUESTIONS

Q1: In what aspect do you think service discovery for mobile computing is different from ordinary service discovery?

A1: The question may be answered from two main aspects. (1) Context awareness: services in mobile computing tend to be dynamic. Their providers may be on the move; thus, service availability must be checked, for example, through periodic polling if there is a local directory server. Besides, service relevance is affected by the mobility of service users and providers, context-aware service matching or reasoning should be provided to enhance user experience. The adaptation to the user device (e.g., PDA, hand phone) may be necessary as well. (2) Security: the environment for users in mobile computing is changing frequently and there are a large number of uncontrolled data exchanges among untrusted entities; thus, service discovery for mobile computing imposes more security, privacy, and trust requirements.

Q2: Summarize the strengths and weaknesses of the illustrated industry approaches.

A2: The industry standards are more focused on practical concerns such as administration control, implementation simplicity, and security. Their aim is usually for service discovery at an enterprise scale or local LAN; therefore, most of them adopt centralized service organization schemes. This is easy to set up and manage. However, from the perspective of system design, industry approaches are not to scale and are not suitable for a dynamic environment. Besides, service matching is known to be keyword-based and category-based, which usually results in low precision and recall as studied in the information retrieval area.

Q3: Summarize the approaches for service description and describe their strengths and weaknesses.

A3: There are two general levels for service description. (1) Syntactic level: attribute–value pair and XML-style representation. This provides the simplest way to describe a service and allows extensibility. The hierarchical relationships among attributes can also be expressed inside the XML document. (2) Semantic level: ontology-based representation. This allows one to achieve a common understanding of a certain concept. Different types of relationships may be expressed among ontology concepts, and, thus, ontology reasoning can be supported. The semantic level has richer expressive power than the syntactic level, but requires more effort for designing and extending the ontology.

Q4: By understanding the different approaches for service organization, which architecture do you think best fits global service discovery?

A4: There is no standard answer to this question. A possible answer can be: The hybrid approach best fits global service discovery. As domains need administration, a centralized approach may be applied, and for dynamic environments, for example, MANETs, the P2P technique can be applied. Overall, the framework is distributed. This allows flexibility in the framework design and caters to different needs. Interoperability may be achieved by transforming messages at different local gateway servers.

Q5: In distributed centralized architecture, what considerations should be taken into account?

A5: The following considerations should be taken into account: (1) discovery of directory servers; (2) creation of backbone for directory servers; and (3) data consistency if the cache of service information is used.

Q6: Describe existing service matching strategies as well as their advantages and limitations.

A6: Service matching strategies used are correlated to the description techniques used for service information. The scope of matching can be at a word level, operation level, and/or process level (for Web services). Word-level matching is usually keyword-based and is simple and fast, but it is mostly done on the syntactical level and may not achieve high accuracy. Semantic matching techniques such as ontology and latent semantic indexing can be used to help match at the semantic level and improve effectiveness. Operation level matching is based on inputs, outputs, and operational parameters for Web services. However, in existing approaches, the final matching is still based on words extracted from the operations, for example, operation name. For Web service composition, process level matching is needed, as the process flow information (i.e., process behavior) also plays a role in determining the similarity score, besides matching at the operation level. However, so far there is no standard solution to evaluate process similarity.

Q7: How can network scale affect the design of the service discovery framework?

A7: For local-scale networks, like the LAN of a home, office, or even company, though routing efficiency is not the primary concern, the effectiveness of searching is rather highly demanded. Designers may choose to apply complex service matching strategies, such as semantic matching over Web service processes, to achieve high accuracy. On the other hand, for global-scale service discovery, which may involve worldwide services, the scalability of the framework is always the major concern. Distributed centralized or decentralized architectures are used to organization services.

Q8: How can network type affect the design of the service discovery framework?

A8: Conventional service discovery frameworks are designed for wired networks; thus, they neglect characteristics and issues in wireless networks. For instance, for MANETs, two practical issues are limited topology knowledge due to wireless communication range and changing topology due to node mobility. Therefore, distributed service organization is more suitable, and context-aware service matching is necessary. For WSNs, it is resource constrained; hence, optimization techniques may be applied (which are as discussed in Section 2.2).

Q9: Why do you think context awareness is crucial to service discovery for mobile computing?

A9: (1) Context can be used to help service selection. By considering contexts of users and services (e.g., location, distance, and preference), relevant services can be selected to improve user experience. (2) Context can be used to help service organization. The organization architecture may take into consideration contexts of services (e.g., proximity and content) to achieve high efficiency when doing query routing. (3) Context can be used to help application adaptation such as device user interface adaptation and discovery protocol adaptation.

References

1. *Wikipedia*. Service definition. http://en.wikipedia.org/wiki/Service_(economics). Last accessed on June 14, 2009.
2. Marin-Perianu, R., Hartel, P., and Scholten, H. A classification of service discovery protocols. *Internal Report*, Department of Electrical Engineering, Mathematics, and Computer Science, University of Twente, Netherlands, 2005.
3. Zhu, F., Mutka, M.W., and Ni, L.M. Service discovery in pervasive computing environments. *IEEE Pervasive Computing*, 4(4), 2005, 81–90.
4. Hosseini-Seno, S.A., Budiarto, R., and Wan, T.C. Survey and new approach in service discovery and advertisement for mobile ad hoc networks. *International Journal of Computer Science and Network Security (IJCSNS)*, 7(2), 2007, 275–284.
5. Ververidis, C.N., and Polyzos, G.C. Service discovery for mobile ad hoc networks: A survey of issues and techniques. *IEEE Communications Surveys and Tutorials*, 10(3), 2008, 30–45.
6. Meshkova, E., Riihijärvi, J., Petrova, M., and Mähönen, P. A survey on resource discovery mechanisms, peer-to-peer and service discovery frameworks. *Computer Networks*, 52(11), 2008, 2097–2128.
7. OASIS. UDDI version 3.0.2. 2005. http://www.uddi.org/pubs/uddi_v3.htm. Last accessed on June 14, 2009.

8. Sun Microsystems. Jini version 2.1. 2005. http://java.sun.com/products/jini/2_1index.html. Last accessed on June 14, 2009.
9. UPnP specifications. http://www.upnp.org/resources/specifications.asp. Last accessed on June 14, 2009.
10. Bluetooth specifications. https://www.bluetooth.org/spec/. Last accessed on June 14, 2009.
11. Service location protocol (SLP). http://www.ietf.org/rfc/rfc2165.txt. Last accessed on June 14, 2009.
12. Salutation. http://salutation.org/. Last accessed on June 14, 2009.
13. Apple Inc. Bonjour technology. *White Paper*. 2007. http://images.apple.com/macrosx/pdf/MACOSX_Bonjour_TB.pdf. Last accessed on June 14, 2009.
14. Hermann, R., Husemann, D., Moser, M., Nidd, M., Rohner, C., and Schade, A. DEAPspace: Transient *ad hoc* networking of pervasive devices. In *Proceedings of the 1st ACM International Symposium on Mobile Ad Hoc Networking & Computing*, Boston, Massachusetts, pp. 133–134, August 11, 2000.
15. Hagemann, S., Letz, C., and Vossen, G. Web service discovery—Reality check 2.0. In *Proceedings of the Third International Conference on Next Generation Web Services Practices*, Seoul, Korea, October 29–31, 2007.
16. Jeckle, M., and Zengler, B. Active UDDI—An extension to UDDI for dynamic and fault-tolerant service invocation. In *Revised Papers from the Node 2002 Web and Database-Related Workshops on Web, Web-Services, and Database Systems*, Erfurt, Germany, pp. 91–100, October 7–10, 2002.
17. Zhou, C., Chia, L.T., Silverajan, B., and Lee, B.S. UX—An architecture providing QoS-aware and federated support for UDDI. In *Proceedings of the 2003 International Conference on Web Services (ICWS 2003)*, Las Vegas, Nevada, pp. 171–176, June 23–26, 2003.
18. Apache River. http://incubator.apache.org/river/RIVER/index.html. Last accessed on June 14, 2009.
19. DNS-based service discovery (DNS-SD). http://files.dns-sd.org/draft-cheshire-dnsext-dns-sd.txt. Last accessed on June 14, 2009.
20. Czerwinski, S.E., Zhao, B.Y., Hodes, T.D., Joseph, A.D., and Katz, R.H. An architecture for a secure service discovery service. In *Proceedings of the 5th Annual ACM/IEEE International Conference on Mobile Computing and Networking*, Seattle, Washington, August 15–19, 1999.
21. Balazinska, M., Balakrishnan, H., and Karger, D. INS/Twine: A scalable peer-to-peer architecture for intentional resource discovery. In *Proceedings of the First International Conference on Pervasive Computing*, Zürich, Switzerland, August 26–28, 2002.
22. WSDL-2.0 Part 1: Core language. http://www.w3.org/TR/wsdl20/. Last accessed on June 14, 2009.
23. OWL-S: Semantic markup for Web services. http://www.w3.org/Submission/OWL-S/. Last accessed on June 14, 2009.
24. WSMO: Web service modeling ontology. http://www.w3.org/Submission/WSMO/. Last accessed on June 14, 2009.
25. WSDL-S: Web service semantics. http://www.w3.org/Submission/WSDL-S/. Last accessed on June 14, 2009.
26. Sun, C., Lin, Y., and Kemme, B. Comparison of UDDI registry replication strategies. In *Proceedings of the IEEE International Conference on Web Services*, San Diego, California, June 6–9, 2004.

27. Lightweight Directory Access Protocol (LDAP): Directory information models. http://tools.ietf.org/html/rfc4512. Last accessed on June 14, 2009.
28. NOMAD Deliverable 3.5: Service discovery middleware, European project NOMAD (Integrated Networks for Seamless and Transparent Service Discovery), February. 2004.
29. Du, Z., Huai, J., and Liu, Y. Ad-UDDI: An active and distributed service registry. In *6th VLDB International Workshop on Technologies for E-Services*, Seoul, Korea, September 12–15, 2006.
30. GICS definitions. http://www.msci.com/equity/GICS_map2005.xls. Last accessed on June 14, 2009.
31. Hu, T. Hsin-ting, and Seneviratne, A. Autonomic peer-to-peer service directory. IEICE/IEEE Joint Special Section on Autonomous Decentralized Systems. *IEICE Transactions on Information and Systems* (*Institute of Electronics, Information, and Communication Engineers*), E88-D(12), 2005, 2630–2639.
32. Sahin, O.D., Gerede, C.E., Agrawal, D., Abbadi, A.E., Ibarra, O.H., and Su, J.W. SPiDeR: P2P-based Web service discovery. In *Proceedings of the Third International Conference on Service Oriented Computing*, Amsterdam, The Netherlands, pp. 157–170, December 12–15, 2005.
33. Stoica, I., Morris, R., Karger, D., Kaashoek, M.F., and Balakrishnan, H. Chord: A scalable peer-to-peer lookup service for Internet applications. In *Proceedings of the 2001 Conference on Applications, Technologies, Architectures, and Protocols for Computer Communications*, San Diego, California, 2001.
34. Tchakarov, J.B., and Vaidya, N.H. Efficient content location in wireless ad hoc networks. In *IEEE International Conference on Mobile Data Management (MDM'04)*, Berkeley, California, January 19–22, 2004.
35. Tsai, H.W., Chen, T.S., and Chu, C.P. Service discovery in mobile ad hoc networks based on grid. *IEEE Transactions on Vehicular Technology*, 58(3), 2009, 1528–1545.
36. Klein, M., König-Ries, B., and Obreiter, P. Service rings—A semantic overlay for service discovery in ad hoc networks. In *Proceedings of the 14th International Workshop on Database and Expert Systems Applications*, Prague, Czech Republic, September 1–5, 2003.
37. Verma, K., Sivashanmugam, K., Sheth, A., Patil, A., Oundhakar, S., and Miller, J. METEOR-S WSDI: A scalable P2P infrastructure of registries for semantic publication and discovery of web services. *Information Technology and Management*, 6(1), 2005, 17–39.
38. Gong, L. JXTA: A network programming environment. *IEEE Internet Computing*, 5(3), 2001, 88–95.
39. Maamar, Z., Yahyaoui, H., and Mahmoud, Q.H. Dynamic management of UDDI registries in a wireless environment of web services: Concepts, architecture, operation, and deployment. *Journal of Intelligent Information Systems*, 28(2), 2007, 105–131.
40. Gnutella specification 0.6. http://rfc-gnutella.sourceforge.net/src/rfc-0_6-draft.html. Last accessed on June 14, 2009.
41. Ratnasamy, S., Francis, P., Handley, M., Karp, R., and Schenker, S. A scalable content-addressable network. In *Proceedings of the 2001 Conference on Applications, Technologies, Architectures, and Protocols for Computer Communications*, San Diego, California, August 27–31, 2001.

42. Rowstron, A.I., and Druschel, P. Pastry: Scalable, decentralized object location, and routing for large-scale peer-to-peer systems. In *Proceedings of the IFIP/ACM international Conference on Distributed Systems Platforms*, Heidelberg, Germany, November 12–16, 2001.
43. Zhao, B.Y., Kubiatowicz, J.D., and Joseph, A.D. Tapestry: An Infrastructure for fault-tolerant wide-area location and routing. *Technical Report*, UMI Order Number: CSD-01–1141, University of California at Berkeley, 2001.
44. Acosta, W., and Chandra, S. Unstructured peer-to-peer networks—Next generation of performance and reliability [Poster]. In *24th Annual Joint Conference of the IEEE Computer and Communications Societies (INFOCOM 2005)*, Miami, Florida, March 13–17, 2005.
45. Banaei-Kashani, F., Chen, C.-C., and Shahabi, C. WSPDS: Web services peer-to-peer discovery service. In *Proceedings of the International Conference on Internet Computing*, Las Vegas, Nevada, pp. 733–743, June 21–24, 2004.
46. Banaei-Kashani, F., and Shahabi, C. Searchable querical data networks. In *Proceedings of the International Workshop on Databases, Information Systems and Peer-to-Peer Computing in Conjunction with VLDB'03*, Berlin, Germany, September 7–8, 2003.
47. Cohen, E., and Shenker, S. Replication strategies in unstructured peer-to-peer networks. *SIGCOMM Computer Communication Review*, 32(4), 2002, 177–190.
48. Klein, M., Kognig-Ries, B., and Obreiter, P. Lanes—A lightweight overlay for service discovery in mobile ad hoc networks. In *3rd Workshop on Applications and Services in Wireless Networks (2003)*, Berne, Switzerland, July 2–4, 2003.
49. Kang, E., Kim, M.J., Lee, E., and Kim, U. DHT-based mobile service discovery protocol for mobile ad hoc networks. In *Proceedings of the 4th International Conference on Intelligent Computing: Advanced Intelligent Computing Theories and Applications—With Aspects of Theoretical and Methodological Issues*, Shanghai, China, September 15–18, 2008.
50. Schmidt, C., and Parashar, M. A peer-to-peer approach to web service discovery. *World Wide Web*, 7(2), 2004, 211–229.
51. Pirró, G., Talia, D., Trunfio, P., Missier, P., and Goble, C. ERGOT: Combining DHTs and SONs for semantic-based service discovery on the grid. *Technical Report*, in CoreGRID Project, 2008.
52. Tang, C., Xu, Z., and Mahalingam, M. pSearch: Information retrieval in structured overlays. *SIGCOMM Computer Communications Review*, 33(1), 2003, 89–94.
53. Zhu, Y., and Hu, Y. Semantic search in peer-to-peer systems. In *Handbook of Theoretical and Algorithmic Aspects of Ad Hoc, Sensor, and Peer-to-Peer Networks*, J. Wu (ed.), pp. 643–664, Boca Raton, FL: Auerbach Publications, 2006.
54. Traversat, B.M., Abdelaziz, M., and Pouyoul, E. *Project JXTA: A Loosely-Consistent DHT Rendezvous Walker*. Palo Alto, California: un Microsystems, Inc., 2003.
55. Arabshian, K., and Schulzrinne, H. GloServ: Global service discovery architecture. In *First Annual International Conference on Mobile and Ubiquitous Systems: Networking and Services (MobiQuitous 2004)*, Boston, Massachusetts, pp. 319–325, August 22–26, 2004.
56. Huang, L. A P2P service discovery strategy based on content catalogues. In *20th CODATA International Conference*. Beijing, China, pp. 23–25, October 23–25, 2006.
57. Unger, H., and Wulff, M. Cluster-building in P2P-community networks. In *Proceedings of the International Conference on Parallel and Distributed Computing Systems* (*IASTED PDCS 2002*), Cambridge, Massachusetts, pp. 680–685, November 4–6, 2002.

58. Ghader, M., Olsen R.L., Genet M.G., and Tafazolli R. Service management platform for personal networks. In *14th IST Mobile and Wireless Communications Summit*, Dresden, Germany, June 19–22, 2005.
59. Bachlechner, D., Siorpaes, K., Lausen, H., and Fensel, D. Web service discovery—A reality check. In *Proceedings of the 1st Workshop: SemWiki2006—From Wiki to Semantics, Co-Located with the Third Annual European Semantic Web Conference (ESWC'06)*, Budva, Montenegro, June 11–14, 2006.
60. Sakaryan, G., Wulff, M., and Unger, H. Search methods in P2P networks: A survey. In *Innovative Internet Community Systems: 4th International Workshop (IISC 2004)*, Guadalajara, Mexico, pp. 59–68, June 21–23, 2004.
61. Lv, Q., Cao, P., Cohen, E., Li, K., and Shenker, S. Search and replication in unstructured peer-to-peer networks. In *Proceedings of the 16th International Conference on Supercomputing*, New York, June 22–26, 2002, 84–95.
62. Salton, G., Wong, A., and Yang, C.S. A vector space model for automatic indexing. *Communications of the ACM*, 18(11), 1975, 613–620.
63. Platzer, C., and Dustdar, S. A vector space search engine for Web services. In *Proceedings of IEEE European Conference on Web Services (ECOWS 2005)*, Växjö, Sweden, pp. 62–75, November 14–16, 2005.
64. Porter stemming algorithm. 2004. http://www.tartarus.org/martin/PorterStemmer/. Last accessed on June 14, 2009.
65. Sajjanhar, A., Hou, J., and Zhang, Y. Algorithm for web services matching. In *Advanced Web Technologies and Applications: 6th Asia-Pacific Web Conference (APWeb 2004)*, Hangzhou, China, pp. 665–670, April 14–17, 2004.
66. Hao, Y., and Zhang, Y. Web services discovery based on schema matching. In *Proceedings of the 30th Australasian Conference on Computer Science—Volume 62*, Ballarat, Victoria, Australia, pp. 107–113, January 30–February 2, 2007.
67. Klein, M., and Bernstein, A. Toward high-precision service retrieval. *IEEE Internet Computing*, 8(1), 2004, 30–36.
68. Bansal, S., and Vidal, J. M. Matchmaking of web services based on the DAML-S service model. In *Proceedings of the Second International Joint Conference on Autonomous Agents and Multiagent Systems*, Melbourne, Victoria, Australia, pp. 926–927, July 14–18, 2003.
69. Pathak, J., Koul, N., Caragea, D., and Honavar, V.G. A framework for semantic web services discovery. In *Proceedings of the 7th Annual ACM International Workshop on Web Information and Data Management*, Bremen, Germany, pp. 45–50, October 31–November 5, 2005.
70. Levenshtein distance algorithm. http://www.merriampark.com/ld.htm. Last accessed on June 14, 2009.
71. WordNet. http://wordnet.princeton.edu/. Last accessed on June 14, 2009.
72. Yu, S.J., Liu, J.W., and Le, J.J. DHT facilitated Web service discovery incorporating semantic annotation. In *Discovery Science: Proceedings of the 7th International Conference (DS 2004)*, Padova, Italy, pp. 363–370, October 2–5, 2004.
73. Fu, Y., Jin, T., Ling, X., Liu, Q., and Cui, Z. A multistrategy semantic web service matching approach. In *Proceedings of the 2007 International Conference on Convergence Information Technology*, Korea, pp. 253–256, November 21–23, 2007.
74. Resnik, P. Semantic similarity in a taxonomy: An information-based measure and its application to problems of ambiguity in natural language. *Journal of Artificial Intelligence Research*, 11, 1999, 95–130.

75. Martino, B.D. An ontology matching approach to semantic web services discovery. In *Frontiers of High Performance Computing and Networking—ISPA 2006 Workshops*, Sorrento, Italy, pp. 550–558, December 2006.
76. Kotis, K., and Lanzenberger, M. Ontology matching: current status, dilemmas and future challenges. In *Proceedings of the 2008 International Conference on Complex, Intelligent and Software Intensive Systems*, Barcelona, Spain, pp. 924–927, March 4–7, 2008.
77. Latent semantic analysis. http://lsa.colorado.edu/. Last accessed on June 14, 2009.
78. Wu, C., Chang, E., and Aitken, A. An empirical approach for semantic web services discovery. In *Proceedings of the 19th Australian Conference on Software Engineering*, Perth, Western Australia, pp. 412–421, March 26–28, 2008.
79. Dong, X., Halevy, A., Madhavan, J., Nemes, E., and Zhang, J. Similarity search for web services. In *Proceedings of the 30th International Conference on Very Large Data Bases—Volume 30*, Toronto, Canada, pp. 372–383, August 31–September 3, 2004.
80. Heß, A., and Kushmerick, N. Machine learning for annotating semantic web services. In *AAAI Spring Symposium on Semantic Web Services*, Palo Alto, California, March 22–24, 2004, 60–67.
81. Patil, A.A., Oundhakar, S.A., Sheth, A.P., and Verma, K. Meteor-s web service annotation framework. In *Proceedings of the 13th International Conference on World Wide Web*, New York, pp. 553–562, May 17–20, 2004.
82. Bruno, M., Canfora, G., Penta, M.D., and Scognamiglio, R. An approach to support Web service classification and annotation. In *Proceedings of the 2005 IEEE International Conference on E-Technology, E-Commerce and E-Service (Eee'05) on E-Technology, E-Commerce and E-Service*, Hong Kong, pp. 138–143, March 29–April 1, 2005.
83. Grigori, D., and Bouzeghoub, M. Service retrieval based on behavioral specification. In *Proceedings of the 2005 IEEE International Conference on Services Computing—Volume 1*, Orlando, Florida, pp. 333–336, July 11–15, 2005.
84. Wombacher, A. Evaluation of technical measures for workflow similarity based on a pilot study. In *Proceedings of the 14th International Conference on Cooperative Information Systems* (*CoopIS'06*), Montpellier, France, *Lecture Notes in Computer Science*, Vol. 4275, pp. 255–272, November 1–3, 2006.
85. Bae, J., Liu, L., Caverlee, J., and Rouse, W.B. Process mining, discovery, and integration using distance measures. In *Proceedings of International Conference of Web Services (IEEE ICWS 2006)*, Chicago, Illinois, pp. 479–488, September 18–22, 2006.
86. Minor, M., Tartakovski, A., and Bergmann, R. Representation and structure-based similarity assessment for agile workflows. In *Proceedings 7th International Conference on Case-Based Reasoning*, Belfast, Northern Ireland, pp. 224–238, August 13–16, 2007.
87. Xiao, Y., Dong, H., Wu, W., Xiong, M., Wang, W., and Shi, B. Structure-based graph distance measures of high degree of precision. *Pattern Recognition* 41(12), 2008, 3547–3561.
88. Bae, J., Caverlee, J., Liu, L., and Yan, H. Process mining by measuring process block similarity. In *Business Process Management Workshops (BPM 2006)*, Vienna, Austria, pp. 141–152, September 4–7, 2006.
89. Yang, R., Kalnis, P., and Tung, A. K. Similarity evaluation on tree-structured data. In *Proceedings of the 2005 ACM SIGMOD International Conference on Management of Data*, Baltimore, Maryland, pp. 754–765, June 14–16, 2005.
90. Niemegeers, I.G., and Heemstra De Groot, S.M. From personal area networks to personal networks: A user oriented approach. *Wireless Personal Communications*, 22(2), 2002, 175–186.

91. El-Khatib, K., Hadibi, N., and Bochmann, G.v. Support for personal and service mobility in ubiquitous computing environments. In *Proceedings of International Conference on Parallel and Distributed Computing (EuroPar 2003)*, Klagenfurt, Austria, pp. 1046–1055, August 26–29, 2003.
92. Park, K.L., Yoon, U.H., and Kim, S.D. Personalized service discovery in ubiquitous computing environments. *IEEE Pervasive Computing*, 8(1), 2009, 58–65.
93. Kozat, U.C., and Tassiulas, L. Service discovery in mobile ad hoc networks: An overall perspective on architectural choices and network layer support issues. *Ad Hoc Networks*, 2(1), 2004, 23–44.
94. Sailhan, F., and Issarny, V. Scalable service discovery for MANET. In *Proceedings of the Third IEEE International Conference on Pervasive Computing and Communication (PerCom 2005)*, Kauai Island, Hawaii, pp. 235–244, March 8–12, 2005.
95. Campo, C., Munoz, M., Perea, J.C., Marin, A., and Garcia-Rubio, C. PDP and GSDL: A new service discovery middleware to support spontaneous interactions in pervasive systems. In *Proceedings of the Third IEEE International Conference on Pervasive Computing and Communication (PerCom 2005)*, Kauai Island, Hawaii, pp. 178–182, March 8–12, 2005.
96. Hassanein, H., Yang, Y., and Mawji, A. A new approach to service discovery in wireless mobile ad hoc networks. *International Journal of Sensor Networks*, 2(1/2), 2007, 135–145.
97. Garcia-Macias, J.A., and Torres, D.A. Service discovery in mobile ad hoc networks: Better at the network layer? In *Proceedings of the 2005 International Conference on Parallel Processing Workshops*, Oslo, Norway, pp. 452–457, June 14–17, 2005.
98. Gerla, M., Pei, G., Lee, S.J., and Chiang, C.C. On-Demand Multicast Routing Protocol (ODMRP) for mobile *ad hoc* networks, WAM Lab, UCLA, 1998.
99. Ververidis, C.N., and Polyzos, G.C. Routing layer support for service discovery in mobile ad hoc networks. In *Proceedings of the Third IEEE International Conference on Pervasive Computing and Communication (PerCom 2005)*, Kauai Island, Hawaii, pp. 258–262, March 8–12, 2005.
100. Varshavsky, A., Reid, B., and Lara. E.D. The need for cross-layer service discovery in MANETs. *Technical Report CSRG-492*, University of Toronto, Canada, 2004.
101. Merlin, C.J., and Heinzelman, W.B. X-lisa: A cross-layer information-sharing architecture for wireless sensor networks. *Technical Report*, University of Rochester, 2006.
102. Zhao, N., and Sun, L. Research on cross-layer frameworks design in wireless sensor networks. In *Proceedings of the Third International Conference on Wireless and Mobile Communications (ICWMC 2007)*, Guadeloupe, French Caribbean, March 4–9, 2007.
103. Chilamkurti, N., Zeadally, S., Vasilakos, A., and Sharma, V. Cross-layer support for energy efficient routing in wireless sensor networks. *Journal of Sensors*, 2009, Article ID 134165, 9 pages.
104. CLAWS Project. http://www.researchportal.be/en/project/a-cross-layer-framework-for-heterogeneous-wireless-sensor-networks-claws–(UA_21944)/. Last accessed on June 14, 2009.
105. NanoIP. http://www.cwc.oulu.fi/nanoip/. Last accessed on June 14, 2009.
106. WBXML. http://www.w3.org/TR/wbxml/. Last accessed on June 14, 2009.
107. Madden, S., Franklin, M.J., Hellerstein, J.M., and Hong, W. TAG: A tiny AGgregation service for *ad hoc* sensor networks. *SIGOPS Operating Systems Review*, 36(Special Issue), 2002, 131–146.

108. Schiele, G., Becker, C., and Rothermel, K. Energy-efficient cluster-based service discovery for ubiquitous computing. In *Proceedings of the 11th Workshop on ACM SIGOPS European Workshop*, Leuven, Belgium, September 19–22, 2004.
109. Baldauf, M., Dustdar, S., and Rosenberg, F. A survey on context-aware systems. *International Journal of Ad Hoc Ubiquitous Computing*, 2(4), 2007, 263–277.
110. Abowd, G.D., Dey, A.K., Brown, P.J., Davies, N., Smith, M., and Steggles, P. Towards a better understanding of context and context-awareness. In *Proceedings of the 1st International Symposium on Handheld and Ubiquitous Computing*, Karlsruhe, Germany, pp. 304–307, September 27–29, 1999.
111. Broll, G., Hußmann, H., Prezerakos, G.N., Kapitsaki, G., Salsano, S. Modeling context information for realizing simple mobile services. In *Proceedings of the 16th IST Mobile & Wireless Communications Summit*, Budapest, Hungary, pp. 1–5, July 1–5, 2007.
112. Hong, C.S., Kim, H.S., Cho, J., Cho, H.K., and Lee, H.C. Context modeling and reasoning approach in context-aware middleware for URC system. *Proceedings of World Academy of Science, Engineering and Technology*, 26, 2007, 129–133.
113. Klan, F. Context-aware service discovery, selection and usage. In *18th GI-Workshop on the Foundations of Databases*, Wittenberg, Saxony-Anhalt, pp. 95–99, June 6–9, 2006.
114. Strang, T., and Linnhoff-Popien, C. A context modeling survey. In *Workshop on Advanced Context Modelling, Reasoning and Management, UbiComp 2004—The Sixth International Conference on Ubiquitous Computing*, Nottingham, England, September 7–10, 2004.
115. Kindberg, T., Barton, J., Morgan, J., Becker, G., Caswell, D., Debaty, P., Gopal, G., et al. People, places, things: Web presence for the real world. *Mobile Networks and Applications*, 7(5), 2002, 365–376.
116. Zhu, F., Mutka, M., and Ni, L. Splendor: A secure, private, and location-aware service discovery protocol supporting mobile services. In *Proceedings of the First IEEE International Conference on Pervasive Computing and Communication (PerCom 2003)*, Fort Worth, Texas, pp. 235–242, March 23–26, 2003.
117. Lee, C., and Helal, S. A multi-tier ubiquitous service discovery protocol for mobile clients. In *International Symposium on Performance Evaluation of Computer and Telecommunication Systems (SPECTS'03)*, Montreal Canada, July 20–24, 2003.
118. Balken, R., Haukrogh, J., Jensen, J.L., Jensen, M.N., Roost, L.J., Toft, P., Olsen, R.L., and Schwefel, H. Context-sensitive service discovery experimental prototype and evaluation. *Wireless Personal Communications*, 40(3), 2007, 417–431.
119. Broens, T., Pokraev, S., Sinderen, M.V., Koolwaaij, J., and Costa, P.D. Context-aware, ontology-based service discovery. In *European Symposium on Ambient Intelligence (EUSAI)*, Eindhoven, The Netherlands, November 8–11, pp. 72–83.
120. Ganter, B., Stumme, G., and Wille, R. Formal concept analysis: Theory and applications. *Journal of Universal Computer Science*, 10(8), 2004, 926.
121. Cuddy, S., Katchabaw, M., and Lutfiyya, H. Context-aware service selection based on dynamic and static service attributes. In *IEEE International Conference on Wireless and Mobile Computing, Networking and Communications (WiMob2005)*, Montreal, Canada, pp. 13–20, August 22–25, 2005.
122. Doulkeridis, C., Loutas, N., and Vazirgiannis, M. A system architecture for context-aware service discovery. In *International Workshop on Context for Web Services (CWS'05)*, Paris, France, July 5, 2005.

123. Spanoudakis, G., Mahbub, K., and Zisman, A. A platform for context aware runtime web service discovery. In *IEEE International Conference on Web Services (ICWS2007)*, Salt Lake City, Utah, pp. 233–240, July 9–13, 2007.
124. Yang, C.Y. Ontology based context-aware web services description and discovery. Master's thesis, Network Learning Technology department, National Central University, Taiwan, 2006.
125. Qi, Y., Qi, S.Y., Zhu, P., and Shen, L.F. Context-aware semantic web service discovery. In *Third International Conference on Semantics, Knowledge and Grid* Shan Xi, China, pp. 499–502, October 29–31, 2007.
126. Al-Masri, E., and Mahmoud, Q.H. A context-aware mobile service discovery and selection mechanism using artificial neural networks. In *Proceedings of the 8th International Conference on Electronic Commerce: the New E-Commerce: Innovations for Conquering Current Barriers, Obstacles and Limitations to Conducting Successful Business on the Internet*, Fredericton, New Brunswick, Canada, pp. 594–598, August 13–16, 2006.
127. Raverdy, P., and Issarny, V. Context-aware service discovery in heterogeneous networks. In *Proceedings of the Sixth IEEE International Symposium on World of Wireless Mobile and Multimedia Networks (WOWMOM'05)*, Taormina/Giardini Naxos, Italy, pp. 478–480, June 13–16, 2005.
128. Raverdy, P., Riva, O., de La Chapelle, A., Chibout, R., and Issarny, V. Efficient context-aware service discovery in multi-protocol pervasive environments. In *Proceedings of the 7th International Conference on Mobile Data Management (MDM'06)*, Nara, Japan, p. 3, May 10–12, 2006.
129. MUSDAC platform. http://wwwrocq.inria.fr/arles/download/ubisec/index.html. Last accessed on June 14, 2009.
130. Zhu, F., Mutka, M., and Ni, L. Facilitating secure ad hoc service discovery in public environments. *Journal of Systems and Software*, 76(1), 2005, 45–54.
131. Trabelsi, S., Pazzaglia, J., and Roudier, Y. Secure web service discovery: Overcoming challenges of ubiquitous computing. In *Proceedings of the 4th European Conference on Web Services (ECOWS'06)*, Zurich, Switzerland, pp. 35–43, December 4–6, 2006.
132. Cotroneo, D., Graziano, A., and Russo, S. Security requirements in service oriented architectures for ubiquitous computing. In *Proceedings of the 2nd Workshop on Middleware For Pervasive and Ad Hoc Computing*, Toronto, Ontario, Canada, pp. 172–177, October 18–22, 2004.
133. Leung, A., and Mitchell, C.J. A service discovery threat model for ad hoc networks. In *Proceedings of the International Conference on Security and Cryptography (SECRYPT 2006)*, Setubal, Portugal, pp. 164–174, August 7–10, 2006.
134. Device Security and Security Console v.1.0, http://www.upnp.org/standardizeddcps/security.asp. Last accessed on June 14, 2009.
135. Bluetooth SIG. Specification of the Bluetooth system—Core and profiles. v.1.1, 2001. Last accessed on June 14, 2009.
136. Service location protocol, version 2. http://www.ietf.org/rfc/rfc2608.txt. Last accessed on June 14, 2009.
137. Trabelsi, S., Roudier, Y., and Pazzaglia, J. Service discovery: Reviewing threats and security architectures. *Technical Report*, Institut Eurecom, France, 2007.
138. Bagga, W., and Molva, R. Policy-based cryptography and applications. In *Proceedings of the 9th International Conference on Financial Cryptography and Data Security*, Roseau, The Commonwealth of Dominica, pp. 72–87, February 28–March 3, 2005.

139. Cardoso, R.S., Raverdy, P.G., and Issarny, V. A privacy-aware service discovery middleware for pervasive environments. In *IFIPTM'07: Joint iTrust and PST Conferences on Privacy, Trust Management and Security*, Moncton, New Brunswick, Canada, pp. 59–74, July 30–August 2, 2007.
140. Zhu, F., Mutka, M., and Ni, L. PrudentExposure: A private and user-centric service discovery protocol. In *Proceedings of IEEE International Conference on Pervasive Computing and Communications (PerCom2004)*, Orlando, Florida, pp. 329–338, March 14–17, 2004.
141. Zhu, F., Zhu, W., Mutka, M.W., and Ni, L. Expose or not? A progressive exposure approach for service discovery in pervasive computing environments. In *Proceedings of the Third IEEE International Conference on Pervasive Computing and Communications (PerCom 2005)*, Kauai Island, Hawaii, pp. 225–234, March 8–12, 2005.
142. Bloom, B.H. Space/time trade-offs in hash coding with allowable errors. *Communications of the ACM*, 13(7), 1970, 422–426.
143. Dingledine, R., Mathewson, N., and Syverson, P. Tor: The second-generation onion router. In *Proceedings of the 13th USENIX Security Symposium*, San Diego, California, August 9–13, 2004, p. 21.
144. Cardoso, R.S., Ben-Mokhtar, S., Urbieta, A., and Issarny, V. EVEY: Enhancing privacy of service discovery in pervasive computing. In *Proceedings of the 2007 ACM/IFIP/USENIX International Conference on Middleware Companion*, Newport Beach, California, p. 27, November 26–30, 2007.
145. Cardoso, R.S., and Issarny, V. Architecting pervasive computing systems for privacy: A survey. In *Proceedings of the Sixth Working IEEE/IFIP Conference on Software Architecture (WICSA'07)*, Mumbai, India, p. 26, January 6–9, 2007.
146. Wang, Y., and Vassileva, J. Toward trust and reputation based web service selection: A survey. In *International Transactions on Systems Science and Applications (ITSSA) Journal, Special Issue on New Tendencies on Web Services and Multi-agent Systems (WS-MAS)*, 3(2), 2007, 118–132.
147. Jøsang, A., Ismail, R., and Boyd, C. A survey of trust and reputation systems for online service provision. *Decision Support System*, 43(2), 2007, 618–644.
148. Allard, J., Chinta, V., Gundala, S., and Richard, G.G. Jini meets UPnP: An architecture for Jini/UPnP interoperability. In *Proceedings of the 2003 International Symposium on Applications and the Internet (SAINT 2003)*, Orlando, Florida, pp. 268–275, January 27–31, 2003.
149. Guttman, E. Service location protocol: Automatic discovery of IP network services. *IEEE Internet Computing*, 3(4), 1999, 71–80.
150. Miller, B. Mapping salutation architecture APIs to Bluetooth service discovery layer. *Bluetooth Consortium 1.C.118/1.0*, July 1, 1999. http://www.afn.org/~afn48922/downs/wireless/1C11800.pdf. Last accessed on Nov 11, 2011.
151. IP mobility support for IPv4. http://tools.ietf.org/html/rfc3344. Last accessed on June 14, 2009.

Chapter 3

Interactive Context-Aware Services for Mobile Devices

Anala Aniruddha Pandit and Anup Kumar

Contents

3.1 Introduction to Context-Aware Mobile Services

Mobile computing devices, more powerful by the day, have become a part of human outfits. Intense competition and customer demands are driving companies to innovate services and products. New applications are being developed and shipped into the market on a daily basis. Response time to customer demand is the driving force. These applications are of two types: developer-defined static applications, where the user does not have any control over the features, and dynamic applications, where the behavior of the application is modified on the basis of context. Moreover, applications that are dynamic in nature can be further classified into two categories: one where the user input is required for the application to alter the behavior and the other where the application modifies itself on the basis of context. Thus, context is extremely important for interactive computing, but has been poorly utilized [1]. Communication between humans is not just through their rich linguistic exchange but also through various other gestures, their understanding of day-to-day situations, and other implicit environmental factors. All these factors widen the conversational bandwidth [2]. A similar benefit is not shared by the human–machine interaction. Computers cannot understand the implicit meanings and gestures unless there are additional devices gathering this information for the machine. Until now, the area of artificial intelligence had been recognized only as an area of research. It is only in recent times that the concept of artificial intelligence is used for practical application development. When applications in mobile devices are able to take advantage of the *context*, the human–machine interaction will be more rich, powerful, meaningful, and ever pervasive.

3.1.1 Current Market Scenario

Although the mobile market is advancing in leaps and bounds, coupled with the extreme market competition, it has become necessary for vendors to create a host of applications that cater to the instantaneous personal needs of an individual. Personalization or individualization is becoming the key factor to capture the customer. Context awareness, a state where the device is aware of the situation in which it is used, is an idea that has gained an increasing amount of attention in recent years. According to the Gartner report, "*Context-aware computing is defined as the concept of leveraging information about the end user to improve the quality of the interaction. Emerging context-enriched services will use location, presence, social attributes, and other environmental information to anticipate an end user's immediate needs, offering more-sophisticated, situation-aware and usable functions*" [3:1]. Thus, a context-aware device can evaluate the situational conditions and alter/ adapt its behavior according to the circumstances. Mobile handheld devices, currently seen in almost everyone's hand (in fact, a mobile phone in one hand and a personal digital assistant, like an iPad, in the other is the order of the day), constitute an interesting platform for context awareness. Context-aware application development is extremely complex as the applications can be used in a variety of situations and user preferences may be different, making it necessary to alter/adapt in different ways under similar/different circumstances. The user is mobile, thus changing the user's current context rapidly. The increased complexity and growing number of features challenges the intuitive and easy use of mobile devices. It is perceived that context awareness may offer solutions to more efficient use of mobile applications and services. Flexibility and user control are of extreme importance. So, these services are more valued by the customer when they feel that they have a greater control over how the application should modify itself with reference to context. A survey conducted by Häkkilä [4], as part of her dissertation work, reveals that the strongest fear toward the use of these applications was that of "spamming" and "personalization and filtering" of context-sensitive push messaging. The most common services identified by Häkkilä [4] were:

- Information regarding restaurants
- Hospitals

- Public transport timetables
- Entertainment schedules in current locations
- Maps and charts while moving through unknown terrains, museums, and so on

Apart from these, users also valued situation-aware reminders and notes [4]. For example, assume that you are traveling to work and there is a traffic jam (a common scenario). It will be welcome if you receive a message on your mobile device specifying an alternative route to your work place where traffic is less [5]. Similarly, a reminder of your spouse's birthday or your anniversary while passing a florist shop is extremely welcome on an otherwise busy day. Advances in ubiquitous computing or pervasive computing are enabling mobile device providers to create many such novel applications that we will discuss later in the chapter. However, the real value of these applications will be determined only when the developed context-aware applications are used in *real time* and are *reusable* and *scalable*. Also, though personalization is appreciated, market penetration to such services will happen only when privacy and security issues are addressed effectively. Gartner predicts that by 2012 context information will be a $12 billion market with at least two global context providers with more than 100 million subscribers each. Most enterprises will use context-aware mobile devices to provide value addition to their existing services to increase the targeted customer base and customer intimacy [3].

3.1.2 Defining Context

Application developers can make effective use of context when the meaning of *context* is understood completely and the various dimensions that are associated with it. Various definitions of context have been described in the literature. Context-aware computing can be traced back to when Weiser published his seminal paper "The computer for the 21st century" [6:78]. Weiser had predicted in 1991 that computers for the twenty-first century will become ubiquitous in human life. Rather than a human being going to a computer and communicating with it through some complex jargon in everyday language, the goal will be to *conceive a new way of thinking about computers, one that takes into account human works and allows the computers themselves to vanish into the background.* Wireless networks, RFID and so on and supporting software will help the ubiquitous computers become invisible. The earliest reference to context awareness was given by Schilit and Theimer in 1994 [7]. They discussed issues related to disseminating location-based information to interested clients, where they defined location objects as anything available at the location—person, printer, or any other device and/or changes to those devices. Additional dimensions of identification include the time of day, temperature, season, and so on, and have been used while defining context awareness [8,9]. Dey [10] has gone beyond this definition in terms of physical parameters (location and orientation, people around the user, date, and time) by incorporating the user's emotional state and focus of attention. Synonymous words have been used by various authors (based on the project in hand) to define context. However, these definitions do not indicate the way in which context can be utilized. The most commonly accepted definition of context is given by Dey and Abowd [2:3–4], where they define context as "*Any information that can be used to characterize the situation of an entity. An entity is a person, place or object that is considered relevant to the interaction between a user and an application, including the user and the application themselves.*" This definition seems to be the most realistic one as it uses only those components of context that are relevant to the application at that instant rather than a list of objects that may or may not contribute to the application. For example, an application that lists the five star hotels in a neighborhood requires only the current location of the user and need not use the temperature or the season or the person(s) around the user to determine the context. However, if the application is to suggest a place to eat a seasonal dish or cuisine available only at specific times in a year, at a particular location, then the location as well as the season are necessary to recommend a place. This definition of context takes care of the implicit and the explicit requirements for eliciting a response from the application on the basis of context.

3.1.3 Defining Context-Aware Computing

Context awareness can be defined as "a state of the device that is aware of circumstances and adapts itself and responds suitably." Figure 3.1 shows the

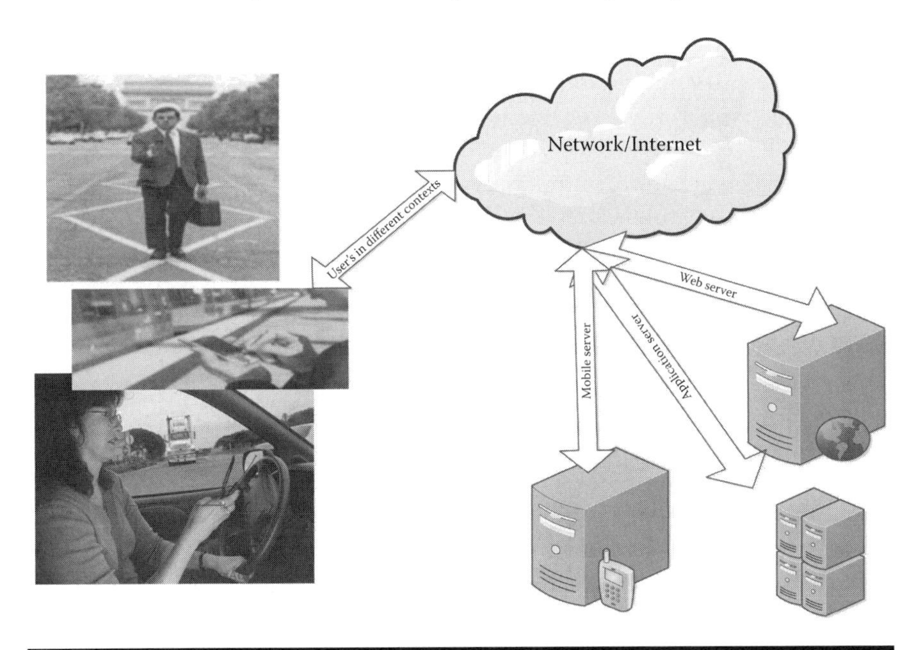

Figure 3.1 Mobile user's context-aware scenario.

communication over the Internet of various mobile devices with the servers. One can find various synonyms for the word in the literature: adaptive, responsive, reactive, situated, environment-directed, and so on. However, one needs to keep in mind that a set of sensors provide the context. The response time of the sensors is also important as a delay in relaying the context will reduce the utility of these applications considerably. Context-aware computing was first defined by Schilit and Theimer [7:22] as "*software that adapts itself according to its location of use, the collection of nearby people and objects as well as the changes to these objects over time.*" This definition implies that software uses the context and also adapts, altering its behavior by itself, with reference to the context. Many such definitions are comprehensively compiled in Dey and Abowd's technical report published in 1999 [2]. In the same report, Dey and Abowd [2:6] have defined a "context-aware system" as the "*one that uses context to provide relevant information and/or services to the user, where relevancy depends on the task the user has in hand.*" This is a comprehensive definition as it takes into account the relevancy of the context. Consider the following example: assume that a person with a context-aware-service-enabled device is walking through a superstore. Many things around the person can be included as "the context." If the designed application provides a list of supersaver items, then it will only take into consideration the person's current shopping list, or give suggestions on the basis of previous purchases made, either currently (from the contents of the person's shopping basket at that moment) or in the past or purchases made by customers with similar profiles (collaborative filtering). This definition, thus, takes into account only the relevancy of the context and response in terms of providing relevant information or service. Although this application would consider the type of response/reaction, it is limited only to where the application provides context-aware service. It does not adapt itself to the context. To illustrate, when a person moves from one time zone to another, the mobile device not only senses the location and displays the relevant time of that location but also modifies the calendar/meeting settings as per the new clock.

As a result of the developments in wireless technology and location-based systems, context-aware systems have reached a new height. Recent advances in telecom, wireless networks, and m-commerce have opened a plethora of applications that are multidimensional in nature. So, apart from modifying the time and calendar, and suggesting nearby restaurants on the basis of one's likes and dislikes, the application will also inform one of the nearest ambulance service in the locality if one happens to be in an area where an accident has taken place. Service providers and handset providers have the capability of storing and analyzing the customer's need/behavior. Whereas handset providers can build in applications that can store customer transactions, service providers have a greater advantage as they are and can remain aware of a customer's location at all times. This initiates a need for a new definition of context-aware systems as outlined below:

> Context-aware systems are those that take advantage of stored knowledge and/or the relevant context information to deliver the relevant information/service while adapting itself to provide newer services based on user needs and demands, using this stored knowledge to ensure "user ecstasy" rather than just "user satisfaction."

Comparing this definition with the ones described in the literature, we ensure that the definition does not eliminate primitive context-aware applications such as redirecting the print command to the nearest printer or switching "off" the ringer of the mobile on entering a conference room or displaying a relevant portion of public transport timetable on the basis of the time of day and the destination. This definition also encompasses the new set of applications that modify their behavior on the basis of context and those that use stored knowledge. For example, a service provider has a record of the movement of the subscriber and the services used by the subscriber or the purchases/money transactions made by the user through the mobile at various locations. Using association mining, collaborative filtering, and content filtering, the provider can predict the required services at various locations and also send prior information with regard to requirements of the customer on a travel path.

3.2 Benefits and Limitations of Context-Aware Mobile Services

The progress in the development of miniaturized, inexpensive, multipurpose sensors that are incorporated in handheld mobile devices opens up the possibility of the ability to perceive, correlate, and utilize the context of the environment and the user. This can make context-aware services a reality. However, the benefits of context-aware mobile systems will be judged using various factors, most importantly utility and the timeliness of the service provided [11]. Other factors include the format in which the service is being provided, the reliability (because of the accuracy of information provided by sensors) of the service [12], the trust the user has in the service provider, the speed at which the service is provided (bandwidth limitations), the value addition to the existing knowledge of the user (if the service is trivial because the user is able to perceive it himself/herself in the context as (s)he is physically present there), clarity of context, and so on. Limitations include, most importantly, limitations of sensors (validity of sensor data and algorithms used) [13], decision on what context is relevant at a particular instance of time, and, of course, the privacy and security concerns of the user. The application developer may also need to define the inputs that can be gathered from the external environment and the inputs that would be necessary to be given by the user. We will discuss them in detail in subsequent sections.

3.2.1 Why Are Context-Aware Systems Becoming Popular?

The development of context-aware systems, especially in the mobile environment, is still in the experimental stage. However, the interest in this field has increased exponentially because these systems bring tremendous amount of flexibility. With miniaturization, the technological support required is also increasing. The "sensors" necessary for judging the "context" as well as the processing capability of the mobile devices have grown rapidly and seen tremendous improvement in the services in recent years [14]. Another important aspect that one cannot ignore is the fact that the world's paradigm has shifted from "I need it" to "I want it." Once the basic amenities are available, people look to get services at the very instant when "I want it." Also, with the advent of the Internet, the world has become a global market. A customer, sitting in front of the computer with Internet facility, is able to get what he wants with the click of a button, at the best price available, from anywhere in the world. Hence, to match the paradigm shift mentioned above and the current market scenario, it has become imperative for the provider to provide the customer a feeling of *ecstasy* rather than only customer *satisfaction*. The need to "lock in" the customers and "lock out" the competition [15] is the single motive, sufficient for service providers to look at providing services that cater to the wants of the customer.

The user, on the other hand, is a winner all the way, provided the services do not invade privacy [16]. These services become extremely important when:

- *The user is treading in unknown territories*: Context-aware services available today are mainly in this domain, whether it is to know about the hotels/restaurants of a specialized category in the vicinity of the user location, stores, movies, or other entertainment programs in the vicinity, location of the nearest possible printer [17–19], and so on.
- *The user requires emergency services*: The location and contact details of ambulance services and/or hospitals in case of accidents or emergency situations; details of police stations in the vicinity and automatic communication with them in case one is a witness of crimes; location of fire stations; incident management, that is, clear the road for traffic as quickly as possible after an incident has happened and ensure safety for emergency services and road users [20], and so on.
- *The user requires assistance at a specific location for a specific purpose*: Information about the current location in a museum and details regarding the artifacts at that location [21]; details of a conference schedule in progress in a particular hall and assistance to the attendee [1]; service in the form of advise to a driver to take a particular route on the basis of his/her current location, destination, and current traffic conditions; telephone call redirection on the basis of the user's location or different service activation on sensing the context of a conference room, meeting room, or canteen [22,23]; defining personal, location-sensitive reminders and notes [4], and so on.

- *Miscellaneous*: A sale going on at a particular location in a store; mobile music experiences [24]; managing patient health-care records using mobile devices [25], and so on.

3.2.2 Types of Sensors

The entire spectrum of context-aware systems is dependent on sensors that will convey the context to the application. Before further discussion it is important to understand the various sensors available and their categorization.

The types of sensors that can be used in mobile devices are increasing everyday. Siewiorek et al. [18] have reported the use of five kinds of sensors (voice microphone, ambient noise microphone, accelerometer, temperature sensor, and visible light sensor) in the development of SenSay, the mobile phone. Yamabe and Nakajima [13] have reported to have successfully incorporated as many as 15 sensors (air temperature sensor, relative humidity sensor, barometer, alcohol sensor, pulse sensor, skin temperature sensor, skin resistance sensor, compass/tilt sensor, three-dimensional linear accelerometer, grip sensor, an ultrasonic range finder, a global positioning system (GPS) receiver, an RFID reader, front/rear cameras, and a microphone) for developing Muffin, a multisensory personal device. However, the sensors used most commonly can be classified into three categories:

- *Motion/Location Sensors*: These sensors assist in finding the location and motion of the user. These are also known as orientation sensors. They include accelerometers, cameras, ambient noise detectors, GPS receivers, RFID readers, ultrasonic range finders, infrared sensors, and so on.
- *Environmental Sensors*: These allow the device to sense the environmental factor as the name suggests. These may include ambient temperature sensor, humidity sensor, barometer, and so on.
- *Physiological Sensors*: These sensors include skin temperature sensors, skin resistance sensors, grip sensors, pulse sensors, and so on.

Among all the sensors mentioned above, the most commonly used are GPS receivers for location identification (outdoors), RFID or ultrasonic range finders, and infrared sensors (indoors). Accelerometers are most popular for gauging motion (sitting, walking, climbing, and so on) as they are inexpensive and also small enough to be embedded in mobile devices without making them bulky. Microphones and cameras help identify the ambient location on the basis of noise levels (recognizing speaking or talking). These are incorporated in most mobile devices available currently in the market. Light and temperature sensors are also used to determine a user's environmental context. The actual sensors included will be on the basis of the applications and the services the provider chooses to incorporate in the device. The number of sensors used by providers in mobile devices will decide the size of such devices. The more the number of sensors, the bulkier the size, reducing its popularity.

3.2.3 Limitations of Existing Infrastructures

As noted earlier, although these are exciting application areas, there are certain technological limitations when one tries to implement these applications in real life. These can be grouped into five types, discussed in detail as follows.

3.2.3.1 Limited Sensor Capability

An important aspect from the providers' point of view is to be able to collate the heterogeneous data from various sensors in real time. The delayed response of the sensors can be a bottleneck in identifying the location of the user, as the context may change, especially when one is working with mobile devices. The context information collected, if dependent on sensors, also faces another difficulty. If the user switches off one of the sensors, the context information may not be accurate. Also, the inaccuracy of the sensor data is likely to give wrong responses. Incorporation of multiple sensors creates an ambiguous situation regarding the context to be extracted. However, in recent decades, there have been reports of a few successful attempts in the development of several context-aware devices. To illustrate, a few examples include: Active Badges [22], SenSay, the context-aware mobile phone [18], the TEA (Technology-Enabling Awareness) approach [26], Muffin, the multisensor device [13], conference assistant [1], teleport [27], tour guide [21], cyberguide [11], and so on.

3.2.3.2 Restrictive Network Bandwidth

Another limitation is network bandwidth. The speed at which information moves over the network is limited by the bandwidth. When the user needs to use context awareness in real time, this can become a real bottleneck. Imagine a situation where the user is video streaming certain information or authentication of certain information is required from the server. The limited bandwidth in such a case can make the application absolutely nonproductive. Also, as the user is mobile, (s)he may be using various connections, such as the wireless local area network (LAN) or Bluetooth or infrared connectivity, at different locations. Services can be disconnected when the user moves from one cell to another and switching takes place. If the context data are being sent to the server for processing owing to limited processing capabilities of mobile devices, this disconnection (permanent connections are not guaranteed) can create problems leading to dissatisfaction [28]. The situation is continuously improving, but there is still a long way to go before any acceptable standards can be reached.

3.2.3.3 Limited Capability of Mobile Devices

For context-aware systems to process large amounts of context data effectively, large memory and processing power are essential. If the processing is done at the server then the speed will be affected. The next limitation arises due to the physical limitations of the mobile devices. This relates to the limited memory of a mobile device, the display space, and the processing power of a device. Although the processing power has improved, the display space is still a limitation. The effective use of display space is a challenge in itself. The information given in the limited space needs to be optimized. In spite of the sufficient memory, if the data are not indexed and properly maintained, retrieval times can be a problem, taking away from the user's feeling of comfort and confidence [29].

3.2.3.4 Rapidly Changing Context

Another limitation is the problem arising as a result of rapid changes in the context and correct interpretation using the received heterogeneous data. When the user is stationary or in a static environment, like sitting on a computer at home and working with specific software or browsing the Internet, providing context-aware services is less challenging. When the user is mobile, the context is changing rapidly. To interpret this context, alignment to the requirement of the user with respect to that context and providing the required services/information will require extremely fast processing to be of any use. Also, the heterogeneous nature of the data gathered by the sensors can lead to ambiguity and complexity of interpretation. The interpretation of certain relative terms like angry or hungry, far and near, can change with situation and application. For example, a "loud voice" needs to be interpreted differently in the conference room, meeting room, or on the road. The "racing pulse" sensed by the pulse sensor when a person is stationary and the person is running is to be interpreted differently on the basis of the activity of the person. The device also needs to work with averages or some predecided computations in case the observations of the sensors vary rapidly. Techniques are required that can help predict the difficulties with perception and interpretation at design time.

3.2.3.5 Trust and Security Requirements

The last category of limitation arises from privacy and security requirements of users. Intuitively, we understand that the user is generally not comfortable with someone "watching over the shoulders" every activity and location of the user. A systematic study on user expectation and requirement has been done comprehensively by Häkkilä [4] as a part of her doctoral thesis work. The findings reported indicate that the user is not comfortable with generalized services and prefers personalized services' thus, the user would be more comfortable with an SMS going to people in their friend's list about their availability in a particular locality. So if the provider gives an option of filtering services (of course, there would be a cost associated depending on requirement), it would be beneficial and desired by the user. Spamming and unnecessary interruptions make users wary of using the many convenient services. An interface to choose the beneficiaries who can share their personal information and the level to which they can share the same is desirable [30]. Thus, the only problem that the user will face is invasion of privacy if proper information is not provided. Also, the issue of receiving the service or information when it has lost relevance needs to be addressed. This may happen due to inaccurate sensor information and network bandwidth limitation (i.e., limiting the amount of data that can be transferred over a network).

3.2.4 Benefits of Context-Aware Systems

The potential benefits of context-aware systems are numerous. The designer can use the knowledge of the context to provide only the required data and information to the user. An example is a context-aware mobile patient record system. If the interface is tailored to suit the mobile nurse's convenience, indicating the location, current activity the nurse is involved in, patients in the proximity, and so on, it would be extremely useful in case of emergency [31,25]. So, in case the user gets the *right* help at the *right* time in the *right'* form, it will spell success for context-aware systems and the service provider. No user will complain if (s)he gets a list of favored brand items on discount in the vicinity of the location that the person is moving through or a suggestion of an alternative route for a destination when the user is stuck in a traffic jam. Once the solutions to the above-mentioned limitations are found, there will be no looking back in this area of application development.

3.3 A Perspective on Contexts in Context-Aware Services

3.3.1 Types of Context-Aware Services/Applications

It is important to understand what qualifies as context in specific scenarios or the application areas and also the type of processing it may go through. The feature categorization of context-aware applications will determine the categories of context. There are basically four dimensions to the categories of application:

- *Selection of information that can be used as context*: Here the selection of the context is based on the proximity of the service required. It could also just be the information needed rather than a choice of service. For example, nurses at a nearby location or restaurants of choice in the proximity or printers in the vicinity are presented and the user chooses one. This has been called "proximate selection" by Schilit et al. [14] and "contextual sensing" by Pascoe [32].
- *Automatic reconfiguration of the context*: The system automatically reconfigures itself on the basis of the context of the user. For example, on recognizing the context of the user, when the user is in conference room, a mobile phone goes on silent mode and calls may be diverted to one's secretary without the

need for one to select this service. It is a system-level technique that creates an automatic binding to an available resource on the basis of the current context. It could also be a process of adding/removing components or altering connections on the basis of the context. There is also a possibility that the results of an application are modified on the basis of the context, as one has seen typically in object orientation.

- *Automatic triggering of action on the basis of context*: Applications that execute commands for the user on the basis of the available context use context-triggered actions. They may be in the form of simple *if–then* rules that specify how context-aware systems should adapt.
- *Tagging of context for future use*: Once the context has been identified, it can be tagged for future use. A variety of applications can use the tagged context.

The categorization taxonomy proposed by Schilit et al. (proximate selection, automatic reconfiguration, contextual information and commands, and context-triggered actions) [14] or Pascoe (contextual sensing, contextual adaptation, context resource discovery, and contextual augmentation) [32] or even, to a major extent, Dey and Abowd (presentation of information, automatic execution of a service, and tagging of information) [2] uses these four basic dimensions to create a taxonomy. However, unlike Schilit et al. [14] and Pascoe [32], Dey and Abowd [2] have not differentiated between service and information.

3.3.2 Types of Contexts

Various kinds of context can be examined. Context information may be permanent or temporary in nature, static or dynamic, and may be manually entered or automatically captured. On the basis of the types of sensors, categories of context can be grouped as follows:

- *Location-Based Context*: Information in this category includes that related to location, physical environment (light conditions, noise levels, traffic conditions, or temperature), and situation (in the canteen, in a conference, in a meeting, or in a movie theater). The initial context-aware applications were built using these contexts as these were minimal and measurable and could be used in available systems without much difficulty.
- *Time-Based Context*: Date and time-based information (day, week, time, season) that can be easily collected from the clock is built into mobile devices. This context is useful in many applications such as health-care applications where reminders can be given for dispensing medication at appropriate times and also requesting stocks at regular intervals.
- *User State-Based Context*: Information in this category includes the user's biological and emotional state, which can be collected by using the physiological

sensors mentioned previously. This context can be extremely useful in emergency situations. For example, alterations in dosage/timing of automatically dispensed medications (e.g., insulin), on the basis of the user's condition, can be an excellent application for saving lives.

- *Application-Based Context*: This type of context generates application- and server-specific information. This information can be used by any service. For example, the traffic density at various locations on a particular route can be used for suggesting alternative routes when there is a traffic jam on a subscriber's route.
- *Network/Device-Based Context*: These consist of network and device parameters and variables. One possible application would be to look at the available bandwidth and reconfigure the information/services that are triggered. For instance, bandwidth utilization while using multiple applications on a mobile device.
- *Context History*: To give the user the experience of "ecstasy," the history of the user as well as the contextual information can be used for creating user profiles. These would be extremely useful in many applications. Of course, the limitations of memory and processing power will need to be addressed before this type of context becomes widely used.

The context can also be bundled in two categories of "physical context" and "logical" context. The physical context, such as location, time, temperature, and so on, would be more directly obtainable, whereas the logical context is a little more difficult to gather as it needs certain processing before one can use it, for example, preferences of the user, relationships, privileges, and so on. Most work has been in the area of physical context processing. Comparatively, there are very few applications in logical context processing as the physical context information has to be sent to the server and connectivity with the server is still in a preliminary stage.

3.3.3 Context Acquisition

As the need for context has become clear, the next issue to be addressed is how to incorporate this context/information in system design. There are two standard ways of obtaining information. One is to pull the required/desired information from the environment and the second is to have the context pushed [33]. Let us look at each one of them and identify implementation issues for each one.

- *Pulling the Context*: This type of information becomes available to the application only when it is requested for. An obvious advantage is that the user will have complete control over the context information requested. Also, "requesting" a certain context, as per the need of the application, could be built into the application. The user then has to be aware of the needs of the application and has to "work" to incorporate the context. If the application is going to automatically pull out the required context information, the bandwidth limitation and costs will come into the picture. In case the required bandwidth is not available, the application may suffer. Also, if the context does not change as frequently as the frequency of "pull" then one will be wasting the bandwidth, without real use. In times of emergencies such as earthquakes or other natural disasters, a large number of consumers may pull out the same information from a supplier, causing severe scalability problems.

- *Getting the Context Pushed*: The other alternative is similar to a "broadcast" facility in networks. The context information will keep getting pushed out at regular intervals to all the users. This simplifies the life of the user (and complexity of application development), in the sense that now the user/application is relieved of the job of getting the contextual information. The utility of the context information is not known unless the user has specifically subscribed for it. It could be possible that a particular context information is irrelevant to a particular user/application. In that case, it is waste of expensive resources in mobile networks. Also, there would be additional need for buffering this context in order to process and check its utility. This increases the cost of the component.

The best approach would then be to combine both these approaches to address all the issues such as bandwidth limitation, cost, frequency of pull, and the utility of context information. The application would request for certain specialized context information (pull) and there could be some information that could keep getting pushed out at regular intervals [34].

3.4 General Framework and Architecture of Context-Aware Services

Building context-aware systems is a challenging task as it involves many parameters that are dynamic in nature. The type of architectures required for supporting context-aware applications will depend on how one approaches the representation of the context. A broad overview of the type of application will also help analyze the architectural requirements. Before we discuss the architectural requirements and generic architectures, an overview of the above-mentioned parameters is given below.

3.4.1 Representation of Context (Context Models)

Early approaches modeled the context from the frame of reference of a single application or a particular genre of application. However, a generic approach is of interest as it will be widely applicable. Contextual information can be provided to/obtained by the application in two ways:

1. The user feeds the information classified as profiled information.
2. Meaningful information is extracted from raw data collected by the sensors, that is, data need to be modeled to reflect the physical entities for interpretation.

Evaluation of the context implies the verification of proposed information generated by modules against a specific context. There are different ways in which context data are modeled. Strang and Linnhoff-Popien [35] give an excellent summary of different context models. These context data models specify the way in which data structures are defined for representing and exchanging the desired context with the system.

- *Key Structure Models*: The simplest of data structures, using key–value pairs to model the context, was suggested and used first by Schilit et al. [14]. In this model, the value of context information (e.g., location information, sound level, and so on) is provided to an application as an environment variable. This model is most widely used in various service frameworks and quite frequently in distributed service frameworks. In such frameworks, the services themselves are usually described with a list of simple attributes in a key–value manner. For instance, the context vocabulary may contain (Temperature, Cold) or (Temperature, Hot) where the environment variable is "temperature" and the context is "cold" or "hot." The service discovery procedure used then operates an exact matching algorithm on these attributes. Note that key–value pairs are easy to manage but lack the capabilities of sophisticated structuring for enabling efficient context retrieval algorithms. Additional attributes such as timestamp, source, and confidence in the accuracy of the context are also of great importance.
- *Markup Scheme Models*: Once systems became Web-based it became natural to represent context as a data structure consisting of a hierarchical context and markup tags (e.g., location: state, city, street, building, and so on). *Profiles* contain typical markup scheme models. This XML-based language or encoding in RDF/S allows accounting for contextual information and dependencies when defining interaction patterns on a limited scale. Various examples cited by Strang and Linnhoff-Popien [35] include CC/PP (composite capabilities/preference profile), UAProf (user agent profile), PPDL (pervasive profile description language), and so on. Owing to the fact that no design criteria and only parts of the language are available to the public, the actual appropriateness and capabilities of this context modeling approach remains unknown.
- *Graphical Models*: UML, the current generic graphical language used to represent objects and their interactions, is found to be suitable to model the

context in multiple scenarios. Another modeling approach can be extended to the object role modeling included in the context.

- *Object-Oriented Models*: Use of object-oriented techniques to model the context gives it all the capabilities of object orientation such as encapsulation, polymorphism, and abstraction. Existing approaches use various objects to represent different types of context. They encapsulate the details of context processing and representation. Access to them is only through well-defined interfaces. An example is the use of objects to represent various inputs from a sensor and the processing of that information necessary to represent the context can be the methods defined for the particular object. The cues in the TEA project are an example of this model [26]. Strang and Linnhoff-Popien [35] have also cited various examples in their seminal paper "A context modeling survey."
- *Logic-Based Models*: It is suitable to use a logic-based system when the contextual information shows a high degree of formality. These systems manage the facts, expressions, and rules used to define the context model. They not only permit updating or deleting existing facts/rules but also permit the addition of new ones. The earliest such approach, of representing context information as formal facts, was published by McCarthy and Buvač [36].
- *Ontology-Based Models*: The nature of context-aware service demands that the context information should be modeled in a platform-independent manner and should support interoperation. Thus, use of ontology in a context model at a semantic level and establishing a common understanding of terms are imperative. The description of the concepts and the relationships between them can be represented as ontologies. This would enable context sharing, reasoning, and reusing in such environments. The simplicity of representation, flexibility, extensibility, genericity, formal expressiveness of ontology, and the possibility of applying the ontology reasoning technique have made these models the most promising instrument in multiple current frameworks for modeling contextual information. This has been concluded in the evaluation of various models by Strang and Linnhoff-Popien [35].

3.4.2 Panoramic View of the Type of Applications

Context-aware applications can be classified into five categories:

- *Location-Aware Applications*: These are applications where the context is basically obtained from the location and the application set provided is also about the location or the services relevant to that location. These applications use either pull or push techniques to gather context and may be made to respond in a specific way on the basis of the context.
- *Time-Aware Applications*: These are applications where the context is basically obtained from time and date and they also respond to the time of day or the

season. This type of context is relatively easy to obtain, from the clock embedded in the device and the simple temperature sensor, and so on.

- *Device-Aware Applications*: In these applications, the context is basically dependent on various devices or the state of the network and other devices that will share the context collected by them.
- *Personalized Applications*: In these applications, the user specifies limits on the contexts that devices should respond to. This means that the user will have to manually customize the conditions that the application is expected to respond to.
- *Adaptable Applications*: These applications/services automatically reconfigure themselves on the basis of the context in the environment and respond to the changing context. These types of systems are still in the development stage. A category of these applications respond to the emotional and physical state of the user along with the environment. Thus, here, context is dynamic in nature, as against being static in earlier applications.

In all the above applications, the context may be explicit or embedded with the service details.

3.4.3 Architectural Requirements

To build context-aware applications relatively easily, designers would find it useful to apply the input handling techniques of application development to context-aware scenarios. The challenges in the development of applications in context-aware scenarios come from the fact that context comes from multiple, heterogeneous, distributed sources, unlike an input given by the user to a system. Also, the confidence in input provided as context is a complicated decision. A set of requirements for the architecture of a framework to support context awareness on mobile devices for specific characteristics of the environment (limited size, processing power, display size, and limited memory) that must necessarily be satisfied are well summarized by Hofer et al. [28]. The basic features desirable for useful architecture are

- *Lightweightedness*: The requirements of the incorporation of context can make the analysis process need higher processing power, higher speed, and higher storage (temporary and permanent) capacity. However, as a result of the limited processing power, in terms of speed and capacity, of the device, the final architecture must be lightweight.
- *Abstraction, Interpretation, and Aggregation*: It is important that the architecture supports aggregation of context about entities in the environment, otherwise it may have a negative impact on both maintainability and efficiency. Aggregation is an abstraction that permits an application to communicate only with a single component for each entity that is of current interest.

- *Robustness*: As the current state does not guarantee continuous connectivity of mobile devices with the server, the architecture should be robust enough to take care of intermittent disconnections with remote sensors/servers in context processing.
- *Metainformation*: The context model needs to store the metadata about various parameters such as distance between remote sensors and devices, their preciseness, and so on, as sensors could provide contradicting information.

More advanced features that could be desirable are

- *Context Sharing*: As a result of the limited capabilities of any one mobile device, it may become necessary to share the context between devices gathering different data. This necessitates separating the concerns of context sensing and the application that will use it.
- *Component Persistence and History*: As one may use this context information in multiple applications, it is necessary for the objects to be persistent. The history of the object states also needs to be maintained.
- *Distribution of a Context Sensing Network*: The architecture must support interoperability of contextual information and heterogeneous application platforms. This is necessary as data are collected from distributed sources and multiple applications need to run on the systems.
- *Extensibility*: As the size of the device has to be small, there is an inherent limitation on the number of slots for adding sensors. This makes it necessary for the architecture to support the connections to remote sensors.

In the current scenario, service-oriented architectures (SOAs) are becoming popular and their requirements will involve the above features as well as those for appropriate service discovery and ontologies.

The above discussion leads to a possible set of components that will be required to cater to any such application. In any system design, if the developer uses the following procedure, which can be termed as *output centric system development*, then the number of iterations necessary in system building and the time for user acceptance tests will be reduced:

1. Identify the outputs expected from the system, their form, the contents of each output, and the users of this output.
2. Identify the inputs required for generating these outputs and their sources.
3. Identify the processes that transform this input to the output.

As the developer has taken into consideration a complete set of input parameters systematically, the gap in the user expectation and the developer's understanding will be considerably reduced, thereby reducing the number of iterations and reworking because of misunderstood requirements. It is a well-known fact that over

30% of software development failures can be attributed to gathering incorrect requirement.

When a system is developed, it could be user-centric or developer-centric. In a user-centric system, the developer has to keep in mind and answer the *5Ws, 1H*: What (is the context), Who (is the beneficiary), Where (can an awareness of the context be exploited), When (is context awareness useful), Why (are context-aware applications useful), and How (does one implement context awareness)—questions regarding the application scenario [37,38]. The developer-centric approach does not necessarily answer all these questions but the system/application is built on the basis of the environment available for development and the simplicity for the developer. Although this is not an ideal methodology, owing to certain constraints, often developers are forced to use this method.

3.4.4 Generic Architecture

The above discussion leads to a possible set of components that will be required to cater to any such applications. To define a generic architecture for a context-aware mobile device, the components (shown in Figure 3.2) that will be necessary to make the service "context aware," flexible and useful include:

1. *Context Providers/Acquirers*: These are sensors or devices responsible for collecting context. These sensors could be *physical* sensors (such as photodiodes, temperature sensors, GPS, and so on), *virtual* sensors (use of retrieved electronic calendars as estimators for position parameters), or *logical* sensors (some combination of physical and virtual sensor data).
2. *Context Aggregator*: This is a module that will take inputs from various context providers/acquirers, aggregate the context, and provide it to a context interpreter. To perform the aggregation, a set of statistical models may be required along with some learning techniques. This layer may be merged in the context inferencer/interpreter in some simple models. However, this extra layer of abstraction will enhance encapsulation and ease the development process and the flexibility available to the developer. Also, it will help improve the network performance when the aggregator can be accessed remotely. Then the client application will not need to access multiple context providers, thus reducing the network traffic.
3. *Context Interpreter/Inferencer*: This module, using the inference engine, provides a "why" to the context. Gathered/interpreted context will be evaluated on the basis of the need of the application (5Ws, 1H).
4. *Context Processor*: This module, based on the repositories and aggregated, interpreted context, generates a list of services required to provide a specific service. The simplest of the models will have the context processing logic embedded in identifying the required service. However, separating the context logic module from other components and the architecture from

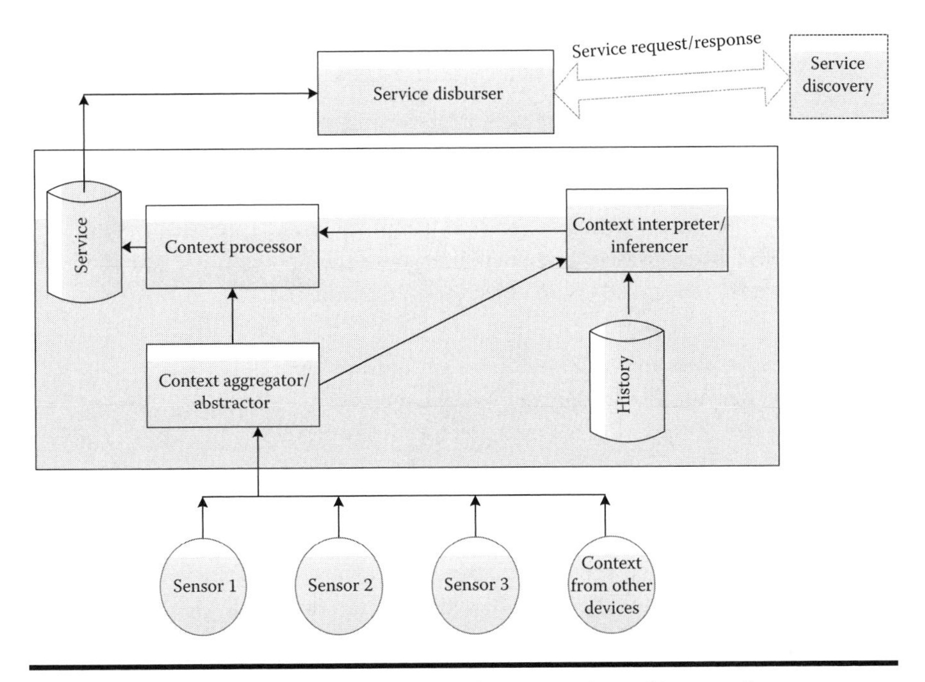

Figure 3.2 Interconnection between various generic architectural components.

supporting service providers will make context processing more generic and also dynamic.

5. *Service Fetcher*: A network may be used to fetch the desired service. The levels of abstraction, and object orientation enhance network performance.
6. *Service Disburser*: This is a component for deploying the required/desired service on the mobile device on the basis of the context.

The functions of components 4, 5, and 6 may be carried out by the server, the network, and the actual mobile device, respectively.

The block diagram shown in Figure 3.2 depicts the interconnections between the basic set of various architectural components. These components are generic in nature. The various architectures reported in the literature are derived from combinations of the above-mentioned components, the priority in their use, and the method of implementation of each. We will discuss a few of them in brief at this point.

In case of the adaptable application/services, there has to be an additional component of an adaptation manager resulting in context-adapted services.

3.4.4.1 Conventional Architectures

Various architectures have been defined specific to particular applications. If studied carefully most of them will be subsets of generic architecture. A comprehensive table of features available in each of these architectures is given at the end of this section. We will review some of the architectures here.

3.4.4.1.1 Context Toolkit

One of the earliest architectures suggested is by Dey et al. [1] in their review submitted to UIST'99. It consists of three main types of components (treated as objects since the context model used is an object-oriented model):

- *Context Widget*: Implements the widget abstraction.
- *Server*: Responsible for aggregation of context.
- *Interpreter*: Responsible for interpretation of context.

Widgets are nothing but standalone pieces of code written with a specific functionality (e.g., an interface widget or a weather widget) and can be embedded into a huge application. They are derived from the idea of code reuse. The level of abstraction thus enhances the ease of application development. Dey et al. [1] created just a first cut of the architecture, and, although not complete, it can give a direction to the designer. Architecture of these context widgets are autonomous in execution by their very nature and can be instantiated independent of each other in their own threads and also executed. These objects can be instantiated all on one device or on multiple computing devices.

The context widgets are similar in nature to the interface widgets and play the role of context collectors. They can be created by specifying the attributes and call backs (types of events used to notify subscribing components) of the widget, the code to communicate with the sensor in question, and when new sensor data are available. Two methods are used: one to validate the data against the current subscription and send relevant data to subscribing components in case of a match and the other to add the data to persistent storage allowing others to retrieve the historical context information.

The context server object plays the role of a context aggregator. This provides additional aggregation and abstraction. The application can communicate with a single object as against the many widgets that collect contextual information. It then becomes necessary to ensure that the server object subscribes to every widget required and acts as a sort of proxy to the application. The creation of the server involves providing the names of the widgets to subscribe to, attributes, call backs, and a condition's object. This is required to ensure that the server receives only the information (from the widget) in which it is interested. Context interpreters, simple or complex, are the objects responsible for implementing the abstraction to permit reuse of interpreters by multiple applications. The state information regarding the context is not maintained across applications and requires a method for interpreting data.

Communication infrastructure plays the role of a service fetcher and a service disburser. The BaseObject class (super class from which all other subclasses are derived and hence inherit the properties, acting both as a client and a server) provides basic communication infrastructure needed for communication with widgets, servers, and interpreters. In particular, it facilitates subscribing and unsubscribing to, querying/polling and retrieving historical content, requesting interpretations, and retrieving object-specific information such as version number and attributes.

The implementation of widgets is carried out using XML and HTTP as they support lightweight integration of distributed components in a heterogeneous environment. Although this architecture defines a broad framework, the developer needs to take care of many more issues while implementing these and this framework can be only used as a starting point.

3.4.4.1.2 Hydrogen Model

To bridge the variation in representations and the progression of Web semantics, ontologies have been created. The abstraction between the application layer and the server becomes important in a multiuser environment where services are shared. Also, when there is data transfer over a network, the system needs to be robust against network disconnections. An architecture variant that takes these into consideration is the *Hydrogen Approach* discussed by Hofer et al. [28].

The key point in this architectural framework is the separation of concerns of interaction between three aspects: physical sensors, storage, and maintenance of context from the application. The topmost layer is the application layer, the middle layer is the management layer, and the bottom layer is the application layer (service fetcher and service disburser of the generic model), as depicted in Figure 3.3. The major benefit comes from the level of abstraction. All the context needs of the application are taken care of by the context server of the management layer (aggregator, interpreter, and inferencer of the generic model), the core context storage component. This also allows sharing of context information with other devices in range.

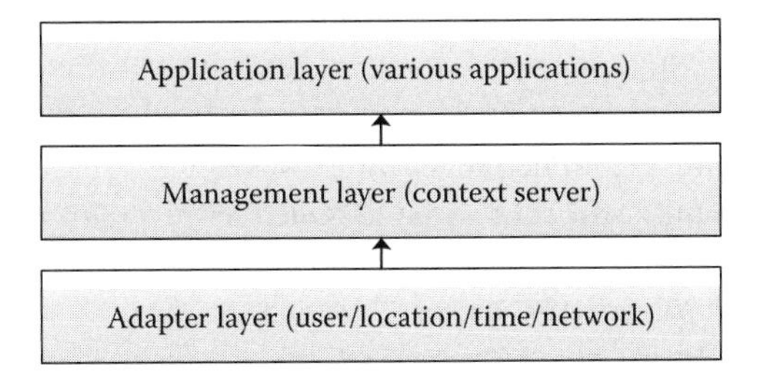

Figure 3.3 Overview of hydrogen context framework architecture.

The adaptor layer combines the roles of context providers and acquirers (physical and possibly logical context) and passes the acquired context to the management layer. The context sever also has embedded in it the methods for applications to retrieve or subscribe to a context (call back of context widgets). The added advantage of having these three services on the same device is that it makes the framework robust against network disconnections. More than one application/service may require the same context. The fact that these contexts are acquired in the adaptor layer allows access to a single context (gathered by a sensor) by multiple applications.

The context server permits two methods of referring to the context. The first one is asynchronous (allowing one to query a specific context from the context server in a pull-based manner). As there is a possibility of change of context during the execution, there needs to be another method—synchronous push access—to ensure that the application is informed about this change.

3.4.4.2 Ontology-Based Architectures

With advances in technology a host of new architectures based on recent concepts of definition, methods of deployment, and use of services are evolving. The ontologies are used for the description of context based on the models to achieve a more expressive and consistent scheme of contextual information, especially in a shared environment where complex context interpretation is a must. A good comparative analysis of various such architectures is presented in Reference [39]. These architectures will be discussed in brief, highlighting their characteristics.

3.4.4.2.1 WASP Architecture

Using OWL, an ontology-based markup language, WASP architecture is based on the semantic Web technology fundamentals. The platform provides services to context providers through the interpreter module. This module collects information for the context model, uses repositories, and exploits the semantic knowledge represented in the form of OWL. Appropriate layers can then be invoked.

3.4.4.2.2 CoBrA Architecture

The second example in this category is CoBrA (Context Broker Architecture). Four key features of this architecture are

- Acquiring context from heterogeneous sources
- Maintaining consistent contextual knowledge
- Enabling knowledge sharing among the agents
- Protecting the privacy of the user

This architecture provides an autonomous agent known as the context broker, which is the most important feature in the multidomain environment.

CoBrA aims to assist devices, services, and agents to become context aware in smart spaces by making a collection of ontologies for modeling contexts. A shared model of the current context and a declarative policy language (wide spectrum language) that users may devise to define constraints on the sharing of private information and protection of resources will be required.

Interested readers may review the study by Chen et al. [40] for a detailed understanding of this architecture.

3.4.4.3 SOA-Based Architecture

As Web services are becoming popular and computing is becoming more community oriented, SOAs are a part of our life. SOAs involve a collection of autonomous services that communicate with each other as needed. The communication can involve either simple data or two or more services coordinating some activity. Some means of connecting services to each other is needed. Another infrastructure suggested by Koskela et al. [41] uses the lightweight Web 2.0 services to provide the user the experience of ubiquitous computing. Figure 3.4 shows a simplified generic architecture for Web SOA for context-aware systems.

Koskela et al. [41] present "a context-aware mobile web 2.0 service architecture that connects user context and community information with web services. This convergence enables the development of device-independent services that are enriched and personalized with user context and community information. Mobile middleware may be needed for efficient delivery of this information from the mobile device to the web services" (p. 41).

3.4.4.4 Multimedia Supporting Architectures

Advances in technology are making it imperative to provide support for multimedia context. The complexity of handling the multimedia context is higher and may

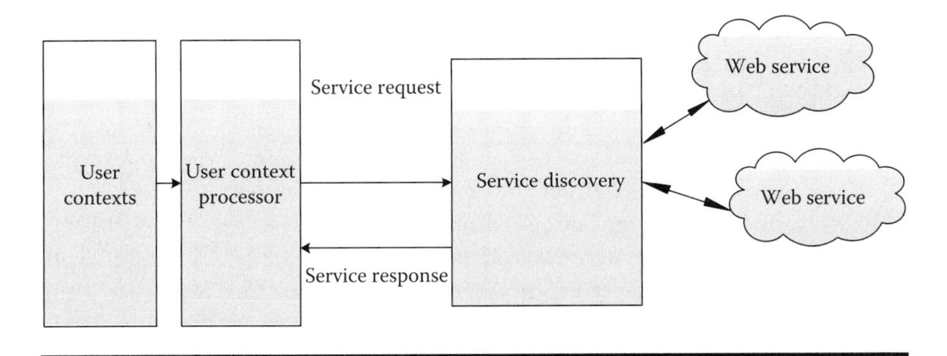

Figure 3.4 Generic framework for Web service architecture.

be reduced by implementing efficient ways of manipulation and storage of this context information. Some recent efforts in this direction are discussed here.

Naguib et al. [42] developed the *QoSDREAM* framework that focuses on supporting the development of context-aware distributed multimedia applications. It has an event messaging component, a data storage component, and a distributed multimedia service component. The framework supports streaming media, but the context is presented by location information only. The framework does not support the mobility of applications (e.g., the applications cannot migrate from one host to another). Besides, it assumes that the network connection is fixed. Mobile devices are used as location sensors only. Davidyuk et al. [43] developed another framework called CAPNET that tries to overcome the shortcomings of QoSDREAM.

Many other variants can be found in the literature, including IrisNet [44], SOCAM [45], and a context awareness substructure (CASS), developed by Fahy and Clarke [46] and by Noltes [47], catering to specific components and altering them in terms of its usage and implementation. Architecture types with embedded or explicit architectures are lucidly explained by Jones [48] in his article on building a context-aware service architecture. Interested readers may look at these sources for further references.

3.4.5 Comparative Evaluation

A comparative evaluation of some of the discussed architectures is shown in Table 3.1, for easy referencing.

3.5 Context-Aware Programming

The challenges of writing the context-aware applications come from the need to dynamically adapt to the changes in the environment and having elements or components that are interoperable in the true sense with the ability to adapt to various hardware platforms as well as to different architectures. This context information comes from a variety of sources and hence may have different formats, units of measurements, and pushed or pulled context data. The second challenge is that the context information may be unreliable and hence may require dynamic manipulation. Also, in advanced applications, the context may be derived/composed from other low-level contexts. Context acquisition and adaptation concerns are most often distributed. This implies that the utilization of usual programming techniques leads to a complex distributed design where high reusability and maintainability are impracticable. This necessitates the development of specific programming models relevant in Context Aware Application Programming Problem (CAAPP) development, discussed in the following sections.

Table 3.1 Comparison of Various Architectures

Features		*Context Toolkit Architecture*	*Hydrogen Context Framework Architecture*	*WASP*	*CoBrA*	*Web 2.0 Service Architecture*	*QoSDREAM (Multimedia Contexts)*
Type of content	Text	√	√	√	√	√	√
	Multimedia	X	X	X	X	√	√
Lightweighted		√	√	√	X	√	X
Interpretation/aggregation		√	√ (raw data only)	√	√	√	√
Abstraction		√	√	√	√	√	√
Inferencer		X	√	√	√		√
Robustness		√	√	√		√	
Context sharing		X	√	√	√	√	√
Component history and persistence		√	X	—	√	√	√
Resource discovery		X	X	√	√	√	√
Security and privacy		—	√	—	√	√	√
Extensibility		√	√	√	√	√	√

Note: √, feature is available; X, feature not supported; — not defined.

3.5.1 Aspect-Oriented Programming-Based Context Implementation

One must remember that programs not only have to perceive the context but also have to adapt to various contexts within the limitations of the platforms on which they run [49]. As there are multiple aspects that need to be taken into consideration, one can treat this scenario as aspect-oriented programming (AOP), where the behavior depends on the context and also the associated actions.

The language constructs will have the ability to scope aspects to certain contexts and refer to the context. Although current AOP does not have this capability, Tanter and Noyé [50] have tried to implement a few context-aware aspects using reflex AOP kernel. *Primitive components*, suggested by Khattak and Barrett [51], work as black boxes in the component-based software development process to give compile time aspect-oriented flexibility to dynamic runtime component context. This approach has proved to score over AspectJ which is the industry standard in AOP. Carton et al. [52] have suggested using a combination of Aspect-Orientated Software Development (AOSD) and a Model Driven Development (MDD) to simplify the challenges of context-aware programming. Many variants are found in the literature. Some of them are summarized here.

The Java aspect component (JAC) [53] is a dynamic AOP framework in Java. An aspect program in JAC is a set of aspect objects that can be dynamically deployed and undeployed on top of running application objects. Besides joinpoint and pointcut core concepts, JAC has a concept named "wrapping methods" that is a code that runs on joinpoints on meeting certain conditions specified in the pointcuts. AOP in JAC is composed of four main parts: the base program, the aspect program, the weaver, and the composition aspect.

3.5.2 General Purpose Context-Based Programming Models

Context-oriented programming (COP) looks at extending the existing traditional object-oriented model to include four new constructs. As per definitions, the context and goals are first-class objects, the open terms are second-class objects, and the stubs are third-class objects. The central idea is to leave gaps that will be filled on the basis of context. As example is to alter the goal of greeting the user in a language based on the location context. The open terms are in the form of gaps that specify the goals and the context and are filled at runtime using the stubs specific to the context. Context-filling thus becomes the distinguishing feature of COP. An excellent and detailed discussion on COP, with various definitions and nuances of the requirements, can be found in the paper by Keays and Rakotonirainy [54] and, more comprehensively, in the paper by Hirschfeld et al. [55].

CSLogicAJ (Context-aware Service-oriented LogicAJ) [56] is a service aspect language that provides context-sensitive aspects. CSLogicAJ is an extension of the

LogicAJ language and thus supports the advice concept, the call pointcut, and logic metavariables.

The Java Context-Awareness Framework (JCAF) aims to create a general-purpose, robust, event-based, service-oriented infrastructure and a generic, expressive Java programming framework for the deployment and development of context-aware applications. It divides the framework into a *Context-Awareness Runtime Infrastructure* and a *Context-Awareness Programming Framework* (or Application Programmer Interface (API)) [57].

Context weaver is a framework by IBM [23], a platform that simplifies writing context-aware applications. In this, all the sources of context are registered so it already has in place a mechanism to deal with different formats and different contexts. They use external sources and composers (programmed entities that compose context information from other sources). The client code is typically written in Java. The user then has to only manage the composition of context from lower-level contexts. The context weaver makes the experience of writing the applications simpler and a pleasurable activity.

3.5.3 Architecture-Based Approach

An architecture-based approach has been described by Sazawal and Aldrich [58] using ArchJava that augments Java with additional syntax for architecture description. The origins of this approach seem to be in VHDL (Hardware Description Language) where connections are defined using ports.

3.5.4 Design Patterns-Based Approach

The most recent efforts in this direction have been made by Nierstrasz et al. [59] where they emphasize that it is not just sufficient for the programs to be model-centric but also context-aware for them to be suitably used in the fast-changing environment and dynamically, efficiently, and effectively adapting to this environments for sustainable long-term evolution. Design patterns may offer a solution for flexibility and adaptability to manipulate the models during development time as well as runtime and adapt the software to the changing context.

3.6 Popular Context-Aware Mobile Services

A service is called a mobile context-aware service if it responds to the situation in the mobile environment by extracting and interpreting the meaning of context and adapts its functionality accordingly. The challenge is to extract and interpret correctly the context (that may even be indirect or deducible) and also for it to be usable by more than one applications with correct interpretation (necessitating abstraction). It is obvious, then, that these applications must also contain certain "intelligence." On the basis of the way the context is used, Khedo [60] has suggested three classes indicating how the application or service uses the context: presenting information and services, executing a service, and tagging captured data. These are discussed in the following:

- *Presenting information and services*: The context information (current location) or response to a query by the user regarding the physical environment is displayed.
- *Automatically executing (adapting) a service*: As the name suggests, the context information displayed by an application, service, or a command is triggered in response to a particular context scenario. For example, the device senses that the user is in a meeting room and the call divert feature to a predefined number is activated. Also, gauging the noise level (or the absence of it) in the room or the location of the room (boss' room), the phone automatically goes into the silent mode.
- *Attaching context information for later retrieval*: Here, the captured data are tagged for further use in applications of class 1 or class 2 or both.

Most of the applications that are known and discussed are based on the passive or active use of location and time parameters. We have already seen earlier that the applications can be location-aware, time-aware, device-aware, personalized, or adaptable. Most of the initial services were mainly location-aware. In fact, to date, maximum applications are in the location-aware area, presenting information category. A discussion on various popular applications is presented in the following sections.

3.6.1 Context (Location/Device/Time)-Aware Applications (Presentation of Information and Services)

- *Active Badge System*: The earliest of the known and popular context-aware applications is the Active Badge system from Olivetti Research Labs [22]. The application locates the wearer of the badge in an office environment. The envisaged use for this badge was to assist the telephone operator to locate the person and forward the call appropriately to the nearest telephone. The system used infrared to locate the badge using infrared sensors at various locations in the building. The location information was passively displayed to the receptionist. Recently, the location context has become active with the help of a PBX that has a digital interface designed for computer-integrated telephony and automatically forwarding the call. There are three major areas of concern with this system:
  1. The system tracked the badge; if the user removed the badge and kept it on the table, the system would give incorrect information about the location of the user.
  2. If the user moved into an area where the sensors were unable to track the user or the sensor failed, the location of the user would not be detected.

3. Privacy of the user was an issue, as users were uncomfortable with their location being tracked every 15 s.

However, if the user was given an option about when, by whom, and how often he could be locatable, the system would be most useful.

- *ParcTab*: The objective of the ParcTab system at Xerox Corporation was to integrate a lightweight palm-sized mobile computer into an office network to enrich the computing environment by emphasizing context sensitivity, casual interaction, and the spatial arrangement of computers. The system is based on palm-sized wireless ParcTab computers (known generically as "tabs") and an infrared communication system that links them to each other and to desktop computers through a LAN. The system is characterized by a combination of parameters as context, the location of the user, the presence of other mobile devices in the vicinity, and the presence of people. The four most popular applications were the accessibility of emails while in motion (fast becoming a popular application on mobile handsets), the weather application—allowing the user to get up-to-date information, the file browser—providing access to text and command files stored in the Unix networking file system, and the tab loader—allowing the user to store information in the tab and use it outside the infrared network area.

  Details of the design can be found in Reference [61]. The bandwidth could become a bottleneck when the number of users increases beyond a certain limit. Also, if the response time of 1–2 s increases, the utility of the device reduces considerably. Of course, the acceptability is dependent on the size, weight, appearance, and response time of applications.
- *Dynamic Ubiquitous Mobile Meeting Board (DUMMBO)*: Brotherton et al. [62], at Georgia Institute of Technology, developed a simple context widget that senses the presence of a user and is able to identify him/her. It relies on a sensor that provides both presence and identity information. The iButtons are the small devices that users snap in a reader to notify their presence. Each button has a unique ID from which the user's identity is derived. This was used in more than one application. The first one was the IN/OUT board presence indicator, where the user, while entering the laboratory area, docks the iButton on the reader indicating that the user is in the laboratory area. Extending use of this further, the iButton reader was mounted on a white board. This whiteboard DUMMBO is an augmented board for impromptu meetings. When multiple users dock their iButtons on the readers, indicating the presence of more than one person, it is assumed that a meeting is beginning and starts capturing the audio and whiteboard drawing that can be used for storing the session.
- *RFID Tracking System for Vehicles (RTSV)*: This system tracks vehicles on the road using RFID [5]. It addresses three major problems: traffic signal timings, congestions on roads, and theft of vehicles. Traffic signaling is made dynamic on the basis of regressions over data archives, containing a detailed

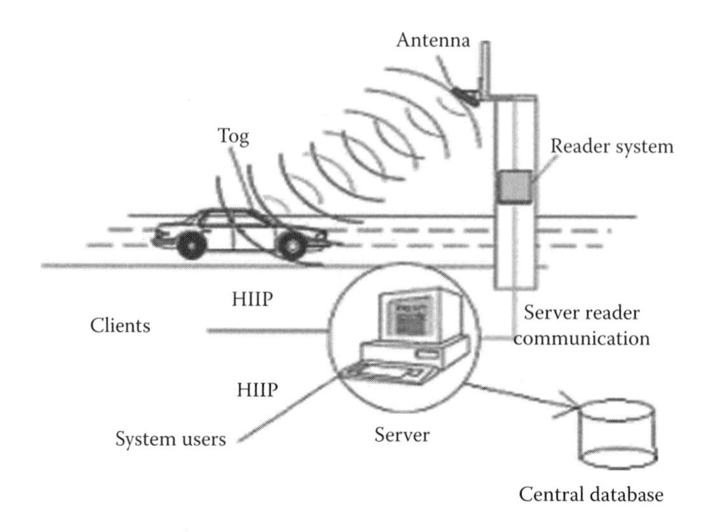

Figure 3.5 Client server architecture for the RFID Tracking System for Vehicles (RTSV).

set of traffic quotient and time. The technique incorporates a simple, unique way to calculate traffic quotient on the basis of the physical dimensions of the road, the nature of traffic, traffic congestion on the road, and so on. Automatic deduction of the amount of toll the user has to pay from a monetarily charged mobile device when the user passes through a toll booth is another feature. The theft of a car is detected using track logs of vehicles. An analysis of congestion forms a key attribute for the traffic signaling system and is used to suggest faster routes to vehicle drivers on their mobile devices and to balance the traffic across various routes. The RTSV requires installing RFID tags on all vehicles and RFID readers at various junctions of the city for tracking. The architecture of the system is shown in Figure 3.5.

3.6.2 Context-Aware Personalized Applications (Presentation of Information and Automatic Execution of Services)

- *CybreMinder*: This is a prototype context-aware tool for sending and receiving reminders with respect to the various contexts associated with the reminder (to oneself or others) using the Context Toolkit infrastructure developed by Dey and Abowd [63]. The difference between CybreMinder and the current methods of reminders, such as post-it notes, alarms, emails, to-do lists, personal information managers, or even human assistants, is the capability of the CybreMinder to define the context (situation) under which the reminder should be delivered and the choice of methodology of receiving this reminder. For example, the user can define the subsituations when this reminder should be served, for instance when the user has entered a particular room, or when the reminder should expire in case the receiver does enter that context during the specified time. The importance can be understood by the user who has been in a scenario where the message that has lost its relevance is delivered on the mobile creating confusion. Also, the receiver may choose to hear the reminders but, if in a meeting room, may just have a beep to indicate that a reminder has arrived. The reminders can be location-based, time-based, colocation-based, or any complex combinations.

- *Tourist Guide*: As mentioned earlier, a vast majority of applications developed using location context are currently available. Especially, when one is moving in an unfamiliar territory, these applications can be extremely useful. One such application tip has been given by Hinze et al. [64] that assists a tourist by showing the location on the basis of the desired context.

  This system, while recommending, not only considers current location and current time when the user is taking assistance but also personal background (location interest, means of travel, user interest based on profile created from history, and so on). The system considers the sights context, such as their location and membership in semantic groups. Recommendations that can be displayed as maps are based on user profiles and feedback. The system is implemented as a client server architecture, supporting mobile as well as desktop clients with appropriate interfaces.
- Other applications of similar nature to assist the visitor's tour through physical and information space are the GUIDE system [65], the information guide for the museum visitor [66], exhibition tourists [67], TramMate [68], or the shopping assistant [69].

3.6.3 Context-Aware Personalized Adaptable Applications (Presentation of Information, Automatic Execution of Services, and Tagging of Information for Later Retrieval)

- *Conference Assistant*: The conference assistant is a complex prototype that assists people attending a conference [70]. This tool is a complex combination of context awareness and ubiquitous computing. The application starts after the completion of registration; the user is provided with the application that will work on a mobile computer. When the user enters the conference area, it automatically displays a copy of the conference schedule, showing the multiple tracks of the conference, including both paper tracks and demonstration tracks, highlighting the tracks of interest to the user. When the user enters the room of interest, the details of the presentation in the room appear as a thumbnail. As the user is informed by the assistant whether the event is being recorded, the user can then appropriately make one's own notes. During the question session, the user can bring up the exact slide from one's own computer for all to see on the topic additional clarification is necessary. This ensures that the user, the presenter, and the rest of the attendees are on the same page. The user can retrieve the entire day's work from one's system and review the day's proceedings. In a similar fashion, a presenter can use a third application to retrieve information about his/her presentation. The application displays a timeline of the presentation, populated with events about when different slides were presented, when audience members arrived and left the presentation (and their identities), the identities of audience members who asked questions, and the slides relevant to the questions.

The conference assistant exploits four capabilities of context awareness: *contextual sensing* (location of rooms with respect to current location of the user, identity of the presenter in the room of interest, location of colleagues), *contextual adaptation* (provision of uniform interface, creation of space for entering notes on new slides after saving the earlier slide notes, and so on.), *contextual resource discovery* (temporary binding to the ongoing presentation in the room), and *contextual augmentation* (all notes made by the user are augmented with the current context data supporting retrieval using context-based retrieval). Thus, this is a complete application that exploits all the four context-aware capabilities suggested by Pascoe [32].

- *Mobile WARD*: The mobile WARD is a context-aware mobile prototype that automatically keeps track of contextual information such as physical location of the patients and staff, upcoming appointments, a pending list of tasks, doctor's visits, direct access to reading details about each patient's history, and access to enter new measures and readings on a patient's current status. It assists the nurses in conducting procedures in the wards [25].
- *SWFIA Context-Aware Wireless Emergency Response Device*: Using the technology developed by Appear, the Context Company, a wireless emergency response solution has been deployed on the IQ Appear platform [71]. The system alerts are disseminated to various components that would be useful for diverting emergencies. For example, if a fire alarm is set off, a message goes to the emergency response center and also to the nearest relevant response team members along with information maps for locating nearest ducts. This has improved capabilities to respond quickly and rapidly to emergency situations and to assist the security person to make a quicker decision.
- *SenSay Mobile Phone*: An adaptive application that responds to the changing physical environment and physiological states of the user is SenSay (Sensing and Saying)—a context-aware mobile phone that adapts to the context [18]. This phone combines various sensors such as voice microphone, ambient noise

microphone, accelerometer, temperature sensor, and visible light sensor to collect the information. The sensor data are passed on to the decision module that analyses the data using the predefined set of rules and specifies the state of the device. It also augments data from calendars, address books, and to-do lists. It is then able to provide services such as manipulating ringer volume, vibration, and phone alerts to the extent of shutting off the ringer when the user is busy in an uninterruptible task and thus eliminating unwanted interruptions (information from the to-do list and calendar). It can also actively inform the user of an incoming call on the basis of urgency. It utilizes the idle time to make call suggestions based on user call history. Thus, developers are looking at integrating more functionality in the phone.

- *Personalized Adaptive Content System for Context-Aware Mobile Learning*: This application looks at personalizing the needs of the user in the mobile environment. Material irrelevant to the learner's preference or contextual information needs to be deleted to cater to smaller displays, lesser memory, and processing powers of mobile devices [72].

  Learners may send questions (including text message, photos, and video/audio captured by cameras or recorders embedded in mobile devices) by mobile email or MMS to participate in the discussion through the mobile device. The results have shown that users feel comfortable and convenient to use this easy-to-use interface.
- *Incident Management*: In this futuristic application, the Ministry of Transport, Public Works and Water Management in the Netherlands has taken an initiative to effectively and efficiently use context awareness to perform the task of incident management [72]. The system will use geoinformation, geoservices, and context awareness to carry out the incident management process in an effective and efficient way. In the Netherlands, there are sectors of integrated water management which include sustainable water management, flood protection, traffic safety, and establishment and maintenance of reliable transport infrastructures over land and air. Incident management in its entirety involves the following two tasks: clearing the road traffic as quickly as possible after an incident has occurred and ensuring safety for emergency services and road users. The geoinformation infrastructure based on the OGC (Open GeoSpatial Web Consortium) Web services architecture has enabled broad geospatial sharing throughout the organization in a cost-effective manner. The architecture is modular and scalable and has three layers consisting of presentation, services, and data. The work done has provided a preliminary business case that can be used for further development of context-aware services for fieldwork in the Netherlands.

Additional applications, classified on the basis of whether or not they are Web-based, are also given in the paper by Truong and Dustdar [73].

3.7 Future Research Issues for Context-Aware Services

Development of context-aware services is still in its early stages although the concept goes back about two decades. Research has been exploratory or demonstrative in nature. Dey, a pioneer in this field, has developed several applications [10,63,70] demonstrating the potential that this field holds. However, to make these applications more useful and user friendly, there are several research challenges. The major hurdles in the development of applications that can be widely used can be divided into nine types.

- *Definition of Context*: As seen in Section 3.1, there are many types of definitions of context and context awareness. Each application developer has chosen a particular dimension of the definition and worked on it. With many new sensors that are being developed each day and the need of the user to have services at one's fingertips, the cost of providing these services is spawning additional definitions. So, the first step would be to come up with a comprehensive definition that would be useful in enhancing the development of all encompassing applications that are universally acceptable.
- *Context Collection, Aggregation, and Abstraction*: The second step would be to identify what represents the context and how to acquire it. This is extremely important to round up the scene/context accurately. There are many parameters in our surroundings that we humans process as a part of context. For the system to recognize what represents context and what represents noise is extremely difficult in a static scenario unless "intelligence" is embedded into the system (self-learning devices). In identifying and capturing the context, the developer may have to use many context acquisition components. Current systems are relatively static and only the predefined parameters are acquired, using specific sensors, location trackers, vision, and so on. These, after capturing, need to be represented appropriately and passed on for further processing. The second challenge here is to aggregate the appropriate components of these gathered contexts to give a comprehensive response to a particular context. The gathered context may be in different, noncompatible formats, making it more difficult to use context information effectively [74].
- *Semantic Interoperability*: In context-aware service scenarios, the data are collected from various contexts in different formats. In an international conference, when every attendee speaks a different language, there has to be a common language so all the participants will be able to comprehend and create collective intelligence which is important. Similarly, in this situation there is need for a common representation of knowledge. OWL [75] is one such language commonly used for formal modeling of contexts. It permits defining the various contexts and their interrelationships providing advantages of expressiveness, knowledge sharing, logic inference, knowledge reuse, and extensibility.

- *A Generic Framework/Architecture*: As described in brief earlier, multiple architectures can be utilized under different circumstances, namely, Context Toolkit, Hydrogen framework, WASP architecture, and so on. Each of these architectures was developed to overcome certain shortcomings of the earlier models. In some cases, the wheel is also reinvented, making it necessary to resolve the problems all over again. A common generic framework would avoid this situation. Also, when we are talking of autonomous services, it is worthwhile to use the required service as needed and as much needed. Today, this indicates the development of a generic framework using SOA. Although there has been some effort in this direction [41], substantial work needs to be done to consolidate this framework and make it simple enough for a general application developer.
- *Suitable Modeling Language*: The context is gathered from various sources and thus the development platform is distributed in nature. Also, as the same context may be necessary for different services/applications at different times, the components have to be reusable. It is obvious that the software paradigm to be used has to be object oriented in nature as all the characteristics of an object-oriented paradigm are necessary in this scenario. AOP or its variants are being used for the development of such systems.

In current times, the software industry is shifting from the development of software to the development of services. This optimally utilizes the developed services on an "as you need, when you need" basis. These services are autonomous in nature, supporting the user without much interaction with the devices. A software architecture that uses the service approach is called SOA. The main challenges in this area are:

1. *Service Description/Discovery*: Determining and developing new services that the user will need requires the development of generic software methodologies and techniques to support context-driven adaptation of the mobile device. The second challenge here is how to discover these services that match user requirements.
2. *Service Monitoring*: These services need to be monitored for their behavior and appropriateness based on user requirements. Feedback on acceptability and user acceptance of the same also needs to be monitored.
3. *Service Composition*: The need for the development of new services is continuous. Thus, one has to look into the process of creating the same using the existing services and the composition of the same.

Some efforts have been made in this direction in the Amigo project [76] and the MADAM project [77]. They address the above-mentioned issues, but there are limitations with respect to lack of interoperability between devices and the intuitive operation of the same. Work done by Pauty et al. [77] in the CoDAMoS (Context

Driven Adaptation of Mobile Services) and CROSLOCIS (Creation of Smart Local City Services) projects undertaken at the Department of Computer Science, Leuven, Belgium successfully addresses at least some of these issues, including self-adaptation based on user context (e.g., change the language of the greeting on the basis of user location), resource discovery, service state capture, and transfer, and also the architecture required to support these kinds of services.

In future, one can then expect to have services that will be context-aware and adaptable (migration to the desired hosts) and will work on a generic architecture (with clear business and value chains) that is semantically interoperable, secure, and preserves the privacy of an individual. The barriers to achieving these goals are reducing each passing day. With the development of new adaptable services, the generic SOA, the generic framework, and the programming languages specifically taking care of special needs of context awareness, the enhanced capabilities of mobile devices (display as well as memory and processing capabilities), improved quality of service with respect to networks and bandwidths, these barriers will soon be overcome. These are indicators of context-aware services coming out of the laboratories and entering into the realm of the real world.

References

1. Dey, A., Salber, D., Futakawa, M., and Abowd, G.D. An architecture to support context aware application. *GVU Technical Report GIT-GVU-99-23*, Georgia Institute of Technology, 1999. Submitted to the *12th Annual ACM Symposium on User Interface Software and Technology (UIST '99)*, Asheville, USA, June 1999.
2. Dey, A., and Abowd, G.D., Towards a better understanding of context and context awareness. *GVU Technical Report GIT-GUV-99-22*, Georgia Institute of Technology, 1999. Submitted to the *1st International Symposium on Handheld and Ubiquitous Computing (HUC '99)*, Karlsruhe, Germany, June 1999
3. *Gartner Symposium, ITxpo*, Sydney, Australia, November 17–19, 2009. http://www.gartner.com/it/page.jsp?id=1229413
4. Häkkilä, J., Towards usable context-aware mobile handheld applications. PhD research in progress (on human-computer interaction with context-aware mobile handheld devices), Information Processing Laboratory, University of Oulu, Finland, 2008, pp. 169–174.
5. Pandit, A.A., Talreja, J., and Mundra, A. RFID tracking system for vehicles (RTSV). In *IEEE Proceedings of the 1st International Conference on Computational Intelligence, Communication Systems and Networks* (*CICSyN 2009*), Indore, India, pp. 160–165, July 23–25, 2009.
6. Weiser, M. The computer for the 21st century. *Scientific American*, 256(3), 1991, 94–104.
7. Schilit, B., and Theimer, M., Disseminating active map information to mobile hosts. *IEEE Network*, 895, 1994, 42–47.
8. Brown, P.J., Bovey, J.D, and Chen, X., Context aware applications from the laboratory to the market place. *IEEE Personal Communications*, 4(5), 1997, 58–64.

9. Ryan, N., Pascoe, J., and Morse, D. Enhanced reality fieldwork: The context aware archaeological assistant. In L. Dingwall, S. Exon, V. Gaffney, S. Laflin, and M. Van Leusen, editors, *Computer Applications in Archeology*, British Archaeological Reports International Series 750. Oxford: Archaeopress, 1999.
10. Dey, A.K. Context aware computing: The CyberDesk project. In *Proceedings of the AAAI1998 Spring Symposium on Intelligent Environments*, Technical Report, SS-98-02, pp. 3–13, March 23–25, 1998.
11. Abowd, G., Atkeson, C., Hong, J., Long, S., Kooper, R., and Pinkerton, M. Cyberguide: A mobile context aware tour guide. *ACM Wireless Networks* 3(5), 1997, 421–433.
12. Van Laerhoven, K., Villar, N., and Gellersen, H.-W., Multilevel sensory interpretation and adaptation in a mobile cube. In *Proceedings of the Third Workshop on Artificial Intelligence in Mobile Systems (AIMS) (UbiComp 2003)*, Memo no. 82, Seattle, Washington, pp. 111–117, October 12–15, 2003.
13. Yamabe, T., and Nakajima, T. Possibilities and limitations of context extraction in mobile devices: Experiments with a multi sensory personal device. *International Journal of Multimedia and Ubiquotous Engineering*, 4(4), 2009, 37–52.
14. Schilit, B., Adams, N., and Want, R. Context aware computing applications. In *Proceedings of IEEE Workshop on Mobile Computing Systems and Applications*, Santa Cruz, California, pp. 85–90, December 8–9, 1994.
15. Parker, C., and Case, T. *Management Information Systems: Action and Strategy*, 2nd edition. New York: McGraw Hill Book Company, 2002.
16. Hinze, A. and Qiu, Q. Trust- and location-based recommendations for tourism. *OTM Conferences*, 1, 2009, 414–422.
17. Priyantha, N.B., Chakraborty, A., and Balakrishnan, H. The cricket location support system. In *Proceedings of 6th ACM International Conference on Mobile Computing and Networking (ACM MOBICOM)*, Boston, Massachusetts, August 6–11, 2000, pp. 32–43.
18. Siewiorek, D., Smailagic, A., Furukawa, J., Moraveji, N., Reiger, K., and Shaffer, J. SenSay: A context-aware mobile phone. *Proceedings of 2nd International Semantic Web Conference (ISWC2003)*, Sanibel, Florida, pp. 248–249, October 20–23, 2003.
19. Li, C., and Willis, K. Modeling context aware interaction for wayfinding using mobile devices. *Proceedings of the 8th Conference on Human-Computer Interaction with Mobile Devices and Services (MobileHCI'06)*, Vol. 159, Helsinki, Finland, September 12–15, 2006, pp. 97–100.
20. Steenberuggen, I.J.G.M., Grothe, M.J.M., Beinat, E., and Wagtendonk, A.J. The value of context aware information systems for incident management. www.imsummit.org/joomla/images/.../Steenbruggen_Grothe.doc
21. Hinze, A., and Buchanan, G. Context awareness in tourist information systems: Challenges for user interaction. *Paper presented at International Workshop on Context in Mobile HCI at the 7th International Conference on Human Computer Interaction with Mobile Devices and Services (Mobile HCI '05)*, Salzburg, Austria, September 19–22, 2005.
22. Want, R., Hopper, A, Falcao, V., and Gibbons, J. The Active Badge location system. *ACM Transactions on Information Systems*, 10(1), 1992, 91–102.
23. Cohen, N.H., Black, J., Castro, P., Ebling, M., Leiba, B., Misra, A., and Segmuller, W. Building context-aware applications with context weaver. *IBM Research Report RC 23388*, October 22, 2004.

24. Seppänen, J., and Huopaniemi, J. Interactive and context-aware mobile music experiences. In *Proceedings of the 11th International Conference on Digital Audio Effects (DAFx-08)*, Espoo, Finland, September 1–4, 2008. http://www.acoustics.hut.fi/dafx08/papers/dafx08_23.pdf
25. Kjeldskov, J., and Skov, M.B. Supporting work activities in healthcare by mobile electronic patient records. In *Proceedings of 6th Asia Pacific Conference on Computer–Human Interaction (APHCI 2004)*, Rotorua, New Zealand, pp. 191–200, June 29–July 2, 2004; *Lecture Notes in Computer Science*, Vol. 3101. Berlin: Springer-Verlag.
26. Gallersen, H.W., Schmidt, A., and Beigl., M. Multi-sensor context awareness in mobile devices and smart artifacts. *Mobile Network Applications*, 7(5), 2002, 341–351.
27. Brown, M. Supporting user mobility. *International Federation for Information Processing*, 41(4), 1996, 315–330.
28. Hofer, T., *Schwinger, W., Pichler, M., Leonhartsberger, G., Altmann, J.*, and Retschitzegger, W. Context-awareness on mobile devices—The hydrogen approach. In *IEEE Proceedings of the 36th Hawaii International Conference on System Sciences (HICSS-36)*, Big Island, Hawaii, January 6–9, 2003.
29. J. Kjeldskov, J., and Paay, J. Indexical interaction design for context aware mobile computer systems. In *Proceedings of the 18th Australia Conference on Computer-Human Interaction: Design: Activities, Artefacts and Environments (OZCHI 2006)*, Sydney, Australia, pp. 71–78, November 20–24, 2006.
30. Häkkilä, J. and Isomursu, M. User experiences of location-aware mobile services. *Proceedings of the 17th Australia conference on Computer-Human Interaction: Citizens Online: Considerations for Today and the Future (OZCHI 2005),* Computer-Human Interaction Special Interest Group (CHISIG) of Australia (ISBN:1-59593-222-4), Canberra, Australia. November 23–25, 2005.
31. Kjeldskov, J., and Paay, J. Just-for-us: A context aware mobile information system facilitating sociality. In *Proceedings of the 7th Conference on Human-Computer Interaction with Mobile Devices and Services (Mobile HCI 2005)*, pp. 23–30, Salzburg, Austria, September 19–22, 2005.
32. Pascoe, J. Adding generic contextual capabilities to wearable computers. In *Proceedings of 2nd International Symposium on Wearable Computers* (*ISWC 1998*), Pittsburgh, Pennsylvania, pp. 92–99, October 19–20, 1998.
33. Cheverst, K., Mitchell, K., and Davies, N. Exploring context aware information push. In *Personal and Ubiquitous Computing*, 6(4), 2002, 276–281.
34. Schmidt, A. Chapter 9: Interactive context aware systems interacting with ambient intelligence. In G. Riva, F. Vatalaro, F. Davide, and M. Alcañiz, editors, *Ambient Intelligence: The Evolution of Technology, Communication and Cognition, Towards the Future of Human–Computer Interaction.* Amsterdam, The Netherlands: IOS Press, 2005, pp. 159–178.
35. Strang, T., and Linnhoff-Popien, C. A context modeling survey. *Workshop on Advanced Context Modelling, Reasoning and Management as part of UbiComp 2004—The Sixth International Conference on Ubiquitous Computing*, Nottingham, England, September 7, 2004.
36. McCarthy, J., and Buvač. Formalizing context (expanded notes). In S. Buvač and Ł. Iwánska, editors, *Working Papers of the AAAI Fall Symposium on Context in Knowledge Representation and Natural Language*, Menlo Park, California, pp. 99–135, November 8–10, 1997. http://www.formal.stanford.edu/buvac/

37. Morse, D., Armstrong, S., and Dey, A.K. The what, who, where, when, and how of context-awareness, Workshop abstract in the *Proceedings of the 2000 Conference on Human Factors in Computing Systems (CHI 2000)*, The Hague, The Netherlands, p. 371, April 1–6, 2002.
38. Oh, Y., Lee, S., Woo, W. User-centric integration of contexts for a unified context aware applications model. In *ubiPCMM 2005*, The Seventh International Conference on Ubiquitous Computing, Tokyo, Japan, pp. 9–16. http://uvr.gist.ac.kr/ubiPCMM05/.
39. Anagnostopoulos, C., Tsounis, A., and Hadjiefthymiades, S. Context awareness in mobile computing environments: A Survey. *Wireless Personal Communications*, 42(3), 2006, 445–464.
40. Chen, H., Finin, T., and Joshi, A. An intelligent broker for context-aware systems (Poster Paper). In *5th Annual Conference on Ubiquitous Computing*, Seattle, Washington, October 12–15, 2003. http://www.cs.umbc.edu/ ~ finin//papers/ubicomp03-poster.pdf.
41. Koskela, T., Kostamo, N., Kassinen, O., Ohtonen, J., and Ylianttila, M. Towards context-aware mobile Web 2.0 service architecture. In *Proceedings of International Conference on Mobile Ubiquitous Computing, Systems, Services and Technologies*, Papeete, French Polynesia, pp. 41–48, November 4–9, 2007.
42. Naguib, H., Coulouris, G., and Mitchell, S. Middleware support for context-aware multimedia applications, *Proceedings of the 3rd IFIP WG 6.1: International Working Conference on Distributed Applications and Interoperable Systems DAIS 2001*, Krakow, Poland, pp. 9–22, September 17–19, 2001.
43. Davidyuk, O., Riekki, J., Rautio, V., and Sun, J. Context aware middleware for mobile multimedia applications. *Proceedings of 3rd International Conference on Mobile and Ubiquitous Multimedia (MUM 2004)*, College Park, Maryland, pp. 213–220, October 27–29, 2004.
44. Nath, S., Ke, Y., Gibbons, P.B., Karp, B., and Seshan, S. IrisNet: An architecture for enabling sensor-enriched Internet service. *Technical Report IRP-TR-02-10*, Intel Research, Carnegie Mellon University, Pittsburgh, December 2002.
45. Gu, T., Pung, H.K., and Zang, D.Q. A middleware for building context aware mobile services. *Proceedings of IEEE Vehicular Technology Conference (VTC 2004-Spring)*, Milan, Italy, pp. 2656–2660, May 17–19, 2004.
46. Fahy, P., and Clarke, S. CASS—A middleware for mobile context aware applications. In *Workshop on Context Awareness at the Second International Conference on Mobile Systems, Applications, and Services (MobiSys 2004)*, Boston, Massachusetts, pp. 304–308, June 6–9, 2004.
47. Noltes, J. An architecture for context aware mobile Web browsing. In *Proceedings of the 9th Twente Student Conference on IT (TSConIT)*, University of Twente, Enschede, The Netherlands, 8pp, June 23, 2008.
48. Jones, K. Building a context aware service architecture, December 12, 2008. http://www.ibm.com/developerworks/architecture/library/ar-conawserv/index.html?ca = drs
49. Tanter, E., Gybels, K., Denker, M., and Bergel, A. Context-aware aspects. In *Proceedings of the 5th International Symposium on Software Composition*, Vienna, Austria, pp. 227–242, March 25–26, 2006.
50. Tanter, E., and Noyé, J. A versatile kernel for multi-language AOP. In *Proceedings of the 4th ACM SIGPLAN/SIGSOFT Conference on Generative Programming and Component Engineering (GPCE 2005)*, volume 3676 of *Lecture Notes in Computer Science*, Tallinn, Estonia, pp. 173–188, September 29–October 1, 2005.

51. Khattak, Y., and Barrett, S. Primitive components: Towards more flexible black box AOP. In *Proceedings of the 1st International Workshop on Context-Aware Middleware and Services (CAMS'09), affiliated with the 4th International Conference on Communication System Software and Middleware (COMSWARE 2009)*, Dublin, Ireland, pp. 24–30, June 26, 2009.
52. Carton, A., Clarke, S., Senart, A., and Cahill, V. Aspect-oriented model-driven development for mobile context-aware computing. In *First Workshop on Software Engineering of Pervasive Computing, Applications, Systems and Environments (SEPCASE 2007), at ICSE'07*, Minneapolis, Minnesota, p. 5, May 20–26, 2007.
53. Pawlak, R., Seinturier, L., Duchien, L., Florin, G., Legond-Aubry, F., and Martelli. L. JAC: An aspect-oriented distributed dynamic framework. *Software: Practice and Experience*, 34(12), 2004, 1119–1148.
54. Keays, R., and Rakotonirainy, A. Context-oriented programming. In *Proceedings of the 3rd ACM International Workshop on Data Engineering for Wireless and Mobile Access (MobiDE'03)*, San Diego, California, pp. 9–16, September 19, 2003.
55. Hirschfeld, R., Costanza, P., and Nierstrasz, O. Context-oriented programming. *Journal of Object Technology*, 7(3), 2008, 125–151.
56. Rho, T., Schmatz, M., and Cremers, A.B. Towards context-sensitive service aspects. *Workshop on Object Technology for Ambient Intelligence and Pervasive Computing, in Conjunction with the European Conference on Object-Oriented Programming (ECOOP '06)*, Nantes, France, July 3–7, 2006. http://sam.iai.uni-bonn.de/projects/csi/
57. Bardram, J. The Java context awareness framework (JCAF)—A service infrastructure and programming framework for context-aware applications. In *PERVASIVE 2005—Proceedings of Third International Conference on Pervasive Computing*, Munich, Germany, pp. 98–115, May 8–13, 2005.
58. Sazawal, V., and Aldrich, J. Architecture-centric programming for context aware configuration. *Proceedings of the Workshop on Self-Healing Systems (WOSS '03)*, 2003. http://archjava.fluid.cs.cmu.edu/papers/
59. Nierstrasz, O., Denker, M., and Renggli, L. Model-centric, context-aware software adaptation. In B.H. Cheng, R., Lemos, H. Giese, P. Inverardi, and J. Magee, editors, *Software Engineering for Self-Adaptive Systems*, volume 5525 of *Lecture Notes in Computer Science*. Berlin/Heidelberg: Springer-Verlag, 2009, pp. 128–145.
60. Khedo, K.K. Context-aware systems for mobile and ubiquotous networks. In *Proceedings of IEEE Fifth International Conference on Networking and the International Conference on Systems (ICN/ICONS/MCL'06)*, Mauritius, p. 123, April 23–29, 2006.
61. Want, R., Schilit, B.N., Adams, N.I., Gold, R., Petersen, K., Goldberg, D., Ellis, J.R., and Weiser, M. Chapter 2: The ParcTab ubiquitous computing experiment. In T. Imielinski and H.F. Korth, editors, *Mobile Computing*. Boston, Massachusetts: Kluwer Academic Publishers, 1996.
62. Brotherton, J., Abowd, G.D., and Truong, K. 1999. Supporting capture and access interfaces for informal and opportunistic meetings. Technical Report GIT-GVU-99-06. Georgia Institute of Technology, GVU Center. Atlanta, GA. Available at: ftp://ftp.cc.gatech.edu/pub/gvu/tr/1999/99-06.pdf
63. Dey, A.K., and Abowd, G.D. CybreMinder. A context-aware system for supporting reminders. In *Proceedings of the 2nd International Symposium on Handheld and Ubiquitous Computing (HUC2K)*, Bristol, UK, pp. 172–186, September 25–27, 2000.

64. Hinze, A., Malik, P., and Malik, R. Interaction design for a mobile context-aware system using discrete event modelling. In *Proceedings of the 29th Australasian Computer Science Conference*, Hobart, Australia, pp. 257–266, January 16–19, 2006.
65. Davies, N., Cheverst, K., Mitchell, K., and Friday, A. Caches in the air: Disseminating tourist information in the GUIDE system. In *Proceedings of Second IEEE Workshop on Mobile Computing Systems and Applications*, New Orleans, Louisiana, pp. 11–19, February 25–26, 1999.
66. Bederson, B.B.. Audio augmented reality: A prototype automated tour guide. In *Proceedings of Conference on Human Factors and Computing Systems (CHI '95)*, Denver, Colorado, pp. 210–211, May 7–11, 1995.
67. Oppermann, R., and Specht, M. A context-sensitive nomadic exhibition guide. In *Proceedings of Second International Symposium on Handheld and Ubiquitous Computing (HUC 2000)*, Bristol, UK, pp. 127–142, September 25–27, 2000.
68. Kjeldskov J., and Howard, S. *Envisioning Mobile Information Services: Combining User- and Technology-Centered Design which Guide the Users*, Vol. 3101/2004. Berlin/Heidelberg: Springer-Verlag, 2005.
69. Asthana, A., Cravatts, M., and Krzyzanowski, P. An indoor wireless system for personalized shopping assistance. In *Proceedings of IEEE Workshop on Mobile Computing Systems and Applications*, Santa Cruz, California, pp. 69–74, December 1994.
70. Dey, A.K., Futakawa, M., Salber, D., and Abowd, G.D. The conference assistant: Combining context-awareness with wearable computing. In *Proceedings of the 3rd International Symposium on Wearable Computers (ISWC '99)*, San Francisco, California, pp. 21–28, October 20–21, 1999.
71. http://www.appearnetworks.com/archives/appear-announces-support-for-intel-umpc-devices
72. Zhao, X., Anma, F., Ninomiya, T., and Okamoto, T. Personalized adaptive content system for context-aware mobile learning. *International Journal of Computer Science and Network Security*, 8(8), 2008, 153–161.
73. Truong, H.-L., and Dustdar, S. A survey on context-aware Web service systems. *International Journal of Web Information Systems*, 5(1), 2008, 5–31.
74. Dimakis, N., Soldatos, J., Polymenakos, J., Fleury, P., Curín, J., and Kleindienst, J. Integrated development of context-aware applications in smart spaces. *IEEE Pervasive Computing* 7(4), 2008, 71–79.
75. W3C Web Ontology Working Group. The Web ontology language: OWL, 2099. http://www.w3.org/2001/sw/WebOnt/
76. The Amigo project: Ambient intelligence for the networked home environment. 2009. http://www.hitech-projects.com/euprojects/amigo/publications/IST-004182%20Amigo-IP%20short%20project%20description.pdf
77. The MADAM project: Mobility and ADaptation enabling Middleware. http://www.ist-madam.org/
78. Pauty, J., Preuveneers, D., Rigole, P., and Berbers, Y. Research challenges in mobile and context-aware service development. *Proceedings of Workshop of Future Research Challenges for Software and Services (FRCSS 2006)*, Vienna, Austria, pp. 141–48, April 1, 2006.

Chapter 4

Mobile Data Services with Location Awareness and Event Detection in Hybrid Wireless Sensor Networks

Liang Hong, Yafeng Wu, and Sang H. Son

Contents

4.1 Introduction

Data aggregation is one of the most important services provided by wireless sensor networks (WSNs) in many applications [1–3]. In some of these applications, mobile sensors are introduced to the WSN to monitor the moving objects, resulting in the need for mobile data services. Some data service queries that are involved in both stationary and mobile sensors cannot be efficiently answered by existing approaches as mobility brings new challenges for data services. Specifically, we consider a class of mobile data service queries that periodically collect and aggregate the data within a specific area around the mobile sensor nodes that detect the event. Such queries require continuous event detection as well as dynamic location awareness on the fly, and hence need specialized optimization techniques to be processed efficiently. An example query can be: "Give me the average percentage of air pollutants within 100 m around the persons whose blood oxygen levels are lower than 90%, sampled every 10 min in the next 20 h." This query is triggered by the event that mobile sensors' data satisfy the user-defined predicate. Multiple mobile sensors may become centers of various target areas (called *query areas*), and they are defined as *reference sensors*. The set of reference sensors may keep changing because different mobile sensors may detect the event every sample period. As a result, query areas change according to the set of reference sensors as well as their locations. We categorize this type of queries as Event-based Location Aware Query (ELAQ), which provides a fine granularity mobile query model that only collects and aggregates the data from the sensor nodes in the query areas instead of the whole sensor network. It is necessary to mention that the ELAQ generalizes many typical data service queries in WSNs. For example, a typical aggregation query is an ELAQ in which query areas are fixed, and a location aware query is an ELAQ without event detection and aggregation.

We propose techniques to process ELAQs in hybrid WSNs. A typical ELAQ process for mobile data services can be described as below. Stationary sensors are first built into a tree structure with the base station as the root. The user's query is then sent to the base station that disseminates the query to each mobile sensor by the sensor tree. After receiving the query, mobile sensors keep monitoring the event at every sample period. Mobile sensors that detect the event become reference sensors and start to propagate the query to the stationary sensors in the query areas. The stationary sensors that receive the query start to collect and aggregate the required data. They adaptively choose routes to transmit aggregation results back to the base station according to current query areas. Such adaptation greatly reduces the transmission cost and query response time. Note that there are possibly multiple concurrently running queries issued to the base station. They are optimized both at the base station and at the mobile sensors to reduce the processing cost. In conclusion, the main contributions are as follows:

- We define a general type of mobile data service query with event detection and location awareness: ELAQ, which, to the best of our knowledge, cannot be efficiently answered by existing works.
- We design a *proxy route* (defined in Section 4.5) for efficient in-network query dissemination to mobile sensors and propose an adaptive proxy selection method to reduce the updating cost of the proxy route when the mobile sensors change their proxies.
- In-network query propagation and Location-based In-network Aggregation (LIA) techniques are proposed to efficiently process ELAQs. Simulation results show that they significantly reduce the query processing cost compared to other techniques.
- To optimize multiple queries, we design a *two-level multiquery optimization* algorithm that filters hidden queries at the base station and instantly rewrites concurrently triggered queries at each mobile sensor.
- We conduct theoretical analyses of the processing cost of ELAQs and an extensive evaluation of the ELAQ processing performance against state-of-the-art techniques. ELAQ exhibits superior performance in terms of energy efficiency, query latency, and query accuracy under various network conditions and outperforms all compared techniques.

The rest of this chapter is organized as follows. Section 4.2 presents the challenges of the research. Section 4.3 discusses related work. Section 4.4 gives details of the system model and query definition. Section 4.5 presents a set of techniques for processing ELAQs. In Section 4.6, we analyze the theoretical cost of processing ELAQs. Section 4.7 proposes a two-level multiquery optimization algorithm. The performance of the ELAQ model is evaluated through extensive simulation in Section 4.8. Section 4.9 concludes the chapter.

4.2 Challenges of Mobile Data Services in Hybrid WSNs

There are several challenges in processing ELAQs owing to sensors' mobility and distributed topology of the sensor networks. The first challenge is that of accessing mobile sensors in a robust and efficient manner in hybrid WSNs. Although ZebraNet [4] has addressed the problem of tracking and accessing mobile sensors in WSNs, its system model and design goals differ from those in our system. We need to design a scheme for efficiently disseminating queries to mobile sensors through stationary sensor infrastructure while reducing the total cost. The second challenge comes from dynamic query areas. Query areas continuously change with the locations of mobile sensors and the event detected each sampling period. This contributes to the challenges of how to efficiently determine query areas and propagate the query to all the sensors in the corresponding query areas. To acquire up-to-date query areas, the traditional centralized methods [2,3,5,6] incur significant transmission overhead and latency. The third challenge is efficient data aggregation in continuously changing query areas. Prior works on in-network aggregation [1–3] focus on the aggregation in static query areas and cannot efficiently aggregate data over dynamic query areas. The fourth challenge is the process of multiple concurrent queries that may cause message collisions and system overload. Multiquery optimization should consider the characteristics of ELAQs. In addition, the sophisticated event detection mechanism is also a challenge that is opened for future work.

4.3 Relate Work

The ELAQ model, to the best of our knowledge, is the first mobile data service scheme with location awareness and event detection for hybrid WSNs. Nevertheless, our work is inspired by a number of related research efforts.

Processing location queries over moving objects has been extensively studied in both centralized and distributed database systems [7,8]. However, these solutions cannot be directly applied to process data service queries in WSNs with constrained resources and distributed topology.

In previous work, query processing in WSNs can be classified into two types: the solutions that consider or do not consider the sensor mobility. Typically, data aggregation query processing approaches collect and aggregate data from stationary sensors within a fixed area. A well-known example is *acquisitional query processing* for sensor networks based on TinyDB [2]. The *Tiny aggregation service* [1] gives an in-network solution to answer aggregation queries over stationary sensors where sensor data are aggregated in the network and sent back through the routing tree to the base station. However, the tiny aggregation service can only answer simple and declarative aggregation queries. In Reference [9], a nearly optimal aggregation tree is built in a distributed manner to aggregate data from a sparse set of nodes in a WSN. A recent work [3] proposes a *two-phase self-join* scheme to efficiently process self-join queries for event detection in sensor networks. Recently, mobile sensors have been widely deployed in many applications where sensor mobility results in new challenges in query processing and should be carefully addressed. However, the above approaches cannot efficiently process mobile data service, as they fail to consider sensor mobility.

Several recent works take sensor mobility into account in query processing in WSNs, such as CNFS [10], CountTorrent [11], and ICEDB [12]. However, these approaches either lack the in-network optimization [12] or introduce additional overhead in query processing by maintaining a partial map of the network [10] and informing all the network nodes of the aggregate query results [11]. Moreover, their system models and query types are different from ours. Thus, these approaches are not efficient for ELAQ processing.

Finally, multiquery optimization in WSNs has been studied in some recent works. In MQO [13], communication cost is optimized on the basis of complexity analysis. In TTMQO [6], the base station aggregates incoming queries into synthetic queries on the basis of the cost model to save the total cost of query processing. However, these methods perform query aggregation in a simple way and are limited in considering sensor mobility as well.

4.4 Preliminary for Data Services

4.4.1 Hybrid WSN System Model

We assume each stationary sensor has a unique identifier and could be aware of its own location through positioning devices (e.g., GPS) or localization techniques such as in Reference [14]. A low-level mechanism, such as beacon messages, enables each stationary sensor to know its neighbors. Current velocities and locations of mobile sensors are measured using GPS. In Figure 4.1a, the whole area is divided into several subareas; each subarea has a base station that is responsible for injecting the user's queries to the sensor network and sending the query results back to the user. The white circles are the stationary sensors and the gray circles are the mobile sensors. In this chapter, we leave the issues of inter-base station communication as open work and discuss the situation in one subarea.

Figure 4.1b shows an application scenario of the system model. The small circle is the mobile body sensor's transmission range and the large circle is the query area. The body sensor keeps monitoring the running person's health status. When it detects the person's blood oxygen level is lower than a given threshold, it triggers and propagates a query to the gas sensors within the query area. Those gas sensors aggregate the air pollutant data and report the results back to the base station.

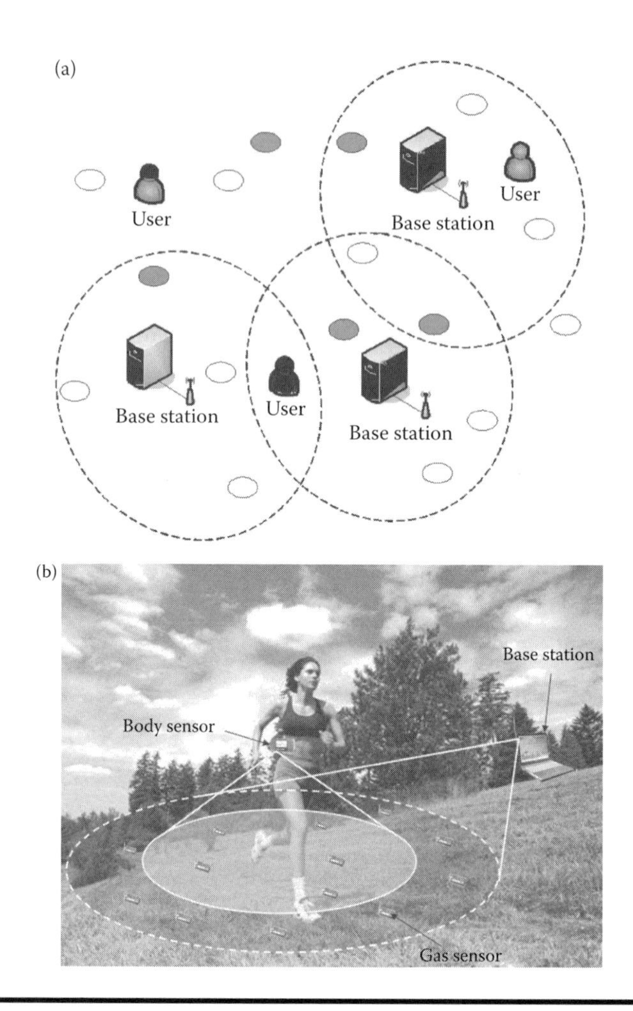

Figure 4.1 (a) System model. (b) Application scenario.

4.4.2 Query Definition

ELAQs are mobile data service queries with location awareness and event detection, and have the following format:

```
SELECT [aggregation operators] S1.att1 ... S1.atti
FROM Sensor1 AS S1
INSIDE (Radius, (SELECT S2.loc
FROM Sensor2 AS S2))
START ON EVENT1 (S2.atti)
STOP ON EVENT2 (S2.atti)
SAMPLE PERIOD Sp
FOR Duration
```

where aggregation operators include SUM, AVG, MAX, MIN, and COUNT; att_i is the data type of the sensor, $1 \leq i < n$, where n is the maximum number of data types a sensor can sample; $EVENT_1$ represents the triggering conditions of the query and $EVENT_2$ represents the stopping conditions of the query; the sample period means that mobile sensors should detect the event at each sample period for a specific duration. As presented in Reference [7], most of the location constraints can be expressed in terms of INSIDE, and thus INSIDE is considered in the query. We focus on the processing of the ELAQs with the aggregation operator AVG, because such ELAQs are the most challenging query type. It models various aggregation operators in such a way that it queries not only the data but also the number of sensors in the query area. In other words, we propose a generalized mechanism that can also efficiently process the ELAQs with other aggregation operators including SUM, MAX, MIN, and COUNT.

4.5 Processing ELAQs

The ELAQ is designed in a distributed and concurrent way to reduce the transmission cost and the query response time. Stationary sensors are first built into a tree structure with the base station as the root, similar to the approach in Reference [5]. As a result, each stationary sensor's level in the tree is proportional to the transmission hops from the base station to itself, which minimizes the number of paths to be traversed in query dissemination. Each stationary sensor maintains a minimum bounding rectangle (MBR) that bounds the maximum extents of all its child nodes' location in the tree [15]. The overlapping areas among MBRs are minimized to decrease the number of messages for accessing sensor nodes. Each stationary sensor also maintains a list of neighbor stationary sensor nodes including each neighbor's ID and location. The above information is updated by each stationary sensor using periodical beacon messages. Note that the topology of stationary sensors does not change frequently, so the cost of maintaining such information is comparatively low. ELAQ processing is divided into four main steps: query dissemination, event detection, query propagation, and data aggregation. We illustrate these steps in the following subsections.

4.5.1 Dissemination to Mobile Sensors

To reduce the cost, the base station disseminates the query to mobile sensors using a tree structure instead of flooding, as in Reference [16]. Each mobile sensor registers to a stationary sensor, called *proxy*, to access the sensor network. The base station assigns each incoming query a unique QID and converts it into the query item (QID, attribute list, aggregation operators, events, radius, sample period, duration). The base station then disseminates the query item to each mobile sensor along a route from the base station to the mobile sensor's proxy (called *proxy route*). Each stationary sensor in the proxy route caches the next hop to the mobile sensor's proxy, and it only needs to unicast the query item to the next hop, which greatly reduces the number of dissemination messages. However, the mobile sensor changes its proxy frequently owing to mobility. As a result, the proxy route should be updated by adding the mobile sensor's proxy information to the new proxy route and deleting the out-of-date proxy information from the obsolete proxy route. We argue that update messages can be reduced if mobile sensors properly select their new proxies.

An extreme situation may occur during the period that the obsolete proxy route has not been deleted immediately after a mobile sensor changes its proxy. A query may be disseminated along the obsolete proxy route to the old proxy that has already lost contact with the mobile sensor. In this case, after the old proxy receives the update message, it sends the query along the old proxy route to the node that caches the next hop to the new proxy. Then the query is disseminated along the new proxy route to the mobile sensor. To avoid message losses, each sensor in the proxy route overhears its next hop. If a parent node does not overhear the query transmitted by the next hop, it assumes the next hop has not received the query, and retransmits the query.

4.5.2 Adaptive Proxy Selection Algorithm

Each mobile sensor checks the connectivity with its proxy by sending periodical handshake messages to its proxy. When a mobile sensor is disconnected with its proxy, it selects a new proxy from its neighbor stationary sensors.

Definition 1: Dissemination Path

Consider the base station's ID is S_0, a stationary sensor S_i's *dissemination path* is ($S_0, \ldots S_i$), which is the order of traversing sensors from the base station to S_i through the tree.

Definition 2: Tree Distance

Consider the stationary sensor node S_i's dissemination path is ($S_0, \ldots S_k, \ldots S_i$) and S_j's dissemination path is ($S_0, \ldots S_k, \ldots S_j$), where S_k is the common node of the two dissemination paths, the *tree distance* between S_i and S_j is $|S_k, \ldots S_i| + |S_k, \ldots S_j| - 2$. $|S_k, \ldots S_i|$ is the number of sensor nodes on the path.

In Figure 4.2a, the dissemination paths for sensor nodes 5 and 6 are (0, 1, 3, 5) and (0, 1, 4, 6), respectively. Therefore, the tree distance between them is 4. When

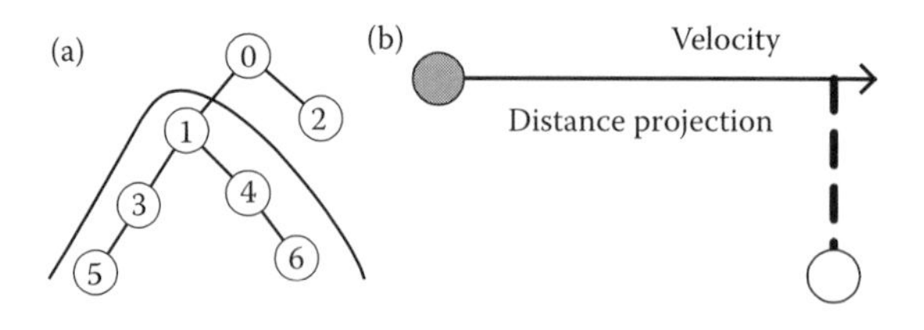

Figure 4.2 Various distances: (a) Tree distance; (b) distance projection.

a mobile sensor changes its proxy (e.g., from 5 to 6), the proxy route should be updated by relaying the mobile sensor's information from the new proxy to the common node (node 1 in Figure 4.2a). Then the obsolete proxy route from the common node to the old proxy is deleted. Therefore, the number of update messages during each proxy change depends on the tree distance between the new and old proxies. We observe that the total update cost for a mobile sensor is affected by the number of proxy changes in the query's duration and the average tree distance during movement. The total update cost can be calculated as

$$C_u = C_{tr} \sum_{i=1}^{N_{pc}} \mathrm{TD}_i \tag{4.1}$$

where C_{tr} is the transmission cost of each message, N_{pc} is the number of proxy changes in the query's duration, and TD_i is the tree distance of the ith proxy change. We propose two proxy selection algorithms to minimize C_u: maximum distance projection (MDP) aims to reduce N_{pc} whereas minimum tree distance (MTD) aims to minimize TD_i.

As shown in Figure 4.2b, the distance between a mobile sensor (the gray circle) and a stationary sensor is projected on the mobile sensor's velocity vector. This projection is called *distance projection*. In an MDP algorithm, each mobile sensor registers to the neighbor stationary sensor that has a maximum distance projection in its velocity direction. By selecting a proxy with MDP, the mobile sensor may stay in the proxy's transmission range for a longer time, which reduces N_{pc} in the query's duration. When a mobile sensor starts to change its proxy, it broadcasts its current location and velocity to its neighbor stationary sensors. Each neighbor stationary sensor calculates its distance projection to the mobile sensor locally and sets a timer that is inversely proportional to the distance projection. The neighbor with MDP first sends its distance projection to the mobile sensor and is selected as the mobile sensor's new proxy. Hereafter, the mobile sensor replies with a register message to the new proxy. Other neighbor stationary sensors overhearing this message will not send their distance projections to the mobile sensor to reduce the transmission cost.

In an MTD algorithm, each mobile sensor registers to the neighbor stationary sensor that has a minimum tree distance with its old proxy. Each stationary sensor keeps a dissemination path that can be obtained in the building procedure of the sensor tree by recording the IDs of the sensors along the traversing path from the base station to itself. Therefore, obtaining a dissemination path does not introduce extra cost. When a mobile sensor needs to change its proxy, it broadcasts its old proxy's dissemination path to the neighbor stationary sensors. Each neighbor stationary sensor calculates its tree distance from the old proxy and sets a timer proportionally to the tree distance. The neighbor node with the shortest tree distance first sends the tree distance to the mobile sensor and is selected as the mobile sensor's new proxy. However, multiple neighboring sensors may have the shortest tree distance from the old proxy at the same time. The tie is broken by selecting the proxy that has the minimum distance with the mobile sensor (MTD&MD) or selecting the proxy that has the maximum distance with the mobile sensor (MTD&MDP). We rely on the MAC layer to avoid collisions between neighbor sensors with the same tree distance.

We have compared MDP, MTD&MD, and MTD&MDP with the typical minimum distance (MD) algorithm [17] by simulation in GloMoSim where the MAC layer protocol is set to IEEE 802.11. In the MD algorithm, each mobile sensor selects the nearest stationary sensor as its new proxy. In our simulation, 200 stationary sensor nodes and 3 mobile sensor nodes are uniformly distributed in a square area as the mobile sensors' mobility patterns use *random way point* with a pause time of 30 s. The simulation time is set to 40 min and each mobile sensor node checks the connection with its proxy every 10 s. We design two experiments with different node densities and sensor transmission ranges.

In the first experiment, we set the simulation area to 600 m × 600 m and each mobile or stationary sensor's transmission radius to 100 m. From Figures 4.3a and 4.4a, we can observe that MDP results in fewer proxy changes and update messages. The advantage becomes more obvious when the maximum speed of mobile sensors increases, because the MDP algorithm prolongs each mobile sensor's resident period within its proxy's transmission range, while reducing the number of proxy changes in the query's duration. In the second experiment, we change the simulation area to 1000 m × 1000 m and each sensor's transmission radius to 200 m. However, we can observe from Figure 4.4a and b that although MDP reduces the number of proxy changes, MTD&MDP results in the minimum number of update messages compared to that of MDP. As the sensor's transmission radius increases from 100 to 200 m, each mobile sensor stays within its proxy's transmission range for a longer time, which results in less saving of the number of proxy changes by MDP. We can observe from Figure 4.3a and b that the maximum saving of proxy changes by MDP decreases from 49 (Figure 4.3a) to 17 (Figure 4.3b). As depicted in Equation 4.1, the total update cost for a mobile sensor is determined by two factors: the number of proxy changes and the tree distance every proxy change. If the saving of the number of proxy changes by MDP has decreased to a certain degree, the saving of the number of update messages (equal to tree distance) every proxy change becomes a dominant factor that determines the total update cost. That is the reason MTD&MDP becomes the optimal algorithm as each sensor's transmission radius increases to 200 m. Note that MDP, MTD&MDP, and MTD&MD always outperform the MD algorithm that requires frequent updates and excessive update messages. The query's duration is usually much longer than the duration in the simulation, so the optimization effects are significant.

We learn from the above evaluations that different system scenarios may lead to different optimal proxy selection algorithms between MDP and MTD&MDP. To decide which algorithm is optimal for specific system scenarios, a training process can be conducted after the sensor tree is built. During the period of training, each

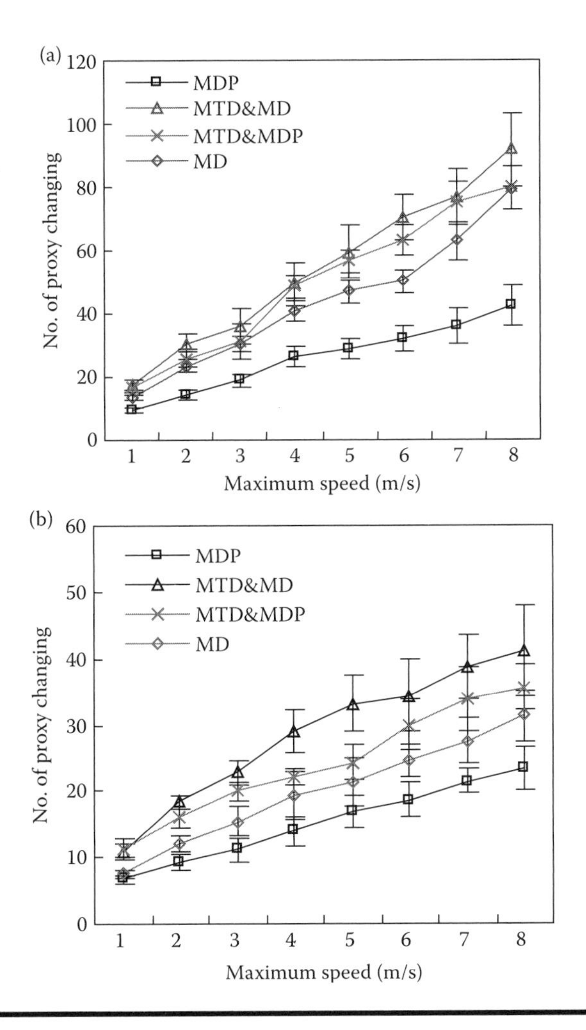

Figure 4.3 Number of proxy changes versus maximum speed: (a) 600 m × 600 m; (b) 1000 m × 1000 m.

mobile sensor runs both algorithms and records the respective total tree distance (equal to the total update messages). The algorithm that has fewer total update messages is chosen as the optimal proxy selection method.

4.5.3 Event Detection

Each mobile sensor performs event detection locally in a distributed and concurrent way. After receiving a query item, each mobile sensor keeps sensing the required data and checking the event every sample period for a specific duration. Once the event is detected, the mobile sensor becomes the reference sensor, and its current location

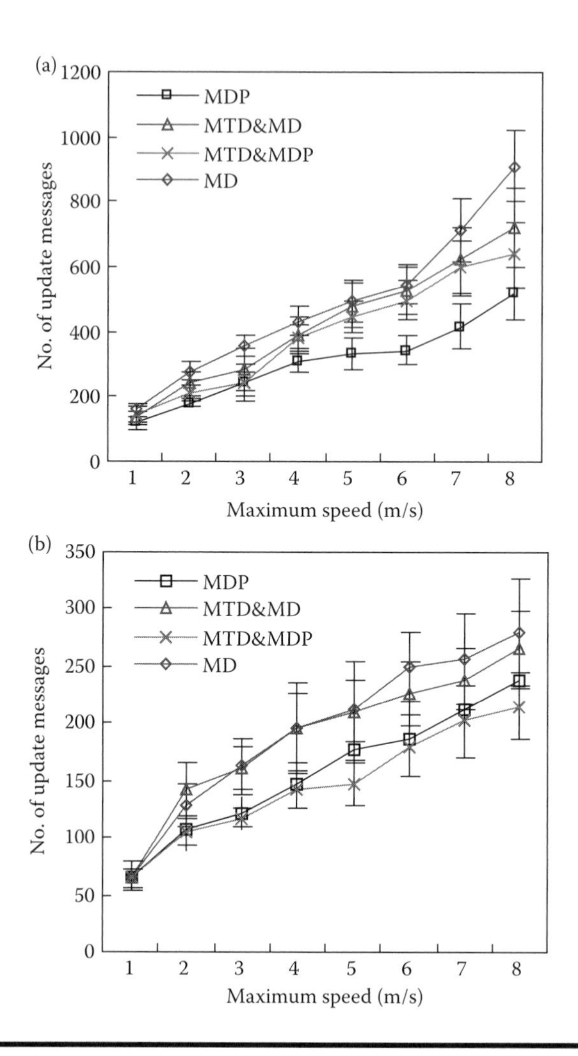

Figure 4.4 Number of update messages versus maximum speed: (a) 600 m × 600 m; (b) 1000 m × 1000 m.

becomes the center of the circle query area. Our approach saves communication cost compared to the approach where the base station is used for event detection [18].

4.5.4 In-Network Query Propagation

After a query is triggered by an event, the reference sensor propagates the *triggered query*. The query message carries the reference sensor's ID and current location, query ID, query radius, and attribute list to the stationary sensors in the query area. If the query radius is shorter than the reference sensor's transmission radius, the triggered query only needs to be broadcast to the sensor nodes in the query area by one message. Otherwise, the triggered query should be propagated to all the sensor nodes in the query area in a multihop fashion. In centralized methods, queries are propagated from the base station to the sensors in the query area by flooding [6] or tree-structured indexes [2,3,5], which wastes propagation messages and time. We propose an in-network query propagation approach for stationary sensors to decide whether or not to propagate the query to their neighbors locally.

Each reference sensor first broadcasts the triggered query to its neighbor stationary sensors. Assume a stationary sensor node S receives the triggered query. If S's parent node is not in the reference sensor's transmission range, it forwards the triggered query to its parent node, because its parent node cannot receive the query directly from the reference sensor. S's parent node can help propagate the triggered query to S's siblings that are isolated from other sensors in the query area. If S is not a leaf node, it broadcasts the triggered query to both its parent node and its child nodes that are out of the reference sensor's transmission range, and their MBRs overlap the query area. However, using the above approach, multiple child nodes may send the triggered query to the same parent node. To save the propagation messages, S overhears its neighbor nodes. If S finds that one of its siblings has sent the triggered query to its parent nodes, S will not transmit the query to its parent node. Note that the triggered query will be eventually sent back to the base station. If the base station does not receive the triggered query from one of its immediate child nodes, S_i, while it finds that S_i's MBR overlaps the query area, the base station will inject the triggered query to S_i's subtree. The reason is that some child nodes of S_i may be in the query area but may not receive the triggered query.

For example, in Figure 4.5, reference sensor 7 broadcasts the triggered query to its neighbor sensors 5 and 6. Sensors 5 and 6 send the triggered query to their parent nodes 2 and 4, respectively. Sensor 6 also sends the triggered query to its child node 8 because node 8's MBR overlaps the query area. Sensors 2 and 4 send the triggered query to their parent nodes 1 and 0, respectively. Sensor node 1 finds that node 3 is in the query area, so it sends the triggered query to node 3.

Theorem 1

Using in-network propagation, if a triggered query is received by at least one stationary sensor, it will be propagated to all the stationary sensors in the query area.

Proof: Assume sensor node S_i receives the triggered query. In our approach, the triggered query is sent to the sensor node's parent after being received. For any sensor node S_j in the query area, the triggered query will be sent from S_i to the lowest common ancestor of S_i and S_j in the tree, S_k. As S_k's MBR overlaps S_j's MBR, S_k will send the triggered query to S_j, which guarantees the correctness of Theorem 1.

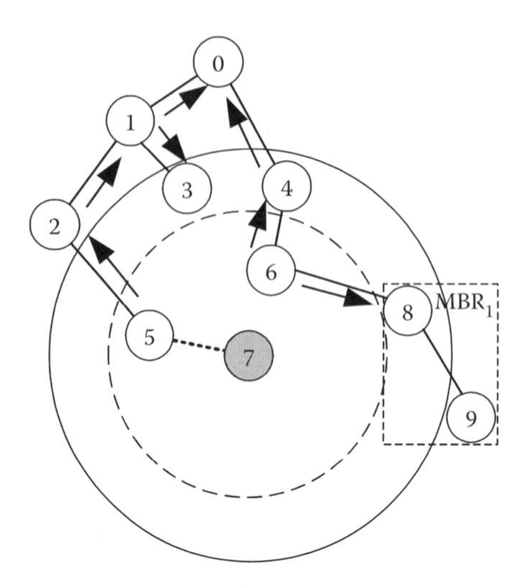

Figure 4.5 An example of propagation of triggered query.

4.5.5 Location-Based In-Network Aggregation

In generic in-network aggregation algorithms [1–3], each sensor collects and sends the data back through a routing tree rooted at the base station. Each sensor also performs data aggregation for the data flowing through it. When query areas keep changing, many messages will be wasted if parent nodes are not in the query areas, but they cannot perform local aggregation. Each sensor in the query area should transmit at least one message to send its data back to the base station. However, sensors out of the query area do not need to send any messages if they have no relaying data toward the base station. Therefore, the transmission cost can be reduced by excluding the sensors that are not in the query area to perform the aggregation.

We propose an LIA algorithm to enable each sensor to dynamically choose the next hop to transmit data on the basis of the reference sensor's current location. A sensor transmits data to its parent node in either of the two situations: (1) its parent node is located in the query area; (2) no neighbor nodes with equal or higher level are in the query area, excluding the neighbor nodes that transmit data to this sensor. Otherwise, the sensor node transmits data to the neighbor node with the highest level in the query area. If there is more than one such neighbor node, the sensor node chooses the nearest one. As a result, fewer sensor nodes are involved in the aggregation and more data are aggregated locally. Each nonleaf sensor node should wait until it has heard from its child nodes in the query area before aggregating and forwarding the results. In the results, the number of sensors has been contained. Several sensor nodes in the same level may transmit data to each other at the same time. To avoid such collisions, each nonleaf sensor's wait interval is set to be proportional to the number of its child nodes in the query area. Because sensor nodes with more child nodes in the query area may aggregate more sensors' data locally, they should wait for other sensors' data. As each node already has a neighbor list with each neighbor node's ID and location, it can easily calculate the distance between each of its child nodes' locations and the reference sensor's current location. If such a distance is shorter than the query radius, the current child node is in the query area. After scanning the whole neighbor list, the sensor can have the number of its child nodes in the query area. Suppose the maximum depth of the sensor tree is d, the nonleaf sensor node is in level l and has i child nodes in the query area ($0 \leq i \leq n$), where n is the maximum number of child nodes in the tree. If each sensor's sensing and processing time is T_{sp} and transmission time is T_{tran}, the wait interval T_{wait} can be calculated as

$$T_{wait} = (d - l)(2 \times T_{tran} + T_{sp}) + i \times T_{tran} \tag{4.2}$$

T_{wait} takes the longest hops $d - l$ into account and the nonleaf sensor node forwards the triggered query to the leaf nodes in the query area, which then transmit their data back to it. The in-network query propagation and LIA algorithms require each stationary sensor in the query area to have up-to-date knowledge of which neighboring sensors are within the mobile sensor's transmission range or the query area. The stationary sensor can get such knowledge locally by scanning its neighbor list and calculating the distance between current neighbor sensor and the up-to-date location of the mobile sensor. In the following, we give an example to illustrate the LIA algorithm.

As is shown in Figure 4.6, sensor 0 is the base station and sensor 7 is the reference sensor that registers to sensor 5. Sensor 5 in the query area first sends its readings to sensor 6 because it has fewer child nodes than sensor 6. Sensor 8 also transmits its readings to sensor 6. Sensor 3 has fewer child nodes than sensor 6 although its level is equal to sensor 6. Therefore, it transmits its readings before sensor 6. Sensor 3 chooses

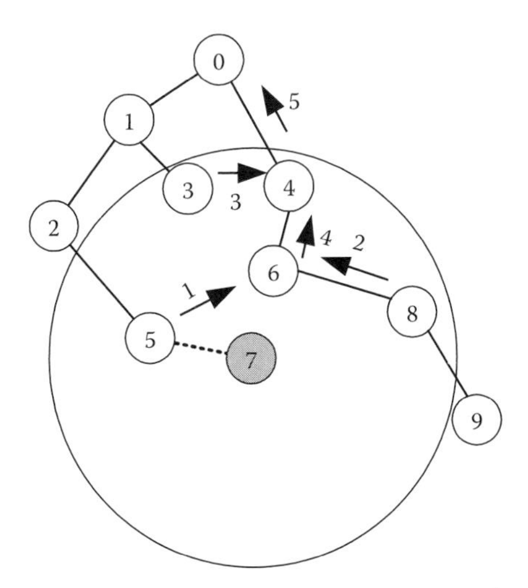

Figure 4.6 Example of the location-based in-network aggregation algorithm.

sensor 4 as the next hop because its parent node is not in the query area whereas its neighbor node sensor 4 is in the query area. When the wait interval expires, sensor 6 calculates the aggregation result from sensor 5, sensor 8, and itself, and transmits the result, including the number of sensors that participated in the aggregation (i.e., three sensors), to sensor 4. Once sensor 4 receives the readings from sensor 3 and the aggregation result from sensor 6, it aggregates the received data with its local readings and sends the final result back to its parent node, the base station.

4.6 Cost Analysis

In the following, we give a theoretical analysis of the cost of ELAQ processing (cost(Q)), that is, the total transmission overhead of dissemination (diss(Q)), propagation (prop(Q)), and aggregation (agg(Q)):

$$\text{cost}(Q) = \text{diss}(Q) + \text{prop}(Q) + \text{agg}(Q) \tag{4.3}$$

An ELAQ Q can be decomposed into attribute set Q.att, event $Q.e$, sample period Q.sp, radius $Q.r$, and duration $Q.d$. Assume N stationary sensor nodes and N_m mobile sensor nodes are uniformly distributed in a circle area with radius $R_{\max}$ and the base station at its center.

Q's dissemination cost includes the cost of query dissemination as well as proxy change. Assume that the average query arrival time is t seconds per query; then diss(Q) is

$$\text{diss}(Q) = C_{\text{tr}}\left(\frac{Q.d}{t}\sum_{j=1}^{N_{\text{m}}}(l(M_{\text{j}})+1) + \sum_{i=1}^{N_{\text{pc}}} TD_i\right) \tag{4.4}$$

where $l(M_j)$ is the level of mobile sensor M_j's proxy.

The probability density of any location within the circle area is $1/\pi R_{\max}^2$. Therefore, the distance R between the mobile sensor (the center of the circle in Figure 4.7) and the base station (point b in Figure 4.7) has the following probability distribution:

$$P(R \le r) = \int_0^{2\pi}\int_0^r \frac{1}{\pi R_{\max}^2} r'\,dr'd\theta = \frac{r^2}{R_{\max}^2}$$

Thus, the probability density function of R is

$$f_R(r) = \frac{\partial P}{\partial r} = \frac{2r}{R_{\max}^2}, \quad 0 \le r \le R_{\max}$$

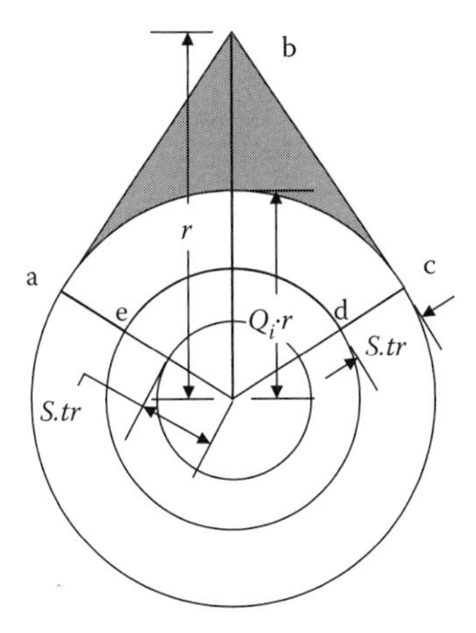

Figure 4.7 Theoretical model.

If $Q.r$ is shorter than the mobile sensor's transmission radius, prop(Q) is 0. Otherwise, prop(Q) includes the cost of propagation within the query area and propagation from *boundary nodes* to the base station. Boundary nodes are the sensor nodes in the query area whose parents are not in the query area. The number of propagation messages within the query area can be approximated to the number of sensors in the ring area between the mobile sensor's transmission range and the query area. As shown in Figure 4.7, the number of propagation messages from boundary nodes to the base station can be regarded as the number of sensor nodes in the shadowed area *abc*. The number of propagation messages out of this area is relatively small. The expectation of the area *abc* can be computed as

$$\begin{aligned} E(\text{Area}_{\text{abc}}) &= \int_{Q.r}^{R_{\max}} \left(Q.r\sqrt{\text{r}^2 - Q.r^2} - Q.r^2 \arccos\frac{Q.r}{r} \right) \frac{2r}{R_{\max}^2} \mathrm{d}r \\ &= \frac{Q.r}{3}\sqrt{R_{\max}^2 - Q.r^2}\left(\frac{Q.r^2}{R_{\max}^2} + 2\right) - Q.r^2 \arccos\frac{Q.r}{R_{\max}} \\ &\le Q.r\sqrt{R_{\max}^2 - Q.r^2} - Q.r^2 \arccos\frac{Q.r}{R_{\max}} \end{aligned}$$

Given the node density $N/\pi R_{\max}^2$ and the sensor's transmission radius $S.$tr, the propagation cost of Q is then

$$\begin{aligned} &\text{prop}(Q) = C_{\text{tr}}\ p(Q.e)\ (Q.d/Q.sp)(N/\pi R_{\max}^2) \\ &(E(\text{Area}_{abc}) + \pi(Q.r^2 - S.tr^2), \quad Q.r > S.\text{tr} \end{aligned} \tag{4.5}$$

where $p(Q.e)$ is the probability that the event occurs during a sample period.

Q's aggregation cost includes the cost of aggregation within the query area and transmitting aggregation results. Aggregation results are aggregated in several boundary nodes at the highest level in the query area before they are transmitted back to the base station. Thus, most of such boundary nodes are in the ring area of *acde* in Figure 4.7 with S.tr as its width. The expectation of the area *acde* is

$$\begin{aligned} E\left(\text{Area}_{\text{acde}}\right) &= \int_{Q.r}^{R_{\max}} \frac{2r}{R_{\max}^2}\left(Q.r^2 - (Q.r - S.\text{tr})^2\right)\arccos\frac{Q.r}{r}\,\text{d}r \\ &= \left(2Q.rS.\text{tr} - S.\text{tr}^2\right)\left(\arccos\frac{Q.r}{R_{\max}} - \frac{Q.r}{R_{\max}}\sqrt{1 - \frac{Q.r^2}{R_{\max}^2}}\right) \\ &< \frac{\pi}{2}\left(2Q.rS.\text{tr} - S.\text{tr}^2\right), \quad Q.r > S.\text{tr} \end{aligned}$$

$$\begin{aligned} E\left(\text{Area}_{acde}\right) &= \int_{Q.r}^{R_{\max}} \frac{2r}{R_{\max}^2} Q.r^2 \arccos\frac{Q.r}{r}\,\text{d}r \\ &= Q.r^2\left(\arccos\frac{Q.r}{R_{\max}} - \frac{Q.r}{R_{\max}}\sqrt{1 - \frac{Q.r^2}{R_{\max}^2}}\right) \\ &< \frac{\pi}{2} Q.r^2, \quad Q.r \le S.\text{tr} \end{aligned}$$

The transmission ranges of all the boundary nodes that have aggregation results can cover the ring area *acde* with little overlap. Thus, the number of these boundary nodes is

$$N_b = \left\lceil \frac{E(\text{Area}_{acde})}{\pi S.\text{tr}^2} \right\rceil$$

The expectation of the average level of these boundary nodes can be simply calculated as

$$\begin{aligned} E\left(l_{\text{avg}}\right) &= \left\lfloor \int_0^{R_{\max}} \left(\frac{r - Q.r}{S.\text{tr}}\right)\frac{2r}{R_{\max}^2}\,\text{d}r \right\rfloor + 1 \\ &= \left\lfloor \frac{2R_{\max} - 3Q.r}{3S.\text{tr}} \right\rfloor + 1 \end{aligned}$$

In the LIA algorithm, each sensor node in the query area sends its data by one message. The number of sensor nodes in the query area is $NQ.r^2/R_{\max}^2$. Thus, agg(Q) is

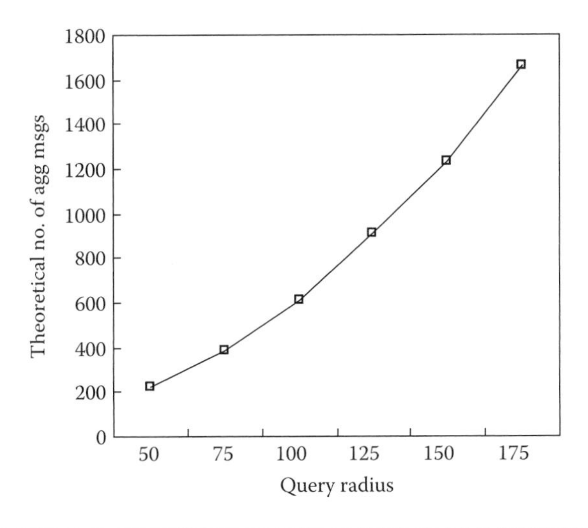

Figure 4.8 Theoretical number of aggregation messages versus query radius.

$$\text{agg}(Q) = C_{\text{tr}}\ p(Q.e)(Q.d/Q.\text{sp})(NQ.r^2/R_{\max}^2 + N_b\ E(l_{\text{avg}})) \tag{4.6}$$

The cost analysis indicates that our techniques reduce the redundancy messages in processing ELAQs. From Equation 4.4, we learn that dissemination cost is primarily decided by the mobile sensors' mobility patterns and by system scenarios. The theoretical upper bounds of the cost of propagation and aggregation can be obtained from Equations 4.5 and 4.6. This can be used to estimate energy consumption in the WSNs. The simulation results in Section 4.8 match well with the theoretical cost. For example, in Figure 4.8, the theoretical number of aggregation messages is quite close to that in the simulation.

4.7 Multiquery Optimization

In many applications, multiple concurrent queries may be issued to the base station. Processing multiple queries in an uncooperative manner will lead to bandwidth contention and transmission collisions. Thus, it is necessary to perform multiquery optimization to reduce the total processing cost. Existing multiquery optimization algorithms [6,13] rewrite the incoming queries at the base station in a static way and do not consider ELAQs' unique properties such as distributed event detection and continuously changing query areas. We propose a two-level multiquery optimization algorithm that filters *hidden queries* at the base station and instantly rewrites multiple concurrently triggered queries into one synthetic query at each mobile sensor.

4.7.1 Optimization at the Base Station

A base station maintains a list of running queries. When an ELAQ Q_i arrives at a base station, our algorithm first scans the query list and compares each part of incoming query Q_i with the corresponding part of each running query Q in the query list. Assume a stationary sensor S's attribute set is S.att and the remaining duration of a running query Q is Q.rd. Q_i is a hidden query and need not be injected to the sensor network if it satisfies all the following hidden rules: (1) $Q_i.\text{att} \subseteq Q.\text{att} \subseteq S.\text{att}$, (2) $Q_i.e = Q.e$, (3) $Q_i.r = Q.r$, (4) $Q_i.\text{sp} \mid Q.\text{sp}$, and (5) $Q_i.d \leq Q.\text{rd}$. Rule (2) guarantees that Q_i's reference sensor set is the same as Q's reference sensor set. By adding rule (3), Q_i's query areas are the same as Q's query areas. Rule (4) means Q_i's sample period is divided by Q's sample period. These hidden rules guarantee Q_i's result set is totally included in Q's result set and can be filtered out by the base station. If no query satisfies all the hidden rules with Q_i, Q_i is added to the list of running queries and disseminated to mobile sensors. In a special case that Q_i satisfies rules (1) through (4) with Q, Q_i will be rewritten to Q_i' with new a duration $(Q_i.d - Q.\text{rd})$. Q_i' will not be injected to the sensor network until Q finishes, because it is possible that when Q_i' is waiting, another incoming query includes Q_i''s results and makes it unnecessary to inject Q_i'. Optimization at the base station on the basis of hidden rules can minimize duplicate access to the sensor network.

4.7.2 Online Query Rewriting Algorithm

Hidden rules are strict for many incoming queries and optimization at the base station does not exploit the similarities among queries at a fine granularity. We apply an *online query rewriting* algorithm at each mobile sensor after queries are disseminated. Compared to rewriting queries at the base station [6,13], our algorithm can easily decide which queries are triggered every sample period, and thus can intelligently rewrite the queries.

Each mobile sensor keeps a list of running queries. After a mobile sensor receives an incoming query Q_i, it sets its own sample period at the greatest common divisor of the sample periods of all the queries. Q_i's start time is set to be divisible by the mobile sensor's sample period. In this way, various queries will sample at the same time, and hence share the event detection. Although introducing such little delay will make the first sampling period start later, for a continuous query, this extra latency is acceptable. In each sample period of the mobile sensor, if multiple queries are triggered at the same time, the mobile sensor instantly rewrites these queries into a synthetic triggered query Q_{syn}. Q_{syn}'s attribute list is the union set of all the triggered queries' attribute list, and Q_{syn}'s query radius is set to the longest query radius of all the triggered queries. To maintain the semantic correctness, Q_{syn} should contain the reference sensor's ID and each triggered query's ID that is mapped to the corresponding query radius. Then, the mobile sensor propagates Q_{syn} to the sensors in the query area instead of propagating the triggered queries separately. Each sensor that receives Q_{syn} sends its data and the ID list of the triggered queries, whose query areas it is in, to the next hop sensor by LIA algorithm. The ID list helps the next hop sensor aggregate multiple results for multiple triggered queries. Rewriting the triggered queries in this way will reduce the cost because the algorithm minimizes the redundant messages of processing multiple ELAQs while only introducing a little computation cost in the mobile sensor and no extra transmission cost.

4.8 Performance Evaluation

We evaluate ELAQ performance using our techniques by simulation in GloMoSim [19]. The implementation of ELAQ includes query dissemination, in-network query propagation, LIA, and two-level multiquery optimization. We use the same setting as in Section 4.5 where the total area is 600 m × 600 m and each sensor's transmission radius is 100 m. The workload query has a sample period of 120 s and duration of 12,000 s. The probability that the event occurs is set to 10%. Each data point in the figures has a 90% confidence interval that comes from the average result of 10 runs.

Dissemination in hybrid WSNs has been studied in References [16,17,20]. However, Reference [20] focuses on interest dissemination from the mobile sink to neighbor stationary sensors, which is opposite to the dissemination direction in our system. Moreover, the flooding approach in Reference [16] is not efficient compared to scalable energy-efficient asynchronous dissemination (SEAD) [17]. Thus, we compare our in-network query dissemination approach with SEAD. In SEAD, the mobile sink selects its nearest neighbor as its access node and sends a join query to the source, and then the source disseminates the data to the access node that delivers the data to the mobile sink. We evaluate the total number of dissemination messages, including the number of query dissemination messages and update messages caused by proxy change. The average query arrival frequency is set to 5 min per query; thus there are around 40 queries being disseminated to the sensor network in 12,000 s. As shown in Figure 4.9, our approach reduces the total dissemination cost. The number of query dissemination messages is proportional to the average level of proxy sensor (called access node in Reference [17]), making the number of query dissemination messages of these two approaches quite close to each other. However, as discussed in Section 4.5, compared to the minimum projection distance method used in our approach, the minimum distance proxy selection method used in SEAD will result in more update messages. Note that total dissemination costs are largely affected by the number of update messages during proxy change when the query arrival rate is not very high, so it is determined by the proxy selection method in most cases. As the maximum speed of mobile sensors increases, the saving becomes more obvious.

As query propagation is a unique procedure in ELAQ processing, no other appropriate technique can be compared. As we set the number of sample periods to 100 and the

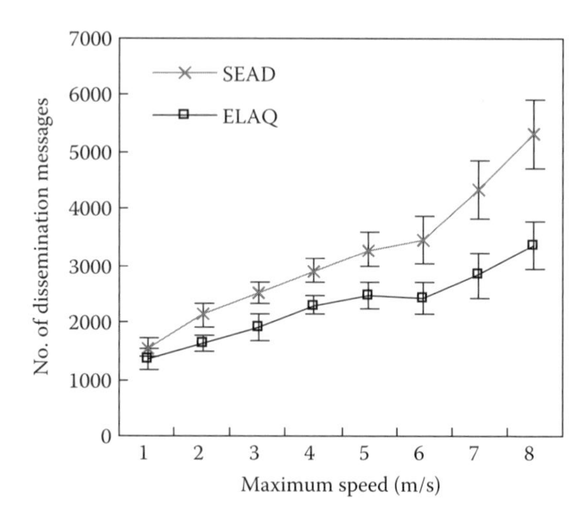

Figure 4.9 Number of dissemination messages versus maximum speed.

probability that the event occurs to 10%, the query is triggered around 10 times at each mobile sensor. We evaluate the total number of propagation messages in the query's duration. If the query radius is shorter than the mobile sensor's transmission radius, the number of propagation messages is 0 because sensors in the query area can receive the query from the mobile sensor directly. Otherwise, we can see from Figure 4.10 that the number of propagation messages increases when the query radius increases.

We compare the LIA algorithm with the generic in-network aggregation algorithm [1–3] that each sensor aggregates the data of its child nodes in the tree structure. We set the maximum speed of mobile sensors to 1 m/s. Figure 4.11 shows

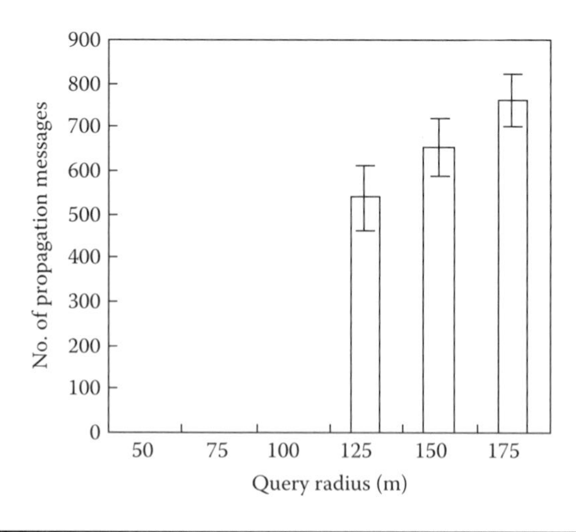

Figure 4.10 Number of propagation messages versus query radius.

that the LIA algorithm always outperforms the generic in-network aggregation algorithm in the number of aggregation messages. According to the results, the saving of messages increases and then decreases as the query radius increases. The most saving occurs around a query radius of 75. As the query radius increases, more sensors are involved in the aggregation, and thus the saving of messages also increases. However, when the query radius is large enough so that most of the sensors' parent nodes are in the query area, these sensors send their data to their parent nodes instead of dynamically choosing the destination. In this situation, only a part of the sensors in the query area save messages by using the LIA algorithm. Message saving degrades if the query radius increases above a certain threshold. For a continuous query, the total saving will be significant when the query's duration is long enough in large-scale sensor networks. It is interesting to note that the dissemination cost forms the main part of the ELAQ processing cost, which reveals that the mobility is a dominant factor that affects the total cost.

After evaluating each technique separately, we compare the accuracy of query results with that in TinyDB [2]. The accuracy is computed by dividing the error (i.e., deviation value) by the true value of the query results. As approaches in Reference [2] cannot answer ELAQ properly, we apply our in-network query dissemination and propagation approaches to TinyDB. The only difference is in aggregation: the generic in-network aggregation algorithm is used in TinyDB whereas the LIA algorithm is used in our scheme. We evaluate the accuracy by measuring the percentage of average error and maximum error. Figure 4.12 shows that query results are more accurate in our scheme than in TinyDB, because fewer sensors are involved in the aggregation, reducing the possibility of message losses. As the query radius increases, both the average error and maximum error increase

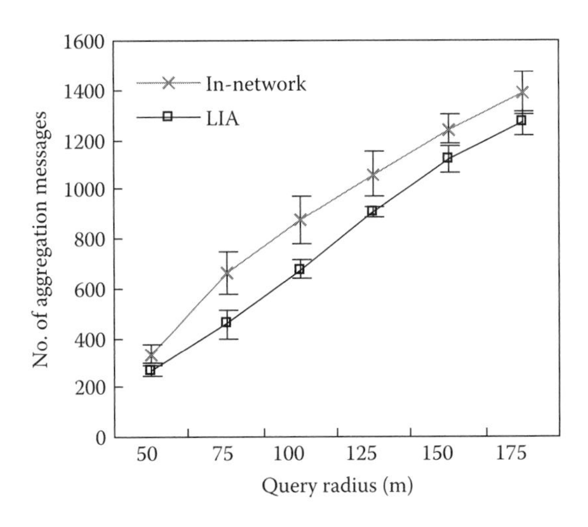

Figure 4.11 Number of aggregation messages versus query radius.

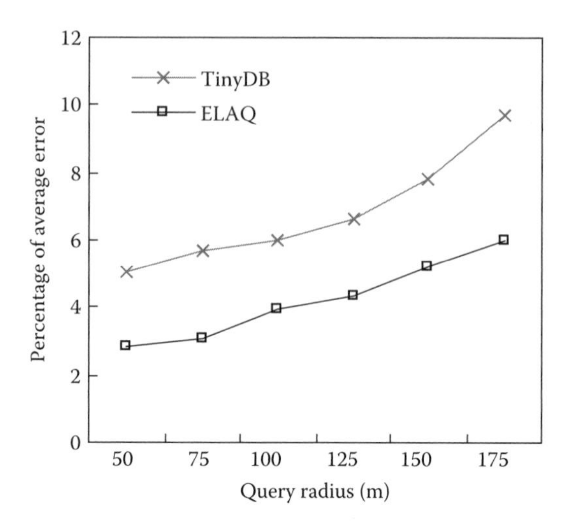

Figure 4.12 Average error versus query radius.

because message losses increase if more sensors are involved in query propagation and aggregation. In fact, the accuracy of query results is affected by messages losses in query dissemination, query propagation, and data aggregation. As discussed in Section 4.5, several techniques are used to avoid message losses. However, in the hierarchical tree structure, a single node failure can result in the aggregated data of a sensor node and its neighbor nodes being lost, making the maximum errors highly variable in Figure 4.13.

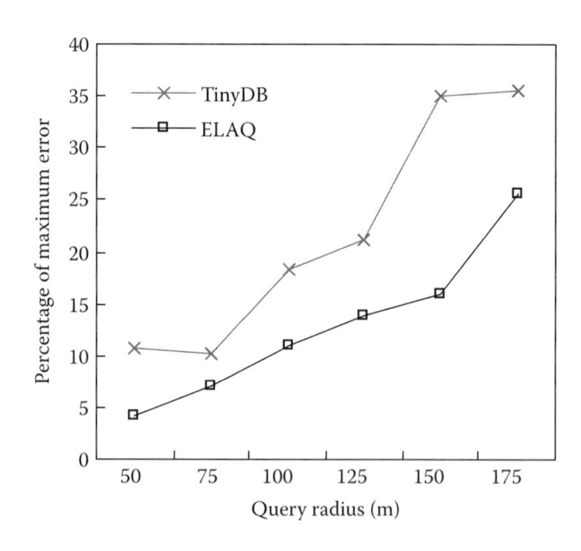

Figure 4.13 Maximum error versus query radius.

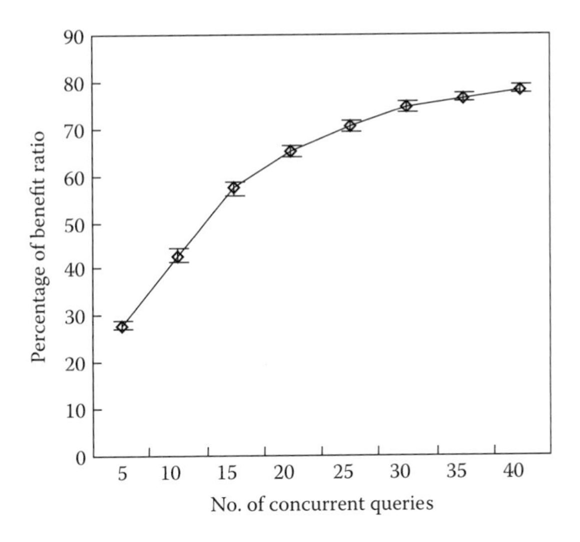

Figure 4.14 Benefit ratio versus number of concurrent queries.

We evaluate the benefit ratio of the two-level multiquery optimization algorithm with the number of concurrently running queries. The benefit ratio is computed by dividing the sum of message savings by the sum of every query's messages. The workloads are set to 100 queries with an average arrival frequency of 60 s per query. These queries randomly choose their attribute lists (lights, temperature), event probability (from 10% to 90%), query radii (from 25 to 175 m), and sample periods (from 50 to 500 s). We vary the average duration to control the average number of concurrent queries. All the queries' sample periods and durations are divided by the smallest time unit of 1 s, and their aggregation operators are set to AVG. As shown in Figure 4.14, the benefit ratio increases significantly from around 28% to 78% as the number of concurrently running queries increase from 5 to 40. This is because greater sharing can be exploited among more running queries.

4.9 Conclusions

In this chapter, we have defined a general query type—ELAQ—that exists in a wide range of data-centric applications in hybrid WSNs. Existing approaches cannot efficiently answer ELAQs and provide mobile data services with location awareness and event detection. Given the challenges of answering ELAQs, we proposed techniques (i.e., in-network query dissemination and propagation, LIA, and two-level multiquery optimization) to efficiently process ELAQs. We also theoretically analyzed the ELAQ processing cost. Cost analysis and experimental results show that these techniques can reduce the processing cost compared to state-of-the-art works. These techniques can also be efficiently applied to typical query types in sensor networks.

There are two open issues for efficiently processing ELAQs. The first issue is designing a new event detection mechanism to accurately detect sophisticated events, which considers not only the mobile sensors' data but also their neighbor sensors' data. The second issue is how to support continuous event prediction to save more query processing costs.

QUESTIONS

Q1: What make ELAQs unique from the typical query types (e.g., aggregation query, location aware query) in WSNs?

A1: ELAQs generalize many typical data service queries directed to sensors moving in WSNs. ELAQs periodically collect and aggregate the data within the specific area around the mobile sensor nodes that detect the event. It requires continuous event detection as well as dynamic location awareness on the fly, which makes ELAQs unique from other typical query types. Specifically, a typical aggregation query is an ELAQ in which query areas are fixed, and a location aware query is an ELAQ without event detection and aggregation.

Q2: Give another application of ELAQ except for the application in this chapter.

A2: Assume the mobile sensor attached to the car keeps monitoring the car's oil status. When it detects the oil is about to be exhausted, it triggers and propagates the query to the stationary sensors that monitor the gas prices of the gas stations within the reachable area. Those sensors aggregate the gas price and report the gas station with the lowest gas price back to the base station.

Q3: What are the main challenges for processing ELAQs?

A3: The challenges are: (1) Accessing mobile sensors in a robust and efficient manner in hybrid WSNs. (2) Determination of continuously changing query areas. (3) Efficient data aggregation in continuously changing query areas. (4) Processing multiple concurrent queries that may cause message collisions and system overload.

Q4: In this chapter, we only discuss the processing of ELAQs with the aggregation operator AVG. Can our techniques process ELAQs with other aggregation operators such as MAX, MIN, and COUNT? If yes, are there any special optimization techniques for other aggregation operators?

A4: Yes, the proposed techniques can process ELAQs with aggregation operators such as SUM, MAX, MIN, and COUNT. As AVG requires not only the data but also the number of sensors in the query area, ELAQs with the AVG operator are usually the most challenging query type compared to ELAQs with other aggregation operators. Special aggregation techniques can be proposed by utilizing the other aggregation operators. For example, the COUNT operator only requires the number of sensors in the query area. The leaf nodes in the query area need not send messages because their parent nodes can calculate the number of their child nodes in the query area locally using the neighbor lists. (This is not the only answer; other reasonable answers can also be accepted.)

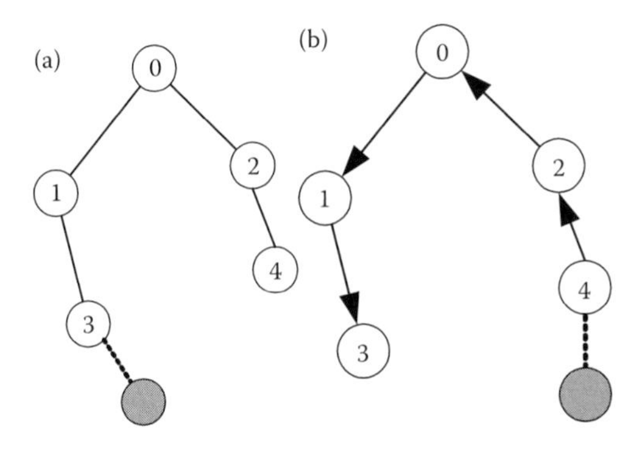

Figure 4.15 Example for proxy change and update of the proxy route: (a) Before proxy change; (b) after proxy change.

Q5: Can you give an example of proxy change and an update of the proxy route?

A5: For example, assume node 0 is the base station, and the mobile sensor's (gray circle) proxy is node 3 in Figure 4.15a. When the mobile sensor moves to the location in Figure 4.15b, it changes its proxy to node 4. The update messages initiate from the new proxy node 4 to its parent node 2, then to node 0, node 1, and finally to node 3. The new proxy information is added to node 4, updated at the base station, and the obsolete information is deleted at nodes 1 and 3.

Q6: In this chapter, we present an extreme situation during query dissemination that the obsolete proxy route has not been deleted immediately after a mobile sensor changes its proxy. Please give an example and illustrate how to disseminate the query to the new proxy.

A6: We use the same scenario as in question 5. If the base station disseminates the query along the obsolete proxy route to the old proxy node 3, node 3 sends the query back to the base station through node 1. Then the base station disseminates the query through the new proxy route to the new proxy node 4.

Q7: Why do MTD and MDP methods reduce the update cost of the proxy route?

A7: As depicted in Section 4.5, the total update cost for a mobile sensor is affected by the number of proxy changes in the query's duration and the average tree distance between the mobile sensor's old and new proxies. The MTD method tries to minimize the tree distance between the new and the old proxy with each proxy change, and the MDP method tries to reduce the number of proxy changes within a specific duration. Thus, they can reduce the update cost, which is proved by the simulations.

Q8: Why is query propagation necessary in ELAQ processing?

A8: If the query radius is longer than the reference sensor's transmission radius, the mobile sensor cannot broadcast the triggered query to all the sensor nodes in the query area by one message. The triggered query should be propagated to all the sensor nodes in the query area in a multihop fashion.

Q9: How does the LIA algorithm reduce the aggregation cost compared to the generic aggregation algorithms?

A9: We observe that each sensor in the query area should transmit at least one message to send its data, whereas the sensor out of the query area does not need to send any messages except relaying the aggregated data to the base station. The LIA algorithm reduces the number of sensors that participate in the aggregation and are not in the query areas by enabling each sensor to dynamically choose the next hop to send its data. Therefore, the LIA algorithm can reduce the aggregation cost compared to the generic aggregation algorithms in which each sensor always sends its data to the parent node.

Q10: Typical multiquery optimization algorithms optimize the incoming queries at the base station or in the network. Why does our multiquery optimization algorithm use a two-level scheme to optimize the incoming queries both at the base station and at the mobile sensors?

A10: Such design is motivated by the ELAQ's unique characteristics: the incoming query first arrives at the base station and then it is disseminated to the mobile sensors by which the query is triggered. Mobile sensors can easily decide which queries are triggered every sample period, and thus can intelligently rewrite the queries at a fine granularity. Compared to the typical algorithms, rewriting at the mobile sensors can exploit more sharing between different queries.

References

1. S. Madden, M. J. Franklin, J. M. Hellertein, and W. Hong, TAG: A tiny aggregation service for ad hoc sensor networks, in *Proceedings of the Fifth ACM Symposium on Operating System Design and Implementation* (OSDI), Boston, Massachusetts, December 2002.
2. S. R. Madden, M. J. Franklin, J. M. Hellerstein, and W. Hong, TinyDB: An acquisitional query processing system for sensor networks, *ACM Transactions on Database Systems*, 30, 122–173, 2005.
3. X. Yang, H. B. Lim, M. T. Ozsu, and K. L. Tan, In-network execution of monitoring queries in sensor networks, in *Proceedings of the ACM SIGMOD International Conference on Management of Data*, Beijing, China, 2007.
4. P. Juang, H. Oki, Y. Wang, M. Martonosi, L.-S. Peh, and D. Rubenstein, Energy efficient computing for wildlife tracking: Design tradeoffs and early experiences with ZebraNet, in *ASPLOS-X 2002: Proceedings of the 10th International Conference on Architectural Support for Programming Languages and Operating Systems*, vol. 37, no. 10, October 2002, pp. 96–107.
5. A. Soheili, V. Kalogeraki, and D. Gunopulos, Spatial queries in sensor networks, in *Proceedings of 13th International Symposium on Advances in Geographic Information Systems* (*GIS 2005*), Bremen, Germany, November 2005.
6. S. Xiang, H. B. Lim, K.-L. Tan, and Y. Zhou, Two-tier multiple query optimization for sensor networks, in *Proceedings of the 27th International Conference on Distributed Computing Systems* (*ICDCS 2007*), Toronto, Canada, 2007.

7. S. Ilarri, E. Mena, and A. Illarramendi, Location-dependent queries in mobile contexts: Distributed processing using mobile agents, *IEEE Transactions on Mobile Computing*, 5, 1029–1043, 2006.
8. B. Gedik, K.-L. Wu, P. S. Yu, and L. Liu, Processing moving queries over moving objects using motion-adaptive indexes, *IEEE Transactions on Knowledge and Data Engineering*, 18(5), 651–668, 2006.
9. J. Gao, L. Guibas, N. Milosavljevic, and J. Hershberger, Sparse data aggregation in sensor networks, in *Proceedings of the 6th International Conference on Information Processing in Sensor Networks* (*IPSN 2007*), Cambridge, Massachusetts, April 25–27, 2007, pp. 430–439.
10. H. Huang, J. H. Hartman, and T. N. Hurst, Efficient and robust query processing for mobile wireless sensor networks, in *Proceedings of the Global Telecommunications Conference* (*GLOBECOM* 2006), San Francisco, California, November 27–December 1, 2006, pp. 1–5.
11. A. Kamra, V. Misra, and D. Rubenstein, CountTorrent: Ubiquitous access to query aggregates in dynamic and mobile sensor networks, in *Proceedings of the 5th International Conference on Embedded Networked Sensor Systems* (*ACM SenSys 2007*), Sydney, Australia, November 2007, pp. 43–57.
12. Y. Zhang, B. Hull, H. Balakrishnan, and S. Madden, ICEDB: Intermittently-connected continuous query processing, in *IEEE 23rd International Conference on Data Engineering* (*ICDE 2007*), Istanbul, April 15–20, 2007, pp. 166–175.
13. N. Trigoni, Y. Yao, A. Demers, J. Gehrke, and R. Rajaraman, Multi-query optimization for sensor networks, in *IEEE International Conference on Distributed Computing in Sensor Systems* (*DCOSS 2005*), Marina del Rey, California, June 30–July 1, 2005, pp. 307–321.
14. R. Stoleru, J. A. Stankovic, and S. Son, Robust node localization for wireless sensor networks, in *Proceedings of the 4th Workshop on Embedded Networked Sensors* (*EmNets 2007*), ACM, New York, June 25–26, 2007, pp. 48–52.
15. A. Guttman, R-trees: A dynamic index structure for spatial searching, in *Proceedings of the ACM SIGMOD International Conference on Management of Data*, Boston, Massachusetts, June 1984, pp. 47–57.
16. F. Ye, H. Luo, J. Cheng, S. Lu, and L. Zhang, A two-tier data dissemination model for large-scale wireless sensor networks, in *Proceedings of International Conference on Mobile Computing and Networking* (*MOBICOM 2002*), Atlanta, Georgia, September 2002, pp. 148–159.
17. H. S. Kim, T. F. Abdelzaher, and W. H. Kwon, Minimum-Energy Asynchronous Dissemination to Mobile Sinks in Wireless Sensor Networks, in Proceedings of the First International Conference on Embedded Networked Sensor Systems (*SenSys 2003*), Los Angeles, California, November 5–7, 2003, pp. 193–204.
18. D. J. Abadi, S. Madden, and W. Lindner, REED: Robust, Efficient Filtering and Event Detection in Sensor Networks, in *Proceedings of the 31st International Conference on Very Large Data Bases* (*VLDB 2005*), Trondheim, Norway, August 30–September 2, 2005, pp. 769–780.
19. UCLA Parallel Computing Laboratory, GloMoSim: Global mobile system simulator. Accessed http://pcl.cs.ucla.edu/projects/glomosim/ on Jan 31, 2012.
20. Y. Wu, L. Zhang, Y. Wu, and Z. Niu, Interest dissemination with directional antennas for wireless sensor networks with mobile sinks, in *Proceedings of the 4th ACM Conference on Embedded Networked Sensor Systems (Sensys 2006)*, Boulder, Colorado, October 31–November 3, 2006, pp. 99–111.

Chapter 5

Service Collaboration Protocols in Mobile Systems

Jingli Li and Bin Xie

Contents

5.1 Introduction

To enable ubiquitous service for mobile users, the Internet and various wireless networks collaboratively interact with each other in such a way that a group of mobile devices such as laptops, personal digital assistants (PDAs), and smart phones could interactively perform a service and they receive the Internet services anywhere and anytime. For such a purpose, extensive research has been conducted for developing flexible, efficient, scalable, user-friendly collaborative applications and services in mobile systems. A number of mobile middleware protocols have been proposed to facilitate collaborative activities and corresponding architectures to support these services. Figure 5.1 illustrates an application example in the distributed enterprise network environment where service collaboration allows flexible mobile services for enterprises. As shown in Figure 5.1a, the distributed enterprise network environment could span across several local Internet or wireless network domains that are located at companies (e.g., Company A, Company C), public domains (e.g., airport), and enterprise factory, or other sites. Figure 5.1b depicts more details of the infrastructure (e.g., servers A, B, wireless access points (AP)) and mobile users involved in a collaborative application that benefits sales operations. We envision the example interactions of service collaborations subsequently.

Suppose a traveling salesman in Company A receives a request from a potential customer from Company B when he is currently at an airport. The salesman uses his PDA to reserve/notify a group of people in Company A for a meeting. The meeting attendees are the salesman, the manager, the marketing clerk, the design engineer, the quality engineer in Company A, and the customer in company B. On receiving confirmation of everyone's availability through his mobile devices or laptops, the manager initializes a video conference and sets up a shared workspace. Everyone joins the meeting online. The manager first introduces the company, and the marketing and sales clerks introduce the company's wireless product. The quality engineer in Company A takes a picture of operating test equipment and shares the digital photo with the customers. The design engineer also exhibits the electronic version product brochure and manual. The customer then reviews the brochure and marks up some items for discussion. Owing to

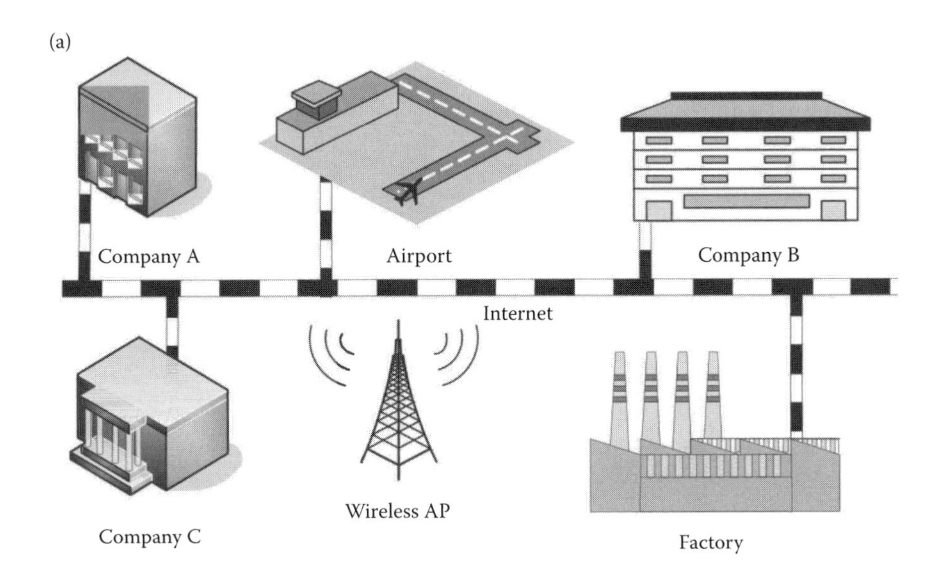

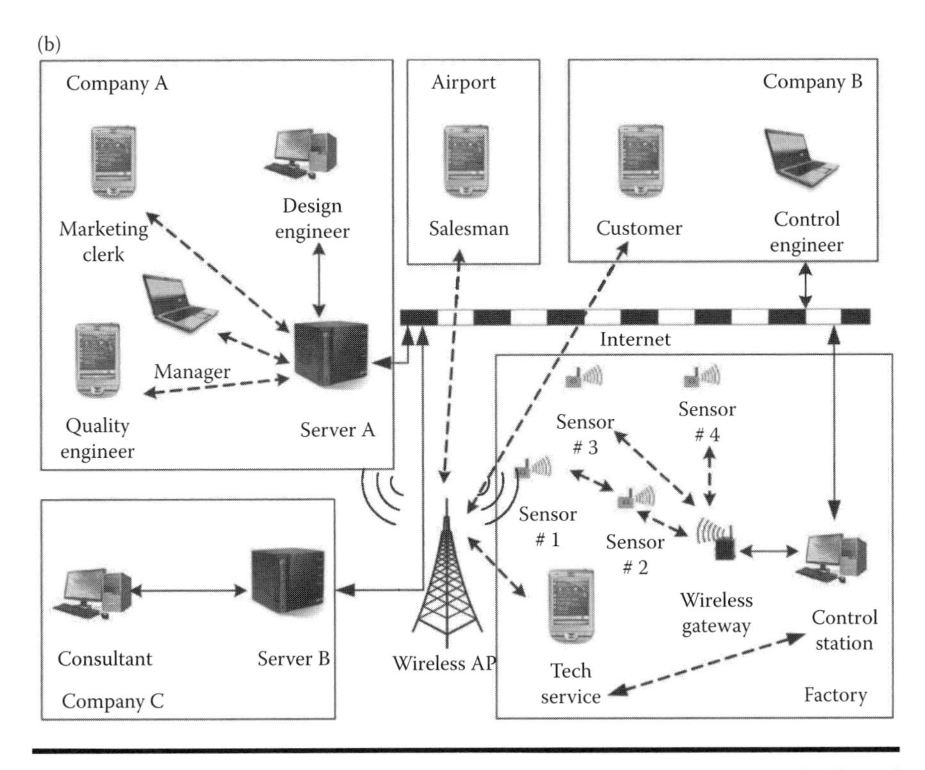

Figure 5.1 A service collaborative application for enterprises. (a) Distributed enterprise network environment. (b) Service collaboration among enterprises.

issues about technical details, the customer can instantly invite a quality control engineer in his company to join the conference and the design engineer in Company A can invite a consultant in Company C to join the conference. The consultant retrieves the relevant data from server B and shares it with everyone. The manager wants to demonstrate to the customer how to operate the wireless products in the field. Thus, he invites the technical service people in his company to join the meeting. After this, he leaves the conference temporally. The technical service person is working on a customer site at a factory as the wireless products (e.g., multiple wireless-enabled sensors and a gateway) of Company A are implemented in this industrial field. He uses his PDA to collect device data and process information in a security-enabled system. He then records a video of the operating devices and everyone watches the real-time video. After a group discussion, the customer is interested in the product. The salesman logs into the database in server A to check inventory, and uses his PDA to create a price quote. The service ends after the customer has reviewed the quote, signed it, and placed an order. In the above application example, service collaboration allows a set of service processes (e.g., conference meeting, production database access, data transfer) to run over one or more machines and mobile devices to interactively access the heterogeneous network and databases, and communicate with each other. In the above example, we can see that service collaboration offers the following capabilities to users:

- *Communication with Mobility*: The communication for a group of users (e.g., the salesman, manager, and customer) is the basic requirement for service collaboration. The customer at the airport can access application functions, and resources in wireless networks, such as cellular or WiFi networks, ensure communication. Wireless communication allows the services (e.g., accessing the database, participating in the conference discussion, browsing the video) to smoothly migrate once a network is disconnected and an alternative network is available.
- *Interoperability and Transparency*: It supports reliable data sharing/accessing and information transmission/reception, spanning different platforms if necessary. It effectively provides interoperability to support distributed architectures in the mobile environment while the operating complexity is effectively hidden from users. Service collaboration allows smooth interactions among various services with minimal coordination cost from users' perspective. Therefore, users only need to focus on what they want to do without considering the implementation of the services even when a service is deployed over Web servers, application servers, and database servers from a laptop or from a moving mobile terminal.
- *Service Initiation and Publication*: Service collaboration also has the ability to initiate a conference service with unicast and multicast communications. The conference offers interoperable service for multiple users and supports

impromptu services. In the conference, users share and access reliable data in a distributed and synchronized way. In the above application, users can initiate a number of services (e.g., conference, inventory checking, and personal communication) sequentially or simultaneously without interrupting each other. The system can provide flexible services for both synchronous and asynchronous services. In addition, service collaboration also enables fast initialization and easy reconnection/rejoining to services.

- *Context-Aware and Scalability*: The service collaborating system is able to deliver services with location-aware, synchronization-aware, device-aware, and connection-aware properties. The scalability means the service is scalable to the number of users and devices that participate in single or multiple services at the same time.

In addition to the above basic issues, some other requirements such as security and efficiency are necessary in service collaboration. In this chapter, we investigate the current progress in developing service collaborating middleware protocols that aim to offer different service collaborations in different network environments. We first illustrate the network environment for supporting the mobility of users when they access collaborative services by different network connections. In all these network environments, the mobility and mobile network environment renders some unique design problems in service collaboration. Second, we depict service collaboration in two types of structures: service collaboration with infrastructure and service collaboration without infrastructure. The infrastructure established in collaborative applications has the Internet or wireless backbone (e.g., cellular) to facilitate service collaboration. On the other hand, applications in the mobile *ad hoc* network (MANET) have no infrastructure in service collaborations. For each type of structure, we illustrate some mobile service middleware protocols that allow us to understand how to implement service collaboration for different applications.

The rest of this chapter is organized as follows. Section 5.2 discusses the network environment and design issues for service collaboration. Section 5.3 illustrates current state-of-the-art collaborative service models and their supported applications. Opening issues are further discussed in Section 5.4. Finally, Section 5.5 concludes the chapter.

5.2 Network Environment and Design Issues for Service Collaboration

Service collaborations for users occur in two general types of network environments: wired network with mobility and wireless network with mobility, where users are mobile and could access the services with mobility. In the first network type, users directly connect to the Internet by a wired connection. In the second network type, a specific wireless network (e.g., on-campus network) is the backbone for service collaboration and Internet access.

- *Wired Network with Mobility*: At first, service collaborations need to be supported on the Internet or computer networks with mobility support in such a way that a mobile device such as a laptop is able to ubiquitously access the rich Internet services. In this case, the mobile device connects to the Internet with cables. To provide flexible service capability, collaborative protocols such as multimedia middleware should be associated with the mobile IP (e.g., mobile IPv4 and IPv6). The mobile IP allows roaming of the mobile device with Internet configuration. On the other hand, the middleware monitors the context change when the mobile device moves from one subnet to another and allows the mobile device to transparently adapt to the new network service environment. Therefore, users only concentrate on what services they want, instead of how to configure the network and service environments, for example, searching and subscribing to a service.
- *Wireless Network with Mobility*: In addition to cable (e.g., wired connection to the Internet), wireless communication has been an indispensable way for users to connect and subscribe to a variety of Internet services. Wireless infrastructure is the Internet gateway with a wireless access point such as the base station of the cellular networks. Wireless communication technologies typically include wireless WANs (wide area networks, e.g., 3G or beyond cellular networks), wireless MANs (wireless metropolitan area networks, e.g., IEEE 802.16), WLANs (wireless local area networks, e.g., IEEE 802.11a/b/e/g/n), and WPAN (wireless personal area networks, e.g., Bluetooth piconets). Service collaborations can be deployed in a wireless network without infrastructure support, such as point-to-point and multipoint communication. Motivated by shared common interests, the collaborative operations occur anywhere and anytime, using mobile computing devices with wireless communication capabilities. It allows data (text, graphics, image, speech, and video) synchronization/exchange between multiple mobile devices. Moreover, the activities and interactions in the mobile group are dynamic and spontaneous.

In addition to the ability to operate independently, the wired network and these wireless networks should be effectively used by a group of collaborators to pursue a common goal and offer their underlying benefits [1]. In terms of the network environment, some unique design issues for service collaborations in developing mobile collaboration applications are:

- *Hardware Limitation*: For carry-on handheld or pocket devices, small size and lightweight are the major concerns for the comfort of customers. Unless there is a big breakthrough for hardware, these aspects only allow battery-powered mobile devices to have small display screen, constrained computation capability, short battery life, and limited storage space. According to this, applications may prefer to choose modest algorithms, unattended coordination/management mechanisms, and special information display patterns to fit into the small screen size without jeopardizing device functionality and user accessibility.

- *Device Heterogeneity*: As users expect the accessibility to the increased collaboration support anytime and anywhere, the platform can reside in both stationary and mobile devices of miscellaneous types. Devices such as desktop computers, data/system servers, and personal notebooks have a file system, rich hardware resources (large amounts of memory, a powerful central processing unit—CPU, constant power supply), and both wired and wireless communication capabilities, whereas other devices such as global positioning system (GPS) and PDAs have only wireless communication capabilities and constrained hardware resources. Functionalities for server and client devices may vary on different device types to achieve optimized resource allocation and best performance. For example, sophisticated services and computation should be only implemented in devices that are equipped with large storage capacities and computational resources to undertake data analysis and presentation.
- *Network Characteristics*: The underlying networks for collaborative services are distinguished by characteristics such as communication range, data transmission rate, delay, network topology, and so on. These features will influence the function of collaborative services in many aspects. For instance, if a video conference is operating between 3G mobile phones and WLAN-based notebooks, the quality of the video may be compromised due to the bandwidth restriction.
- *Intermittent Connectivity*: Unstable wireless connectivity is a common characteristic of collaborative applications. The network coverage may be unavailable in a short term as users are mobile and they may move outside the radio range. Weak connectivity caused by interference can make a user temporarily unavailable to communicate with others during the collaboration process. In mobile collaborative systems, tasks are initialized dynamically and spontaneously, and users can choose to join or leave a group randomly. Collaborative services must be able to deal with irregular activity and handle both intended disconnection/reconnection or accidental disconnection.
- *User Mobility*: Mobility is a double-edged sword for collaborative services. On one hand, it is one of the main characteristics and advantages of mobile services. On the other hand, it renders complexity in designing a collaborative service. As users are on the move to carry out their activities, they change locations randomly or constantly for different purposes. In location-aware applications, information about user locality and proximity is required to be accurately updated as it is pertinent in determining an operation and the participants. In MANET, users' mobility changes the network topology and a good routing mechanism is required for frequent evaluations of better paths between a consumer and a provider.

Service collaboration paves the way and eases the interaction for industrial/academic applications such as entertainment, social network, business cooperation, field device monitoring, construction management, traffic control, disaster relief, health care, and so on. For example, a business transaction may involve the cooperation of different parties. In construction management, builders may frequently exchange messages, data, or drawings on the construction site. Field engineers in the process control industry use mobile devices to connect to engineering data and collect information for field instrument/device diagnosis and maintenance. Certain employees in large global enterprises perform daily work (sending/receiving message or email, participating video conference, sharing business documents) using a PDA or laptop anytime and anywhere as required. A fire brigade arrives at the scene of a fire, initiates a dynamic formation of the emergency response group, and works together with laptops and PDAs. The diversity of mobile collaborative applications increases the complexity of protocol development and it is hard to satisfy various requirements of services using a unified design. Mobile collaborative applications are generally case specific because of different scenarios and various goals. In other words, a generic architecture cannot meet the vast variety of possible collaboration and interaction patterns. Collaborative services can be context oriented, location oriented, object oriented, service oriented, message oriented, and so on. With these observations, the following sections illustrate the service collaboration protocols according to the types of network environment.

5.3 Collaborative Architectures and Protocols

According to their structures, we divide the mobile service collaborating protocols into two categories at a high level as shown in Figure 5.2. The first category, infrastructure-based service collaboration, includes the Mobile Collaboration Architecture (MoCA) [2,3], the Yet Another Collaboration Environment (YACO) [4], the Supporting Mobile Collaboration Services Platform (SMCSP) [5], ActiveCampus [6,7], the Service-Oriented Architecture Framework for Collaborative Services (SOA-CS) [8], the Service-Oriented Architecture for Mobile Collaboration (SOA-MC) [9,10], and a Java framework Sync [11]. The Internet and wireless infrastructure are available for service or communication collaboration. The second category is a set of mobile service collaborating protocols that are infrastructureless. Most of these protocols are deployed in the MANET with the Internet and wireless infrastructure such as base stations. They are the Allocation and Group-Aware Pervasive Environment (AGAPE) framework [12], the Communication and Coordination Patterns (CCP) established in References [13–15], the collaborative peer-to-peer platform Proem [16], and the location-aware event-based middleware Scalable Timed Events and Mobility (STEAM) [17–19].

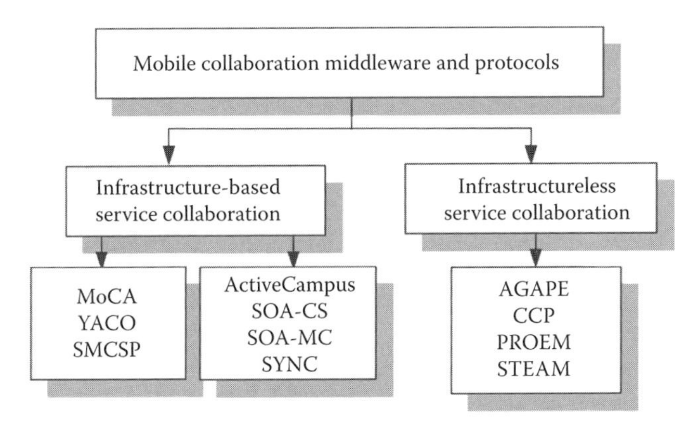

Figure 5.2 Classification of service collaboration schemes.

5.3.1 Infrastructure-Based Service Collaboration

5.3.1.1 Mobile Collaboration Architecture

Sacramento et al. [2,3] present a MoCA design as a mobile collaborative middleware for context-aware collaborative applications for mobile users. The middleware allows a mobile user to roam in the wireless network (e.g., 802.11 wireless networks or cellular) while accessing various services that are implemented by the Internet infrastructure (e.g., an application server). In particular, MoCA consists of client and server application programming interfaces (APIs), core service stack, and an object-oriented framework for instantiating customized application proxies as illustrated as below. Figure 5.3 shows these components as described by Sacramento et al. [2,3].

- *Client and Server APIs*: For a collaborative application, the client and server APIs support synchronous or asynchronous, message-oriented or sharing-oriented, communication. In addition, the monitor that is a daemon executing on each mobile terminal collects data and sends data to the context information service (e.g., step 4 in Figure 5.3).
- *Core Service Stack*: The core service stack has a design of four service components: discovery service, location information service, context information service, and configuration service. Figure 5.3 shows these parts in the core service stack. The core service stack allows MoCA to monitor and infer the context of the mobile terminal. The basic functionalities of these four components are as follows. The discovery service stores the information of application or service that has registered with the MoCA middleware. The location inference service infers users' geographical location. The context information service is a distributed service used to process the mobile terminal's context

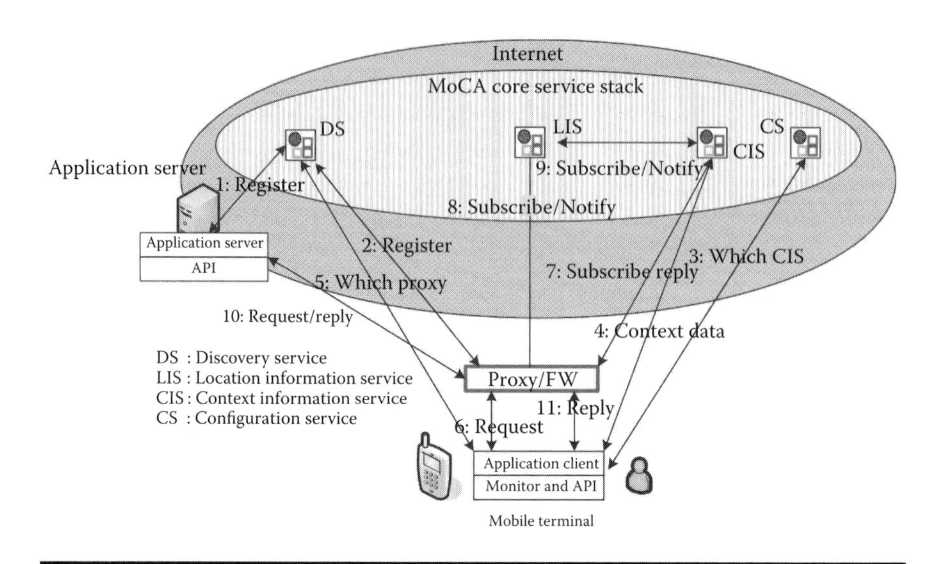

Figure 5.3 Interaction sequence chart for a collaborative application. (Adapted from V. Sacramento et al., *IEEE Distributed Systems Online*, 5(10), 2, 2004; V. Sacramento et al., In *Proceedings of the 13th IEEE International Workshops on Enabling Technologies: Infrastructure for Collaborative Enterprises (WETICE '04)*, Modena, Italy, pp. 109–114, June 14–16, 2004.)

information, and functions on the nodes of the wired network. The configuration service stores and manages configuration information.

- *Object-Oriented Framework*: The ProxyFramework (i.e., Proxy/FW in Figure 5.3) is a design of the object-oriented framework. Its goal is to allow instantiating customized application proxies. The ProxyFramework offers the application developer a simple mechanism to access context information. Furthermore, the proxy can distribute the application-specific processing among the server and its proxies if a task (such as data compression, protocol conversion, user authentication) requires significant processing effort.

Figure 5.3 also illustrates the MoCA interaction sequence chart among these above components for a collaborative application. We use the example provided in References [2,3] to illustrate the interactions for implementing a collaborative application. At first, the application server has to register itself at the discovery service (i.e., step 1 in Figure 5.3). In this process, the discovery service records the name and the properties of the collaborative services available at the application server. In the same way, each proxy of the application receives a copy of the service information (i.e., step 2 in Figure 5.3). Hereafter, the mobile client can query the discovery service to ascertain how to access the collaborative service from its visiting network (i.e., the closest proxy). The monitor in the mobile client first connects to the configuration service to obtain the knowledge of context information service (i.e., step 3). Then, it periodically sends its context information (e.g., local resource and radio signals) to the context information service (i.e., step 4).

Once the mobile client discovers a desirable collaborative service from the discovery service (i.e., step 5), the client starts sending requests to the application server. In this process, the ProxyFramework is responsible for forwarding all the requests (i.e., step 6) after applying the application-specific adaptations. Suppose the proxy subscribes with an interest expression (i.e., step 7), for example, {"FreeMem < 10KB" OR "APChange = True"}. The context information service then generates a notification for the subscription. When the context information service receives a device's context information (from the corresponding monitor), it checks whether this new context evaluates any stored interest expression to true. If so, the context information service generates a notification and sends it to all proxies that have registered interest in this change of the device's state.

The context information service enables the mobile terminal to receive its application while it roams from one access point to another. To support mobility, the application requires the location information of the mobile terminal. The client registers with the location information service (i.e., step 8). This subscribes to the context information service (i.e., step 9) to receive periodic updates of the mobile terminal's radio signals. The location information service uses these signals to infer the device's location and send the corresponding notification to the application proxy. When the application server receives the client's request, the request is processed and the application service sends a reply back to some or all proxies (i.e., step 10). The proxies compress or filter the reply, or perform other processes on the reply according to the new state of the corresponding mobile terminal's wireless connection. The implementation of the context-specific adaptation depends on the collaborative application's specific requirements. For example, if the proxy knows that the quality of a mobile device's wireless connectivity has dropped below a certain threshold, it could temporarily store the server's replied data in a local buffer. Furthermore, the proxy can determine when and how to deliver data to the client, on the basis of other contextual information, such as the device's location or proximity (i.e., step 11).

MoCA offers generic and extensible infrastructure for context-oriented services and collaborative applications development. The implementation of two context-aware collaborative applications is discussed in this framework: W-Chat (wireless chat) and NITA (notes in the air).

W-Chat is a chatting program that can diffuse the connectivity status of chatroom participants. It utilizes the connectivity information from the context information service so that the reconnected user can obtain all messages posted during the disconnection time. The proxy intercepts the messages between the client and the application server and has a local message buffer for each client. W-Chat's client and the proxy can synchronize the reconnected users' states. A special feature of W-Chat is the collaboration awareness. Suppose one client moves outside the wireless coverage, a status icon will appear on the mobile terminal to indicate this user is disconnecting from other users. The MoCA design achieves efficiency as it effectively reduces W-Chat's complexity.

NITA is a location information service-based message retrieval service. In NITA, users post text messages and files to a symbolic region and define how long the message will be available and who is the authorized recipient. The client can search for available NITA servers, their regions, and visible users in the regions. Receivers can set their visibility to others and choose the messages they want to receive and when to display the message. NITA proxy queries the location information service about the area's structure, registers interest in its clients' location, and manages clients' profile to make the system scalable.

5.3.1.2 Yet Another Collaboration Environment

The YACO [4] framework is a design based on SIENA (Scalable Internet Event Notification Architectures) [20,21] and MOBIKIT [22]. SIENA is a distributed and content-based framework for event notification services to maximize both expressiveness and scalability. MOBIKIT is a mobility service toolkit based on proxies to support service for mobile publish/subscribe applications. The integration of them in the YACO offers an event-based system for mobile collaborative services among corporate domains. Considering an enterprise network, the YACO allows cooperation and sharing of expertise between coworkers across local network domains where the mobility frequency of coworkers across domains becomes high in the enterprise operations. In the enterprise, the servers (e.g., file servers, database servers) are located at different domains over the company network, but they should be able to offer services to users in spite of their physical domains. In each domain, there are stationary users such as desktops and workstations used for database management and concurrent versions systems. Employees working at any location with their PDAs or notebooks are mobile (e.g., across the domain) while they want to stay connected with the network to continue their work. The example collaborative services offered in such an enterprise environment include messaging, distributed search of artifacts, file transfer service, user discovery, mobile and stationary users, additional entities, and so on.

On the enterprise network, YACO consists of various communication and application servers and a number of peers (i.e., mobile and stationary users). YACO servers provide communication services by means of SIENA access points and MOBIKIT proxies. Every peer has the same software architecture to access the YACO standard service from the application servers. Figure 5.4 shows the YACO software architecture that consists of the following components:

- *YACO API*: YACO provides the API functions for accessing the local repository, distributed search, and message publishing and receiving. The application servers are logically located above the YACO API component such that the user can access the services by invoking the API functions.

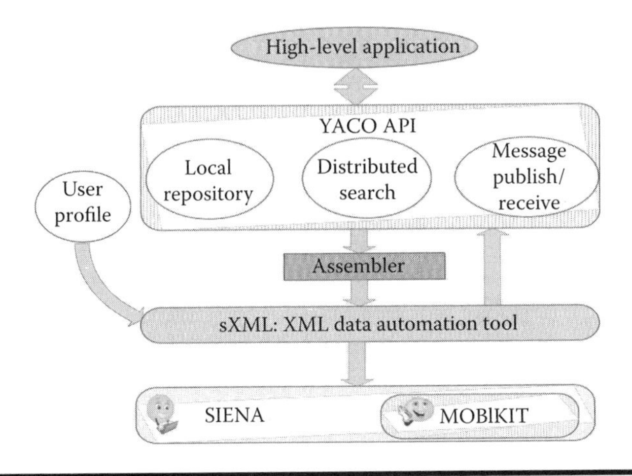

Figure 5.4 YACO architecture.

- *Assembler*: An assembler component assembles information and passes it to the sXML [23] module.
- *sXML and XML*: The sXML module provides the way to communicate between YACO and SIENA. Moreover, it uses an extensible markup language (XML) definition for user and document profiles.
- *SIENA*: The SIENA layer is for sharing artifacts and exchanging messages.
- *MOBIKIT*: A MOBIKIT component provides mobility support to mobile peers. The MOBIKIT enables clients to disconnect and reconnect to the system without losing interesting events. Using the MoveOut operation in MOBIKIT, a client can inform a server of its disconnection. A reconnected client can invoke the Moveln operation, which replays all the events missed during the disconnected period. MOBIKIT increases the reliability for high-mobility users.

To provide flexibility for applications, YACO uses a publish/subscribe system as the communication protocol, which enables users to be location unaware [24]. Publish/subscribe is an asynchronous messaging mechanism where the publishers (i.e., senders) of messages are not programmed to send their messages to specific subscribers (i.e., receivers). This decoupling of publishers and subscribers allows for high scalability and a more dynamic network topology. This is different from the traditional way in which the sender specifies the intended receiver of the message (e.g., the name or the IP address of the destination). On the contrary, the published messages could be classified (e.g., according to content) without knowing what subscribers there may be. By subscription of interest message types, the subscribers only receive the messages of interest with no concern of who are the publishers. YACO uses SIENA for its scalability, simplicity, and light weight and MOBIKIT for its mobility support to mobile peers. In SIENA, each publisher posts notifications through its SIENA access point, and the subscribers use their SIENA access points to subscribe to the filtered notifications. The filters in SIENA use a set of operators to evaluate the attributes of notifications and to find the match of the interest and the posted content. In MOBIKIT, the mobility service is implemented by a stationary component named mobility proxy objects. These objects run at the SIENA access points of the publish/subscribe system. For a disconnecting client, a MoveOut function in its local mobility interface will transfer the stored subscriptions to its mobility proxy. The proxy subscribes and stores all the messages in a dedicated buffer. On reconnection of the network, the client uses a Moveln function to pass the address of its original proxy to a local proxy. In this way, the buffered messages can be transferred to it without message loss. In YACO, users refer to each other by the user ID instead of the host address. This offers the flexibility for mobile users to move from one domain to another by getting temporary addresses instead of their fixed network addresses.

5.3.1.3 SMCSP: A Multiagent-Supported Adaptive Mobile Collaborative Service Platform

SMCSP [5] is developed to support collaborative services for multiple mobile users. It uses the flexible layered configuration structure, the multiagent technology, and the context-aware module to increase adaptability in the dynamic network environments. Furthermore, its service collaboration is implemented according to cooperative semantic granularity, cooperative task partition, and task assignment.

Table 5.1 shows the layered configuration structure that consists of the following five components from the higher application layer to the lower layer in that order:

- *Application Layer*: This layer consists of a service management module and a number of collaborative services such as the mobile electronic commerce service, the mobile collaboration police service, and the mobile entertainment service. All these collaborative services are managed by the service management module. The services can come from the SMCSP or can be obtained from a third party. If they are obtained from a third party, the service profile, service model, and service grounding need to be registered before they join the platform.
- *Mobile Collaborative Application Supporting Layer*: This layer has the functionality of collaborative task partition and allocation.
- *Optimized TCP/IP Protocol Layer*: This TCP/IP (Transmission Control Protocol/Internet Protocol) protocol layer is to alleviate the network traffic and provide effective information transmission.

Table 5.1 Structure of Supporting Mobile Collaboration Services Platform

Layer	*Function*
Application layer	E-commerce, mobile collaboration police service, mobile entertainment, stock
Mobile collaborative application supporting layer	Task partition and allocation
Optimized TCP/IP protocol layer	Information transmission and traffic alleviation
Trusted infrastructure layer	Security mechanism
Communication network layer	Physical layer for both wired and wireless devices

- *Trusted Infrastructure Layer*: Security mechanisms are implemented in this layer.
- *Communication Network Layer*: This is the physical layer for both wired and wireless devices.

According to personal interest, a group of mobile users can form a dynamic collaborative group structure. To implement this in a mobile cooperative system, users are divided into several groups first. As shown in Figure 5.5, there are three groups

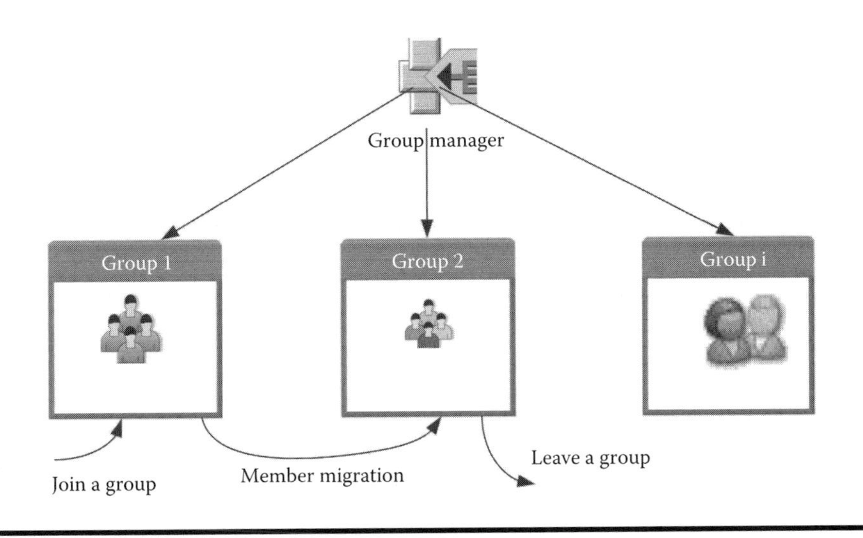

Figure 5.5 Dynamic collaborative group structure.

for three different interests. Each group has multiple users and each user is represented by an agent in the mobile cooperative system. A member can join or leave a specific group. In SMCSP, the cooperative semantic granularity is divided into the following three collaboration levels:

- *Group Collaboration Level*: This level indicates the groups that include multiple users who carry on cooperative work for the common goal.
- *Member Collaboration Level*: This level indicates users in a specific group.
- *Realization Collaboration Level*: This level allows the collaboration among multiple agents, which implements the system function.

The cooperative mechanism of a multiagent plays an important role in the SMCSP. In the multiagent system (MAS), each mobile client is represented by a mobile collaboration agent that interacts with the MAS through the message agent and the graph information agent. The partitioned mobile collaboration agent platform management level and application service level in MAS consist of various agents. There are the infer agent, knowledge agent, user information agent, collaboration management agent, directory of file agent, and ontology language agent. SMCSP further implements an agent management system to manage the runtime states of all agents in the MAS system. SMCSP considers a hybrid network system with both wired and wireless devices such as mobile client terminal units and platform system servers. In this prototype, the MAS that supported mobile collaboration services was implemented to represent each mobile client terminal unit and platform system server. Third-party service can also run on the platform system servers. SMCSP aims to be simple, adaptable, extensible, and easy to implement; it supports collaborative services with desirable performance. For this purpose, this application consists of the following three modules: (i) mobile and collaborative management module to perform task division, assignment, and coordination; (ii) group management module for group construction and member management; and (iii) context-aware module for collaborative context information management.

5.3.1.4 ActiveCampus Context-Aware Infrastructure

ActiveCampus [6,7] is a collaborative service system at the University of California, San Diego, as a suite of personal services for sustaining an educational community. On the on-campus wireless network, ActiveCampus uses a context-aware infrastructure to provide location-based services for academic communities. For easy administration of the system and simple applications for the end-user devices, which have limited computation capability and resources, ActiveCampus adopts a centralized client–server approach. In this approach, the server supports all internal functions whereas the end-user devices only sense the context and provide a user interface. This approach greatly reduces the computation load and relieves the program development burden on sensing and display devices. Moreover, the centralization approach achieves task allocation and component organization easily. ActiveCampus architecture consists of two parts: the interfaces to external components and the internal system architecture. The interfacing component on the user side runs on the end-user device to capture and transmit data. The interfacing component on the other side runs on the server to receive data and organize it into a database. Figure 5.6 shows the application that can capture user location and display it on a map. Note that in Figure 5.6, the solid arrows represent the calls and the dashed arrows denote the event notifications. As illustrated in Figure 5.6, the internal component architecture is composed of the following five layers:

- *Device Layer*: Components residing in devices of this layer connect to ActiveCampus through XML.
- *Environment Proxy Layer*: The proxy component transfers data (e.g., wireless signal/map information) between devices and ActiveCampus's internals through XML or HTML (Hyper Text Markup Language).
- *Situation Modeling Layer*: This layer synthesizes the situation of an individual entity from multiple available data sources. It interprets the sensor data to

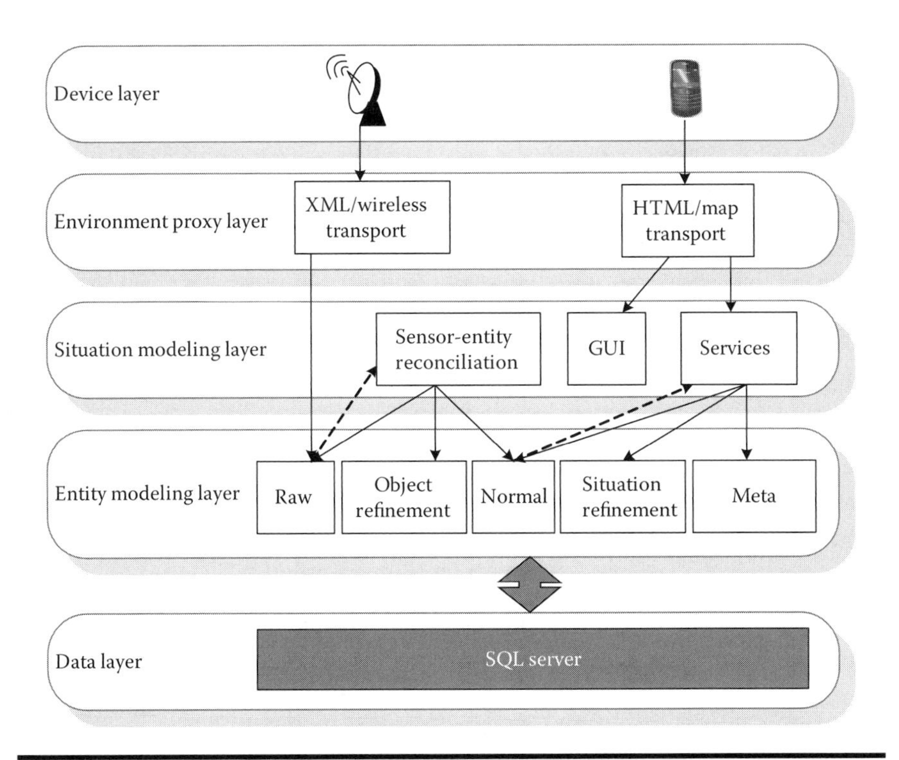

Figure 5.6 Active campus architecture.

internal forms and uses entities to represent them. It also consolidates entities for service execution and graphical user interface display.

- *Entity Modeling Layer*: This layer represents entities such as users and sensors in multiple forms. Raw external representations such as a dynamically assigned network address are first mapped to normal forms such as a permanently assigned unique integer through object refinement. Then, the normal entities are mapped to their rich presentation such as a screen name through situation refinement. Object refinement integrates data from multiple sensors and other sources to determine attributes such as position and kinematics. Situation refinement interprets the relationships among the objects and events in the context of the operational environment.
- *Data Layer*: This layer is used for data storage and retrieval. In this application, the SQL (Structured Query Language) database is established.

To meet the goals of extensibility, integration, and performance, additional architectural constraints and mechanisms are implemented in ActiveCampus to maintain a high level of integration and performance while decoupling services from modeled entities and services. These mechanisms include the following:

- *Entity Bloat*: Entity normalization and situation modeling methods are used to divide the entity object representation into a collection of context index identifiers. On one hand, four rules are used in entity normalization. Rule number one, only intrinsic data are part of the object. Alternate representations for intrinsic data must be stored separately. Rule number two, intrinsic data are represented in a compact normal form, which provides an index for fast retrieval, by alternate representations. Rule number three, all contexts are also standardized into indexable normal form representation. Rule number four, contexts are mapped using their indexed forms. By obeying these rules, new services can be added in a bumpless transaction. The representation and its mapping from the service are independent, which makes it reusable by other services without further coupling services. On the other hand, situation modeling processes data from multiple entities in a context to discern or present the situation of an entity. To manage different forms of entities, the sensor–entity reconciliation component in Figure 5.6 mediates the complex sensor–entity relationship by receiving change notification from the sensor, and propagates updates to appropriate entities.
- *Service Coupling/Decoupling*: A registration mechanism along with interface rules are used for service decoupling and introspection interaction is adopted for service coupling. Services are structured as follows. All services are designated to run on behalf of a subject, which denotes entity or user. Most services can also run on the accessible context of an entity called object. To support service cooperation, all service objects are required to support a compatible method and a renderID method. The compatible method performs a

type check to determine whether the types of the subject and object can support the service. The renderID method renders the target service's activation button without specifying the representation of that button.

- *Cache*: To improve performance, a scoped, context-indexed data cache is implemented. ActiveCampus applications can tolerate inconsistent and stale data. For example, it is not necessary to report data such as user's location and their buddy's location at the same time. So the data can be cached. Caching is performed through the memorization of method calls into the entity modeling layer. If a service invokes changes to the data, it must update the cache.

ActiveCampus makes life in university more convenient and helps everyone in campus stay connected. An example of this is the new student orientation. The map service in ActiveCampus will provide an outdoor or indoor map of the new student Sarah's vicinity, with buddies, sites, and activities overlaid as links at their location. The buddies service shows colleagues and their locations, organized by their proximity. With these services, Sarah can easily send a message to the buddy or quickly find her way to a nearby buddy.

5.3.1.5 SOA-CS: A Service-Oriented Architecture Framework for Collaborative Services

SOA-CS [8] is a generic coordination service framework in a distributed scenario that uses the service-oriented architecture (SOA) framework for collaborative services. SOA-CS organizes a number of services together and makes them available as collaborative applications. As another SOA-based system, SOA-CS is a distributed structure with built-in loosely coupled, extendible, flexible systems to integrate well with existing legacy systems. It has the functions of describing and publishing service, discovering a service, and consuming or interacting with a service. For example, three users can access a collaborative service at the same time as shown in Figure 5.7. Each user collaborates with other users through the user interface as

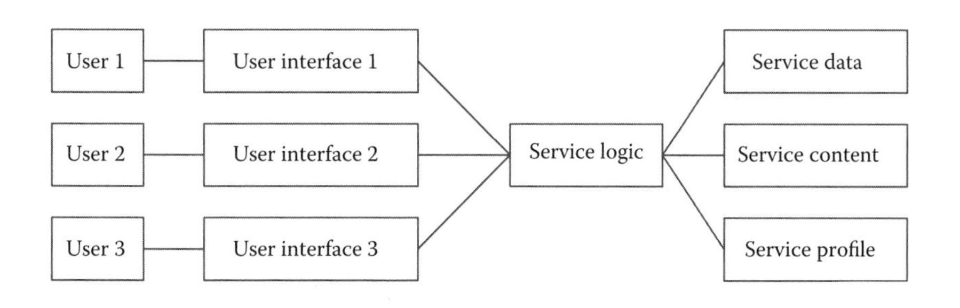

Figure 5.7 Architecture of a collaborative service.

shown in Figure 5.7. The service logic in Figure 5.7 is a special program that defines the dynamic behavior and functions of a service. As shown in Figure 5.7, SOA-CS decomposes a service by the following three components:

- *Service Data*: status, variables, and parameters used in execution of the services.
- *Service Content*: the data that are the product of service usage.
- *Service Profile*: the settings information that is related to users and devices.

The SOA-CS framework consists of three layers: the top layer is the application layer; the bottom layer is the service layer that is composed of various services; the middle layer is called the collaborative services layer. This layer has the following components:

- *Locking Mechanism*: It prevents data corruption during knowledge and resource sharing.
- *Presentation Control Mechanism*: This component collects input, assigns operation permission of a shared application to only one user at a time, and controls the output.
- *User Presence Management Component*: It implements registration, deregistration, login, logout, and options for users to subscribe or unsubscribe.
- *Organization Management Component*: This component allows users to add, remove, and assign rights to participants, and allows responsible users to define different collaborative activities.
- *Communication Control Mechanism*: This manages the communication between collaborators.

A collaborative application can be built by composing or by orchestrating these basic services and other selected services. Two general examples are shown in Figure 5.8. In the general example on the upper part of Figure 5.8, collaborative application X has a main function that invokes service A and service B clients. These clients call their respective services. An application is the chat room (application X), where there is a virtual space for users to share voice, text, multimedia, and so on. During an online chat between Tom and Mary, Tom (client A) uses a text processor (service A) to edit text to present to Mary, whereas Mary (client B) uses a photo editor (service B) to make an image to share with Tom. In the general example on the lower part of Figure 5.8, collaborative application Y is an orchestration of service C, service D, and service E. An application is the international Web conference scheduling process. The scheduling application (application Y) first invokes a user management service (service C) to determine who will present the conference, then it checks the user availability for a best time through the calendar service (service D), and after that, it sends out a meeting schedule and outline to all participants using the mail service (service E).

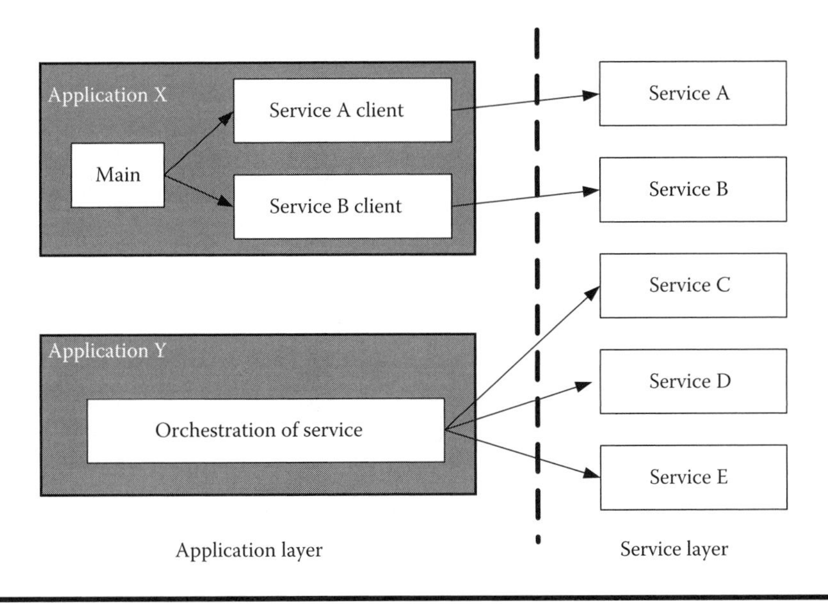

Figure 5.8 SOA-CS collaborative services.

5.3.1.6 SOA-MC: A Service-Oriented Architecture Framework for Mobile Collaboration

SOA-MC is a lightweight SOA-based architecture [9,10] for mobile devices. It aims to achieve efficient mobile data exchange, while being timely, robust, and easily accessible to service-oriented architecture. For this purpose, it addresses the transparency between the connected, occasionally connected, and disconnected modes. Considering the intermittent wireless connection, SOA-MC offers services for asynchronous connections between back-end systems and devices to maintain functions in the occasional and disconnected modes.

To promote business communication to a new level by means of architectural flexibility, lightweight design, and enhanced support, the SOA-MC architecture (as shown in Figure 5.9) consists of the following components:

- *Services*: Services such as email and instant messaging are wrapped together by means of proxy components to provide core collaboration services. The composite services orchestrate the core services using workflow process scripts.
- *Middleware*: The middleware server provides Web service invocation, data exchange, transformation, and interfacing with the users adopting components such as service invocation broker, proactive cache, XML compression, connection management, and staging database. In detail, it uses a compressed XML format to transfer and store data, which increases the

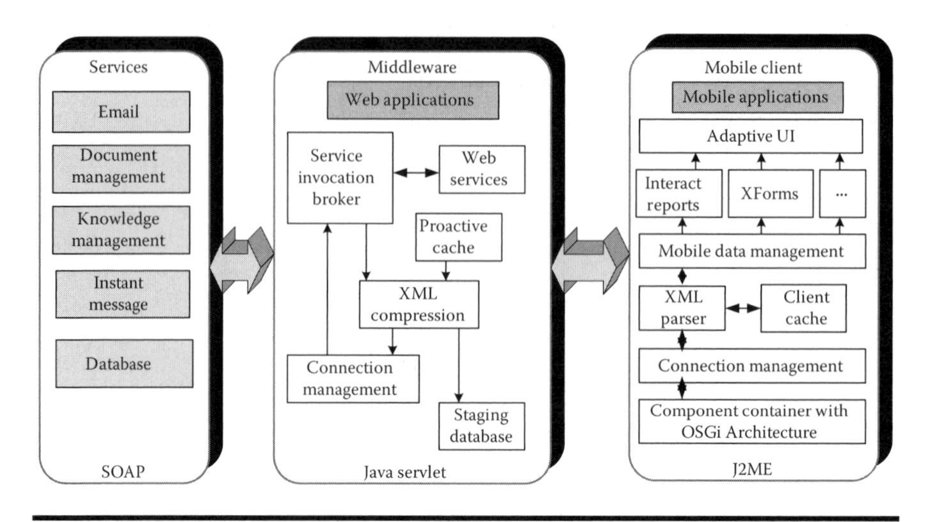

Figure 5.9 SOA-MC collaborative framework.

efficiency of SOA-MC services by using XML compression for objects. It also performs proactive data push on the basis of the client's service invocation history.

- *Mobile Client*: The client is supported by multiple mobile application components based on J2ME [25] technology. SOA-MC develops an approach that uses J2ME midlet that asynchronously exchanges information with the server using highly compressed XML format and client cache, under mobile data and connection managements. It is proved that the client application supports a complex interaction without requiring a powerful smart phone. To achieve a light application, neither the business logic nor the user interface forms are hard-coded in the client application. On the contrary, abstractions of open industry standards such as XForms are implemented in the client application for easy modification or augment at a low cost. The open services gateway initiative (OSGi) [26] framework is used in the mobile collaboration client. Services in the OSGi registry are special proxy components that bridge remote Web service invocations as local service calls to the mobile application components. With the OSGi-based architecture, new components can be easily created in the mobile client and integrated with Web services and features available on the server side and on the mobile devices, such as the phone camera, Bluetooth, GPS, and so on.

To provide flexibility for service composition, the collaborative services can be classified into the following two general types according to the service complexity as shown in Figure 5.10:

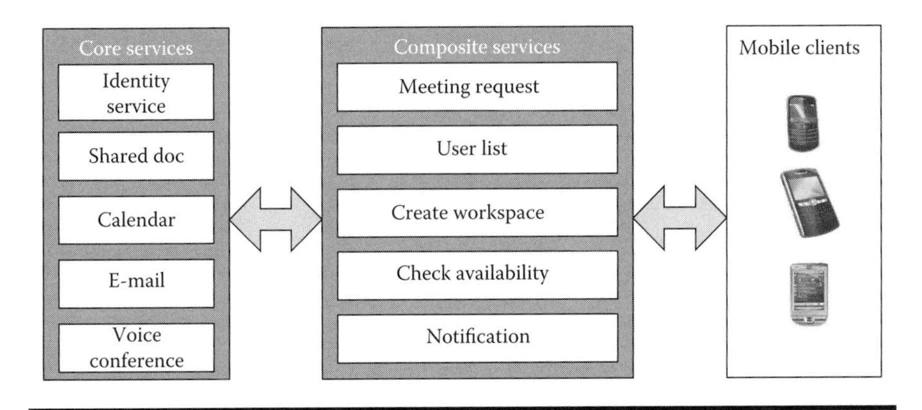

Figure 5.10 SOA-MC collaborative services example: A business meeting.

- *Core Services*: These services provide a generic agreed interface to business logic of common building blocks of people's interaction. The core services can be identity service, calendar, document sharing, voice conference, instant message, email, and so on.
- *Composite Services*: These services account for complex collaboration and interaction patterns. The composite services can be user list generation, document creation, participant selection, updates checking, editing, task list creation, meeting creating, document publishing, agenda creation, notification, and so on.

The core and composite Web services are distributed across the client and an auxiliary proxy server to implement the business logic. For a specific service, SOA-MC can integrate the core services and composite Web services to construct a complex service. Figure 5.10 shows an example of how a business meeting is processed. In the meeting planning stage, the meeting organizer uses the MeetingRequest service in his cell phone to request a meeting and setup a shared workspace. The UserList service provides a list of potential participants. Then, the Calendar service is invoked to check the availability of participants. In addition, the Notification service distributes the meeting invitations to participants using the Email service. The meeting-related documents are then sent to or shared with participants. The organizer will then set up voice and multimedia conference services. Even though asynchronous, synchronous, and parallel interactions are included in this complex scenario. By using the above flow, the results in References [9,10] show that the application can run well on regular cell phones.

5.3.1.7 Sync: A Java Framework for Mobile Collaborative Applications

Sync [11] is a Java framework based on object-oriented replication. It allows users to access and update the local replicas. The centralized asynchronous synchronization mechanism in Sync allows clients to connect and synchronize their changes at different times. Sync offers four levels of synchronization-aware classes. From low to high levels, they are default policies, table-based declarations, method overriding, and complete class declaration. Sync allows programmers to define conflicts and specify conflict resolution on the basis of the application's structure and semantics. To guarantee performance on low-bandwidth connections, Sync enables applications to share changes at a granularity as small as basic-type updates. To provide flexibility, Sync allows programmers to alter the synchronization behavior of Sync's replicated classes by overriding the method or creating completely new replicated classes by inheriting from the replicated class.

The Sync framework comprises the following five models, as shown in Figure 5.11a:

- *Basic System Model*: Figure 5.11b shows the basic system model. In this model, the client handles all user interactions. The client side consists of the user interface, code of application, and data that are represented as objects. The user interface component interacts with the replicated object through object methods, and the replicated object interacts with the Sync replication client through Sync methods. The server side consists of an application object replica and an interface for users to access external resources. Operations such as triggering a software system rebuild or sending and receiving email run on this side. The synchronization server provides a reliable resource sharing in an asynchronous mode by all clients.
- *Programming Model*: In this model, programmers can define the replicated object using a small set of predefined Sync-replicated classes or subclasses derived from them. By this approach, programmers effectively expose the structure of

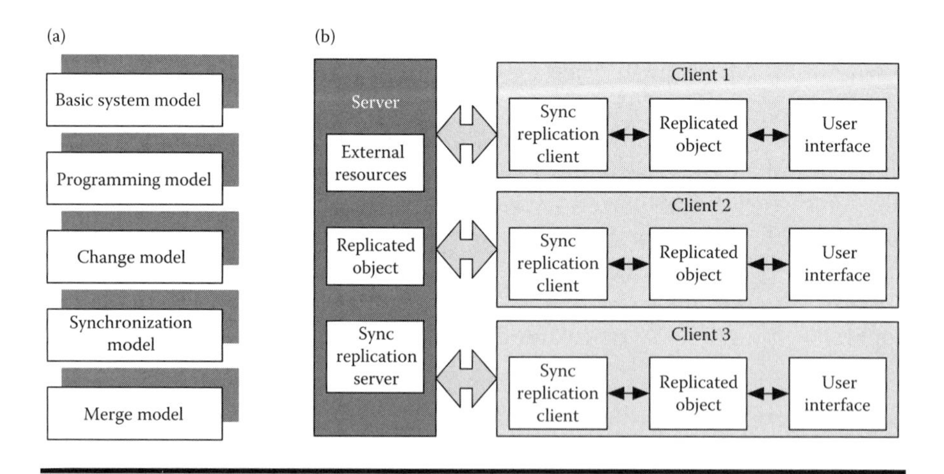

Figure 5.11 Sync framework and the basic system model. (a) Sync framework. (b) Sync basic system model.

their data to the Sync replication and synchronization system. Characterized by this feature, Sync can offer automatic, fine-grained change detection and merging. At the client side, a model–view–controller paradigm is adopted. The user interface implements a Java's observer interface and the replicated object extends Java's observable class. At the server side, a Sync server interface allows the application to respond to synchronization notifications from Sync.

- *Change Model*: This model specifies how objects represent the way they changed between synchronizations. It features class-specific changes and changes that are hierarchical structures. That is, changes to lower-level structures in a replicated object are reported as changes to upper-level structures. Moreover, changes are reported to the granularity of the basic types. This reduces the transmission data load, speeds communications, and reduces the network charges when the communication is charged by byte.
- *Synchronization Model*: The central synchronization server accepts synchronization requests from remote replicas, and collects and manages changes received by the server. In the current version of Sync, an update occurs when it reaches the server. This fixed rule for resolving conflict sacrifices some flexibility but it requires less memory and it is simpler to implement.
- *Merge Model*: The merge model describes how two sets of changes are merged with an object and their conflicts are resolved. A construct named merge matrix is the core of this model. The merge matrix is class specific, and for each class consists of rows and columns labeled by the operations for that class. With the rows representing the operations of the remote client and the columns representing the operations already committed at the server, the merge matrix entry identified by a particular row and column determines the action the merge procedure will take, such as accepting one operation or the other, or both. Sync populates merge matrices with default actions, but programmers may replace these at runtime through a provided method defined for all merge matrices, or supply their own merge action, to implement an application-specific action.

A Sync-based collaborative service example is introduced in Reference [11]. The example application considers a group of people planning for a conference. In the user interface, there is a newsgroup-like discussion forum and a drawing. Users discuss meeting planning details through the forum. At the same time, they can view the drawing that shows the layout and computer network of the conference room.

5.3.2 Infrastructureless Service Collaboration

5.3.2.1 AGAPE: Allocation and Group-Aware Pervasive Environment Framework

A context-aware group communication service based on the AGAPE framework is proposed in Reference [12] for collaborative application in MANET environments.

AGAPE supports impromptu collaboration in the environment characterized to have unpredictable user/device mobility, frequent disconnection, reconnection, and continuous network topology changes. AGAPE provides context-based point-to-point and multipoint communication patterns for colocated group members. In AGAPE, context is defined as the collection of useful information to characterize the runtime situation of communication entities, such as location, profile, desired collaboration preferences, and properties. It facilitates reliable and adaptable message delivery.

AGAPE formulates the group members dynamically based on their attributes and characteristics rather than simply depending on message recipients' names. In addition, members can schedule the message order or tailor the message format according to the communicating entity context. In the AGAPE communication model, only entities that are members of the same group can interoperate. Each group has a group unique identifier and its profile specifies commonly agreed interests, preferences, activities, and goals. In one group, each member has a unique personal identifier. The membership is dynamic due to user mobility, connection status, network connectivity, and partitions. The AGAPE group model has two entity roles. The managed entity (ME) is group members that implement the support service. On the other hand, the locality manager (LME) collaborates and supports the group management operation on behalf of MEs. Figure 5.12 shows the

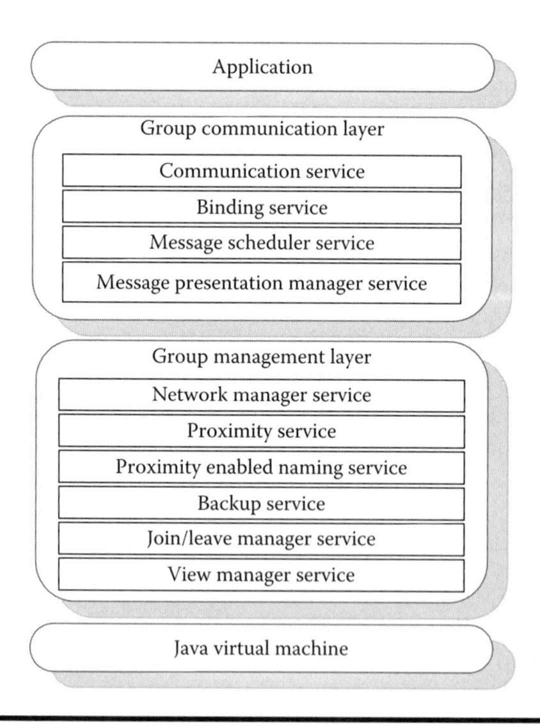

Figure 5.12 AGAPE framework.

AGAPE-specific architecture. As shown in Figure 5.12, on the top of the Java virtual machine, there are AGAPE services organized in two logical layers: the group communication layer and the group management layer. In the group management layer, there are:

- *Network Manager Service*: This service supports the transmission of UDP packets between all AGAPE entity devices and the mobile *ad hoc* network. It allows both point-to-point and multipoint communication.
- *Proximity Service*: This service permits both ME and LME members to advertise their online availability in the locality at regular times.
- *Proximity-Enabled Naming Service*: This service randomly generates unique group identifiers (GIDs) and ME/LME personal identifiers (PIDs) by exploiting a special naming approach. It senses the online advertisement packets of MEs/LMEs and establishes a table that associates each available AGAPE entity with an entry containing its GID/PID, its role (ME or LME) and its IP address.
- *Backup Service*: This service allows LMEs to decide whether they need to distribute context-dependent views. It can reduce unnecessary view propagation; for example, when multiple LMEs belonging to the same group and defining the same locality attempt to disseminate the same view to collocated group members.
- *Join/Leave Manager Service*: This service allows AGAPE entities to join/leave the group and to requalify themselves when their profile information changes. In detail, during the joining phase, all AGAPE entities submit their user/device profiles to the service, if the service decides that the entity is allowed into the group, it then returns to the new entity the group profile and its GID/PID.
- *View Manager Service*: This service allows LMEs to create and disseminate group views to AGAPE group members at regular times. When group members connect or disconnect from the network or when they change access device and/or group profile, AGAPE generates event notifications to report the view changes to members into the locality. These event notifications cause the view manager service to coordinate with the proximity-enabled naming service and to update the group view accordingly by inserting or removing the new/old member.

In the communication layer, there are four services:

- *Communication Service*: This service supports asynchronous unreliable message-oriented communication and implements both context-based any-cast and multicast patterns.
- *Binding Service*: This service supports two binding strategies: early binding and late binding. The early binding strategy determines the set of members

matching the characteristics specified in the searching profile associated with a binding request, that is, the target member set, by keeping into account the availability of members when the binding is created. The late binding strategy determines the target member set dynamically each time a communication message is to be sent.

- *Message Scheduler Service*: This service assigns dynamically a priority to exchanged messages on the basis of the application-specific scheduling preferences. In particular, the message scheduler service associates each user/device/group profile with a priority level and builds a priority table registering these priorities. This service can also discard messages depending on application-specific preferences. For example, messages coming from undesired members will be discarded if the available memory is limited.
- *Message Presentation Manager Service*: This service supports plug-in and filter selection to adapt the content of exchanged messages. In detail, this service extracts messages from the message scheduler service and uses user/device profile information to choose the most suitable tailoring filter among available ones. For example, downscaling operations will be performed to convert images from GIF to JPG format when delivering images to a resource-limited mobile device.

Two middleware prototypes are developed. One is for portable devices with rich computation resources. This version supports LME and includes all AGAPE services. The other is for devices with limited resources. It only supports ME operations and a subset of AGAPE services. Figure 5.13 shows an AGAPE service in a

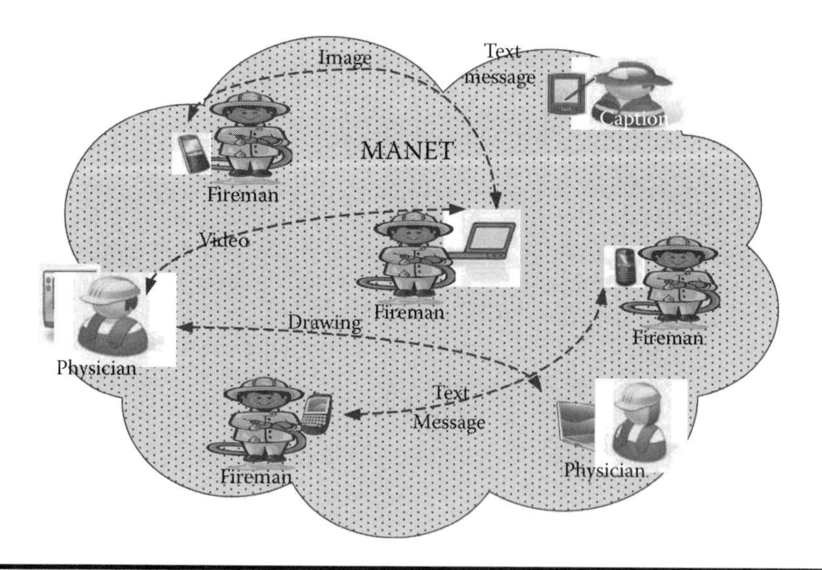

Figure 5.13 A fire scene scenario.

simplified emergency rescue scenario where the firefighters and physicians are cooperating. After a fire brigade arrives at the fire scene, the captain initiates a dynamic formation of the emergency response group on the field. People work together with 802.11b-enabled laptops and PDAs. They have different roles in the operation and they communicate and share instructions or images via message exchange services. The performance of the AGAPE is evaluated from the aspects of the network overhead, bandwidth usage, memory occupation in the worst case, and the ordinary operation scenario. The results indicate its ability to handle communication between transiently collaborating partners with the desirable performance.

5.3.2.2 CPP: Communication and Coordination Patterns for Mobile Collaboration

CPP presents a layered and fully distributed architecture [13] as shown in Figure 5.14. The layered structure aims to support a system with interoperability, scalability, and flexibility. The replicated architecture provides the mobile unit the autonomy in terms of services and data, which well suits a distributed system. As shown in Figure 5.14, the architecture is composed of the following layers:

- *Collaboration Layer*: The mobile collaborative applications reside in this layer.
- *Coordination Layer*: The user/session management, information sharing, and notification services coexist in this layer.
- *Communication Layer*: This layer handles routing, cross layer, and messaging.

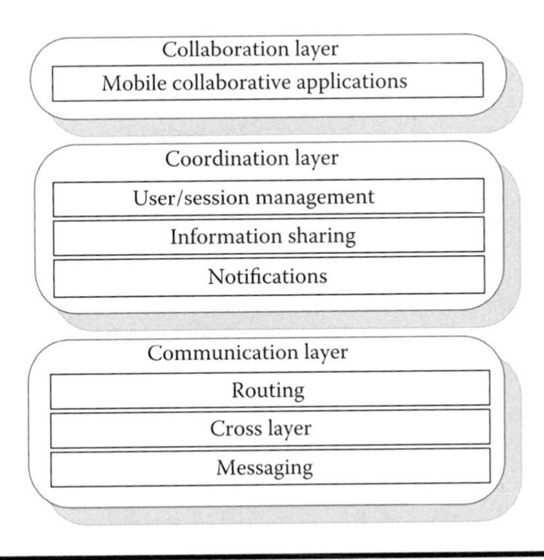

Figure 5.14 A layered and fully distributed architecture.

Table 5.2 Communication Patterns for Mobile Collaboration

Communication Pattern	*Characteristic*
Distributed session management pattern	Low coordination cost, autonomy, and ease to deploy
Work session pattern	Information sharing and low coordination cost
Mobile user list pattern	Awareness of user's reachability and variability of the work context
Cross-layer pattern	Interoperability and awareness of user's reachability
MANET routing pattern	Awareness of user's reachability and easy to deploy

CPP has several communication patterns for mobile collaboration [13] and Table 5.2 shows these patterns described as follows:

- *Distributed Session Management Pattern*: A distributed session management pattern enhances the cooperation in a mesh network or MANET with a dynamic topology. It allows users to establish a network environment including a transport layer (TCP/IP) and a session layer for collaborative interactions using a MANET. It can use synchronous or asynchronous interaction for fully distributed architecture and autonomous mobile units. Regardless of the device type or IP address, it uses a unique virtual identity for each user to link to the real identity. User sessions are operated on a dynamic group of virtual identities.
- *Work Session Pattern*: A work session pattern is to speed up setting up issues in MANET collaboration to ease the interaction and to enhance functions such as information sharing and message delivery. This pattern allows users belonging to the same session to collaborate using a MANET. In this pattern, each work session is modeled as an entity with a session-shared data space and a group of members. Each session member has a private data space, a public data space, locally recorded information for reachable mobile units, and an input/output communication channel. The input/output communication channel is shared by the session members.
- *Mobile User List Pattern*: A mobile user list pattern is proposed for user awareness applications. In particular, this pattern presents who is currently participating in a work session. A list of reachable mobile units is used to record the mobile unit ID, the virtual ID, the user's role, the visibility attribute, and the

neighbors. The list is updated by peer discovery using a typical synchronization process.

- *Cross Layer Pattern*: A cross-layer pattern is proposed to allow information sharing and to provide service interoperability among different devices such as laptops, PDAs, or smart phones. It supports information sharing among layers of a groupware system architecture. The shared data could be a mobile unit list, the MANET topology, and features of the mobile computing devices. In this pattern, groupware service and public data structures on different layers of the architecture are accessed through an API. Important data are shared to guarantee interoperability.
- *MANET Routing Pattern*: A MANET routing pattern is proposed for an ad hoc network to transmit messages between mobile users located at a distance more than the wireless communication range. A routing method is used to convey messages between mobile units located at more than one hop of a distance. In this application, the *ad hoc* gossip multicast method [27] is used as it offers an intermediate solution between the routing and flooding techniques. The delivery strategy is based on an algorithm with the following three phases: potential disconnection detection, the correction phase for ignoring/accepting potential disconnection detection, and the maintenance phase for units in the potential disconnection areas to detect the requesting nodes and set its state accordingly.

These proposed patterns are applied to a middleware platform [14,15] to deal with collaborative issues in disaster relief operations, construction site inspection, and exam managements.

5.3.2.3 Proem: Collaborative Peer-to-Peer Platform for Mobile Ad Hoc Applications

Proem [16] presents a complete solution for collaborative peer-to-peer applications in MANETs and personal area networks. It aims to achieve adaptability, universality, interoperability, extensibility, and platform independence. Figure 5.15 shows a typical Proem network as MANET that is infrastructureless, self-organized, decentralized, and dynamic. In such a MANET, Proem network defines the following terms in design:

- *Peer*: A peer is an autonomous host or device that is part of a peer-to-peer relationship. Examples of peers are the PDAs, laptops, and mobile phones in Figure 5.15.
- *Individual*: An individual is a person who owns and uses one or more peers. Examples are the people in Figure 5.15. Note that a person can own multiple

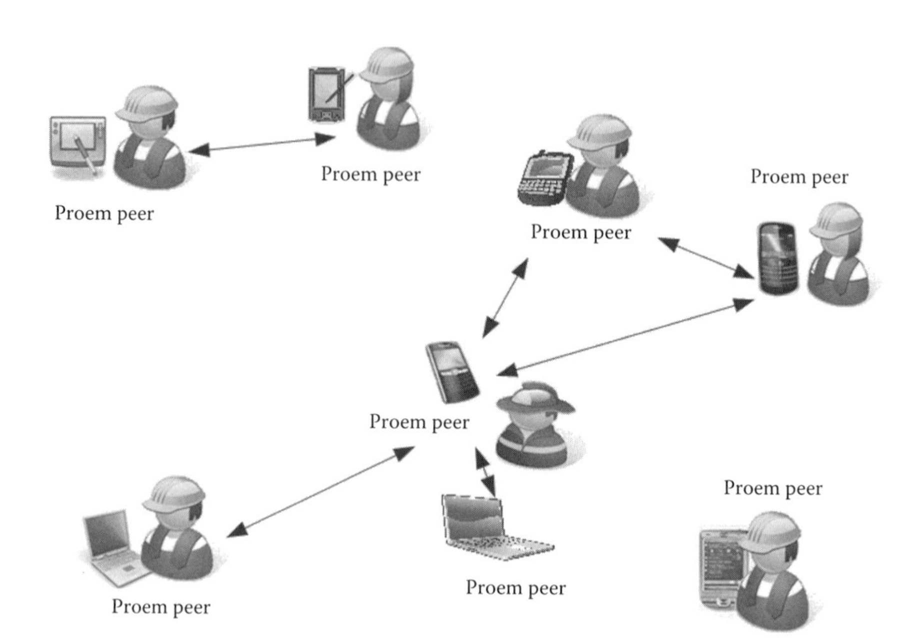

Figure 5.15 Proem network.

peers, for example, the police officer owns a PDA and a notebook, but each peer only belongs to one individual.

- *Data Space*: A data space is a collection of data items that is owned and managed by a set of peers. Data spaces are stored as copies for all peers that share them. An example is the drawing shared by the two engineers in Figure 5.15.
- *Community*: A community is a set of entities such as peers, individuals, data spaces, and even other communities. Communities are used to define access rights to data and functionality to group entities. Each entity can be a member of several communities and each community can contain members of different types. For example, the PDA and notebook owned by the police officer are of a community. People that communicate to the officer are of another community.
- *Name*: All these above four types of entities are identified by names. Each name is unique and only refers to one entity. Each entity is allowed to have one or more names, which may be used for pseudoanonym. An example name for a peer can be proem:peer:001.
- *Profile*: Profile is an XML-based data structure that describes the attributes of entities. An example profile can be the name, address, and email for an individual.

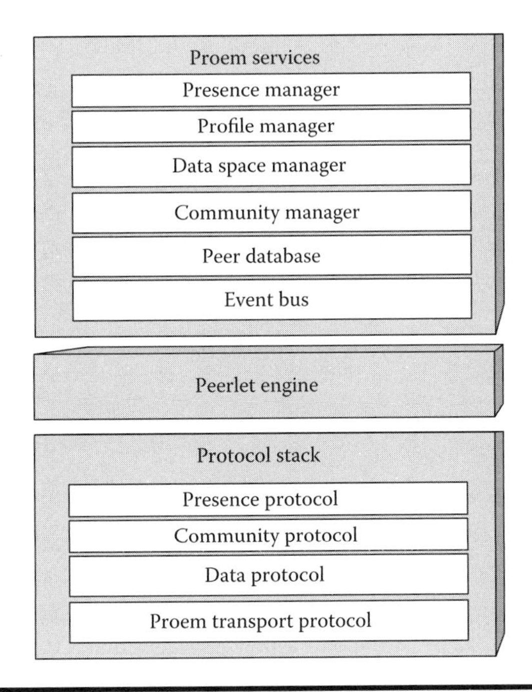

Figure 5.16 Proem architecture.

As shown in Figure 5.16, Proem architecture is composed of three components: the protocol stack, the peerlet engine, and the application services. In the protocol stack, Proem uses the following four communication protocols to define the syntax and semantics of messages that peers can exchange:

- *Proem Transport Protocol*: This is a low-level connectionless asynchronous communication protocol. This Proem protocol can be applied on top of various existing protocols such as TCP/IP (Transmission Control Protocol/Internet Protocol), UDP (User Datagram Protocol), or HTTP (Hypertext Transfer Protocol). Proem passes data between peers (autonomous mobile hosts or devices) in an atomic unit. Messages are encoded as XML documents and then sent from one peer to another. Proem peers can be implemented in any programming language without specific transport protocol.
- *Data Protocol*: The data protocol allows peers to share and synchronize data via data spaces. Data spaces are stored in a replicated fashion on all peers that share them.
- *Community Protocol*: The community protocol assigns and verifies membership for a set of entities.

- *Presence Protocol*: The presence protocol uses profiles as its basic message type. It allows peers to announce their presence and available entities to the network.

To extend its capability, Proem also allows application developers to define application-specific protocols. The peerlet engine manages peerlets. The management tasks include instantiation, execution, and termination. Peerlets are simply structured peer-to-peer applications that are confined by an event-based programming model. Peerlets are designed as drop-in modules and can be added to or removed from the peerlet engine during runtime. Some Proem services are as follows:

- *Presence Manager*: The presence manager discovers directly or indirectly reachable peers within the current network topology and announces a peer's presence.
- *Profile Manager*: The profile manager handles the entities' profile.
- *Data Space Manager*: The data space manager is responsible for data space storage and access control.
- *Community Manager*: The community manager maintains a peer's membership in communities and performs validation check for other peers' community memberships.
- *Peer Database*: The peer database keeps record of one peer's interaction with other peers and allows peerlets to store custom metainformation on peers.
- *Event Bus*: The event bus provides event-based communication among system components and peerlets. Its publish/subscribe model supports anonymous data exchange. System components and peerlets can broadcast availability of data items and subscribe to update events. Events are also used by the present manager to inform peerlets about the presence or absence of a peer.

For rapid deployment of peerlets, a Peerlet Development Kit is used in the proposed design. This kit has an extensive collection of high-level Java APIs. It provides classes for naming, communication, data management, and event handling. The Proem platform is capable of providing high-level support for mobile peer-to-peer application development. On the basis of the proposed platform, students at the University of Oregon developed several impromptu MP3 file-sharing systems successful within a short time frame.

5.3.2.4 STEAM: A Location-Aware Event-Based Middleware

STEAM [17–19] is a location-aware event-based middleware. The system is specially designed for MANETs (e.g., a vehicle *ad hoc* network) that include a large number of mobile application components distributed over a large geographical area. In STEAM, a component can establish a dynamical connection to nearby

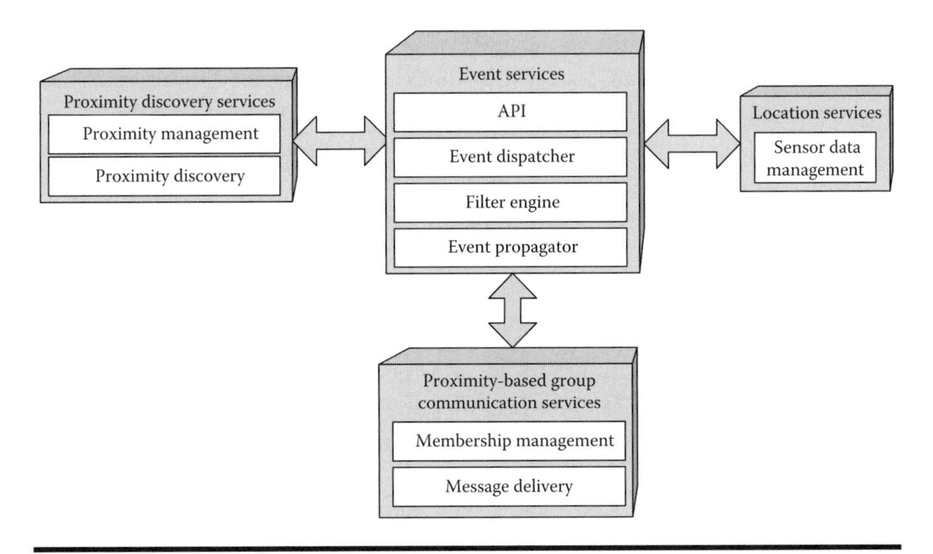

Figure 5.17 STEAM architecture.

components. STEAM supports filtering of event notifications based on both subject and proximity. The design enables location services, event services, proximity-based group communication services, and proximity discover services. Figure 5.17 shows STEAM architecture and the components for these services described as follows:

- *Location Services*: Location services collect sensor data to compute the current geographical location of its host machine and, subsequently, submit the location information to the middleware and to event producers and consumers running on that machine. The designed sensor data management achieves these functions. Example sensor data are the latitude and longitude coordinates collected by a GPS satellite receiver.
- *Event Services*: The STEAM event services consist of API, the event dispatcher, the filter engine, and the event propagator. The middleware only delivers events to consumers who subscribe to them, and when they are located proximate to where events occur until they unsubscribe. A consumer may move from a proximity to another without reissuing a subscription when entering the new proximity. Consequently, subscriptions are transparent inside the middleware and persistent for all proximities. A STEAM event type is defined by specifying a subject, content name values, and attribute list. Moreover, a STEAM event instance is defined in a similar manner by specifying the subject, content parameter values, and attribute list. Subject and content represents functional attributes, whereas a self-describing attribute list represents nonfunctional attributes. The subject defines the name of a

specific event type and the content defines the names and types of a set of associated parameters. STEAM provides three different event filters: subject filters, content filters, and proximity filters. Subject filters match the subject of events allowing a consumer to specify the interested event type. Content filters contain a filter expression that can be matched against the values of the parameters of an event. Proximity filters are location filters that define the geographic scope within which events are relevant and correspond to the proximity attribute associated with an event type.
- *Proximity-Based Group Communication Services*: The proximity-based group communication services are developed as the underlying means for entities to interact. The membership management service and message delivery service are exploited as following. To obtain the group membership, an application component must be located in the geographical area corresponding to the group and interested in the group. To specify a proximity group, STEAM defines a functional aspect and a geographical aspect. The functional aspect represents the common interest of producers and consumers on the basis of the type of information that is propagated among them, whereas the geographical aspect outlines the bounded scope within which the information is valid. In STEAM, subject and proximity to the functional and geographical aspects of proximity groups are mapped, respectively. Furthermore, proximity groups can be either an absolute proximity group that is geographically fixed and attached to a fixed point in space or a relative proximity group that is attached to a moving point represented by a specific mobile node. This concept serves as the basis for the stationary and mobile scopes supported by STEAM's programming model.
- *Proximity Discovery Services*: The proximity discovery services use beacons to discover proximities. On the discovery of the proximity, the associated events will be delivered to subscribers that are located inside the proximity. The proximity discovery services also map discovered proximities to subscriptions and to the underlying proximity-based communication groups. Through these services, the middleware facilitates a host to join a proximity group of interest when it has either a subscription or an announcement and is within the associated geographical scope, and to leave the proximity group on departure from the geographical scope.

The STEAM architecture assumes that mobile components are more likely to interact with each other when they are in close proximity. The system allows event producers to publish specific event types and consumers to subscribe to particular event types. Producers must define event types to specify the functional and nonfunctional attributes of the event for dissemination. Consumers must subscribe to event types and the subscriptions are persistent even if it leaves a certain proximity and enters a new proximity. The event announcement and subscription is only valid in certain geographical area. Therefore, STEAM supports location awareness and

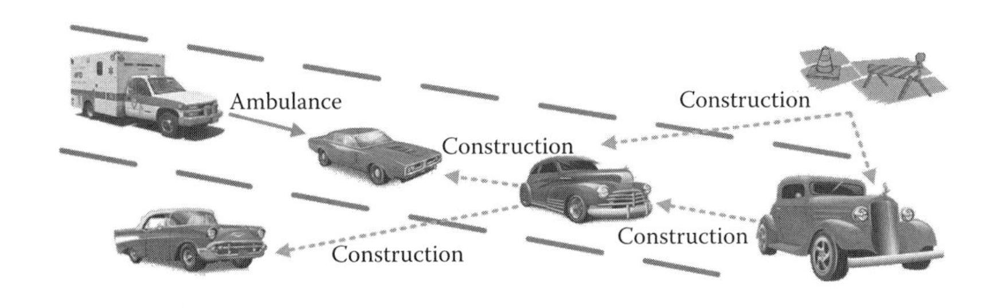

Figure 5.18 Event dissemination scenarios.

mobility. STEAM supports both single-hop and multihop event propagation. The single-hop event propagation performs when the transmission range of the sender covers the entire scope of the proximity. Figure 5.18 shows an example where the ambulance transmits a signal ("Ambulance") to the orange car, as indicated by the red solid arrow. The multihop event propagation is adopted when the proximity exceeds the transmission range of the sender. For example, the road construction messages ("Construction") are propagated from the construction sign to multiple cars, as indicated in Figure 5.18.

Examples of STEAM, as described in References [18,19], are traffic management applications. In these scenarios, there are mobile objects such as cars, buses, fire engines, and ambulances, and stationary objects such as traffic signals and traffic lights. As shown in Figure 5.18, a traffic sign can broadcast a speed limit change due to road conditions to all approaching cars. The ambulance can disseminate its location to nearby cars so that they can get ready to give the right of way to the ambulance. In addition, a crashed car can disseminate an accident notification to approaching cars to ensure safe driving. In STEAM, the producer only propagates events to users that are within a certain geographical area. By limiting the forwarding of event messages, the usage of communication and computation resources is reduced and the radio frequency interference is mitigated. Moreover, more predictable communication and reliable event delivery can be achieved.

5.4 Future Research and Opening Issues

There are many challenges in designing service collaboration protocols and many of these challenges come from the distinguished network environments and application scenarios. According to application requirements, the collaborative services can be context oriented, location oriented, object oriented, service oriented, message oriented, and so on. The challenges include hardware limitation (e.g., small screen of the PDA), device and network heterogeneity, mobility and intermittent connectivity, and others. On the other hand, the development of most of the existing approaches is still at early stages and they are not implemented in the real network environments for applications. Although some of them have been prototyped and limited simulations have been conducted for performance evaluation, these are not enough for practical applications. Therefore, in addition to developing new protocols, continuous research and implementation are required for validating the proposed architectures, frameworks, and service collaboration protocols.

Furthermore, all the current service collaboration protocols are limited in their addressed network environments and applications. It is very hard to develop a generic model and architecture that can work for diversified applications and network environments. These developed protocols are mostly case-specific mobile applications and there is no generic architecture for all applications. As a result, they are highly different even for similar applications. The case-specific architecture also makes it hard to evaluate and compare the protocol performance by a set of general criteria. Therefore, a simple, unified, open, and generic application model that encompasses several major aspects or different goals may be helpful for developers to apply their applications. This may be even more useful for industry if several generic or unified architectures can be standardized such that the diverse applications can be implemented over the framework.

5.5 Conclusion

Mobile service collaboration is a challenging but promising technology that leads the way for industrial/academic interactive applications such as entertainment, social network, business cooperation, field device monitor, construction management, traffic control, disaster relief, health care, and so on. In this chapter, we investigate mobile collaborative service design issues related to hardware limitation, device heterogeneity, network characteristic, intermittent connectivity, and user mobility. On the basis of these design requirements, we present a comprehensive review on the latest research breakthroughs and innovations for service collaborative architectures and protocols, as well as their experiments and applications. Finally, we discuss directions for further improvement and opening issues for future research and development.

QUESTIONS

Q1: Illustrate some key characteristics for mobile collaborative services.

Q2: Describe a service collaborative application that you think is useful in our daily life.

Q3: What are the design considerations/requirements for mobile collaborative services?

Q4: What are the challenging issues in developing mobile collaboration applications?

Q5: How to determine whether an application is infrastructure based or infrastructureless? What are the advantages and disadvantages of using infrastructure based or infrastructureless architecture for service collaboration?

Q6: Explain "context awareness" in service collaboration. Use an example to illustrate how the design can achieve context awareness in service collaboration.

Q7: How to improve the scalability in service collaboration? Please illustrate this by an example.

Q8: What do you think are the future research and development directions for service collaboration protocols?

References

1. B. Xie, A. Kumar, and D. P. Agrawal, Enabling multiservice on 3G and beyond: challenges and future directions, *IEEE Wireless Communications*, 15(3), 66–72, 2008.
2. V. Sacramento, M. Endler, H. K. Rubinsztejn, L. S. Lima, K. Gonçalves, F. N. Nascimento, and G. A. Bueno, MoCA: A middleware for developing collaborative applications for mobile users, *IEEE Distributed Systems Online*, 5(10), 2, 2004.
3. V. Sacramento, M. Endler, H. K. Rubinsztejn, L. S. Lima, K. Goncalves, and G. A. Bueno, An architecture supporting the development of collaborative applications for mobile users, In *Proceedings of the 13th IEEE International Workshops on Enabling Technologies: Infrastructure for Collaborative Enterprises* (*WETICE '04*), Modena, Italy, pp. 109–114, June 14–16, 2004.
4. M. Caporuscio and P. Inverard, Yet another framework for supporting mobile and collaborative work, In *Proceedings of the 12th IEEE International Workshops on Enabling Technologies: Infrastructure for Collaborative Enterprises* (*WETICE '03*), Linz, Austria, pp. 81–86, June 9–11, 2003.
5. Y. Cao, W. Wang, and Z. Qin, A multi-agent supported adaptive mobile collaborative service platform, In *Proceedings of the 3rd International Conference on Natural Computation*, Haikou, China, pp. 331–335, August 24–27, 2007.
6. W. G. Griswold, R. Boyer, S. W. Brown, and T. M. Truong, A component architecture for an extensible, highly integrated context-aware computing infrastructure, In *Proceedings of the 25th International Conference Software Engineering* (*ICSE '03*), Portland, Oregon, pp. 363–372, May 3–10, 2003.
7. W. G. Griswold, R. Boyer, S. W. Brown, T. M. Truong, E. Bhasker, G. R. Jay, and R. B. Shapiro. ActiveCampus—sustaining educational communities through mobile technology, *Technical Report CS2002-0714*, UC San Diego, Department of CSE, July 2002.
8. I. Jørstad, S. Dustdar and D. Van Thanh, A service oriented architecture framework for collaborative services, In *Proceedings of the 14th IEEE International Workshops on Enabling Technologies: Infrastructure for Collaborative Enterprise* (*WETICE '05*), Linköping, Sweden, pp. 121–125, June 13–15, 2005.
9. Y. Natchetoi, H. Wu, and Y. Zheng, Service-oriented mobile applications for ad hoc networks, In *Proceedings of the IEEE International Conference on Services Computing*, Honolulu, Hawaii, 405–412, July 7–11, 2008.
10. Y. Natchetoi, V. Kaufman, and Y. Karabulut, Service-oriented architecture for mobile collaboration, *In Proceedings of the International Conference on Collaborative*

Computing: Networking, Applications and Worksharing, New York, pp. 371–375, November 12–15, 2007.

11. J. P. Munson and P. Dewan, Sync: A Java framework for mobile collaborative applications, *IEEE Computer*, 30(6), 59–66, 1997.
12. D. Bottazzi, A. Corradi, and R. Montanari, Context-awareness for impromptu collaboration in MANETs, In *Proceedings of the Second Annual Conference on Wireless On-demand Network Systems and Services* (*WONS '05*), St. Moritz, Switzerland, pp. 16–25, January 19–21, 2005.
13. R. Messeguer, S. F. Ochoa, J. A. Pino, L. Navarro, and A. Neyem, Communication and coordination patterns to support mobile collaboration, In *Proceedings of the 12th International Conference on Computer Supported Cooperative Work in Design*, Xi'an, China, pp. 565–570, April 16–18, 2008.
14. A. Neyem, S. F. Ochoa, and J. A. Pino, Supporting mobile collaboration with service-oriented mobile units, In *Proceedings of the 12th International Workshop on Groupware: Design, Implementation, and Use, Lecture Notes in Computer Science*, Vol. 4154. Berlin / Heidelberg: Springer-Verlag, pp. 228–245, 2006.
15. A. Neyem, S. F. Ochoa and J. A. Pino, Designing mobile shared workspaces for loosely coupled workgroups, In *Proceedings of the 13th International Workshop on Groupware: Design, Implementation, and Use, Lecture Notes in Computer Science*, Vol. 4715. Berlin / Heidelberg: Springer-Verlag, pp. 173–190, 2007.
16. G. Kortuem, J. Schneider, D. Preuitt, T. G. C. Thompson, S. Fickas, and Z. Segall, When peer-to-peer comes face-to-face: collaborative peer-to-peer computing in mobile ad hoc networks, In *Proceedings of the 1st International Conference on Peer-to-Peer Computing*, Linköping, Sweden, pp. 75–91, August 27–29, 2001.
17. R. Meier and V. Cahil, Exploiting proximity in event-based Middleware for collaborative mobile applications, *In Proceedings of the Distributed Applications and Interoperable Systems: 4th IFIP WG6.1 International Conference* (*DAIS '03*), *Lecture Notes in Computer Science*, Vol. 2893. Berlin/Heidelberg: Springer-Verlag, pp. 285–296, 2003.
18. R. Meier and V. Cahill, Location-aware event-based middleware: A paradigm for collaborative mobile applications? In *Proceedings of the 8th CaberNet Radicals Workshop*, Ajaccio, Corsica, France, October 5–8, 2003.
19. R. Meier and V. Cahill, STEAM: Event-based middleware for wireless ad hoc networks, In *Proceedings of the International Workshop on Distributed Event-Based Systems* (*ICDCS/DEBS '02*). Vienna, Austria: IEEE Computer Society, 639–644, July 2–5, 2002.
20. A. Carzaniga, D. S. Rosenblum, and A. L. Wolf, Achieving scalability and expressiveness in an Internet-scale event notification service, In *Proceeding of the 19th Annual ACM Symposium on Principles of Distributed Computing*, Portland, Oregon, pp. 219–227, July 16–19, 2000.
21. A. Carzaniga, D. S. Rosenblum, and A. L. Wolf. Design and evaluation of a wide-area event notification service, *ACM Transactions on Computer Systems*, 19(3), 332–383, 2001.
22. M. Caporuscio, A. Carzaniga, and A. L. Wolf. Design and evaluation of a support service for mobile, wireless publish/subscribe applications, *EEE Transactions on Software Engineering*, 29(12), 1059–1071, 2003.

23. A. Carzaniga and J. Giacomoni. sXML. Available at: http://www.inf.usi.ch/carzaniga/siena/sxml/index.html
24. P. Fenkam, E. Kirda, S. Dustdar, H. Gall, and G. Reif, Evaluation of a publish/subscribe system for collaborative and mobile working, In *Proceeding of the 11th IEEE International Workshops on Enabling Technologies: Infrastructure for Collaborative Enterprises* (*WETICE '02*), Pittsburgh, Pennsylvania, pp. 23–28, June 10–12, 2002.
25. Sun Microsystems, J2ME Web Services Technical White Paper, July 2004.
26. OSGi. Available at: http://www.osgi.org/. Last accessed Nov. 11, 2011.
27. Z. Haas, J. Halpern, and L. Li, Gossip-based ad hoc routing, In *Proceeding of the IEEE Infocom '02*, New York, pp. 1707–1716, June 23–27, 2002.

Chapter 6

Mobile Agents for Mobile Services

Ratan K. Ghosh

Contents

6.1 Introduction

Humans, by nature, continually seek increasing levels of comfort in day-to-day interactions with the environment around them. Environment changes with mobility, which is a key feature of human life. The smartness of environments, like homes, offices, hospitals, shopping malls, streets, or airports, and so on, can be judged by their ability to adapt to the user's requirements in sync with change in environment and to provide efficient customized services any time at any location [1] So, service personalization, which guides user's interactions with the environment, is an important goal of any smart environment. Service personalization in smart spaces is achieved by a powerful convergence model for Service-Oriented Architecture (SOA) realized through mobile agents in conjunction with the Internet and wireless communication. Mobile agents move freely over a network of computers combining various reusable Web services from different peers to satisfy the services sought by a user. Therefore, the key requirement of the convergence model is that mobile agents should be able to dynamically discover, bind, and execute services concurrently and opportunistically in different platforms.

The integration of a mobile agent with a Web service has been a topic of intense research in recent years [2–7]. Both the technologies follow similar approaches to organize services. For example, Web services can be discovered through a platform-independent registry called UDDI (Universal Description, Discovery, and Integration). Similarly, a mobile agent can find other agents through a yellow page or a navigation agent that keeps track of all agents and agent platforms. Web services communicate using predefined standard XML schemas such as SOAP (Simple Object Access Protocol) and WSDL (Web Services Description Language). Likewise, agents can communicate among themselves if the messages are in a previously agreed format and language. Agents may as well understand SOAP and WSDL messages like Web services do. Both agents and Web services provide interoperability between systems with different operating systems and application languages. However, there are also some fundamental differences between the two technologies; notable among these are execution autonomy and nonmutable operation. Agents can execute autonomously, but Web services are passive and have to be invoked explicitly to provide any service. The methods of a Web service are nonmutable. An agent can, however, be programmed to overcome this difficulty by including a fall-back plan for executing another service if the one according to the original plan is unable to fulfill the user's need. So, the two technologies can be integrated harmoniously to provide personalized mobile services to the users.

Another aspect of the convergence model is the ability to handle personal mobility. Limited discrete mobility is possible while availing remote services over wired networks. SOAs are equipped to handle such discrete personal mobility. The ability to communicate wirelessly allows continuous, *always on*, personal mobility to the user. However, wireless networks offer much lower bandwidth and considerably high communication latency compared to wired counterparts. Furthermore, wireless devices, owing to portability, have very little resources. So, there is no easy way to extend SOA designs to work on wireless networks with low bandwidth, high network latency, and resource-starved mobile devices. The minimum requirement of any mobile service is that a user carrying a portable wireless device should be able to avail those services while roaming within the area under the coverage of the network. Using mobile agents, a client's mobile device could create a mobile agent corresponding to the user's request in the fixed network. This agent then contacts a stationary registry agent to find appropriate methods on Web services; it spawns clones that migrate and invoke chosen Web services in parallel. Finally, the clones return with the results to the user's device. The solution seems eminently possible assuming that the user's device has sufficient resources to create, send, and receive an agent. However, apart from having resource scarcity, portable user devices, being driven by small capacity batteries, operate with very low energy. So, these devices are not equipped for extended local computations. Mobile services, therefore, have to be supported by wired infrastructures; and end-user applications are realized through flexibility in relocating heavy computation to wired infrastructure. In other words, flexibility in computation should complement the flexibility in mobility of devices.

Two closely related concepts, namely, code portability and code mobility, are keys to the needs of flexibility in computation. Java and scripting languages such as Tcl, Tk, Perl, Python, and so on, provide code portability, as a code is either semi-compiled in the form of a machine-independent byte-code or directly interpreted from the source. Therefore, programs written in any such language are portable. Code mobility, which allows a small piece of code to move from one computing device to another, provides the required flexibility in computation. It serves several purposes. Important among these are

1. Opportunistic use of resources at the server instead of draining limited resources at the client.
2. Overcoming the problem of an unreliable and low-bandwidth wireless link between the client and the server.
3. Personalization of services through execution of selective sequences of mobile codes at different servers.

The deployment of mobile service, thus, faces significant challenges not only because of the mobility of the user or mobile device but also because of the mobility of the service component. As wireless communication provides only last mile connectivity, the novelty in design of mobile services lies in the ability to handle challenges for extending the fixed infrastructure at the service provisioning time, when necessary, to accommodate user and device mobility.

In this chapter, our aim is to provide an insightful study on mobile agent technology and its importance in the design of mobile services. Section 6.2 deals with general concept of agents, software agents, classification of agents, and agent services.

The topic of discussion in Section 6.3 is the technology behind the mobility of software agents. It deals with code mobility, agent migration, and agent execution environment. Section 6.4 talks about specific application domains where the impact of mobile agent technology can be significant in the design and performance of applications. The focus of discussion in Sections 6.5 through 6.7 is on how the mobility of both computing units and program units can harmoniously complement each other, providing an integrated development environment for mobile applications notwithstanding the network outages and heterogeneity. Section 6.8 explains how mobile agent technology can be used in the development of distributed applications as effectively as client server computing. Section 6.9 deals with security issues arising out of mobile agent computing. The issue of performance is addressed in Section 6.10. Section 6.11 summarizes the chapter. Section 6.12 gives exercises.

6.2 Agents and the Services Offered by Agents

The dictionary meaning of the word *agent* is *the one who carries out certain tasks delegated by someone on his/her behalf.* Thus, an agent is an entity that represents and has power or authority to act on behalf of individuals, groups, institutions, group of institutions, or even a nation. This definition of an agent implicitly refers to only human agents. Most agents we know of are commercial (human) agents who assist in brokering business deals.

However, an agent need not necessarily be a human. It can as well be a hardware agent (e.g., a robot) or a software agent. So, an abstract definition of agent would be: *an entity that has a perception of its environment through sensors and acts upon the environment's requirements from time to time through its effectors.* Our discussion in this chapter concentrates only on software agents; we do not elaborate on human agents or hardware agents.

6.2.1 Software Agents

Loosely speaking, a software agent is essentially a piece of computer code that performs a set of predefined actions on behalf of a client (a program fired by a user), and has the intelligence to autonomously adapt to the requirements of the client. For example, a user interface agent, such as the *MS Office Assistant* that provides help during document preparation, is an adaptive autonomous agent.

The idea of considering software as an agent can be naturally extended to decouple *computing units* from *places of computing.* A computer is considered as a place of computing. A software agent is a logical carrier of computing. Thus, by allowing software agents to move or conduct themselves as mobile agents, a computation unit is effectively decoupled from the place where it is executed. Using several mobile agents concurrently on a network of computers we can create a distributed computation. This approach to distributed computation also matches with the well-known abstraction of a process-based model. In other words, an agent-based approach to distributed computing overcomes the gap between abstraction and implementation.

6.2.2 Taxonomy of Agents

There are several ways to classify agents. A careful study of some of these schemes suggests that they are not very different from one another.

An agent can be classified using one or more of the following criteria: (i) mobility, (ii) interaction capabilities, and/or (iii) agent attributes.

An agent that can migrate from one environment to another or from one host to another is a mobile agent. An agent that cannot move out of its place of computation is referred to as a stationary or a static agent. A mobile agent executes in widely varying environmental settings, and at times in potentially hostile environments. Therefore, a mobile agent must be crafted carefully in order to prevent unnecessary exposure of its client to security attacks. An agent may have a model for internal reasoning to interact and negotiate with the external world including other agents. Certain agents, however, may not have any interaction capabilities except in a predetermined manner and only in response to specific external events. Such agents are termed as reactive agents.

Sometimes agents may be classified on the basis of a set of ideal attributes. In Reference [8] three attributes have been specified for classification of agents. These are:

- *Autonomy*: There may be a number of anticipated as well as unanticipated exigencies wherein an agent may have to take independent decisions on behalf of the clients it represents. So, a framework of operational independence should be provided to an agent.
- *Intelligence*: Agents must have sufficient intelligence to take independent decisions.
- *Adaptability*: Agents operate in hostile guest environments that differ widely. Many external events may expect a response from an agent. Therefore, agents must be equipped to react to such external events/conditions and adapt accordingly.

If an agent has independent capabilities to cooperate with other agents then it is also known as a collaborative agent. However, collaboration excludes the scenario in which an autonomous agent may possibly go out of the creator's control. An agent that has capabilities to adapt to environmental conditions and cooperates with other agents either under control of its creator or autonomously is referred to as a learning agent. Learning subsumes intelligence. An agent that can learn obviously has a model for internal reasoning. If a learning agent is also autonomous then it is an intelligent agent.

Another simple way to classify agents is based on the framework of (i) mobility, (ii) collaboration, and (iii) intelligence. This classification is not very different from the framework discussed earlier. Autonomy empowers agents to move around, seek remote resources for performing computations, and interact with other entities when needed. The ability of an agent to collaborate with other agents comes from the fact that it has interaction capabilities. Furthermore, as an agent moves from one environment to another, the environment around it changes. So, an agent should be able to sense and perceive the dynamicity in its environment and act upon it, or in other words adapt accordingly. The mobility of code and its execution in remote places (computing nodes) essentially constitute *remote programming*. When collaboration among agents is possible, a number of agents can be used for various parts of a single task and can complete the same effectively by executing in parallel. So, the classification of agent can be viewed as a three-dimensional framework [9] depicted in Figure 6.1. The dashed vertical line indicated in the figure represents the notional plot of the mobile agent system that would ideally be used for implementation of mobile services.

There may be many elaborate classification schemes for agents, as the one found in Reference [10]. Basically, all these classifications provide a somewhat fine grain distinction of agents on the basis of a number of attributes as explained below.

- *Communicative*: Ability to communicate is fundamental to cooperation and interaction.
- *Cooperation*: If several agents cooperate to perform a job, then that job can be done quickly and efficiently.

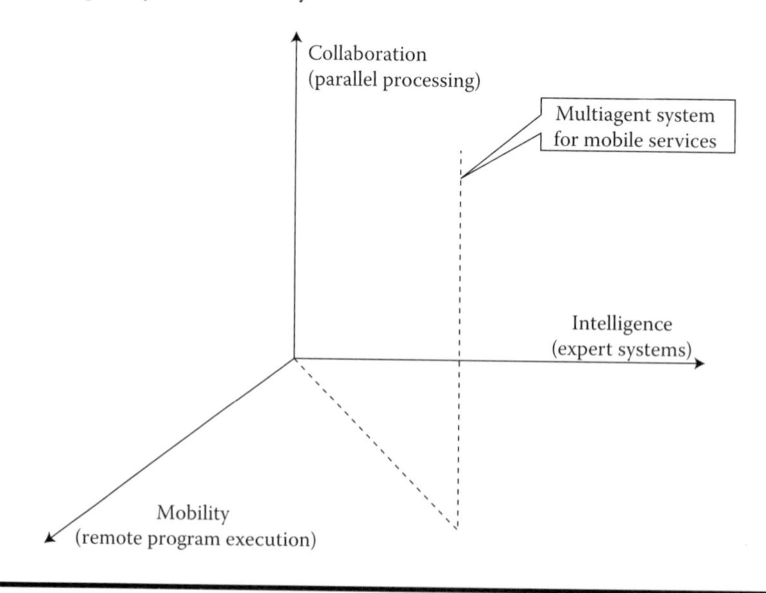

Figure 6.1 The space of multiagent systems.

- *Autonomy*: An agent can follow the goals autonomously without requiring any triggers from the environment or explicit instructions from its controlling entity.
- *Mobility*: An agent can move from one network node to another on its own.
- *Reactivity*: Sensors and actuators enable agents to react.
- *Proactivity*: An agent can take initiative on its own to accomplish the task.
- *Character*: An agent may have a personality and an emotional state; it can represent the person it is representing. In other words, it is a virtual person.
- *Learning*: An agent should modify its internal reasoning capabilities on the basis of previous experiences. That is agents should learn and adapt themselves to the environments in which they execute.

Such an elaborate attribute classification does not add much to the understanding of an agent computing system. These attributes are easily derivable from the basic notions of the agent framework.

6.3 Mobile Agent Technology

There have been considerable efforts in building agent systems [11] and experimenting with them. Most mobile agent systems [12–14] are Java based whereas others use different scripting languages. The choice of Java for agent programming is driven by a number of reasons. The notable among these are that it enforces security through strong-type safety, and sandbox execution environment. Static-type checking in the form of byte code verification is used to check the safety of a downloaded code. Some dynamic checking is also performed during runtime. With sandbox execution environment, it is possible to isolate memory and method accesses, as well as to maintain mutually exclusive execution domains [15]. Aglets [16], Ajanta [17], Concordia [18], Voyager [19], Mole [20], and Odyssey [21] are some of the examples of Java-based mobile agent systems.

Non-Java-based agent platforms use scripting languages such as Tcl, Perl, Python, Telescript, and so on. Telescript [22] and Safe Tcl [23] are specially designed to support development of mobile agent systems for commercial applications. Safe Tcl was used in the early development of the Agent Tcl [24] system. It uses a padded cell concept, whereby a second-level *safe* interpreter prevents harmful commands from being executed by the main Tcl interpreter. Telescript uses two key concepts, namely, *places* and *agents*. The virtual locations occupied by agents are known as places. Apart from built-in support for autonomous process migration, it provides for a complex four-level security model that enforces safe execution of mobile codes in foreign platforms.

6.3.1 Code Mobility

The idea of executing code on a remote machine is not new. When a printable file is dumped on a network printer, it essentially constitutes the act of dispatching an agent with limited functionality (of printing) from the user's computer to the printer. In that effect, postscript is a rudimentary form of executing code on a remote computer. When a *servlet* located at a remote machine gets executed in response to a user's mouse click on a hyperlink, it is equivalent to an agent performing a job, namely, executing a servlet for retrieving the dynamic contents of the relevant link. Similarly, an applet is an agent dispatched by the remote machine to perform a user-requested job at the client. Thus, any form of remote procedure call can be considered as an agent for a specific task.

A piece of software consists of three elements, namely, (i) data, (ii) code, and (iii) program stack. The mobility support may be *weak* or *strong* depending on the mobility of these three elements of a program unit. Strong mobility is represented by process migration involving mobility of an executing or a suspended process from one computer to another. It offers a number of capabilities such as dynamic load distribution, fault resilience, and locality of data access in a distributed system. Weak mobility refers to transfer of code from a server to a client or vice versa.

Code migration may be initiated either by the receiver or by the sender. A client sending a query to a database server is an example of sender-initiated code migration, because the execution at the server is triggered by the client's request. In most situations involving a *sender-initiated code migration*, the receiver would check for authenticity of the sender. So preregistration of clients with a server needed for a client (sender)-initiated code migration. When a client wants to execute some code available in a remote machine, then it becomes a receiver. A Java applet is an example of a *receiver-initiated code migration*. The receiver can be anonymous in a receiver-initiated code migration. The three remote execution models, we are familiar with are *client–server*, *code on-demand*, and *remote evaluation*.

In the client–server model, both program and stack are static and these resources are provided independently for the execution of parts in the client and the server, respectively. The client sends data that will be used as parameters for the procedures to be executed at the server. The data are the only mobile entity in sender-initiated weak code mobility. Figure 6.2 illustrates the client–server type of code mobility. The client sends the request for some service that is communicated in the form of a function call. The code for the function resides at the server that is executed locally. The response from the server is communicated as a result to the client.

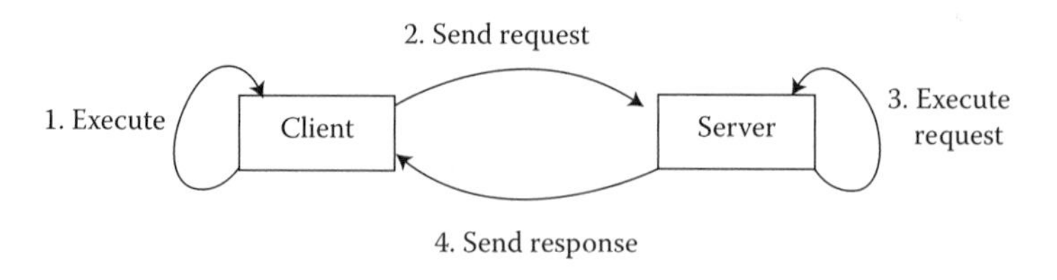

Figure 6.2 Code mobility in the client–server model.

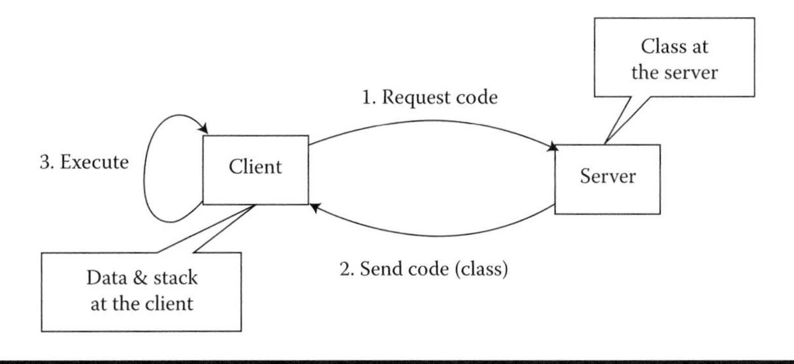

Figure 6.3 Mobility in code on demand.

The code-on-demand represents receiver-initiated weak mobility of codes. This paradigm is suitable for clients with limited capabilities. A Java applet is a popular example of code on-demand computing. Although the code is mobile, both stack and data are static being located at the client that provides the resources for execution. Figure 6.3 depicts the technique for code-on-demand-type mobility. The execution of code takes place in the client. The server essentially serves as a code repository providing specific codes on demand.

Remote method evaluations (RMEs) demonstrate the third type of mobility support. An application's code is divided into two parts. One part is executed in a client and the other part in a server. The part executed in the server constitutes RMEs. RMEs belong to sender-initiated weak mobility. SQL and postscript printing are examples of this type of mobility. Here again, the data and the program stack are static and the code is mobile. However, the static parts are located at a service provider's system rather than at the client. Figure 6.4 demonstrates code mobility in RME.

The type of code migration that allows transportation of limited data required for state initializations and storing of results along with the code is viewed as programming with mobile objects (MOs). The idea of using MO-based programming mainly originates from load-balancing type of applications. Figure 6.5 illustrates

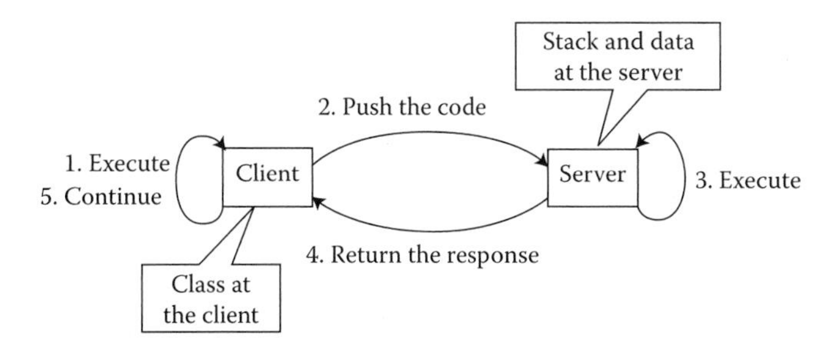

Figure 6.4 Code mobility in remote method evaluation.

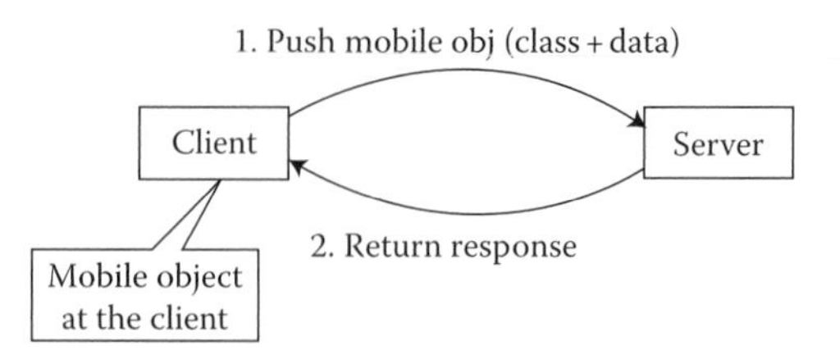

Figure 6.5 Mobile object.

the concept of programming with MOs. A mobile agent system can be viewed as an extension of the idea of programming with MOs. It consists of a collection of MOs, where each MO hops through a sequence of sites performing parts of computation at each site. The code and the state of methods are carried by MOs through migration sequences. Figure 6.6 depicts how a mobile agent system supports code mobility. An agent is initially created at a client and dispatched to a server. The code moves through a sequence of servers. Each server provides an execution environment for the agent to execute at the server. An agent autonomously transfers itself to other server sites seeking more services to accomplish its goal. Finally, it returns to the client with results. So, mobile agent systems demonstrate examples of strong sender-initiated code migration capabilities. Table 6.1 summarizes the mobility of elements of programs in the five remote execution paradigms discussed above. Note that programming with MOs is not shown explicitly, as it is subsumed by mobile agent programming.

As a mobile agent system allows the dispatch of a program for execution on a remote computer, it appears to have evolved out of client–server computing and remote evaluation. An agent can transport itself from one remote site to another and try to complete its task by availing services offered at different sites on a network.

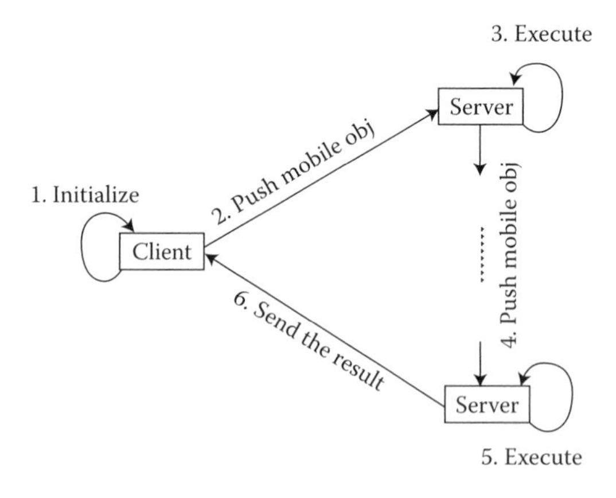

Figure 6.6 Code mobility involving mobile agents.

Table 6.1 Classification of Code Mobilities

		Direction of Mobility		
Paradigm	*Mobilty Type*	*Local Host*	*Remote Host*	*Execution At*
Client/ server	Weak sender initiated		Code	Remote host
			Stack	
		Data ⇒		
Code on demand	Weak receiver initiated	⇐	Code	Local host
		Stack		
		Data		
Remote method evaluation	Weak sender initiated	Code ⇒		Remote host
			Stack	
			Data	
Mobile agent	Strong sender initiated	Code ⇒		Remote host
		Stack ⇒		
			Data	

Therefore, mobile agent-based computing is a uniform and extremely powerful paradigm for distributed computing that introduces a neat decoupling of units of computing from the platform of computing. Such an approach can orchestrate computing in a way where agents seek services from computing platforms that guarantee best performance on a distributed network. However, the major hurdle in the use of mobile agents appears to be the lack of infrastructural support for security, fault tolerance, and transaction for remote execution of code [25].

6.3.2 Agent Execution Environment

Agent programs require an *agent execution environment* (AEE) that provides an interface in the form of agent application programming interfaces (APIs). Using the agent APIs, user applications may pass appropriate parameters to its agents and dispatch them, seeking required services at different servers over a network. On the server side, the AEE allows the two entities, namely, the server (the agent platform) and the client (the agent), to interact in a meaningful way. On the client side, AEE also provides an interface to the agents for interacting with the application program at the client to return the results.

The fundamental role of AEE is to bind to the user interface libraries at the client for collecting information directly from input devices such as the keyboard or for displaying information directly on the screen. The AEE must also bind operating system functions such as memory management, clock, file systems, and so on. So, in principle, there is no need for an application to run at the client [26]. As the AEE looks symmetrical at either ends—the server and the client—there is uniformity in the agent-based computing paradigm unlike the client–server model.

6.3.3 Execution and Migration

The parameters that an application passes to the agent through AEE are used for the agent's state initialization. Once the initialization is over and the agent has been launched, it becomes autonomous. Autonomy does not mean absolute freedom. The agent still has to behave within an expected framework. The most important among the expected behaviors is the dispatch of code for execution at a remote destination. So, at some point of time, an agent may execute an instruction whose effect may be either (i) to suspend the current agent process or (ii) to create a new child agent process. In response to this, the current state of the suspended process or the new process is saved. The state includes (i) process state and (ii) stack, heap, and all external references.

The saved data are marshalled into a machine-independent message format before the message is sent to the intended destination. The message may be addressed explicitly to one of the following

1. A final destination
2. A postoffice function that performs address resolution before sending the message to the final destination
3. An intermediate station that routes the agent on the basis of its content

The message is handed over to a message subsystem and routed directly or indirectly to the destination server, where it is delivered by the server's message subsystem to the AEE. In the AEE, the received message is reconstituted into an executable, and a process or a thread is dispatched. The execution continues at the next instruction in the agent program.

Figure 6.7 illustrates a generic system architecture for mobile agent systems. The interaction with the agent system is through a console. It provides a symmetric

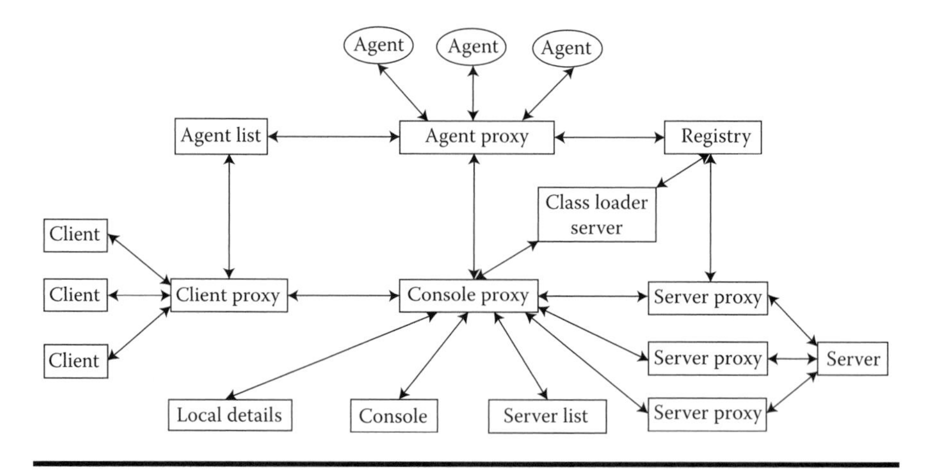

Figure 6.7 Agent system architecture.

picture for execution of the remote agent and of locally created agents. The clients represent a bunch of remote hosts whose agents seek service from the local host. The client proxy manages all the remote requests in a coordinated fashion. It creates an entry into a system-wide agent list before handing over the agent to the agent proxy. The agent proxy provides the actual environment for agent executions. The agents created by the local host also seek service from a remote host through the local agent proxy in the symmetric way. The local registry provides directory service. The agent proxy uses the registry to find the class loader server for servicing requests from the remote agent. Server proxies represent remote servers where locally created agents are transferred. One proxy is created for managing all the agents that have moved to the same server. The server list, which is dynamically created, has the details of the available server for agent services. Local details constitute the information about this host's agent service, and these details are provided to remote agent platforms when needed.

When the agent executes at the server, it passes the information received from the client application to the server methods. The agent in turn may receive other information from server applications. At the completion of this stage, the agent (i) may be terminated, or (ii) may be suspended at the server, waiting for some event to be delivered from a server application, or (iii) may once again migrate either by forking a new child process or by suspending and migrating itself. If the agent becomes resident at the server waiting for an event, it may also repeatedly satisfy specific services desired by the user. Figure 6.8 demonstrates how a mobile agent operates over its lifespan. If the agent migrates multiple times, it can either (i) return to its originating client or (ii) continue at another server or another client. This means that the agent may be able to perform recovery actions and visit another server if it determines that either the required service is not available or the service is unsatisfactory. In such a scenario, the agent may decide to visit another server on the basis of data received from the current server or the itinerary initially provided by the client.

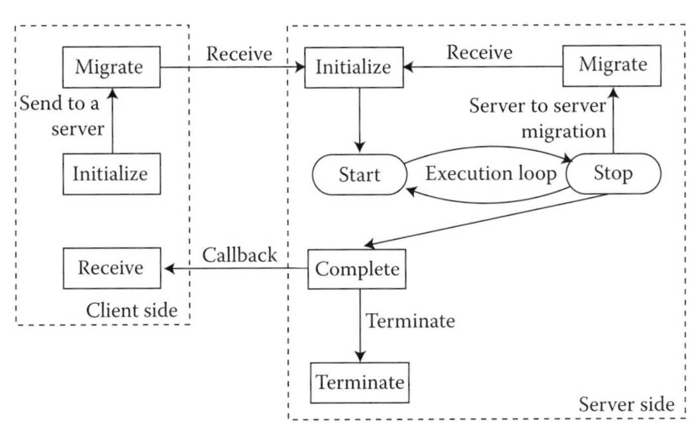

Figure 6.8 Life cycle of a mobile agent.

6.4 Application Domains

Although the potentials of mobile agent computing in many application domains are widely accepted, not many prominent applications have yet been built. Thus, the emphasis of most research projects in mobile agent technology is still on identifying applications where mobile agent-based implementations are found to perform better than client–server or other distributed computing models. Some representative application domains where this technology can provide significant performance gains could be as follows [27]:

- *E-commerce*: The most promising and effective use of mobile agent technology seems to be in the area of e-commerce. Agents acting on behalf of a client can approach appropriate service providers for specific service requirements such as gathering business information over the Internet and can collaborate with other agents for information sharing in order to strike a favorable deal on behalf of the client.
- *Personal Assistance*: A mobile agent can act as an assistant by performing tasks on behalf of a user. The mobile agent can migrate to remote network nodes and continues to represent the creator and negotiate fixing meeting schedules, booking travel tickets, or other business exchanges.
- *Telecommunication Network Services*: Value-added services in telecommunication networks require dynamic network reconfiguration capabilities and user customization. With mobile agent technology it is becomes easy to carry out reconfiguration and customization while keeping the network glued together for ongoing services.
- *Information Gathering*: An obvious application of mobile agents is information gathering. As a mobile agent interacts with other agents and autonomously executes at different agent platforms, it can gather information that is not known beforehand. It can accumulate certain knowledge on its own by interacting with other agents, and may decide to visit places where information of substance may be available. For example, the use of mobile agents has been proposed for sensor data gathering [25,28].
- *Information Dissemination*: Dissemination is the reverse process where information is distributed to all network nodes from the source. One important application could be to disseminate software updates/patches. Agents can bring new software components or installation procedures directly to the consumer's personal computer and autonomously update the software [16].
- *Watch-Dog Application*: Any watch-dog application monitoring occurrences of certain events can be implemented using mobile agents. For example, we can place a network management agent that detects traffic congestion at different nodes in a network by monitoring the patterns of network traffic. It can alert a network administrator when congestion is detected or when an

unusual pattern of data transactions is detected. This could be important especially in a network environment where data communication is broadcast oriented.

- *Parallel Processing*: Parallel processing is an obvious application domain for mobile agent technology. A mobile agent can clone itself and dispatch the clones to distributed computing nodes. Taking advantage of idle processing powers of distributed nodes, these agents can then execute a task in parallel without affecting any important ongoing activity at those nodes. The parallel processing application based on mobile agents can also scale up easily.
- *Pervasive Computing*: It is an exciting domain of applications that remains largely unexplored even today due to many challenges. The research in pervasive computing involves creating computer systems that can sense and interpret the actions or even moods of users under its environment. Solutions for many of the challenges encountered in pervasive computing can be met through code mobility [29].

6.5 Mobile Computing with Mobile Agents

Mobile agents and mobile computing represent two complementary concepts of mobility. Mobile computing enables user mobility as well as terminal mobility. So, it facilitates access of network services anytime and anywhere. The end-user runs mobile computing applications over mobile devices, such as personal digital assistants, cell phones, laptops, and so on. These devices are portable and equipped with wireless communication interfaces. Mobile computing allows some mobile nodes to disconnect from a collaborative distributed computation, move out physically, and reconnect to the network in order to resume their participation at a later stage. Mobile computing is, thus, a specialized form of distributed computation that may or may not use mobile agents. Mobile agent computing, on other hand, refers to the mobility of program units across a network of computers. Mobile agents do not require continuous connectivity. An autonomous agent can be injected into a fixed network from a mobile terminal. This agent would then start accessing services even when the mobile terminal loses connectivity. With location awareness, the agent can deliver results to the mobile terminal upon reconnection. The mobile agent platform needs to be extended by adding a top layer that provides mobility services [30].

There are several reasons as to why mobile agents are particularly suitable for development of mobile applications. Important among these are

1. Mobile agents can support disconnected operations on mobile devices.
2. Mobile agents allow dynamic adaptation and easy integration of heterogeneous environments. An agent-based middleware can, therefore, support new innovative location-aware and context-aware mobile applications.

3. Mobile agents can respond to dynamic changes in network and computation loads by providing a flexible framework for developing distributed applications where threads of computation are decoupled from the places where the threads execute.
4. Mobile agents enforce security police in an effective manner through agent infrastructures at service providers' ends.

6.6 Disconnected Operation

The fragility of wireless connection and severe constraints on the availability of bandwidth force mobile devices to intermittently disconnect from the network either voluntarily or involuntarily. A mobile agent, representing a computing unit for a client at a remote machine, can continue the computation on the client's behalf. The only requirement is that the client device must have an active connection, for a sufficient time, enabling it to inject a mobile agent into the fixed network. The results can be delivered by the mobile agent to the client when it reconnects later. Thus, mobile agents present ideal technology for writing applications for thin mobile clients such as cellular phones, personal digital assistants, automotive electronics, and military equipment.

However, there is some technical difficulty in allowing agents to jump to and from a mobile device due to the intermittent nature of wireless connectivity. The agents may not be able to exactly synchronize their attempts with the brief period of network connectivity enjoyed by the client device. To overcome this difficulty, a stationary agent called a docking agent [31] can be placed for each mobile terminal in the fixed network. Essentially, a docking agent works as an interceptor on behalf of a mobile host, and intercepts all connection requests to that host. As shown in Figure 6.9, the interceptor has a counterpart running in the mobile terminal too. Many interceptors may be hosted in a single machine on a fixed network. The interceptor on the fixed host puts

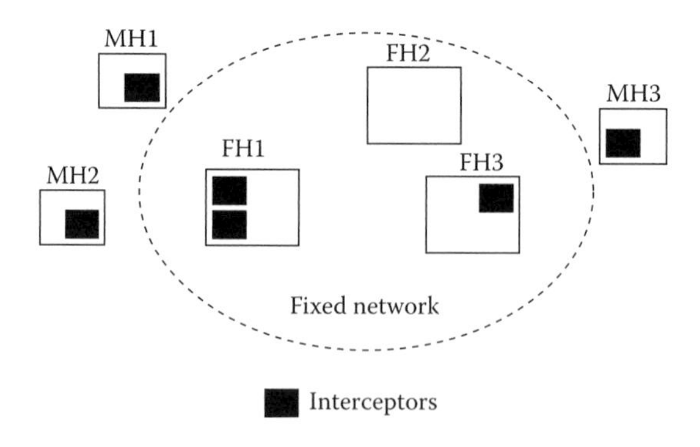

Figure 6.9 Placement of interceptor agents for mobile hosts.

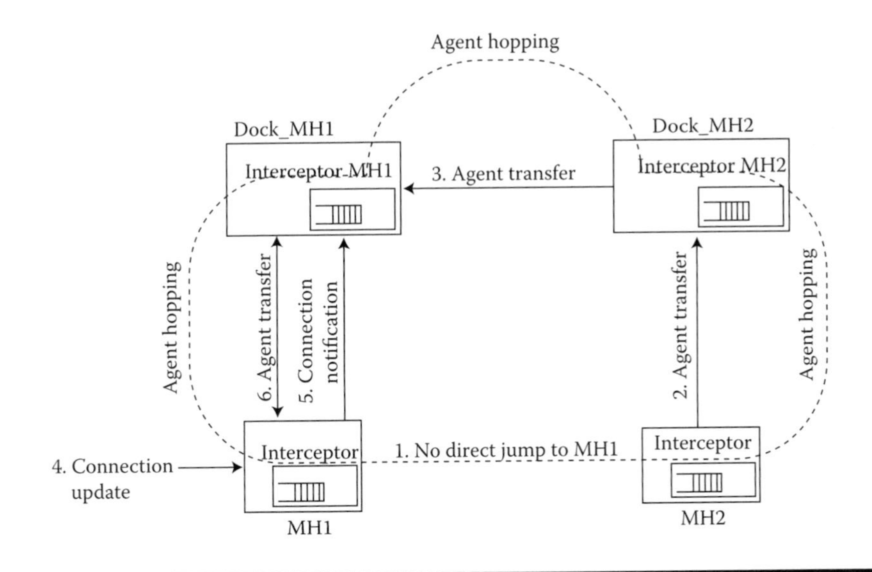

Figure 6.10 Hopping of mobile agents.

any agent trying to execute a jump in the waiting mode when the concerned mobile host does not have an active connection with the network.

Figure 6.10 illustrates the hopping scenarios for a mobile agent [31]. Initially, the agent from MH2 is transferred to its own docking station. From there the agent hops to the docking station of MH1, where it stays in waiting mode. After establishing a connection with MH1 (following a connection update), the waiting agent from MH2 can be transferred to MH1 from dock MH1. There may also be some waiting agents at MH1. These agents are also transferred between MH1 to dock MH1 while the time connectivity between the two remains alive. In the reverse situation when an agent attempts to jump out from a disconnected host, the local interceptor at the host saves the state of the agent and continually polls network to check if it is physically connected to the network. Once connection is established the waiting agent is immediately transferred to the fixed network.

To realize the agent hopping, as explained above, the interceptor must be able to detect the network connection [31]. This is aided by a set of network-sensing tools. These tools gather network status information that additionally helps smarter agents to plan their movement sequences in advance, given the estimates of network latencies to move between sites. Before an agent can move out of a client device, it needs to know whether the client's mobile has an active physical connection with the wired network. A tool that pings the broadcast address on the local subnet can be used for this purpose. If a response to a ping is received within a short interval, the network connection is considered to be active. The standard ping can be used to determine whether or not a specific host is reachable from the network. A traffic monitoring agent at each site can keep track of the amount of data it has recently sent. The monitoring data can be provided to any mobile agent on query. A mobile agent can, therefore, predict the expected bandwidth on the basis of past experience on the volume of traffic handled by the site. If no monitoring data are available, the traffic agent can contact neighboring sites to collect traffic information.

6.7 Heterogeneous Environment Integration

It is the responsibility of mobility middleware to keep track of a mobile terminal's location. Taking advantage of location awareness, mobile applications can apply specific optimizations according to the properties of a network connection and the characteristics of the mobile device [30], whereas mobile agents can continue computations on behalf of the user, leveraging the advantage of a fixed network for the best performance. Additionally, agents provide dynamic personalization by following the user's movements. An agent may interact with other agents to accomplish a task delegated by the client. An agent may autonomously decide to move to a different host if the computational load of a host platform is too high.

Mobile agents do not have dependencies either on the host or on the underlying transport layer [26]. The fundamental idea on which they function is code mobility. Some of the important requirements for supporting code mobility across network computers are

- A common application language for writing agent codes
- Persistence of agent processes (requires transfer of execution state along with its code when an agent migrates)
- Communication mechanism between agent hosts
- Security mechanisms to protect agents from agent platforms and vice versa

These requirements would make mobile agent systems eminently suitable for heterogeneous environments. However, many agent systems, such as Aglets [16], Grasshopper [32], Odyssey [21], and so on, differ widely in their architectures and implementations. Therefore, interoperability cannot be realized without a standardization of agent technology [33]. One level of difficulty in agent interoperability arises due to the need for translation between agent communication languages. Another level of difficulty is presented by the need to preserve semantic contents of parameters among applications. Two standardizations having distinct origins were proposed for agent interoperability.

- *Mobile Agent System Interoperability Facility* (*MASIF*) [34]: It was developed by the object management group (OMG) and defines interfaces for CORBA-based interoperability between heterogeneous mobile agent systems.

- *Foundation of Intelligent Physical Agents* (*FIPA*) [35]: It was developed originally by a Swiss nonprofit group that later became a group under the IEEE Computer Society Standard Organization. FIPA concentrates on agent communication to achieve interoperability as opposed to MASIF that focuses on agent mobility.

A mobile agent-based mobility middleware would require some extensions to the architecture of an agent platform. These extensions have mostly evolved out of adaptability requirements of applications to heterogeneity of hosts, networks, and user environments [30,36]. The underpinning of these requirements is that the system must somehow let the mobile applications become aware of the context in which they execute. A user's mobility rapidly changes the context as it not only changes a user's location but also brings the user into an entirely different computing environment.

In Reference [30], the authors propose extensions that provides for a mobility layer on top of the core layer for mobile agent services. The layer supports the following four basic mobility services:

1. Mobility-enabled naming services (MNSs)
2. User's virtual environment (UVE)
3. Mobile virtual terminal (MVT)
4. Virtual resource manager (VRM)

The overall layered architecture of mobile agent-based middleware for supporting mobile services could be as illustrated by Figure 6.11 [30].

MNSs are used to translate high-level names of entities into globally assigned unique identities. It may also maintain current locations of the entities. The naming service comes under tremendous pressure because of the mobility of entities. Typically, two different services, namely, directory and discovery services, are used

Layer for mobility services			
Mobility naming services	User virtual environment	Mobile virtual terminal	Virtual resource manager
—	—	—	—
Core mobile agent services (identification, transportation, etc.)			
Heterogeneous distribution system			

Figure 6.11 Mobile agent-based mobility middleware. (Adapted from Bellavista, P. et al., *IEEE Computer*, 73–81, 2001.)

for improving performance and scalability. Whereas a discovery service provides information about service availability on a local scale to the clients, the main responsibility of a directory service is to permit authorized clients to locate services through a hierarchical query mechanism. With replication and caching, a directory service can provide excellent performance.

UVE allows a user to connect to the Internet from different locations. It works like a personal and logical docking port for a user. The user's personal information is stored in a profile at the fixed host where the user registers first. UVE may replicate this profile at several other UVE servers at different locations. By leveraging a directory service, UVE can transparently map the requests for a user's profile to the nearest UVE server. Thus, in spite of the user's mobility, the performance remains unaffected and services become highly available. Mobile agents greatly simplify the dynamic distribution of UVE information. Agents save a user's session state and restore it by transporting to the new location.

MVT supports physical movements of a mobile device to different locations by allowing continuity of its local execution while preserving the state of interactions with the network resources and services. Mobile agents are used to trace mobile terminals. Using care-of addresses [37], discovery and directory services, the location of a mobile terminal can be traced. An agent at a fixed location acts as proxy for any mobile device and forwards undelivered messages to it, whereas a discovery service keeps track of the mobility of the device within a network locality. The directory service makes the mobile device visible to other authorized entities that may want to communicate with the former. MVT interacts with VRM to provide necessary resources from time to time. It also supports a disconnected operation through mobile agents using a procedure similar to the one explained in Section 6.6. MVT allows secure interoperability by exploiting the agent security tools and compliances to standards such as Internet protocols and CORBA compliance.

Mobile agent-based VRM allows for modification and migration of system resources at runtime to deal with complex resource management. VRM provides agent wrappers to standardize interface and access control. With agent wrapping it is also possible to enforce established mobile agent security policies, to control accesses, and to monitor resource accesses via logs.

6.8 Distributed Computing

Traditional approaches to distributed computing are based on a client–server model. Although it is able to achieve scalability by defining a hierarchical structure on the capabilities of networked computers, the client–server model cannot respond to dynamic change in network and computational loads. Redistribution of load in a dynamic environment is made possible through code migration techniques. Code migration addresses the issue of load redistribution through a twin strategy that could: (i) reduce network load and latency by moving computation to place of data and (ii) solve the problem of resource scarcity through relocation of computation.

Therefore, it is appropriate to consider mobile agent computing as just another technique for distributing computing resources, which is not very different from traditional distributed computing paradigms.

6.8.1 Advantages over Client–Server Systems

According to Pleisch [27], owing to dynamic load redistribution, mobile agent computing offers several advantages over a client–server model. Some of these are:

- *Communication Latency and Bandwidth*: It is always beneficial to move computation to the place of data rather than moving data, especially if the available bandwidth is low and the amount of data is large. A mobile agent exploits the locality of data in computation not only to reduce communication latency but also to decrease bandwidth requirements.
- *Offline Processing*: Because of its roaming nature, mobile computing devices often experience network outages for long time. A mobile device can delegate responsibilities to an agent to act on its behalf and perform requisite actions at the servers besides collecting the results. The results may additionally be postprocessed by the agent and a consolidated message carrying the result may be sent to the mobile device. So the entire task of computation can be carried out effectively offline as far as the mobile client is concerned.
- *Improved Response Time*: If the mobile agent is carefully designed, it can exploit locality information very effectively. The response time of a mobile client can be greatly reduced, because the mobile agent acts as a surrogate of the remote client and provides instant response on its behalf.
- *Large Asynchronous Request Granularity*: When mobile agent computing is used, a remote client can send many requests within a single agent. This type of interaction is also ideal when there is a low bandwidth, or high communication latency, or both.
- *Dynamic Load Distribution*: Mobile agents include code to take preemptive actions in order to adapt to environmental changes. For example, when an agent determines that a heavily loaded server is unable to provide the required service, it reacts autonomously either by migrating to an alternative server or by returning to the originating host.
- *Protocol Encapsulation*: Mobile agents can be very effective in encapsulating the protocols of the legacy applications running at the servers. As remote clients interact through mobile agents using a proprietary protocol, they do not have to bother about the protocol of the legacy application at the server.

The advantages of mobile agent computing over traditional client–server computing as discussed above, thus, make the former a suitable technology for many applications in diverse domains.

6.9 Agent Security

The most difficult challenge in the widespread use of mobile agent technology has been the inability to fully address security concerns in remote execution of code. Security threats arising out of execution of the code originating from a remote client machine can be of various forms. Broadly, these security threats can be classified into three types:

1. Disclosure of information
2. Denial of information
3. Corruption of information

A mobile agent system is modeled in terms of servers (agent platforms) and agents. Consequently, the security scenario can be examined from three different perspectives:

- *Attack on Servers by Agents*: The attack by an agent may be in the form of masquerading, denial of services, and unauthorized accesses.
- *Attack on Agents by Other Agents*: Agent-to-agent attacks can belong to any of the three categories of attacks mentioned above. In addition, nonrepudiation is another security concern when agents are allowed to conduct transactions among themselves. The basic premises on which an agent society works are trust, truthfulness, and confidentiality. Thus, a compromise in any aspect entails a collapse of the whole agent society.
- *Attack on Agents by Servers*: The server or the computing platform itself can be nontrustworthy. The attack by a nontrustworthy server may again be in the form of masquerading and denial of services. In addition, a server may induce agents to modify its behavior and eavesdrop on the communication between interacting agents.

All countermeasures conform to a set of baseline assumptions regarding the agent environment. Qualitatively, we may view these assumptions under trust and encryption. An agent has to trust the home platform where it gets created and starts execution. The home platform and other equally trusted platforms are nonvulnerable to various security attacks, and they behave nonmaliciously. Finally, there is public key infrastructure that manages public key cryptography.

6.9.1 Protecting Servers or Agent Platform

To protect servers from unauthorized access by agents or to protect agents inside a server from one another, one simple approach could be to establish separate isolated domains for each agent and platform, and to have all interdomain accesses mediated through a monitor. Such a monitor is referred to as a reference monitor. A *reference monitor* has the following characteristics [15]: (i) it is always invoked, cannot be bypassed and mediates all accesses; (ii) it is tamper-proof; and (iii) it is small enough to be analyzed and tested.

A reference monitor enforces a number of conventional security techniques, such as:

- Isolating processes from one another and from the control process.
- Controlling accesses to computational resources.
- Enforcing application of cryptographic methods to encode information exchanges, and to identify and authenticate users, agents, and server.
- Creating audit trails for security-relevant events that occur at an agent platform.

Some of the techniques for the security of a mobile code and a mobile agent have evolved out of the guidelines for building a reference monitor, and include the following:

- *Software-Based Fault Isolation*: The basic idea is to isolate application modules into distinct fault domains by software. So, untrusted programs written in an unsafe language, such as C, can be executed safely within the single virtual address space of an application. Untrusted machine interpretable code modules are transformed such that all memory accesses are confined to code and data segments within their fault domain. The technique alternatively referred to as *sandboxing*.
- *Safe Code Interpretation*: It does not allow an agent to execute any command that can potentially harm the server. The agent may be allowed to execute some of the less harmful commands after these commands are made safe. For example, an agent cannot execute any arbitrary string as a program segment.
- *Signed Code*: Digital signatures are used for authenticity of an agent, its origin, and its integrity. The agent code can be signed by the creator of the agent, or by the user of the agent, or by a third entity that has reviewed the agent. Typically an agent operates on behalf of an end-user or an organization. Therefore, most mobile agent systems [16,24,38] present the signature of the user as often as needed to ascertain the authority under which the agent is operating.
- *Authorization and Attribute Certificates*: It represents a digital document explicitly listing the permissions that the issuing authority has granted to the

holder for use of a service or a resource that the issuer controls. So the agents can seek attribute certificates prior to the use of any resource at the server and then present these certificates to seek legitimate services at the server.

- *State Appraisal*: It ensures that an agent has not been subverted due to modification in its state information. Appraisal functions determine privileges to be granted to an agent on the basis of certain conditional factors and whether or not any identifiable state invariants are violated.
- *Path Histories*: It represents an authenticated record of the previous platforms visited by an agent. Using this record, a newly visited platform decides whether or not to process an agent and, if so, what resource constraints to apply. The maintenance of a path history requires each agent platform to add a signed entry to the path indicating its identity and the identity of the next platform to be visited, and to supply the complete path history to the next platform.
- *Proof-Carrying Code*: It requires the code producer (e.g., the author of an agent) to formally prove that the program possesses safety properties prescribed by the code consumer (e.g., security policy of the agent platform). The code and the proof are sent together to the code consumer where the safety properties are verified. A safety predicate representing the semantics of the program is generated directly from the native code to ensure that the companion proof actually corresponds to the code.

6.9.2 Protecting Clients or Agents

The countermeasures for protection of an agent platform have directly evolved out of traditional mechanisms used by trusted hosts and emphasize active prevention measures. In contrast, those for the protection of agents are designed as detection mechanisms and serve as deterrents for the attackers. The reason for this approach stems from the fact that an agent when executing under a foreign environment is completely susceptible to that environment. No countermeasure can prevent the occurrence of a malicious behavior; it may only be able to detect the same. Some general-purpose techniques for protecting an agent include the following:

- *Partial Result Encapsulation*: It refers to a kind of miniaudit trail for an agent's operations and itinerary. The information encapsulated depends on the agent's goal, but it typically includes responses to inquiries made or transactions issued at a platform as well each platform visited. Any tampering by malicious hosts can be detected by verification of the encapsulated results when the agent returns to the point of origin or to some intermediate points.
- *Mutual Itinerary Recording*: It can be viewed as a variation of path histories. It allows for recording of an agent's itinerary and tracking by another cooperating agent and vice versa [39]. An agent conveys the information on the last, the current, and the next platforms to the cooperating peer through an

authenticated channel. The peer maintains a record of the itinerary and takes appropriate action when inconsistencies are detected.

- *Itinerary Recording with Replication and Voting*: Instead of having a single copy of an agent to perform a computation, multiple copies of the agent are used. The idea is that a malicious platform may corrupt only a few copies of the agent. With enough replications circulating in the system the validity of computation can be checked through a voting mechanism. On the flip side, it leads to wastage of resources.
- *Execution Tracing*: It requires each platform to create and retain a nonrepudiable log or a trace of operations performed by the agent while it was resident there, and to submit a cryptographic hash of the trace upon conclusion. The technique also defines a secure protocol to convey agents and associated security-related information among the various parties involved, which may include a trusted third party to retain the sequence of trace summaries for the agent's entire itinerary.
- *Environmental Key Generation*: It calls for construction of agents in such a way that on encountering an environmental condition (e.g., string match in search) it generates a key. This key is used to unlock some executable code [40]. The environmental condition is hidden through either a one-way hash or a public key encryption of the environmental trigger. The technique ensures that a platform or an observer of the agent cannot uncover the triggering message or provoke action by directly reading the agent's code.
- *Computing with Encrypted Functions*: It represents a strategy that allows a mobile code to safely compute cryptographic primitives, such as a digital signature, even executing under an untrusted computing environment, and this can be done without requiring any interactions with the home platform. The trick in using such a strategy lies in inventing an appropriate encryption scheme that can transform arbitrary functions as desired. The approach requires distinction between a function and a program that implements the function.
- *Obfuscated Code*: The strategy behind this technique is to scramble the code in such a way that no one will be able to get a complete understanding of its function (i.e., specification and data), or to modify the resulting code without detection.

6.10 Performance

Whether a particular paradigm of computing is better or worse largely depends on application-specific requirements. A single computing paradigm cannot be appropriate for all application scenarios [41]. So, the performance of one computing paradigm can be compared with another with respect to special functional requirements of a certain class of applications. That apart, a computing paradigm may be implemented in multiple ways, differing in execution environments, topologies, load distribution among computing entities, properties related to programming languages, security considerations, and so on. However, considering that mobile agent computing combines the advantages of the client–server model and the remote evaluation paradigm, a comparison of the three paradigms would provide insight into the nature of applications where one is preferred over the other two.

One fundamental advantage of mobile agents over the traditional client–server model is in reduction of network load. However, mobile agents suffer from several drawbacks arising out of technicalities involved in code migration. For example, mobile agents could be ideal for information filtering when data size is relatively large compared to the amount of filtered information. In contrast, for simple distributed information systems, downloading documents via agents could be very expensive, as it has to carry a large baggage (all documents found so far, its code, and its current state) when the agent migrates from one server to another. However, using the traditional client–server computing model, the download can be accomplished with less load on network traffic. Once a document has been located, the client can download the same immediately from the corresponding server.

Indeed, a concrete expression for network traffic load generated in each case can be obtained as follows [41]. Let there be N servers, each holding D documents. The server supplies certain keywords as header information to identify the documents it holds. The relevant information is assumed to be uniformly distributed among the servers. The important assumptions concerning download request and documents are as follows

1. i: density of relevant information at each server
2. h: size of header information of each document in bits
3. b: maximum size of a document in bits
4. r: size of the request sent from a client to a server

A client–server downloader application is expected to have a search engine component at the client who will connect to N document servers. The search engine at each client will issue D requests for document headers generating Dr amounts of network traffic. After receiving Dh bits of header information, the search engine matches keywords to identify relevant documents, and finally issues $i \times D$ requests to the servers, injecting iDr amounts of network traffic. So, the total traffic for requesting relevant documents alone results in $Dh + Dr + iDr = Dh + (D + iD)r$ bits. After a server gets requests for iD documents, it sends those relevant documents to the client. This implies that iDb bits are sent to the client. Thus, the total network traffic on each server is: $((D + iD)r + Dh + iDb)$. With N document servers, the total load will be

$$T_{CS} = ((D + iD)r + Dh + iDb)N$$

When the same downloading is realized through a mobile agent, we need to consider two other parameters, namely,

1. C_{MA}: the size of the code for a mobile agent
2. s: the size of mobile agent's state, that is, its internal data structures including list of sites and/or log of agent operations

The agent will hop from one server to another carrying relevant documents with it and will return to the user. The network traffic at each hop j consists of the following two components:

- $r + C_{MA}$: network load for sending a mobile agent with download request to one server
- S_j: the size of the state component at the jth hop

Note that the agent will carry downloaded data with it when hopping from one server to another. So, at each hop j, it will carry $\Sigma_1^j iDb$ traffic load due to documents. Therefore, $S_j = s + \Sigma_1^j iDba$. This means the total load due to the jth hop of the agent is

$$T_{MA}^j = r + C_{MA} + s + \sum_1^j iDb$$

Thus, summing over all hops the total load generated by the agent is

$$\begin{aligned} T_{MA} &= \sum_{j=0}^{N}\left(r + C_{MA} + s + \sum_1^j iDb\right) \\ &= \left(r + C_{MA} + s + \frac{N}{2} iDb\right)(N+1) \end{aligned}$$

Comparing the loads generated by the two computing paradigms, it is clear that the client–server model outscores the mobile agent for the downloader application. The reason for a bad performance of the mobile agent is obviously due to the fact that the agent carries downloaded data from each data server as it moves to the next. As opposed to this, in the client–server model, data are sent directly to the client once a relevant download request has been received at the server.

The comparison of performance of the document downloader application with remote evaluation is even more revealing [41]. The remote evaluation offers more flexibility in computation under a basic client–server paradigm by allowing the code for processing a request also to be supplied by the client. So with remote evaluation, the search engine executes in the same site that holds the document. Assuming

C_{REV} to be the size of the search engine's code, the network load generated by the download application is

$$T_{\mathrm{REV}} = (r + C_{\mathrm{REV}} + iDb)N$$

The total traffic purely for download of the relevant documents is $I = iDb$ in all cases. So, the overheads of different computing paradigms are as indicated in Table 6.2.

The discovery and the allocation of resources for a task are two other basic issues that have an impact on the performance directly [42]. The efficiency of resource allocation strategy can be measured by the time to complete resource requests. The completion time is the time between the initiation and satisfaction of a resource request.

Let us consider the client–server paradigm first. Assume the client knows about the location of the server. The client's interaction with the server would consist of marshalling the request in the form of a flat message, transferring the message, and then unmarshalling the message that is received as a reply from the server. Marshalling and unmarshalling will be needed for any message transfer, so this requirement is independent of the client–server system's functioning. The transfer time will depend on the distance between the client and the server and the amount of data transferred. This implies that the transfer time is dependent on the application's requirement and does not relate to the client–server paradigm. So, the part that is related to the operation of the client–server system is the service provided by the server in response to a client's request. As the server gets many requests from different clients, the time to complete a request is governed by the principles of M/M/1 queue. Assuming the arrival rate of request to be λ, and service rate to be ρ, the service completion time of request from a client is [42]

$$T_{\mathrm{CS}} = t_{\mathrm{res}} + \frac{1}{\rho} + \frac{1}{\lambda} + t_{\mathrm{res}}$$

where t_{req} is the time to transfer the request from client, t_{res} is the time to transfer the response from server, and l is the average number of clients in queue at the server.

Table 6.2 Performance Comparison of Three Mobile Code-Based Computing Paradigms

Computing Paradigm	*Total Network Traffic*	*Overhead*
Client server	$((D + iD)r + h)N + 1$	$((D + iD)r + h)N$
Remote evaluation	$(r + C_{\mathrm{REV}})N + I$	$(r + C_{\mathrm{REV}})N$
Mobile agent	$((r + C_{\mathrm{MA}}) + (I/2))(N + 1) + I$	$((r + C_{\mathrm{MA}}) + (I/2))(N + 1)$

Now let us examine the paradigm of mobile agent computing. Assume the client has no *a priori* knowledge about which server has the required resources for the service needed. It sends a mobile agent that moves around in the network and finally reaches the server that has the resource and autonomously executes there. The results are then carried back to the client by the agent. Assuming that the agent visits $k - 1$ other servers before it finds the server with the required resources, the time to complete the request of the client is [42]

$$T_{MA} = kt_{req} + \frac{1}{\rho_k} + \frac{l_k}{\lambda_k} + t_{res}$$

where t_{req} is the maximum time to transfer the request from client to the first server and between two intermediate servers, t_{res} is the time to transfer the response from the kth server, l_k is the average number of clients in queue at the server k, ρ_k is the rate of service at the kth server, and λ_k is the rate of arrival of requests at the kth server.

Let $(1/\rho)_{min}$ denote the minimum service time, $(1/\rho)_{max}$ denote the maximum service time, $(1/\lambda)_{min}$ denote the minimum waiting time in the queue at a server, and $(1/\lambda)_{max}$ denote the maximum waiting time in the queue at a server.

Then, ignoring the request transfer time, the upper bound of difference in service completion times between the client–server and mobile agent is

$$(1/\rho)_{min} - (1/\rho)_{min} + (1/\lambda)_{min} - (1/\lambda)_{max}$$

The request transfer time for a mobile agent dominates that of a client–server model. However, a mobile agent is able to handle the case when the client is unaware of the location of the server. It is expected that k will be small and constant; server capabilities are identical and service-related parameters as well as loads are uniform across servers. Thus, the difference in completion time, as mentioned above, is negligibly small. So the overhead of using a mobile agent will be $(k - 1)t_{req}$. Note, however, that request is related to network traffic load. Two other simple variations of a basic computation scenario involving a client service request for both the client–server and the mobile agent models have also been provided in Reference [42]. However, except for the details, it does not add more to the basic understanding of the framework of performance analysis.

The comparative study of performances of the three paradigms indicates that the agent paradigm does not necessarily lead to efficiency in distributed computation, especially if the agent were to carry substantial amounts of data related to the results of computation. Most distributed applications, implemented using mobile agents that we know of, can as well be implemented, perhaps more efficiently, using either client–server computing or remote evaluation paradigm. So, the reasons for using mobile agents in building distributed applications cannot just be judged on the basis of performance alone. The mobile agent system provides a single, unified framework in which distributed, information-oriented applications can be implemented efficiently and easily, with the programming burden spread evenly across information, middleware, and client providers [43]. If performance is the only criterion then it may be advisable to use a mixture of different paradigms to implement an application. For example, a mixture of agent migration and remote procedure calls can be used to reduce network load. Mobile agents migrate only to a few selected hosts while the remaining hosts are accessed by remote procedure calls [44]. The optimal selection of hosts for agent migration depends on the size of requests, the results, and the quality of links between the hosts. These parameters should be available before deciding the sequence. A mathematical model developed in Reference [44] also takes into account the ability of the agent to compress the results before transmitting back to the client. The agent code may not always be transmitted along with its state. The class corresponding to the agent code is dynamically loaded from the agent's home server when needed. The class download can also be avoided if the required class is already available at the destination agent server. This approach is followed in most Java-based agent systems. So, agent technology should be evaluated from the perspective of convenience as a flexible framework for distributed computation.

6.11 Summary

The topic of discussion in this chapter is centered on examining how mobile agent technology can be leveraged for design of efficient, flexible, and highly personalized mobile services. The important points emphasized in this discussion are: (i) the concept of software and mobile agents, (ii) code migration and related problems, (iii) mobile agents in mobile computing, (iv) performance of mobile agents and other known remote execution paradigms, and (iv) security arising out of code mobility. A comprehensive solution covering all the issues in an agent framework, including security, performance, availability, and so on, is difficult to find. However, mobile agents provide the kind of flexibility that is not possible in other known computing paradigms for distributed computation. The computation flexibility of an agent would be ideal for many application domains where end-users can seek network services anytime and anywhere over portable mobile devices or thin clients. The important bottleneck in the widespread use of mobile agents is a concern for security. An array of techniques for implementing security in agent systems is available. However, not all of them are compatible with one another, nor are they suitable for most applications where performance is the most premium need. Many of the security features can be implemented within the framework of the agent system, whereas a number of them can be applied independently within the context of the application. Though elementary security techniques may be adequate for a number of agent-based applications, many applications would require a more comprehensive set of mechanisms. Moreover, to meet the needs of a specific application, a flexible framework that allows application of a selected subset of mechanisms must exist. The trick, of course, lies in the ability to select a comprehensive baseline of countermeasures that can meet the philosophy of protection by guiding the design of the agent system, fulfill the needs of most applications, include compatible mechanisms, and be extended to include other advanced mechanisms that may be invented subsequently. Clearly, the state of research in mobile agents still belongs to the period where establishing such a baseline requirement calls for more experimentation and experience with alternative design choices, including those involving trade-offs in performance, scalability, and compatibility.

EXERCISES

1. Give a precise definition of code mobility.
2. Identify the key benefits of using a mobile code versus a static code.
3. Explain how MOs may ease the transaction processing scenario involving mobile devices. Design and implement such a transaction system and experiment with it.
4. Consider a Java applet and explain how agent security issues are addressed in the execution of an applet.
5. Provide a flow diagram of generic operations involving code migration from a sender and its execution on a receiver in a network.
6. Give a simplified state diagram for the life cycle of a mobile agent and contrast this with that of a process.
7. Spanning trees are important for many communication protocols such as routing, broadcasting, multicasting, and so on. Design a distributed mobile agent-based algorithm for identifying a spanning tree. The algorithm should not make any assumption about topology or size of the network.
8. Experiment with the *ad hoc* routing algorithm called *antNet* [20] based on the concept of mobile agents using an appropriate simulation platform.
9. Discuss how mobile agents can be used to reduce communication overhead and information redundancy for data gathering operations in a wireless sensor network.
10. In Time Division Multiple Access (TDMA) slot assignment to nodes in a wireless sensor network, no two nodes should be assigned the same slot if they have a common neighbor. Design a simple mobile agent-based graph coloring algorithm for solving the TDMA slot assignment problem.

References

1. I. Marsa-Maestre, M. A. Lopez-Carmona, J. R. Velasco, and A. Navarro. Mobile agents for service personalization in smart environments. *Journal of Networks*, 3(5):30–41, 2008.

2. D. Cooney and P. Roe. Mobile agents make for flexible web services. In *Proceedings of the Ninth Australian World Wide Web Conference* (*AusWeb 2003*), Queensland, Australia, pp. 385–391, July 5–9, 2003.
3. Y. Wang. Dispatching multiple mobile agents in parallel for visiting e-shops. In *Proceedings of the Third International Conference on Mobile Data Management* (*MDM2002*), Singapore, pp. 53–60. IEEE Computer Society Press, USA, January 8–11, 2002.
4. Y. Wang and J. Ren. Building Internet marketplaces on the basis of mobile agents for parallel processing. In *Proceedings of the Third International Conference on Mobile Data Management* (*MDM2002*), Singapore, pp. 61–68. IEEE Computer Society Press, USA, January 8–11, 2002.
5. D. Greenwood and M. Calistia. Engineering web service—Agent integration. In *Proceedings of the IEEE International Conference on Systems, Man and Cybernetics* (*SMC 2004*), The Hague, The Netherlands, Vol. 2, pp. 1918–1925, October 10–13, 2004.
6 W. Zahreddine and Q. H. Mahmoud. An agent-based approach to composite mobile web services. In *Proceeding of the 19th International Conference on Advanced Information Networking and Applications* (*AINA05*), Taipei, China, Vol. 2, pp. 189–192, March 28–30, 2005.
7. F. Ishikawa, N. Yoshioka, Y. Tahara, and S. Honiden. Toward synthesis of web services and mobile agents. In *Proceedings of the AAMAS2004 Workshop on Web Services and Agent-based Engineering* (*WSABE*), New York, USA, pp. 48–55, July 2004.
8. H. S. Nwana. Software agents: An overview. *Knowledge Engineering Review*, 11:205–244, 1996.
9. D. Gilbert, M. Aparicio, B. Atkinson, S. Brady, Ciccarino, B. Grosof, P. O'Connor et al., IBM intelligent agent strategy. *Technical Report, White Paper*, IBM TJW Research Center, 1995.
10. M. Grabner, F. Gruber, L. Klug, and W. Stockner. Agent technology: State of the art. *Technical Report SCCH 49/2000*, Software Competence Center, Hagenberg, Austria, 2000.
11. S. Robles. Mobile agent systems and trust, a combined view towards SEURE sea-of-data applications. PhD thesis, Department of Computer Science, Universitat Autonoma de Barcelona, Spain, 2002.
12. J. Kiniry and D. Zimmerman. A hands-on look at Java mobile agents. *IEEE Internet Computing*, 1(4):21–30, 1997.
13. D. Wong, N. Paciorek, and D. Moore. Java-based mobile agents. *Communications of the ACM*, 42(3):92–102, 1999.
14. R. S. Gray, D. Kotz, G. Cybenko, and D. Rus. Mobile agents: Motivations and state-of-the-art systems. *Technical Report*, Dartmouth College, Hanover, New Hampshire, 2000.
15. W. Jansen and T. Karyiannis. Mobile agent security. *Technical Report Special Publication 800-19*, National Institute of Standards, August 1999.
16. D. Lange and M. Oshima. *Programming and Deploying Java? Mobile Agents with Aglets.* Addison-Wesley Publishing Company, Reading, Massachusetts, 1998.
17. A. Tripathi, N. Karnik, M. Vora, T. Ahmed, and R. Singh. Mobile agent programming in Ajanta. In *Proceedings of the 19th International Conference on Distributed Computing Systems* (*ICDCS'99*). IEEE Computer Society Press, USA, 1999.
18. Mitsubishi Electric ITA. *Concordia: An Infrastructure for Collaborating Mobile Agents.* Mitsubishi Electric ITA, Waltham, Massachusetts, 1997.

19. Object Space, Inc. *Voyager: ORB 3.1 Developer Guide*, 1999. http://www.objectspace.com/products
20. J. Baumann, F. Hohl, M. Straber, and K. Rothermel. Mole—A Java based mobile agent system. *The World Wide Web Journal*, 1(3):123–137, 1998.
21 B. D. Noble and M. Satyanarayanan. Experience with adaptive mobile applications in odyssey. *Mobile Networks and Applications*, 4(4):245–254, 1999.
22 J. E. White. Mobile agents. *Technical Report*, General Magic, Inc., Sunnyvale, California, October 1995.
23. N. Borenstein. Email with a mind of its own: The Safe-Tcl language for enabled mail. In *Proceedings of the IFIP WG 6.5 Conference*, North Holland, Amsterdam, May 1994.
24. R.S. Gray. Agent Tcl: A transportable agent system. In *Proceedings of the CIKM Workshop on Intelligent Information Agents*, Baltimore, Maryland, December 1–2, 1995.
25. H. Qi, Y. Xu, and X. Wang. Mobile-agent-based collaborative signal and information processing in sensor networks. *Proceedings of the IEEE*, 91(8):1172–1183, 2003.
26. C. G. Harrison, D. M. Chess, and A. Kershenbaum. Mobile agents: Are they a good idea? *Technical Report RC 19887 (88465) 12/21/95*, IBM TJW Research Center, Yorktown Heights, New York, 1995.
27. S. Pleisch. State of the art of mobile agent computing—Security, fault tolerance and transaction support. *Technical Report RZ 3125 (#93198) 06/28/99*, IBM Zurich Research Lab, 1999.
28. M. Chen, T. Kwon, Y. Yuan, and V. C. M. Leung. Mobile agent based wireless sensor networks. *Journal of Computers*, 1(1):14–21, 2006.
29. F. Corradini, R. Culmone, and M.R. Di Berardini. Code mobility for pervasive computing. In *Proceedings of the 13th IEEE International Workshop on Enabling Technologies: Infrastructure for Collaborative Enterprises* (*WETICE 2004*), Modena, Italy, pp. 431–432, June 14–16, 2004.
30. P. Bellavista, A. Corradi, R. Montanari, and C. Stefanelli. Mobile agent middleware for mobile computing. *IEEE Computer*, 34(3):73–81, 2001.
31. R. S. Gray, D.Kotz, S. Nog, D. Rus, and G. Cybenko. Mobile agents: The next generation in distributed computing. In *Proceedings of the Second Aizu International Symposium on Parallel Algorithms/Architecture Synthesis*, Aizu-Wakamatsu, Japan, pp. 8–24, March 17–21, 1997.
32. C. Bäumer and T. Magedanz. Grasshopper—A mobile agent platform for active telecommunication. In *IATA '99: Proceedings of the Third International Workshop on Intelligent Agents for Telecommunication Applications, Stockholm, Sweden, August 9–10, 1999*, pp. 19–32. Springer-Verlag, London, 1999.
33. Z. Linda, B. Nadjib, and E. Aouaouche. An architectural model for a mobile agents system interoperability. In H. Labiod and M. Badra, editors, *New Technologies, Mobility and Security*, pp. 555–566. Springer, Dordrecht, The Netherlands, 2007.
34. D. Milojicic, M. Breugst, I. Busse, J. Campbelland, S. Covaci, B. Friedman, K. Kosaka et al., MASIF: The OMG mobile agent system interoperability facilities. In *Proceedings of Second International Workshop, MA'98, Lecture Notes in Computer Science*, Vol. 1447. Springer-Verlag, Berlin / Heidelberg, pp. 50–67, 1998.
35. Foundation for Intelligent Physical Agents (FIPA). FIPA abstract architecture specification, 2000. http://www.fipa.org

36. F. Corradini and E. Merelli. Hermes: Agent-based middleware for mobile computing. Formal methods for mobile computing, *Lecture Notes in Computer Science*, Vol. 3465. Springer-Verlag, Berlin/Heidelberg, pp. 234–270, 2005.
37. C.s Perkins. *Mobile IP: Design Principles and Practices*. Addison-Wesley Longman, Reading, Massachusetts, 1998.
38. J. Tardo and L. Valente. Mobile agent security and Telescript. In *Proceedings of IEEE CompCon Spring '96*, Santa Clara, California, pp. 58–63, 1996.
39. V. Roth. Secure recording of itineraries through cooperating agents. In *Proceedings of ECOOP Workshops*, Brussels, Belgium, pp.297–298, 1998.
40. J. Riordan and B. Schneier. Environmental key generation towards clueless agents. In G. Vinga, editor, *Mobile Agents and Security*, Vol. 1419. Springer-Verlag, Berlin/Heidelberg, 1998.
41. A. Fuggetta, G. P. Picco, and G. Vigna. Understanding code mobility. *IEEE Transactions on Software Engineering*, 24, 1998. http://www.cs.ucsb.edu/~vigna/listpub.html.
42. M. Bakhouya, J. Gaber, and A. Koukam. Observations on client-server and mobile agent paradigms for resource allocation. In *IPDPS '02: Proceedings of the 16th International Parallel and Distributed Processing Symposium*, pp. 257–261. IEEE Computer Society, Washington, DC, April 15–19, 2002..
43. D. Kotz and R. S. Gray. Mobile agents and the future of the Internet. *ACM Operating Systems Review*, 33(3):7–13, 1999.
44. M. Straer and M.s Schwehm. A performance model for mobile agent systems. In *Proceedings of the International Conference on Parallel and Distributed Processing Techniques and Applications* (*PDPTA97*), Vol. 2, Las Vegas, Nevada, pp. 1132–1140. CSREA Press, Athens, June 30–July 3, 1997.
45. B. Barán. Improved antNet routing. *SIGCOMM Computer Communication Review*, 31(2 supplement):42–48, 2001.
46. R. S. Gray, G. Cybenko, D. Kotz, R. A. Peterson, and D. Rus. D'Agents: Applications and performance of a mobile-agent system. *Software: Practice and Experience*, 32(6):543–573, May 2002.
47. D. Johansen, R. van Renesse, and F. B. Schneider. Operating system support for mobile agents. In *Proceedings of the 5th IEEE Workshop on Hot Topics in Operating Systems* (*HOTOS-V*), Orcas Island, Washington. IEEE Computer Society, USA, May 4–5, 1995.

MIDDLEWARE FOR MOBILE SERVICES

Chapter 7

Middleware for Mobile and Pervasive Services

Antoine B. Bagula, Mieso K. Denko, and M. Zennaro

Contents

Communication networks have evolved from small islands of closed networks into mobile pervasive networks that use the Internet as a common platform for all forms of modern communication with an increased appearance of mobile applications, users, and services [1] in a computing world where computing devices are available anywhere but remain invisible. Wireless mobility has enabled a ubiquitous dimension in modern communication networks where mobile users expect to be best connected anywhere and anytime for whatever task they perform [2]. Furthermore, the combination of wireless sensors [3] with radiofrequency identification (RFID) technology [4] is emerging as an important segment of the first mile of Internet connectivity which complements its ubiquitous dimension by allowing the information to be accessed not only anywhere and anytime but also by anyone and using anything. Derived from the Latin world *Ubique*, which means *everywhere*, ubiquitous computing, also called pervasive computing and very often related to ambient intelligence, is a postdesktop model of human–computer interaction that considers a thorough integration of information processing into everyday objects by involving many computational devices and systems simultaneously into everyday activities to achieve specific tasks. This is opposed to the desktop paradigm that considers only a single device designed for a specialized task in a more homogeneous environment. As illustrated by Figure 7.1, a common vision of ubiquitous networking consists of sharing small, inexpensive, and robustly internetworked processing devices that are generally launched into distinctly commonplace ends such as homes, markets, hospitals, streets, and workplaces. These devices are distributed at all scales throughout everyday life to deliver different services in a heterogeneous environment that involves a number of applications, protocols, operating systems, processors, and architectures. Figure 7.1 illustrates the schematic layers of a ubiquitous sensor network (USN) where different layers are used to provide different services to different types of applications in a multitechnology, multidevice, multiprotocol platform. It reveals (1) a sensor networking layer (the bottom layer) where sensor and RFID devices are launched into the environment to sense what is happening and report to sink nodes via USN bridges; (2) a USN access networking layer (the second layer) where the combination of USN bridges and sink nodes are used as an access network for the first-mile connectivity of a next-generation network (NGN) of gateways; (3) a USN middleware layer (third layer) that is used as an interface between the NGN and the application layer; and (4) different applications embodied into a USN application layer (the last layer) to perform tasks related to logistics, structural health monitoring, agriculture control, disaster surveillance, military field surveillance, and disaster/crisis management. The ubiquitous networking scenario

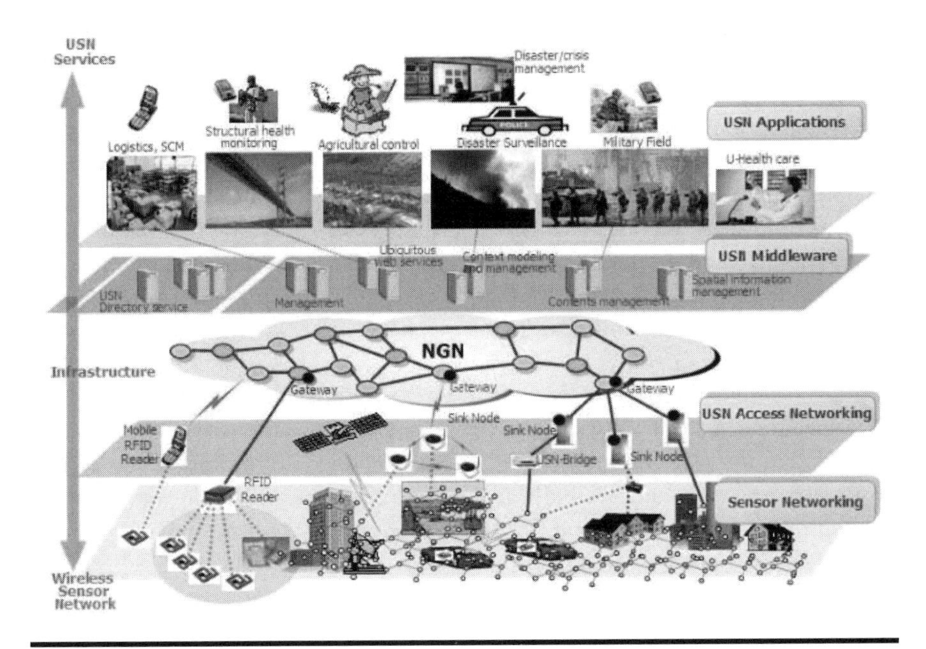

Figure 7.1 Schematic layers of a ubiquitous sensor network.

illustrated by Figure 7.1 is built around a communication platform using (1) different protocols such as WiFi, wireless LAN technology, and code- and time-division multiple access wireless communication protocols; (2) different devices that are based on different processors such as personal digital assistants (PDAs), smart phones, and laptops; and (3) all these protocols and devices being built around different architectures such as centralized, distributed, or peer-to-peer architecture. Such a heterogeneous platform requires a software layer often referred to as *middleware* that resides between programs, operating systems, hardware, and communication platforms to let the different applications and systems work together. The main functions of such a middleware are threefold. First, it aims at hiding the underlying complexity of the environment. Second, it helps in insulating the applications from explicit protocol handling, disjoint memories, data replication, network faults, and parallelism. Lastly, such a middleware hides the heterogeneity of computer architectures, operating systems, programming languages, and networking technologies to facilitate application programming and management. As assumed by Vaughan-Nichols [5], this is done by easing the transformation of markup languages, delivery of content and data, recognition of protocols and devices, incorporation and routing of business logic through enterprise systems, and the adaptation of data formats for compatibility with multiple databases.

7.1 Introduction and Background

Middleware systems for mobile and pervasive computing can be classified using different criterion. On the one hand, building on a perceived dual view between the information technology vendor and the wireless service provider, where the former supports a software infrastructure view of middleware for different types of applications and the latter adopts a narrower approach supporting only few applications, Vaughan-Nichols [5] classifies middleware into (1) device-level middleware, (2) carrier-level middleware, (3) application-integration middleware, and (4) standalone and hosted wireless middleware.

Device-level middlewares are run on mobile devices to enable the execution of local applications even in the absence of wireless connections. Such local applications include games, appointments, calendars, and so on. These device-level middlewares require minimum resources in terms of memory footprint and system processing resources and are built around common cross-platform development environments such as J2ME and the Qualcomms binary runtime environment for wireless connections.

Carrier-level middlewares are used by telecommunication carriers to provide reliable connections between their infrastructure and wireless networks. A project like "Parlay" by the Third-Generation Partnership Project (3GPP) [6] has been built around this type of middleware with the objective of developing commercial carrier-level wireless middlewares.

Usually operating on a server that sits between the mobile device and a traditional application, an *application-integration middleware* enables enterprise application integration by letting traditional programs such as database systems work with wireless applications. An evolution of such a type of middleware is perceived in the transition toward a "mobile phone sensing" paradigm where mobile phones are used as sensors recording what is happening in the environment and reporting the sensor reading for public safety purposes. The Personal Environmental Impact Report (PEIR) [7] is an example of an application illustrating this evolution by providing an online tool that allows citizens to use their mobile phone to explore and share how they impact on the environment and how the environment impacts on them. PEIR uses location data that are regularly and securely uploaded from a mobile phone to create a dynamic and personalized report about the environmental impact and exposure of a mobile-phone user.

A *hosted middleware* is hosted by a wireless service provider for customers. This approach has the advantage of being inexpensive and eliminates the need for customers to manage the software. A *standalone middleware* is one that a company buys and uses on its own systems. It has the advantage of giving users more control over the middleware but is more expensive and requires that the company manages the software.

On the other hand, two other classifications built on cluster-oriented architectures [8] may emerge from a dual view differentiating middleware for mobile computing from middleware for wireless sensor networks (WSNs). Cluster-based architectures have been widely investigated in both MANETs and WSNs to promote resource usage efficiency in large dynamic networks by easing the implementation of data-centric mechanisms and localized algorithms. However, compared with mobile networks that are highly dynamic, WSNs are designed to remain more static and usually consist of stationary nodes. Although this eases the use of clustering to control energy efficiency in WSNs, clustering itself is a highly costly process for mobile networks. Most middleware for WSNs are thus based on layering a clustered architecture above the physical WSN at an affordable price, whereas this is more costly for mobile computing middleware. Middleware solutions have thus followed different classifications depending on whether mobile computing or WSNs are concerned.

7.2 Middleware for Mobile and Pervasive Computing

Middleware for mobile computing may be classified into different types [9–12]). Emmerich [9] laid the foundations of a fourfold classification that ranked middlewares into transactional middleware, procedural middleware, message-oriented middleware, and object-oriented middleware. A similar classification was proposed by Talarian Corporation [10] where homegrown middlewares are added to this fourfold classification and specific emphasis is put on deployment by looking at the middleware market and describing the "smart sockets" as a commercial message-oriented middleware (MoM) solution. Hennadiy [11] revisits this classification by looking at its advantages, disadvantages, and applicability whereas Park et al. [12] apply this classification in the design of real-time and fault-tolerant middleware that looks at the automotive requirements to adopt an MoM for automotive software.

7.2.1 Classification of Middleware for Mobile Computing

Transactional middleware supports the development of systems that allow transactions to run across multiple distributed hosts. A transaction ensures that the operations required will occur either on all hosts in the system or on no hosts in the system. Transactional middleware uses the two-phase commit protocol to implement these transactions. The distributed transaction processing (DTP) protocol defines a programmatic interface for two-phase commit, which is used by most relational database management systems. This allows servers and database management systems to be easily integrated. Transactional middleware supports both synchronous and asynchronous communication between hosts and are a good fit in situations where transactions need to be coordinated and synchronized over multiple databases.

Procedural middleware supports remote procedure calls (RPCs) as its communication model. In this model, communication can be made by using primitives similar to local procedure calls. Specifically, a server exports several procedures for communication and a client invokes these procedures. Then, procedural middleware marshals an RPC into several messages and vice versa. As suggested by Talarian Corporation [10], procedural middlewares are not a good match for enterprise-wide applications requiring high performance and high reliability. They are better for small, simple applications using, primarily, point-to-point communication.

MoM is a specific class of middleware that supports the exchange of general-purpose messages in a distributed (also referred to as uncoupled or loosely coupled) application environment. MoM operates on the principles of message passing and/or message queuing that supports both synchronous and asynchronous interactions between distributed computing processes. MoM ensures message delivery by using reliable queues and by providing the directory, security, and administrative services required to support messaging [13]. MoMs are built around a reliability paradigm that is suitable for applications where the availability of a network or all components is not warranted.

Object-orineted middleware is an evolution from procedural middleware. It adds to middleware the concept of object-oriented programming such as object identification through references and inheritance. This type of middleware enables independent development and distribution of each component as each interaction of components is defined by interfaces. Object middleware evolved from RPCs. The development of object-oriented middleware mirrors similar evolutions in programming languages where object-oriented programming languages such as C ++ evolved from procedural programming languages such as C. The idea here is to make object-oriented principles, such as object identification through references and inheritance, available for the development of distributed systems. Systems in this class of middleware include the Common Object Request Broker Architecture (CORBA) of the Object Management Group (OMG), the latest versions of Microsoft's Component Object (COM), and the Remote Method Invocation (RMI) capabilities that have been available since Java. More recent products in this category include middleware that supports distributed components such as enterprise Java beans. Object-oriented middlewares are suitable for applications whose scalability is not an immediate concern.

These types of middleware have been studied on the basis of some key features that include network communication, coordination, reliability, scalability, and heterogeneity.

7.2.2 Features of Middleware for Mobile Computing

7.2.2.1 Network Communication

In *transactional middleware*, network communication is performed through definition of services offered by server and client components and implementation of both of these components within a transaction. The middleware allows client and server components to reside on different hosts and therefore requests are transported via the network in a way that is transparent to client and server components.

Procedural middleware uses RPCs to support the definition of server components as RPC programs. An RPC program exports a number of parameterized procedures and associated parameter types. Clients that reside on other hosts can invoke those procedures across the network. Procedural middleware implements these procedure calls by marshalling the parameters into a message that is sent to the host where the server component is located. The server component unmarshals the message, executes the procedure, and transmits the marshalled results back to the client, if required. Marshalling and unmarshalling are implemented in client and server stubs, which are automatically created by a compiler from an RPC program definition.

In *MoM*, client components use MoM to send a message to a server component across the network. The message can be a notification about an event or a request for a service execution from a server component. The content of such a message includes service parameters. The server responds to a client request with a reply message containing the result of the service execution.

Object-oriented middleware supports a distributed object request, which means that a client object requests the execution of an operation from a server object that may reside on another host. The client object has to have an object reference to the server object. Marshalling operation parameters and results is again achieved by stubs that are generated from an interface definition.

7.2.2.2 Coordination

Transactional middleware coordination is achieved by allowing client components to request services using synchronous or asynchronous communication, supporting various activation policies, and allowing services to be activated on demand and deactivated when they have been idle for some time. Activation can also be permanent, allowing the server component to always reside in memory.

In *procedural middleware*, RPCs are used as synchronous interactions between exactly one client and one server. Asynchronous and multicast communication is not supported directly by procedural middleware. Procedural middleware provides different forms of activating server components. Activation policies define whether a remote procedure program is always available or has to be started on demand. For startup on demand, the RPC server is started by an `inetd` daemon as soon as a request arrives. The `inetd` requires an additional configuration table that provides for a mapping between remote procedure program names and the location of programs in the file system.

A strength of *MoM* is that this paradigm supports asynchronous message delivery very naturally. The client continues processing as soon as the middleware has taken the message. Eventually, the server will send a message including the result and the client will be able to collect that message at an appropriate time. This achieves decoupling of client and server and leads to more scalable systems. The weakness, at the same time, is that the implementation of synchronous requests is cumbersome as the synchronization needs to be implemented manually in the client. A further strength of MoM is that it supports group communication by distributing the same message to multiple receivers in a transparent way.

The default synchronization primitives in *object-oriented middleware* are synchronous requests that block the client object until the server object has returned the response. However, the other synchronization primitives are supported too. CORBA 3.0, for example, supports both deferred synchronous and asynchronous object requests. Object-oriented middleware supports different activation policies. These include whether server objects are active all the time or started on demand. Threading policies are available that determine whether new threads are started if more than one operation is requested by concurrent clients or whether they are queued and executed sequentially. CORBA also supports group communication through its event and notification services. These services can be used to implement push-style architectures.

7.2.2.3 Reliability

Transactional middleware reliability is achieved by allowing a client component to cluster more than one service request into a transaction using the two-phase commit protocol. This is achieved even if the server components reside on different machines and are implemented by the DTP protocol, which has been adopted by the Open Group.

In *procedural middlewares,* the RPCs are executed with at-most-once semantics. The procedural middleware returns an exception if an RPC fails. Exactly once semantics or transactions are not supported by RPC programs.

MoM achieves fault-tolerance by implementing message queues that store messages temporarily on persistent storage. The sender writes the message into the message queue and if the receiver is unavailable because of a failure, the message queue retains the message until the receiver is available again.

In *object-oriented middleware,* the default reliability for object requests is at-most once. Object middleware supports exceptions, which clients catch in order to detect that a failure occurred during execution of the request. CORBA messaging or the notification service can be used to achieve exactly once reliability. Object middleware also supports the concept of transactions. CORBA has an object transaction service that can be used to cluster requests from several distributed objects into transactions. COM is integrated with Microsoft's Transaction Server, and the Java Transaction Service provides the same capability for RMI.

7.2.2.4 Scalability

Transactional middleware scalability is implemented by allowing transaction monitors to support load balancing and replication of server components, which is often based on replication capabilities that the database management systems provide upon which the server components rely.

The scalability of *procedural middleware* is limited by the RPCs. Unix and Windows RPCs do not have any replication mechanisms that could be used to scale RPC programs. Thus, replication has to be addressed by the designer of the RPC-based system.

MoMs do not support access transparency very well, because client components use message queues for communication with remote components, whereas it does not make sense to use queues for local communication. This lack of access transparency disables migration and replication transparency, which complicates scalability. Moreover, queues need to be set up by administrators and the use of queues is hard-coded in both client and server components, which leads to rather nonflexible and poorly adaptable architectures.

The support of *object-oriented middleware* for building scalable applications is still somewhat limited. Some CORBA implementations support load-balancing, for example, by using name servers that return an object reference for a server on the least-loaded host, or using factories that create server objects on the least-loaded host, but support for replication is still rather limited.

7.2.2.5 Heterogeneity

Transactional middleware heterogeneity is achieved by allowing the components to reside on different hardware and operating system platforms and different database management systems to participate in transactions using the standardized DTP protocol.

Procedural middleware can be used with different programming languages. Moreover, it can be used across different hardware and operating system platforms. Procedural middleware standards define standardized data representations that are used to represent transport of requests and results. The distributed computing environment, for example, standardizes the network data representation for this purpose. When marshalling RPC parameters, the stubs translate hardware-specific data representations into the standardized form and reverse mapping is performed during unmarshalling.

MoM does not support data heterogeneity very well either, as the application engineers have to write the code for marshalling. With most products, there are different programming language bindings available.

Object-oriented middleware supports heterogeneity in many different ways. CORBA and COM both have multiple programming language bindings so that client and server objects do not need to be written in the same programming language. They both have a standardized data representation that they use to resolve heterogeneity of data across platforms. Java/RMI takes a different approach as heterogeneity is already resolved by the Java virtual machine in which both client and server objects reside. The different forms of object middleware interoperate. CORBA defines the Internet InterOrb Protocol (IIOP) standard that governs how different CORBA implementations exchange request data. Java/RMI leverages this protocol and uses it as a transport protocol for remote method invocations, which means that a Java client can perform a remote method invocation of a CORBA server and vice versa. CORBA also specifies an interworking specification to Microsoft's COM.

An assessment of the strengths and weaknesses of MoM reveals that this class of middleware is particularly well suited for the implementation of distributed event notifications and publish/subscribe-based architectures. The persistence of message queues also allows the event notification to be achieved in fault-tolerant ways by allowing components to receive events when they restart after a failure. Some of the weaknesses of MoM include (1) support for at-least-once reliability, where the same message could be delivered more than once; (2) no support for transaction properties, such as atomic delivery of messages to all or no receiver; and (3) only limited support for scalability and heterogeneity. The loosely coupled architecture of MoM provides several advantages over traditional client–server systems. These advantages include a high degree of anonymity, persistent storage, and latency hiding. A *high degree of anonymity* between the message producer and the consumer in a system is an advantage as, for a consumer, it does not matter which producer sent a given message. This, in turn, makes messaging systems very flexible and highly scalable, as entire subsystems can be replaced without the need to replace other parts of the system. *Persistent storage* allows backup of the message transfer medium. It allows asynchronous delivery of messages by enabling the sender and the receiver to connect to the network at different times. This becomes particularly useful when dealing with intermittent connections, such as those in mobile communications. *Latency hiding* allows the client to be freed and continue with other tasks once a request message from the client is transmitted. This differs from the typical implementations of client–server systems where a client is suspended while its request is transmitted to the server, processed by the server, and a reply sent back to the client. With MoM, however, the request and reply phases are fully decoupled. There are two common messaging paradigms supported by current MoM systems: message queuing and publish/subscribe.

- *Message queuing:* The message queuing paradigm is a one-to-one model where a message sent by a message producer gets delivered to one of the interested message consumers. The main difference between this and the publish/subscribe paradigm is that only one of the interested message consumers will receive any one message. This one-to-one message distribution is useful in applications where one-to-one communication is required, but where the identity of the receiving client is less important. An example of a messaging product using the message queuing model is IBM's MQSeries [4].
- *Publish/subscribe*: The publish/subscribe paradigm, also referred to as topic-based messaging, is an asynchronous messaging paradigm whereby the providers of information (publishers) are decoupled from the consumers of

that information (subscribers) using a broker. Subscribers express interest in one or more classes, and only receive messages that are of interest, without knowledge of what (if any) publishers there are. An example of a messaging product using the publish/subscribe model is that of Tibco, Inc. JMS is the MoM standard for Java.

Object middleware provides very powerful component models. They integrate most of the capabilities of transactional, message-oriented, or procedural middleware. However, the scalability of object middleware is still rather limited and this disables use of the distributed object paradigm on a large scale.

7.3 Middleware for Pervasive Computing and WSNs

Several ubiquitous middleware architectures have been proposed by researchers [14,15], but these systems are unfortunately usually based on computation-intensive technologies such as laptops and PDAs that are not supported by other less-powerful mobile devices such as cell phones. Salminen and Riekki [1], for example, propose a lightweight middleware architecture for mobile phones. Although addressing some of the requirements for mobile middleware such as adaptability, interoperability, and context awareness, this architecture, however, does not consider autonomic communication. As described in Reference [16], Hermes is a middleware for mobile computing that uses agents and policies but considers only tools for the design and execution of distributed applications in mobile computing by using service agents. As research on middleware for autonomic communications has gained momentum through the emergence of several communication technologies, research focusing on only middleware for mobile computing has been presented in the literature [17–19]. Among those that focus on autonomic communication, very few deal with building the actual framework for developing ubiquitous applications in autonomic communication. Some studies on middleware [20,21] have focused mainly on autonomic communication in a disruptive environment that is defined by Kyng et al. [20] as a physical location where there is a high probability that some disruption will occur diverging from normal, expected behavior. A comprehensive policy-driven, autonomic solution that guarantees service delivery using multitechnology systems is presented in Reference [22].

It has been found that most of the research on autonomic communication deals with either infrastructure or *ad hoc* networks, discounting the integration of both aspects. This is exemplified for example by the work presented in Reference [23], where an autonomic system developed under the U.S. Army—CERDEC Dynamic ReAddressing and Management for the Army (DRAMA)—is presented. This work deals in great depth with policy-based networking but is based on only mobile *ad hoc* networks (MANETs). A middleware for mobile applications beyond 3G that deals with both *ad hoc* and infrastructure-based networks is presented in Reference [16]. However, it does not consider application requirements, such as session and security requirements, quality of service, and so on, which are important parameters on which the efficiency of the system using the middleware may depend. An innovative policy-driven approach is taken by RASCAL [2], where a middleware communication layer residing between the user applications and the underlying networks intercepts all messages sent from the application and decides, on the basis of a set of policies, the most appropriate actions to be taken. This makes applications in RASCAL remain communication- and protocol-specific. As suggested by Raatikainen [16], RASCAL does not consider device properties. As proposed by Tigli et al. [13], WComp is a ubiquitous middleware model that integrates (1) event-based Web services, (2) a lightweight component-based approach for the design of dynamic composite services, and (3) an adaptation approach using the concept of "aspect of assemble." Although the work presented in Reference [24] is based on a survey model, to the best of our knowledge, it presents the most comprehensive survey of requirements and solutions for ubiquitous software. Using the programming models used by WSNs, Hadim and Mohamed [25] classify middleware for WSNs into two different types: programming support middleware and programming abstractions middleware.

7.3.1 Classification of Middleware for WSNs

7.3.1.1 Programming Support Middleware

Programming support middleware provides systems, services, and runtime mechanisms such as reliable code distribution, safe code execution, and application-specific services. This class of middleware includes five subclasses: (1) virtual machines, (2) modular programming-based middleware, (3) database-based middleware, (4) application-driven middleware, and (5) message-driven middleware. *Virtual machines middleware* such as MATE that runs on TinyOS is based on a flexible virtual machine approach where virtual machines, interpreters, and mobile agents are used in a three-step process in which (1) the system developers write applications in separate small modules; (2) these modules are injected and distributed through the network using tailored algorithms; and (3) the modules are interpreted by the virtual machine. *Modular programming-based middleware* such as Impala is based on the modularity of the applications to facilitate their injection and distribution through the network using mobile codes. *Database-based middleware* such as Cougar is a middleware approach where the whole network is perceived as a virtual database system and an easy-to-use interface is provided to the user to query the sensor network. *Application-driven middleware* such as Milan is based on an architecture that reaches the network protocol stack to let programmers fine-tune the network on the basis of application requirements. *Message-driven middleware* such as Mires provides a communication model using the publish/subscribe mechanism to facilitate the exchange of messages between nodes and based station or sink nodes.

7.3.1.2 Programming Abstractions Middleware

Programming abstractions middleware focuses on the way we view a sensor network and provides concepts and abstractions of senor nodes and sensor data. It includes two subclasses of middleware: (1) global behavior and (2) local behavior middleware. *Global behavior middleware* such as Kairos introduces macroprogramming in sensor networks, a new paradigm where the sensor network is programmed as a whole rather than using low-level software programming to drive individual nodes. This is done using high-level specification to enable automatically generated nodal behaviors to relieve application developers from dealing with low-level concerns at each network node. *Local behavior middleware* such as Abstract regions focuses on the nature of the sensed data generally involving a group of sensor nodes in a specified region.

Some of the most known wireless sensor middleware systems include (1) Mate and MagnetOS that are based on the virtual machine approach; (2) Cougar, SINA, DsWare, and TinyDB that are built around the database abstraction approach; (3) Impala that uses the modular programming approach; (4) Milan that is modeled around the application-driven approach; (5) Mires that uses a message-oriented approach; (6) EnviroTrack and Abstract region that use local behavior; and (6) Kairos that is based on a macroprogramming global behavior. Whereas middleware design and management issues for sensor middleware are described in References [25,26], projects focusing on middleware for sensor networks are presented in References [26,27].

Mate [27] is a synchronous middleware based on the TinyOS operating system that uses a byte code interpreter and mobile active capsules. It aims to provide better interaction and adaptation to the ever-changing nature of sensor networks.

MagnetOS [28] is a middleware that is based on a single system image and the Java virtual machine model. It uses a single system image and object migration. Its main aim is to overcome the heterogeneity of distributed, *ad hoc* sensor networks by exporting the illusion of the Java virtual machine on top of distributed sensor networks.

Impala [29] is an asynchronous middleware based on an autonomic approach that uses mobile agents, binary instructions, and an application finite state machine.

Cougar [30] is based on a database approach that uses a virtual relational database to abstract data types. Using language such as the Standard Query Language, Cougar is very suitable for large sensor collections and offers an easy-to-use database query system for different network operations.

TinyDB [31] is based on an easy-to-use declarative query system that abstracts the network as a virtual database with SQL-like statement and semantic routing tress.

SINA [32] is an easy-to-use query-processing database that uses a spreadsheet database and SQL-like statement and incorporates hierarchical clustering and a base naming scheme.

DsWare [33] is a middleware that is based on the database-based approach and an easy-to-use interface similar to conventional database systems. It provides a set of services so that applications do not have to implement their own application data service, implements event detection, and supports real-time aspects.

Milan [34] is an application-driven middleware that deals with high-level concerns, quality of service requirements, and a network protocol stack.

Mires [35] is an asynchronous MoM that uses nesC programming and runs on TinyOS. It uses active messages in a publish/subscribe environment.

Kairos [36] is a centralized macroprogramming middleware that includes compile time and runtime subsystems on top of a TinyOS operating system. It represents the next step in sensor network programming by adopting a macroprogramming model.

Abstract region [37] runs on top of TinyOS as a suite of general-purpose communication primitives for developing middleware services for sensor network applications rather than a complete middleware. It is built around in-network data aggregation, radio and geographic neighborhood, and local regions, spanning tree methods, and approximates planar mesh.

EnviroTrack [38] is an object-oriented data-centric middleware that runs on top of TinyOS with the objective of achieving environmental tracking.

7.4 Comparison of Middleware for Mobile and Pervasive Computing

Current generation human–computer interaction models using command-line, menu-driven, or graphical user interface-based interaction are inappropriate and inadequate for ubiquitous environments that include contemporary devices such as mobile phones, digital audio players, RFID tags, global positioning system, and interactive whiteboard. To address this problem, modern ubiquitous systems have to tackle challenging issues in terms of system design, modeling, engineering, and user interface design. Some of the basic requirements and characteristics of ubiquitous systems have been studied by different authors in terms of requirements of computing and their effect on system, software and business [24]. They allow software services to be adapted to different kinds of terminals, different networks, dynamically changing contexts and user preferences different kinds of radio capabilities, and dynamically changing conditions through structural and behavioral changes. Whereas structural adaptation enables modification of an assembly of components while preserving its behavioral services, behavioral adaptation defines a sequence of operations to be executed on a particular component. Whereas this adaptability has an impact on software in terms of context awareness, personalization, resource management, and agility, it also affects business in terms of infrastructure, multiprovider, and services.

Heterogeneity is the ability to handle heterogeneous network technologies, operating systems, programming languages, devices and users, and interaction methods. Heterogeneity affects (1) software designs in terms of user interaction support, service access via different network infrastructures, and different service needs for the network and device capabilities and (2) the business in terms of multiactor system infrastructure, integration of multiprovider and multiactor devices, and value network. *Extensibility* allows easy extension and addition of new functionalities to the system. *Scalability* allows a system to grow in the future and extend to higher-load applications or a wider network. *Security* consists of using authentication, authorization, confidentiality, and accountability mechanisms to protect data. It involves reliable transactions, privacy of communication and content, and terminal and user location. It has an impact on software security, reliability, survivability, and location-monitoring and affects business in terms of security and fault-tolerance of multiactor system infrastructure and security and privacy of multiprovider service network. *Reactivity* enables a pervasive or ubiquitous adaptive system to react on context changes by handling some kind of event notification such as a publish/subscribe mechanism. It has an impact on software in terms of computing platforms for users, spontaneity, and dynamic structures and affects business in terms of multiactor system infrastructure and multiprovider and multiactor services.

Mobility is a requirement for all ubiquitous middlewares allowing the creation of applications from mobile devices and a changing context by enabling users to move from one place or terminal to another while still receiving a personalized service. Mobility has an impact on software in terms of actual, virtual, and physical mobility; full mobility and physical device; and personal, session, and service mobility. *Roaming* in multiactor systems and *value network* are two business parameters that are affected by mobility. Discovery allows devices to be discovered dynamically. Updating allows components or services to be updated at runtime. *Interoperability* is the ability of the software to understand the exchanged information.

Similar challenging requirements have been proposed by Hadim and Mohamed [25] to lead the design of WSNs. They include the management of limited power and resources. *Scalability* is the capability of a WSN to growth anywhere and anytime with the growth of applications without affecting network performance. Similar to mobile computing, scalable WSNs are flexible enough to grow with higher-load applications at acceptable performance. *Mobility dynamic* network topology management aims at achieving a robust network operation to adapt to a network-changing environment on malfunctioning, device failure, moving obstacles, mobility, and interference. Such a robust operation is achieved by supporting mechanisms for fault-tolerance, sensor node self-configuration, and sensor node self-maintenance. *Heterogeneity* in WSNs is implemented through low-level programming models that bridge the gap between the hardware technologys raw potential and the application requirements expressed in terms of a broad range of activities such as reconfiguration, execution, and communication.

Table 7.1 assigns a value of "1" or "0" to each of the 10 middleware models depending on whether they meet one of the nine requirements depicted by the first column. Note that the middleware models Gaia, One World, and Homeros are presented in the same column as they meet the same requirements. This table reveals that while the three models outperform the others by meeting all the nine requirements, the SCORPIO, Aura, ExORB, and SATIN models come second but by meeting the same requirements except for Aura while CORBA and DoAmI come third by meeting the same requirements and CORTEX is the least score.

Table 7.2 describes the efficiency of different middleware projects by assigning a value of "3" for fully meeting a requirement, "2" for partially meeting a requirement, and "1" for poorly meeting a requirement. The table reveals that Magnet and EnviroTrack outperform the other projects whereas Cougar comes last. The Abstract region has not been assigned any values as it is considered as only a subset of a middleware.

7.5 Implementation of Middleware for Wireless *Ad Hoc* and Sensor Networks

In this section, we present an overview of middleware for MANETs and discuss sensor network applications. The design and implementation details of a WSN application will also be presented.

7.5.1 A Publish/Subscribe Middleware for MANETs

Denko et al. [39] propose a mobility-aware cross-layer-based middleware (MOBCROSS) for MANETs. This is based on the publish/subscribe approach and the motivation behind this proposal is to overcome the limitations of existing middleware proposed for wireless networks and their use in infrastructureless networks. One of the main reasons for their not being of much help in this environment is that they do not take advantage of cross-layer optimization. Owing to the dynamic nature of the network, asynchronous publish/subscribe message middleware is a suitable communication model for MANET environments. The proposal by Denko et al. [39] provides the following main features:

- *Mobility awareness*: The middleware is aware of node mobility and address changes.
- *Network reconfiguration*: The middleware supports network reconfiguration and adaptation.
- *Message caching*: The middleware is capable of message caching for later delivery.
- *Cross-layer interaction support*: The middleware supports cross-layer information exchange and sharing.

Table 7.1 Mobile Pervasive Computing Middleware Systems

Requirement	*Gaia, One World, Homeros*	*CORTEX*	*CORTEX ExORB*	*Aura*	*CORBA*	*SATIN*	*DoAmI*	*SCORPIO*
Adaptability	1	0	1	0	0	1	0	1
Heterogeneity	1	0	0	1	1	0	1	0
Extensibility	1	0	1	0	0	1	0	1
Scalability	1	1	1	1	1	1	1	1
Security	1	1	1	1	1	1	1	1
Reactivity	1	1	0	1	0	0	0	0
Mobility	1	0	1	1	1	1	1	1
Discovery	1	0	0	1	1	0	1	0
Updating	1	0	1	0	0	1	0	1
Efficiency score	9	3	6	6	5	6	5	6

Table 7.2 Sensor Networking Middleware Systems

Project	*Power Awareness*	*Openness*	*Scalability*	*Mobility*	*Heterogeneity*	*Easy-of-Use*	*Efficiency Score*
Mate	3	3	3	3	2	1	15
Magnet	3	3	3	3	2	3	17
Cougar	2	1	1	1	1	3	9
SINA	3	1	1	1	1	3	10
DsWare	3	2	2	1	1	3	12
TinyDB	3	2	2	2	2	3	14
Impala	3	3	3	3	1	3	16
Milan	2	3	3	1	1	3	13
Mires	3	3	3	2	2	3	16
EnviroTrack	3	3	3	3	2	3	17
Kairos	2	2	2	2	2	2	12

Experimental results based on simulation have indicated that the proposed middleware performs well in terms of reducing overheads and packet delivery.

7.5.2 A Database-Oriented Middleware for WSNs

Besides using laptops and standalone workstation computers, the gateways used in WSNs may be built around smart boards such as the Alix [40] to provide flexibility and database replication at a cheaper price than traditional gateways. Such smart boards are endowed with modern communication interfaces such as Ethernet, WiFI, GPRS, and USB to disseminate sensor readings in different ways and to provide the capability to run MySQL databases with replication on different other boards forming a gateway network as depicted by the higher layer of Figure 7.2. As depicted by Figure 7.2 and implemented in our middleware, the use of smart boards as gateways allows the deployment of hybrid star/mesh networks, referred to as hybrid-SM, where a mesh network of smart boards (gateways) is layered above islands of star WSNs. This section describes the hybrid-SM middleware component as implemented by the WSN management system using Open Source SquidBee [41] and SunSPOT [42] technologies in our testbed.

A middleware is an important component of a WSN. As currently designed, wireless sensor middlewares are tightly coupled to the particular wireless sensor application that uses wireless sensor technology. Such a situation raises issues as the use and number of WSNs spreads. The design of a generalized interface layer that fulfills the required management roles in WSNs will be a major milestone in the adoption of the technology. In WSN middleware approaches, how the network is abstracted is of outmost importance, as it determines how the user who wants access to the network's data will interact with the WSN. The middleware proposed in this section is based on the database abstraction model described earlier as one of the implementation models for the commonly proposed middleware approaches for

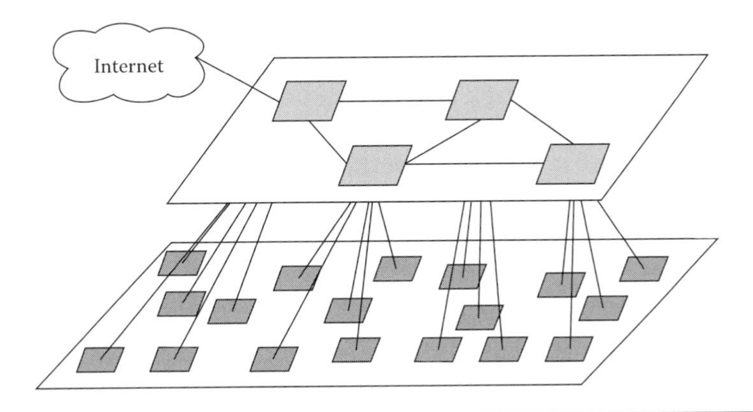

Figure 7.2 Cross-layer-based middleware for mobile *ad hoc* networks.

WSNs. It has been implemented in the environment monitoring testbed using the SquidBee [41] technology to measure environmental parameters such as humidity, light intensity, and temperature at different locations [43]. As described, database abstraction can be loosely defined as an abstraction whereby the WSN is viewed as a logical ordered set of data, stored within a relational database.

7.5.2.1 The Implemented Model

Figure 7.3 depicts the main components of the implemented middleware model for a testbed network using the SquidBee technology. These components include:

- *User*: The user is an abstraction of the service that requires data from the WSN. It queries the WSN using the widely used SQL and receives responses in the standard SQL format.
- *Control/Interface Layer*: This interface monitors the response from the relational database to the user. Trigger conditions can thus be set that will activate upon certain conditions and perform tasks. For example, the timestamp of the data requested is too far in the past from the current time. The layer, in this case, would update the data in the database using the data from the WSN interface. Hence, this layer also can interact with the WSN interface, and use it to update the relational database.
- *WSN Interface*: This is the interface to the actual WSN that transforms sensor readings in terms of voltage values into human-friendly values. Depending on the WSN and the user requests and WSN responses, different formats are used for sensor readings.
- *Relational Database*: The virtual version of the WSN is stored in this database. The structure of the database is dependent on the projected most-common requests for data.

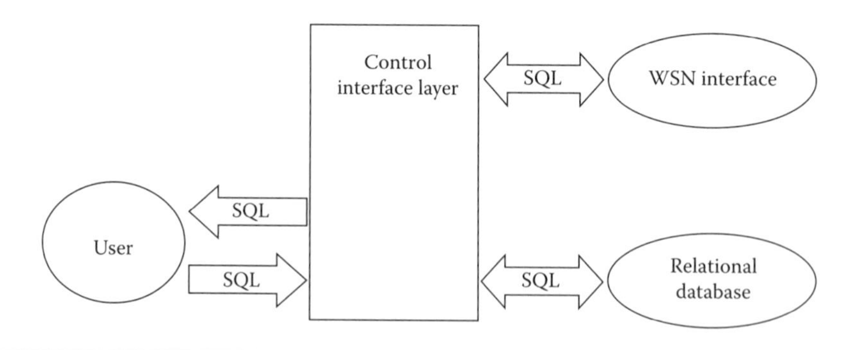

Figure 7.3 Components of the implemented middleware model for a testbed network using the SquidBee technology.

7.5.2.2 The Prototype Used by the Model

As implemented in the SquidBee testbed, the model presented by Figure 7.4 includes the following software components implemented in Python and PHP.

- *WSN Interface*: The interface is programmed in Python and communicated with the Squidbee Gateway over a USB connection. Essentially, it contains a method for requesting readings from the sensors and a method for reading the responses from the sensors. The gateway also wrapped the data into reading objects.
- *Readings*: These are the raw data from the wireless sensors that were wrapped into Python reading objects that contain all of the data sensed along with information such as sensor and network identities.
- *Control Layer*: This is programmed in Python and the script instructs the WSN interface to take readings and wrap them as reading objects. It then passes those reading objects to the database interface. The data reading procedure is time based.
- *Database Interface*: This interface interprets the reading objects and inserts them into the relational database used, in the decided-upon structure.
- *Relational Database*: This is a standard MySQL database used as a backend application.
- *User*: This represents a user. A phpMyAdmin database administration software is used to interact with the relational database.

Some of the strengths of the proposed model have a number of benefits. First, the model has a well-established external interface (SQL). Second, the control of actual data requests sent to the WSN allows for high-level policy setting to control traffic volumes and hence minimize power cost on the WSN itself. Third, given a

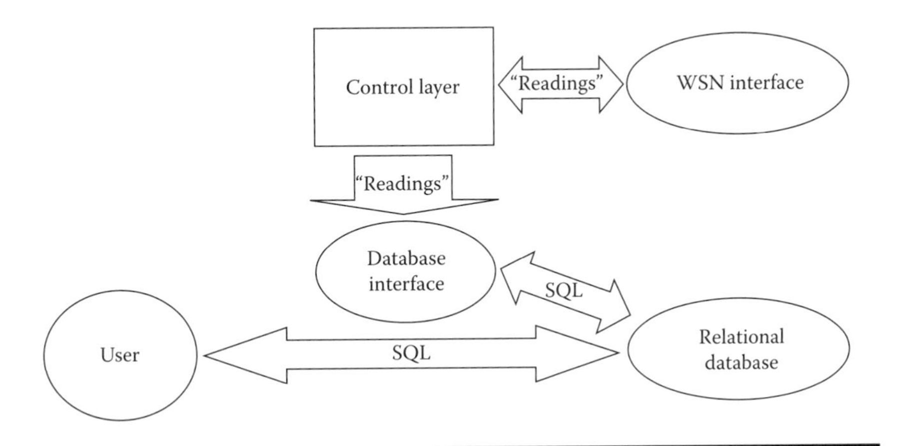

Figure 7.4 Prototype used by the implemented middleware model.

WSN interface, potentially, any WSN can be interfaced with it. Finally, it can facilitate a high volume of external requests fairly easily. The model also has some limitations such as limited input types and the need for more general applications.

7.6 Future Trends and Research Directions

Most existing middleware solutions for mobile and pervasive networks are specific to the network and applications for which they have been designed. This limits their suitability for applications in emerging next-generation mobile and pervasive computing. Thus, the convergence of different communication networks, technologies, and services requires new techniques and approaches for middleware design and implementation. Some of the main research challenges in next-generation mobile and pervasive computing middleware are:

- Owing to the heterogeneity of networks and computing and communication devices, middleware solutions based on one network cannot be directly applied to heterogeneous networks and ubiquitous systems. This requires new research in the design of communication architecture and protocols that dynamically adapts to heterogeneous networks and computing platforms.
- The growing complexity and heterogeneity of networks and computing devices make the design of middleware more challenging. Research efforts addressing autonomic configuration, self-organization, and self-healing supporting fault-tolerance and context awareness are needed to close this gap.
- Security is important to protect user identity and system resources in heterogeneous and pervasive computing environments. Hence, the design of secure and efficient middleware for mobile and resource-constrained pervasive computing is an area of future research. We envision that future research in mobile and pervasive middleware will address these and other similar architectural, algorithmic, and implementation issues while also developing metrics for evaluation and deployment.

7.7 Conclusions

In this chapter, we introduced middleware solutions for mobile and pervasive services. The classification of existing and emerging middleware solutions has been presented. The importance and some distinguishing features of existing solutions, implementation challenges, and their limitations have been discussed. The chapter also presents middleware for MANETs and WSNs. An application scenario using the implementation and deployment testbed for WSN middleware was discussed. The prototype implementation was based on MySQL relational databases and

SquidBee testbed. Both the strengths and weaknesses of the proposed model were outlined to give ideas for further improvement. Finally, a brief discussion on future trends and possible research directions was presented.

References

1. T. Salminen and J. Riekki, Lightweight middleware architecture for mobile phones, In *Proceedings of the 2005 International Conference on Pervasive Systems and Computing (PSC 2005)*, Las Vegas, Nevada, pp. 147–153, June 27–30, 2005.
2. R. Ghizzioli and D. Greenwood, The RASCAL system for managing autonomic communication in disruptive environments, *Proceedings of the 1st IEEE Workshop on Autonomic Communication and Network Management (ACNM '07), as part of the 10th IFIP/IEEE International Symposium on Integrated Network Management*, Munich, Germany, pp. 41–47, May 21–25, 2007.
3. I.F. Akyildiz, W. Su, Y. Sankarasubramaniam, and E. Cayirci, A survey on sensor networks, *IEEE Communications Magazine*, 40(8), 102–114, 2002.
4. ITU-T, Ubiquitous sensor networks (USN), *ITU-T Watch Briefing Report Series*, No. 4, February 2008.
5. S.J. Vaughan-Nichols, Wireless middleware: Glue for the mobile infrastructure, *Computer*, 37(5), 18–20, 2004.
6. 3GPP—A global initiative, The mobile broadband standard, http://www.3gpp.org
7. PEIR project, http://urban.cens.ucla.edu/projects/peir/
8. M. Steenstrup, Cluster-based networks, In C.E. Perkins, *Ad Hoc Networking*, Addison-Wesley, New York, 2001, pp. 75–138.
9. W. Emmerich, Software engineering and middleware: A roadmap, In *Proceedings of the 22nd International Conference on Software Engineering (ICSE 2000), the Future of Software Engineering Track*, Limerick, Ireland, pp. 76–90, June 4–11, 2000.
10. Talarian Corporation, Everything you need to know about middleware: A guide to selecting a real-time infrastructure, 2000, http://searchwebservices.techtarget.com/searchWebServices/downloads/Talarian.pdf, last accessed May 7, 2009.
11. P. Hennadiy, Middleware: Past and present—A comparison, White paper of the University of Maryland, http://userpages.umbc.edu/~dgorin1/451/middleware/middleware.pdf, 2004, last accessed May 7, 2009.
12. J. Park, S. Kim, W. Yoo, and S. Hong, Designing real-time and fault-tolerant middleware for automotive software, In *Proceedings of SICE-ICASE International Joint Conference*, Busan, Korea, pp. 4409–4414, October 18–21, 2006.
13. J.-Y. Tigli, S. Lavirotte, G. Rey, V. Hourdin, D. Cheung-Foo-Wo, E. Callegari, and M. Riveill, WComp middleware for ubiquitous computing: Aspects and composite event-based Web services, *Annals of Telecommunications*, 64(3–4), 197–214, 2009.
14. M. Román, C. Hess, R. Cerqueira, R.H. Campbell, and K. Nahrstedt, Gaia: A middleware infrastructure for active spaces, *IEEE Pervasive Computing*, 1, 74–83, 2002.
15. R. Grimm, One world: Experiences with a pervasive computing architecture, *IEEE Pervasive Computing*, 3, 22–30, 2004.
16. K. Raatikainen, Middleware for mobile applications beyond 3G, In O. Martikainen, K. Raatikainen, and J. Hyvärine, editors, *Smart Networks*, Kluwer Academic Publishers, Dordrecht, The Netherlands, 2002, pp. 3–18.

17. C.-L. Fok, G.-C. Roman, and G. Hacmann, A lightweight coordination middleware for mobile computing, *Lecture Notes in Computer Sciences*, 2949/2004, 135–151, 2004.
18. R. Cheung, An adaptive middleware infrastructure incorporating fuzzy logic for mobile computing, In *Proceedings of the International Conference on Next Generation Web Services Practices (NWeSP 2005)*, Seoul, Korea, pp. 449–451, August 22–26, 2005.
19. C. Mascolo, L. Capra, and W. Emmerich, Mobile computing middleware, *Lecture Notes in Computer Science*, 2497, 20–58, 2002.
20. M. Kyng, E.T. Nielsen, and M. Kristensen, Challenges in designing interactive systems for emergency response, In *Proceedings of the 6th ACM Conference on Designing Interactive Systems*, New York, NY, USA, pp. 301–310, June 26–28, 2006.
21. E.T.N. Margit Kristensen and M. Kyng, IT support for healthcare professionals acting in major incidents, In *Proceedings of the 3rd Scandinavian Conference on Health Informatics*, Aalborg, Denmark, pp. 37–41, August 25–26, 2005.
22. M. Calisti and M.D. Greenwood, Adaptive service access management for ubiquitous connectivity, In *Proceedings of the 1st IEEE Workshop on Management of Ubiquitous Communication and Services (MUCS), as part of the 10th IFIP/IEEE International Symposium on Integrated Network Management*, Munich, Germany, May 21–25, 2007. Available at http://www.whitestein.com/library/WhitesteinTechnologies_Paper_MUCS2007.pdf
23. R. Chadha, Y.-H. Cheng, J. Chiang, G. Levin, S.-W. Li, and A. Poylisher, Policy-based mobile ad hoc network management for drama, In *Proceedings of the IEEE Military Communications Conference (MILCOM 2004)*, IEEE Vol. 3, pp. 1317–1323, October 31–November 3, 2004.
24. E. Niemela and J. Latvakoski, Survey of requirements and solutions for ubiquitous software, In *Proceedings of the 3rd International Conference on Mobile and Ubiquitous Multimedia (MUM 2004)*, College Park, Maryland, pp. 71–78, October 27–29, 2004.
25. S. Hadim and N. Mohamed, Middleware: Middleware challenges and approaches for wireless sensor networks, *IEEE Distributed Systems Online*, 7(3), 1, 2006.
26. K. Römer, C. Frank, P.J. Marron, and C. Becker, Generic role assignment for wireless sensor networks, In *Proceedings of the 11th ACM SIGOPS European Workshop*, Leuven, Belgium, pp. 7–12, September 19–22, 2004.
27. P. Levis and D. Culler, Mate: A tiny virtual machine for sensor networks, In *Proceedings 10th International Conference Architectural Support for Programming Languages and Operating Systems (ASPLOS-X)*, San Jose, California, pp. 85–95, October 5–9, 2002.
28. R. Barr, J.C. Bicket, D.S. Dantas, B. Du, T.W. Danny Kim, B. Zhou, and E.G. Sirer, On the need for system-level support for ad hoc and sensor networks, *Operating Systems Review*, 36(2), 1–5, 2002.
29. T. Liu and M. Martonosi, Impala: A middleware system for managing autonomic, parallel sensor systems, In *Proceedings of the ACM SIGPLAN Symposium on Principles and Practice of Parallel Programming (PPoPP '03)*, San Diego, California, pp. 107–118, June 11–13, 2003.
30. P. Bonnet, J. Gehrke, and P. Seshadri, Towards sensor database systems, In *Proceedings of the 2nd International Conference on Mobile Data Management (MDM '01)*, Hong Kong, pp. 314–810, January 8–10, 2001.
31. S.R. Madden, M.J. Franklin, and J.M. Hellerstein, TinyDB: An acquisitional query processing system for sensor networks, *ACM Transactions Database Systems*, 30(1), 122–173, 2005.

32. C. Srisathapornphat, C. Jaikaeo, and C. Shen, Sensor information networking architecture, In *Proceedings of the International Workshop on Parallel Processing*, Toronto, Canada, pp. 23–30, August 21–24, 2000.
33. S. Li, S. Son, and J. Stankovic, Event detection services using data service middleware in distributed sensor networks, In *Proceedings of the 2nd International Workshop on Information Processing in Sensor Networks (IPSN 03)*, Palo Alto, California, pp. 502–517, April 22–23, 2003.
34. W.B. Heinzelman, A.L. Murphy, H.S. Carvalho, and M.A. Perillo, Middleware to support sensor network applications, *IEEE Network*, 18(1), 6–14, 2004.
35. E. Souto, G. Guimarães, G. Vasconcelos, M. Vieira, N. Rosa, and C. Ferraz, A message-oriented middleware for sensor networks, In *Proceedings of the 2nd International Workshop on Middleware for Pervasive and Ad Hoc Computing (MPAC '04)*, Toronto, Canada, pp. 127–134, October 18–22, 2004.
36. R. Gummadi, O. Gnawali, and R. Govindan, Macro-programming wireless sensor networks using Kairos, In *Proceedings of the International Conference on Distributed Computing in Sensor Systems (DCOSS '05)*, Marina del Rey, California, pp. 126–140, June 20–July 1, 2005; *Lecture Notes in Computer Science*, Vol. 3560, Springer, Berlin/Heidelberg.
37. M. Welsh and G. Mainland, Programming sensor networks using abstract regions, In *Proceedings of the 1st Usenix/ACM Symposium On Networked Systems Design and Implementation (NSDI '04)*, San Francisco, California, pp. 29–42, March 29–31, 2004.
38. T. Abdelzaher, B. Blum, Q. Cao, D. Evans, S. George, J. Stankovic, T. He, et al., EnviroTrack: Towards an environmental computing paradigm for distributed sensor networks, In *Proceedings of the 24th International Conference on Distributed Computing Systems (ICDCS 04)*, Tokyo, Japan, pp. 582–589, March 23–26, 2004.
39. M.K. Denko, E. Shakshuki, and H. Malik, Enhanced cross-layer based middleware for mobile ad hoc networks, *Journal of Network and Computer Applications*, 32(2), 490–499, 2009.
40. PC Engines, Alix2 board. http://www.pcengines.ch/alix2c2.htm, last accessed March 2009.
41. SquidBee. http://www.SquidBee.org, last accessed March 2009.
42. SUNSpots. http://www.SunSPOTworld.com/, last accessed March 2009.
43. A. Bagula, G. Inggs, S. Scott, and M. Zennaro. On the relevance of using open wireless sensor networks in environment monitoring, *Sensors*, 9(6), 4845–4868, 2009.

Chapter 8

Context-Aware Middleware for Supporting Mobile Applications and Services

Wenwei Xue and Hung Keng Pung

Contents

8.1 Introduction

We inhabit a world that is moving toward a profusion of digital devices for assisting the well-being of humans, including the sustainable use of resources in our living environments—both physical and virtual. The physical environment is being digitally equipped and strewn with embedded sensor-based devices, portable and mobile devices, together with conventional computing devices and systems.

Pervasive computing enables the utilities and computing capabilities derived from these environments to be intelligently released and efficiently utilized [1,2]. This exciting vision is also known as *ubiquitous computing*. Though there are research issues common to both pervasive computing and mobile computing, they are indeed two different research areas that do not overlap. More specifically, the research issues of pervasive computing are closely related to those of mobile computing, wireless *ad hoc* and sensor networks, artificial intelligence, software engineering, human–computer interaction, and social computing.

The ultimate goal of pervasive computing is to provide "anytime, anywhere" network connectivity and services away from the traditional desktop model to mobile human users throughout the world in a way that is transparent to the underlying details of hardware and software architectures. *Context awareness* is a key enabler of pervasive computing for the dynamic and continuous adaptation of operations or behaviors in applications with as little explicit human intervention as possible [3,4]. A context-aware application can sense the context in its current environments and react to such changing context automatically. A *context-aware pervasive system* encompasses a set of system software technologies (software and hardware) and services that enable actors (e.g., human and software agents) to interact through context-aware applications running on portable devices (e.g., laptops, personal digital assistants—PDAs, phones, sensors, and actuators) and other computing hardware platforms (e.g., servers and databases) cohesively. The system facilitates the provision of information and assistance for the applications to make appropriate decisions in the *right* manner, at the *right* time, and at the *right* place (*3Rs*). Figure 8.1 sketches

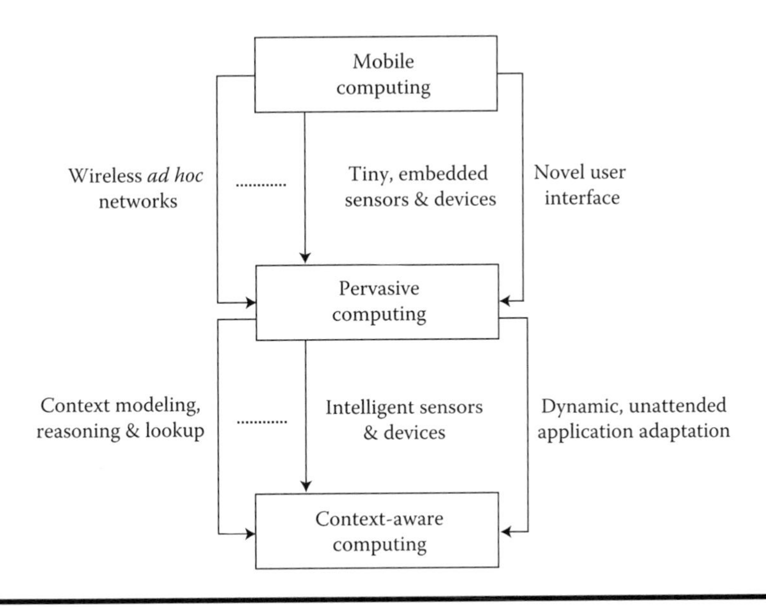

Figure 8.1 From mobile to pervasive context-aware computing.

the evolution of research on future computing paradigms from mobile computing to pervasive and context-aware computing.

Many different definitions of context are reported in the literature. We adopt here the definition given by Dey [5:4–7], in which "context is any information that is used to characterize the situation of an entity involved in the user-application interaction." According to such a definition, we use the two terms "context" and "context data" interchangeably in this chapter. We define a *context source* as any entity described in the definition that can provide one or more kinds of context data, and a *context attribute* as a specific kind of context data. Examples of context sources are sensors, computers, people, vehicles, homes, shops, and offices. Variants of context definitions reported in the literature are surveyed and their categorizations are presented in References [6,7].

Despite being investigated for more than a decade, context-aware systems have not been widely deployed today partly because of the immaturity of the context acquisition technologies, which are not reliable and not "plug-and-play" systems. There is a lack of appropriate software engineering methodology and programming tools to support the development and deployment of applications, as well as a lack of novel and adaptive interfaces for end-users. Furthermore, the sheer diversity of exploitable context data and the potentially large number of heterogeneous sensing technologies for data acquisition work against the real-world deployment of interoperable context-aware systems.

Motivated by its potential benefits of lowering the application development effort and to better application robustness, many recent researches have approached the problem of generic systems support for context-aware computing from a middleware point of view [8]. A *context-aware middleware* is a software system lying between context-aware mobile applications (services) and context sources, excluding operating systems and networking. A service-oriented context-aware middleware should consist of functional components operating as system services to perform context acquisition (such as discovery, lookup, and retrieval) and facilitate their processing (such as context data aggregation and reasoning) for the applications.

A generic context-aware middleware that supports heterogeneous applications in various domains, such as commerce, healthcare, road transportation, and environmental resources management, must have most of the following features:

- *Management of heterogeneous context sources*: The middleware needs to manage numerous context sources in the physical or even virtual environments [9]. These context sources are distributed throughout the Internet, and are likely to be administered by different organizations. As a result, context data may be modeled and represented differently by the respective stakeholders [10]. The middleware should therefore unify these models and provide a single abstract view of context representation over all sources to the applications. Moreover, context sources may dynamically join and leave the middleware system from time to time, especially for those that are mobile such as persons and vehicles. The middleware must shield such dynamics and mobility of context sources to the applications.

- *Real-time context provision*: The middleware should provide a friendly query or programming interface that allows applications to flexibly specify what context attributes are to be acquired from which context sources at all times. A context attribute can represent *static* data that seldom changes or *dynamic* data that changes frequently. Examples of static attributes include *name*, *preference* of a person and *location*, *opening hours* of a shop, and so on. Examples of dynamic attributes include *location*, *mood* of a person and *temperature*, *crowd level* in a shop, and so on. Because a context-aware application requires up-to-date context from the middleware to drive its behavior adaptation, the middleware must implement an efficient context lookup mechanism to promptly obtain such context from corresponding sources. Otherwise the adaptation cannot be fulfilled in real time and may become meaningless to the user. For dynamic context attributes, their values may become obsolete and result in incorrect adaptations if not returned to the application in a timely manner.
- *Support of context reasoning*: A context source can provide many simple context attributes whose values are sampled or fused by multiple types of sensors associated with the source, for example, the *temperature*, *light*, and *noise* in an office. We use the term "sensor" to generally refer to any physical sensing device or logical software program that produces the values of a simple attribute. A context source can also provide deduced context attributes whose values are postprocessed or reasoned from those of simple attributes. We call each of these deduced attributes an *event*. An event may represent a physical phenomenon (e.g., gas leakage or fire emergency) or a human activity (e.g., having a party or watching TV) occurring in the context source (e.g., a smart home). A context-aware middleware should use effective context reasoning techniques centrally at a server and/or distributed in agents of individual sources, to enable intra- and intersource event detection. This makes the context sources "smart" and capable of generating high-level contextual information. The middleware should further provide mechanisms to support event subscription and notification for context-aware applications.
- *Service discovery*: A context source can be a provider of context-aware services as well. The middleware needs to have a service discovery mechanism that organizes the diversified services provided by context sources into different domains and enables the applications to efficiently search the required services. This mechanism can be either integrated into or separated from the mechanism for context data lookup in the middleware.

Earlier research in context-aware systems has focused on location tracking of the human user. A number of location tracking systems have been proposed in the literature, for example, Active Badge [11], Cricket [12], and RADAR [13]. Several location-aware information systems have also been developed, such as comMotion

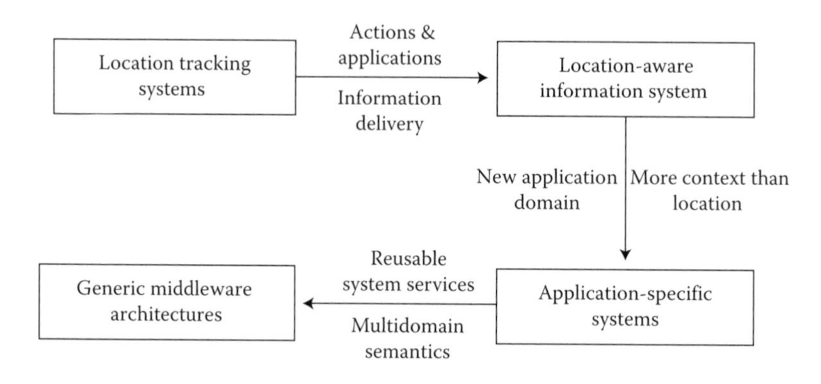

Figure 8.2 From application-specific to middleware-based context-aware computing.

[14], GeoNotes [15], and E-graffiti [16]. As one of the pioneers in context-aware computing, the research group at the Georgia Institute of Technology had developed many application-specific context-aware systems, each of which was tailored for a particular application in a domain. Examples of these systems include Cyberguide [17], CyberDesk [18], Conference Assistant [19], CybreMinder [20], and Smart Floor [21]. Other application-specific context-aware computing systems reported include the guide for tourism [22–24], museum exhibition [25,26], or zoo visit [27] and the assistance for shopping [28], office [29], or hospital [30–32]. These location-based or application-specific approaches to the design of context-aware systems have resulted in vertical solutions that are not interoperable or are incompatible, difficult to integrate or extend for other applications. Figure 8.2 briefly depicts the evolution of context-aware systems from location-based or application-specific systems to generic middleware architectures.

The rest of this chapter is organized as follows. In Section 8.2, we next present a multilayered and system service-oriented architecture (SOA) that abstracts most of the recent context-aware middleware. In Section 8.3, we propose a taxonomy of context-aware middleware and in Section 8.4 we provide a survey of many existing middleware systems according to this taxonomy. Finally, in Section 8.5, we identify and present a number of challenging problems for future research in context-aware middleware and conclude the chapter in Section 8.6.

8.2 Architecture of Context-Aware Middleware

The overall architecture and functionality of a context-aware middleware can be abstracted using a multilayered diagram as shown in Figure 8.3. Each layer contains a number of system services that are functional components interconnected to one another. Such architecture decouples the dependency of the techniques adopted in individual layers. This allows a greater flexibility in the adoption of technique for

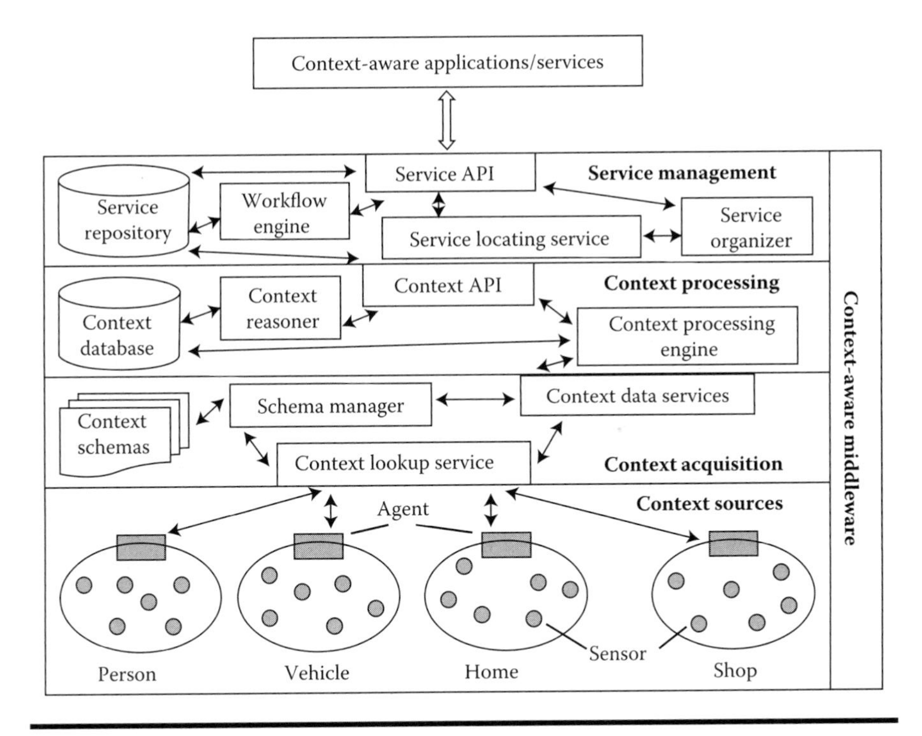

Figure 8.3 Multilayered middleware architecture.

each layer by the system administrator and the utilization of an application programming interface (API) by application developers.

The architectural diagram in Figure 8.3 generally captures the design principles and system organization of existing context-aware middleware, although with minor differences such as in the number of layers and placements of components. It also encompasses most of their key functional components. Thus, Figure 8.3 represents a reasonable abstraction of current context-aware middleware and can be a useful reference model for future research. We describe the layers of the architecture from bottom-up as follows.

8.2.1 Context Source Layer

This layer contains various classes of context sources providing context data for applications. A *context source* encapsulates at least one but most commonly a collection of sensors (include software-sensors) that is installed or attached to people, objects (e.g., vehicles and desks), and an environment (e.g., home, shop, and office). It has an *agent* software module as its representative in the middleware. The agent provides a single uniform context acquisition interface to the users of the context source, such as the upper-layer middleware services as well as the agents of other sources. These agents of context sources can be centrally located at a middleware server or be distributed onto computers corresponding to individual sources.

In the bottom of Figure 8.3, a large circle denotes a context source, whereas the small dots in the circle denote the sensors or software sensors and the solid rectangle sitting on top of each circle denotes the agent of the source.

8.2.2 Context Acquisition Layer

The role of this layer is to acquire context data from heterogeneous sources in the lower layer and provide the data to the upper-layer services with a unified view of data types. The *context lookup service* in the layer dynamically manages the stationary or mobile context sources registered to the middleware. It communicates with the agents of context sources and implements mechanisms to enable the effective organization of context sources and the efficient lookup of a specific source.

The set of *context data services* collect context attribute values from specified sources or subscribe to events of the sources. These data services should support both pull- and push-based context acquisition in an on-demand request/respond style or a continual access manner. Individual services in the set can perform different preprocessing operations on the collected context, such as filtering, aggregation, calibration, and cleaning; other design considerations such as data privacy [33] and quality of context [34] may also be addressed in these services.

A *context schema* is a specific description of the context data provided by a source in terms of a general context model [35,36]. The *schema manager* manages the heterogeneous issues of schemas for different context sources. There are at least two possible approaches in managing heterogeneous context schemas: (i) the first approach mediates centrally the local context schemas of each source into a set of global schemas for the middleware, and (ii) the second approach decentralizes mapping so that a query on a context source's schema can be reformulated in a query on another context source's query. The second approach, such as pair-wise schema mapping, is practical only for a small number of context sources, whereas the first approach is feasible only when the dynamic and autonomous nature of context sources (such as mobile context sources) are taken care by the process of mediation. To support a huge number of context sources in the first approach, the schema manager may define multiple classes of global context schemas for the middleware. For each schema of a context source, the manager then converts it into the unified context schema of one of the schema classes. To adapt to the dynamicity of context sources, the schema manager must update (i.e., modify, create, or purge) the classes of global schemas according to the availability of the context sources.

8.2.3 Context Processing Layer

This layer implements a *context API* for the applications or services on top. The *context processing engine* parses the context data requests received by the API.

Examples of requests are queries, scripts, or messages for function call. For each request, the engine invokes corresponding context data services in the context acquisition layer and performs certain postprocessing on the collected context depending on the specific API implemented.

The *context reasoner* detects application-interested events involving multiple context sources. The *context database* stores historical data acquired by the processing engine as well as training datasets and results of the context reasoner.

8.2.4 Service Management Layer

Any context-aware service provided by a context source or separately by a service provider is maintained in this layer. These services are application-level entities and belong to a different category from the system services in the middleware. An *application-level service* refers to a context-aware service that encapsulates application-oriented processing logic. The service can be used as a building block in the workflow-based service composition of an application [37].

The *service API* allows the applications to search and invoke context-aware services having specified properties through the *service locating service*. An application may also use the API to store service instances in the *service repository* and specify workflows of service composition to be executed by the *workflow engine*. A typical implementation of the service API can be based on Web service standards such as Simple Object Access Protocol (SOAP) [38] and Web Services Description Language (WSDL) [39]. The *service organizer* clusters and interconnects the services to accelerate the discovery process in the service locating service.

8.2.5 Application Layer

The *context-aware applications* in this layer can invoke services using the service API or acquire context data using the context API directly or indirectly through a service. Both these applications and the context sources can be mobile. They are not regarded as part of the middleware in essence.

8.3 Taxonomy of Context-Aware Middleware

On the basis of the architecture in Figure 8.3, we create a taxonomy for context-aware middleware, as shown in Figure 8.4. The taxonomy is defined on the basis of our knowledge and perception of the state-of-the-art middleware systems in context-aware computing research. Each major category in the taxonomy describes a key functional aspect of the middleware; each has several different methods of fulfilling. We describe each of the categories in the taxonomy in the following subsections.

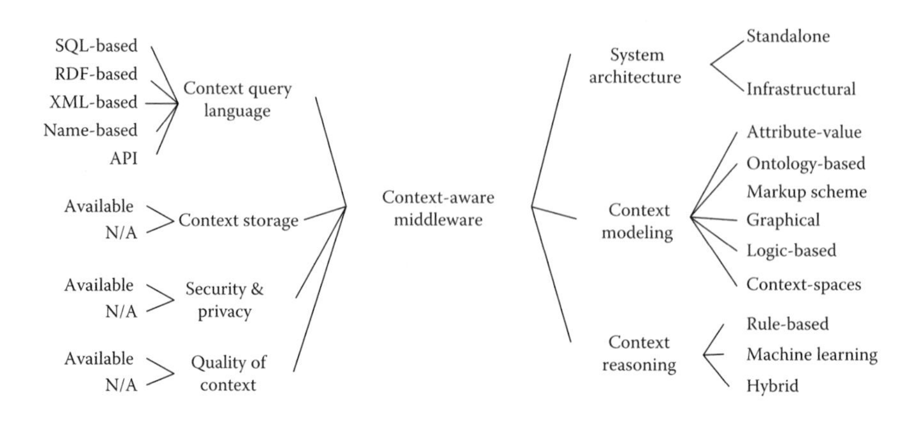

Figure 8.4 Classifying context-aware middleware systems.

8.3.1 System Architecture

In a *standalone* context-aware middleware, all system services and agents of context sources reside in a single resource-abundant *middleware server.* As a result, the context acquisition, lookup and processing operations are all performed in a centralized manner at the server.

In comparison, when a context-aware middleware has as an *infrastructural* architecture [40], its components and agents of context sources reside in multiple computers. These computers providing middleware functionalities are indeed the dedicated servers deployed by the middleware infrastructure providers together with the computers in which agents of context sources reside. A system-wide communication model is therefore needed to enable context exchange among the agents and servers. The model incorporates predefined networking protocols, data formats, system settings, and so on. The agents' computers are organized into distributed network topologies based on the model. A typical topology that is widely adopted in existing systems is *Peer-to-Peer* (P2P) [41,42]. We call such an infrastructure a *context-aware middleware infrastructure.*

A full-fledged infrastructure is, in general, more scalable and reliable than a standalone server. It avoids a single point of system failure and performance bottleneck. As an intrinsic requirement, a context-aware middleware infrastructure must automatically handle the dynamic joining, leaving, and mobility of distributed applications and context sources. The agents in the infrastructure should be able to communicate with one another on the failure of middleware servers and a few other agents.

8.3.2 Context Modeling

Context modeling means using a generic data model to describe how context data are represented and accessed. Various context modeling approaches are applied in current context-aware middleware, including:

- *Attribute Value*: This approach uses simple key–value pairs or relational databases [43] to model context data. The approach is straightforward and easy to implement in practice using commercial DBMS products, such as IBM DB2, Microsoft SQL Server, and Oracle Database. However, the model lacks complex semantic structures and has limited expressiveness.
- *Ontology-Based*: An *ontology* in computer science is a formal representation and shared vocabulary used to model a set of concepts as well as their properties and relationships in a specific domain [44]. It is a knowledge base agreed upon by multiple parties in a domain that provides a general and consistent perception of the domain. Components in an ontology can be objects, classes, attributes, relationships, rules, and so on. A common language in the literature to specify and compose ontology is the *Web Ontology Language* (OWL) [45]. The language is based on the schemas or vocabularies of the *Resource Description Framework* (RDF) [46] and the XML syntax. Earlier ontology languages superseded by OWL include DAML, OIL, and DAML + OIL [47]. An ontology-based context model has very rich semantic expressiveness and is good for logical, rule-based context reasoning. The drawback of the model, in contrast to an attribute-value model, is the complex specification and heavy processing cost of the ontology. This makes the approach less desirable on the resource-limited mobile devices with simple user interfaces.
- *Markup Scheme*: This approach uses hierarchical tags in a markup language, which is usually profile-oriented [10] and extended from XML and/or *RDF Schema* (RDFS) [48], to recursively specify and serialize a number of entities and attributes. It can be regarded as a middle and transitional approach between the attribute-value and ontology-based approaches.
- *Graphical*: An abstract model that is composed and depicted using multiple graphical components is used to represent context data in this approach. The graphical model can be visualized using a diagram in a standard and general notation method such as the *Unified Modeling Language* (UML) [49]. It can also be a proprietary model proposed in a specific middleware system.
- *Logic-Based*: In this approach, context is represented using a set of elements in a logic-based system, including expressions, facts, and rules [10]. As an example, using first-order logic, context data are defined as predicates in a triple form (*subject*, *predicate*, *object*), for example, (*Keith*, *activity*, *eating-Dinner*) and (*KeithOffice*, *event*, *meeting*) [50,51]. The approach is well suited for rule-based context reasoning. An ontology-based context modeling approach combines and evolves from the logic-based and markup scheme approaches.
- *Context Spaces*: Padovitz et al. [52–54] proposed a context spaces model that consists of three basic concepts: context attributes, context states, and context spaces. In the model, every context attribute corresponds to a dimension in a multidimensional data space. A context state is the collection of values for a

subset of context attributes at a specific time. A context space, or situation space, contains a number of regions each of which defines ranges of values for a context attribute. Each context space represents a real-world situation, for example, standing, walking, and running. When the context state of a middleware system varies along time, the model continuously examines whether the current context state satisfies the definitions of certain predefined context spaces. This approach is good for context reasoning like the logic or ontology-based approach. It extends the attribute-value approach and is a variant of the logic-based approach.

More detailed descriptions of all context modeling approaches in our taxonomy, except for the context spaces approach, are available from a previous survey paper written by Strang and Linnhoff-Popien [10].

8.3.3 Context Reasoning

Context reasoning refers to the inference process for the values of high-level, deduced context attributes from those of simple, low-level attributes. Common context reasoning approaches are

- *Rule-Based*: A *rule* is a declarative statement that specifies how the values of several simple or deduced attributes can be combined to derive the value of another deduced attribute. Both logic and ontology-based context models imply the use of rule-based context reasoning. The rules can express implicit semantics embodied in the model language constructs as well as explicit user-defined semantics. A widely used rule-based engine for ontology reasoning is Jena [55].
- *Machine Learning*: This approach applies various machine learning techniques [56], such as the Bayesian networks and utility theory, to identify patterns in context data and use these patterns to infer a context attribute value from a set of values of other attributes.
- *Hybrid*: A middleware system can use multiple rule-based engines and machine learning algorithms as its individual context reasoning blocks. The value of a high-level attribute can be deduced using a single block or a combination of multiple blocks.

8.3.4 Context Query Language

Many context-aware middleware systems use a declarative query language as part of the context API. The language enables the flexibility and user convenience in context acquisition. Haghighi et al. [57] surveyed several categories of query languages that have been used in context-aware systems. Our classification of context query languages contains the following categories:

- *SQL-Based*: This category includes the SQL query language used in relational DBMSs [43] and its variants.
- *RDF-Based*: Query languages over context data represented in RDF format are in this category. Examples are SPARQL [58] and its ancestor RDQL.
- *XML-Based*: A middleware can use an XML query language, such as XQuery [59], to acquire context data stored in XML format.
- *Name-Based*: It is a restricted version of SQL. A query in this category of languages is a conjunction of equality predicates [50,60]. Each predicate contains the name of a context attribute and a constant value for the attribute. The query specifies a condition and every context source that satisfies the condition will return the current values of several context attributes. These values to be returned are indicated as special query variables in the value fields of a few predicates.
- *API*: This category represents the case that a middleware does not have an explicit context query language. In this case, the requests for context acquisition are encapsulated in the context and service APIs of the middleware.

8.3.5 Context Storage

In addition to real-time context, applications may require access to historical data such as the location of a person or names of those in the office an hour ago. This category describes whether a context database is available in the middleware to store historical context data for future processing and probing. Such historical data can be query results or training datasets of the context reasoner. The context database can be distributed and partitioned onto multiple agents and servers in an infrastructural architecture.

8.3.6 Security and Privacy

The security and privacy issue [33,61] of context data access is one of the major concerns of mobile users in a world-wide, pervasive computing environment. This category describes whether this issue has been investigated in the middleware.

8.3.7 Quality of Context

This category describes whether the middleware implements a mechanism to monitor the quality of context data [34] or quality of context-aware services that it provides.

Our taxonomy can be extended with many other major or minor categories, such as context lookup approaches, object-oriented context model [10], and graphical query language [57]. On the other hand, we believe the current taxonomy encapsulates key functional aspects and commonly used technologies in existing context-aware middleware. For example, a standalone middleware server usually uses a centralized approach to context lookup whereas a middleware infrastructure implies a distributed approach. The taxonomy is effective to classify current middleware systems, as we can see in the next section. We have not included any category corresponding to service management in the taxonomy because we find that it is not well separated and often tightly integrated with context data management in most existing middleware.

8.4 Survey of Context-Aware Middleware

To have a better understanding of the open research issues and the progresses in the development of the technologies, we surveyed, studied, and compared state-of-the art context-aware middleware systems. Among the list surveyed, we present, in this section, an overview of the features and principle of operations of each of the 20 representative context-aware middleware systems in a chronological order, and compare them with regard to the taxonomy. Among the shortlisted, a few of the recent middleware systems have not been included in previous survey studies [7,8,62]. To facilitate further reading, references for related work that may of interest to readers but have not been discussed in detail in our short list.

8.4.1 Gaia

Gaia [50] is a distributed context-aware middleware infrastructure. The middleware extends a physical space in the real world into an *active space* by providing a set of basic system services and a component-based framework to assist the application development over multiple coordinated devices in the space.

The context modeling in Gaia is based on first-order logic. A space repository service maintains the metadata about all dynamically joining or leaving hardware devices and software entities in the active space. Applications can issue name-based queries to the space repository to identify context sources of interest. As an example, to locate all projectors in the active space a query *Category* = =“*Device*” and *Type* = =“*Projector*” is specified. Both pull- and push-based context acquisitions are supported by Gaia. Applications can query context data from a context service or subscribe to the event channels in an event manager service.

Gaia allows the application users to make personal storage automatically available in their current locations through a context file system service. This basic service constructs a virtual directory hierarchy. Context types and values are represented as path components in the directories. A security mechanism, Cerberus [61], has been later developed and integrated into Gaia as a core system service. The mechanism combines context awareness with rule-based automated reasoning to enable user identification and authentication as well as the access control of context data and services.

8.4.2 CMF

The *context management framework* (CMF) [63] is a client–server software architecture. Each client in the framework has a resource server that collects sensory context and performs low-level feature extraction, quantization, and semantic labeling over the data. The context manager at a centralized server gathers preprocessed context data from the resource servers and stores it into a context database. Application running at the clients can then request context data or subscribe to context changes through the context manager API. Both the clients and the server run on mobile devices in the framework design. As every client communicates with the server merely but not the other clients, the framework essentially forms a standalone middleware.

The framework uses an ontology to represent context data and a naive Bayesian classifier to reason high-level context. The context manager uses a security service to examine the context data sent by the clients whereas the functional details of the service are not presented in the paper.

8.4.3 CARISMA

Capra et al. [64] developed a context-aware middleware for mobile computing called CARISMA. The standalone middleware server of CARISMA encapsulates the context data that drives the behavior adaptation of a mobile application into an *application profile* in XML. The profile specifies how a set of services provided by the middleware are associated, what policies are used to deliver these services, and when these policies are triggered on context changes for the corresponding application. The profile can be fine-tuned to express what kind of quality of service the application may achieve. An application can use a reflective API to modify its profile in the middleware. No context reasoning service is discussed in CARISMA.

8.4.4 CARMEN

CARMEN [65] is a multilayered middleware infrastructure that facilitates the design and implementation of context-aware Web services in the wireless Internet. It utilizes an underlying secure and open mobile agent platform to implement the shadow proxies that migrate with and provide continual context access for the mobile users or clients in the infrastructure. Context is reflected as two types of metadata stored in the middleware components: profile and policy. The former describes the characteristics of the resources, including users, devices, service components, and sites, and is represented in an XML format. The latter specifies how to bind, migrate, or control the access of resources and middleware components upon user mobility. Context reasoning is not explored.

The low level of the CARMEN architecture contains an event manager that triggers the policies by relevant events, a naming system with identification, directory, and discovery facilities, a monitoring scheme for context changes, and a location mechanism that supports the tracking of mobile devices with heterogeneous wireless communication technologies. The high level includes a metadata manager and a context manager that maintains the current context of all clients.

Corradi et al. [66] illustrated the effectiveness of CARMEN for context-aware service provision via a mobile news service scenario. They subsequently developed a ubiquitous context-based security middleware (UbiCOSM) [66] by adding extra security and user privacy protection mechanisms on top of the low-level CARMEN facilities.

8.4.5 CAPNET

The CAPNET [67] middleware uses a client–server architecture to support the development of context-aware multimedia applications on mobile devices. It allows the storage and retrieval of a wide variety of context data in a database, such as time, available services, and user's location and preferences. Privacy policies are defined in the database for the access control of context data. However, no detailed context modeling and reasoning approaches are presented. Media components of the middleware are also stored in the database. Client applications can use a Jini-based service discovery component to locate the multimedia services and system components.

8.4.6 Mercury

Solarski et al. [68] proposed Mercury, a P2P middleware infrastructure for context-aware provision and adaptation of mobile services. Mercury applies mobile agent technologies to enable the interaction among a network of autonomous and decentralized software entities without requiring the presence of a middleware server. A communication pattern for dynamic service discovery is used in the infrastructure for the entities to publish and subscribe to the service advertisements.

Mercury uses a profile-based markup scheme approach for context modeling and an XML-based query language, XQuery, to acquire context data stored in the XML files of the profiles.

8.4.7 SOCAM

SOCAM [51] is a service-oriented middleware infrastructure that supports the building of context-aware services in multiple domains. The architecture of SOCAM contains three layers: (i) a context sensing layer with physical and virtual sensors in the bottom, (ii) a context middleware layer with context providers, interpreter, and database in the middle, and (iii) a context application layer with application-level services on the top. Java RMI is used for the distributed communication between systems services in the middleware.

The context providers in SOCAM collect low-level context from sensors and the context interpreter uses rule-based logic reasoning to deduce high-level context. Context is represented using an ontology-based model. The ontologies are written in OWL and consist of a general upper ontology describing common concepts in all domains as well as domain-specific lower ontologies describing particular concepts in a domain. Applications can use a *service locating service* to search the context providers and interpreters in wide-area networks.

8.4.8 ACAI

Khedr and Karmouch proposed ACAI [69], a multiagent context-aware infrastructure for supporting spontaneous mobile applications in different domains. Three layers exist in the ACAI architecture: (i) a lower sensing layer for context acquisition, (ii) a middle context service layer for context processing and context-aware service provision, and (iii) an upper application layer that provides the interface between mobile users, applications, and system services. ACAI functions across multiple physical spaces for a context-aware application. These spaces are categorized into the home, current, local, and remote sites of the mobile user of the application.

ACAI uses an ontology context model and a hybrid context reasoner that combines logical, fuzzy, and rule-based reasoning. These two functionalities are implemented by the ontology and reasoner agents in the ACAI agent model, respectively. The context management agent is the kernel of the agent model that administrates the whole ACAI system. A knowledge base repository is used for the storage and retrieval of various information related to the system, for example, context ontologies, policies, and user profiles. The system knowledge base agent provides an interface for probing and updating the repository. The coordinator agent manages data stored in the repository. Each context source in ACAI is associated with a context provider agent.

Two core services are provided by the ACAI system. The first is a context-aware, inference-based service discovery module for applications. The second is a context-sensitive communication protocol that maintains the user sessions with privacy preservation.

8.4.9 JCAF

The *Java context awareness framework* (JCAF) [70] provides a Java-based infrastructure and programming API for developing context-aware applications. The middleware infrastructure is service oriented and uses a P2P network to organize and search context services to be used in the applications. It is also event based, thus allowing the subscription and notification of context events for the applications.

A graphical context model in UML is used in JCAF. The JCAF API is similar to many other traditional Java API specifications, such as JDBC and JMS. The API has been designed to support the measurement of quality of context. A context service with the API allows the storage and access of context attributes. JCAF addresses the security issue using a digital signature-based authentication mechanism for clients via the Java security API.

8.4.10 CHIL

Soldatos et al. [71] developed a multiagent middleware infrastructure in their CHIL research project to accelerate the development of context-aware services. A situation model specified in XML is used for context representation and a common API is provided to acquire context from hardware sensors in the infrastructure. Applications may retrieve historical context data via storage services.

8.4.11 ScudWare

ScudWare [72] is a multiagent and component-based middleware that operates in a smart vehicle space. A context-aware application is decomposed into a number of task units in the ScudWare infrastructure. Each task is performed by a few *semantic virtual agents* (SVAs) running on the mobile devices in the space. An SVA contains metaobjects, each of which is a service having one or more CORBA components. A semantic interface protocol is used to handle the dynamic joining, leasing, self-update, P2P, or multicast-based discovery among the SVAs in task cooperation.

The middleware contains a semantic context management service to acquire and fuse context data from various sensors. Wu et al. [72] built an ontology in OWL for context modeling and reasoning. Another adaptive component management service is responsible for the built-time package, assembly, deployment, and allocation as well as the run-time migration, replacement, update, and vibration of the distributed components for metaobjects. The underlying adaptive communication environment used by ScudWare provides both security service for SVA cooperation and quality-of-service control.

8.4.12 W4

Castelli et al. [73] proposed a simple W4 model to represent context data collected by agents on mobile user devices from various sensors in the physical world. The model abstracts a piece of context data as a quadruple of attribute-value pairs—Who, What, Where, When. The authors further designed and implemented a middleware infrastructure based on the W4 model to support context-aware mobile application and services. We still use W4 to denote this middleware.

The W4 middleware infrastructure consists of multiple software agents that sense and generate context data in the format of W4 tuples, the application agents that utilize the data, and several publicly accessible shared Web tuple spaces that store the data. The tuple spaces are implemented using a database in the middleware prototype. A query engine is implemented as a system component to convert name-based W4 queries into SQL queries over the database. Context reasoning, quality, and security issues are not explored.

8.4.13 SOA

Kim and Choi [74] described an SOA [74] for context-aware middleware in a smart home. The authors separated the context information ontology that models ordinary context data from the context metadata ontology that models quality of context as well as other additional information such as context dependency, sources, and categories. SPARQL is used as the context query language in the middleware and a rule-based reasoning engine is used on the basis of an ontology repository. A centralized context discovery service maps context data to its corresponding context provider services.

8.4.14 CMSANS

Santos et al. [75] presented a service-oriented middleware that integrates a context management service (*CMS*) with an awareness and notification service (*ANS*) [75]. We denote this middleware as *CMSANS*. CMS provides an ontology model for context sources to represent and publish their data. Client applications can use SPARQL to query CMS and locate the sources having required data. ANS implements a rule-based reasoning engine that takes monitoring rules submitted by client applications and notifies the applications when the context-based conditions in the rules are satisfied. ANS subscribes to the context data required for the evaluation of every rule from the corresponding sources through CMS. It stores the rules and ontologies used for reasoning in a knowledge repository.

8.4.15 ACoMS

ACoMS [76] is an automatic context data management system for pervasive computing based on graphical, application-oriented context models. These models are able to incorporate quality of context information as quality indicators. The system architecture consists of a standalone middleware server and a number of sensors producing context data.

ACoMS provides the dynamic self-configuration of available sensors and runtime replacement of faulty sensors. This enables the fault-tolerant context acquisition in applications. An application can request or subscribe context data from the sensors using name-based queries.

8.4.16 ECORA

Padovitz et al. [77] proposed an agent-based hybrid architecture called ECORA with a design focus on facilitating unified context modeling and reasoning under uncertainty in context-aware applications. The middleware infrastructure of ECORA integrates a centralized context server providing reasoning services with multiple mobile agents having reasoning capabilities to improve the system reliability and flexibility. The authors used the context spaces model proposed in their previous work [52–54] to represent context data. Historical context is stored in a database at the context server for future inference. On the basis of the context model, the authors applied utility theory in machine learning for context reasoning in uncertain contextual conditions. Security issues are not investigated in ECORA as the architecture assumes the mobile agents only run on host computers with authorization.

8.4.17 Solar

Solar [60] is a data-centric context-aware middleware infrastructure. It uses an application-level P2P service overlay based on a *distributed hash table* (DHT) [41,42] to interconnect several *Planets* that are host computers of service providers. A Planet is both a context source and an application client. It provides a set of key services, including context acquisition from sensors as well as operator-oriented context processing and dissemination, to enable the stream-based context data publishing and subscription in the infrastructure. As peers, the Planets can be flexibly added or removed from the P2P network to adjust the system performance.

Solar uses the concept of *operator* to represent a software component that takes data pushed from one or more context sources or downstream operators, performs certain filtering on the input data, and outputs the processed data to another upstream operator. An operator runs on a Planet and the operators can be interconnected into an acyclic *operator graph* for the context dissemination for an application over the P2P overlay.

An attribute-value context model is used in Solar with a lightweight name-based context query interface for applications. The operators do not have context reasoning and storage capability, as suggested by Chen et al. [60]. The authors described two main services in Solar, resource discovery and data dissemination, and evaluated their performance using a campus-wide deployment. The second service has considered how to satisfy the quality of context requirements of applications.

Researchers of Solar have also studied security model for context-sensitive authorization [78] and privacy protection of user's digital footprints [79] on top of the Solar architecture using policies.

8.4.18 CoWSAMI

CoWSAMI [80] is a middleware infrastructure that enables interface-aware mobile applications in context-aware computing. An application can issue SQL queries to acquire context data that is modeled using relational views from the dynamically appearing sources in its surroundings. A context source in the infrastructure publishes its *context relation* via a number of standard Web services based on SOAP and WSDL. The tuples of attribute-value pairs in a context relation can be stored in the device memory of the context source. CoWSAMI allows the dynamic discovery of context sources around the current location of a user application when the user moves. The discovery is based on multiple update policies for different services and is realized using two system services—*Naming&Discovery* and *ContextSerivceDiscovery*.

8.4.19 CADeComp

Ayed et al. [81] proposed CADeComp, a platform-independent component middleware that lies between a context-aware middleware and the applications. CADeComp enables context-adaptive deployment of component-based applications on the devices of mobile users. A component-based application consists of a set of independent and interconnected software components. A component can have properties that describe security policies or quality of context concerns.

Context data and component discovery queries in CADeComp are both specified in XML. A deployment service provider presents repositories to store the component packages of applications as well as the metadata to drive application adaptation.

8.4.20 CAMPH

CAMPH [82] is a context-aware middleware for pervasive homecare. The middleware consists of four layers from bottom-up: (i) physical space layer, (ii) context data management layer, (iii) service management layer, and (iv) application layer. Although CAMPH is presented for the healthcare domain in the paper by Pung et al. [82], the infrastructure is general and applicable to multiple domains.

A *physical space* in CAMPH abstracts a context source. Each physical space is equipped with a *physical space gateway* (PSG) that provides the mere data communication interface between the space and all external parties, including the context-aware applications, the middleware server, and the other PSGs. The PSG acts as the agent of the context source and is a software program that can run on any computer associated with the physical space. Example classes of physical spaces in CAMPH include persons, homes, clinics, and hospitals.

The context data management layer provides an SQL-based query interface for applications to acquire context data from physical spaces in a pull- or push-based manner. The layer contains an adaptive, lightweight context schema matcher that integrates the local context schemas from various physical spaces into a set of global schemas at the middleware server. An application can issue context queries either to the middleware server on the basis of the global schemas or to the individual PSGs on the basis of their local schemas. CAMPH uses a server-side query processor as

Table 8.1 Classification of Selected Context-Aware Middleware Systems

Middleware	*System Architecture*	*Context Modeling*	*Context Reasoning*	*Context Query Language*	*Context Storage*	*Security and Privacy*	*Quality of Context*
Gaia	Infrastructural	Logic-based	Rule-based	Name-based	Available	Available	N/A
CMF	Stand-alone	Ontology-based	Machine learning	API	Available	Available	N/A
CARISMA	Stand-alone	Markup scheme	N/A	API	N/A	N/A	Available
CARMEN	Infrastructural	Markup scheme	N/A	API	Available	Available	N/A
CAPNET	Stand-alone	N/A	N/A	API	Available	Available	N/A
Mercury	Infrastructural	Markup scheme	N/A	XML-based	Available	N/A	N/A
SOCAM	Infrastructural	Ontology-based	Rule-based	RDF-based	Available	N/A	N/A
ACAI	Infrastructural	Ontology-based	Hybrid	RDF-based	Available	Available	N/A
JCAF	Infrastructural	Graphical	N/A	API	Available	Available	Available
CHIL	Infrastructural	Markup scheme	N/A	API	Available	N/A	N/A
ScudWare	Infrastructural	Ontology-based	Rule-based	API	N/A	Available	Available
W4	Infrastructural	Attribute value	N/A	Name-based	Available	N/A	N/A
SOA	Stand-alone	Ontology-based	Rule-based	RDF-based	Available	N/A	Available
CMSANS	Stand-alone	Ontology-based	Rule-based	RDF-based	Available	N/A	N/A
ACoMS	Stand-alone	Graphical	N/A	Name-based	N/A	N/A	Available
ECORA	Infrastructural	Context spaces	Machine learning	API	Available	N/A	N/A
Solar	Infrastructural	Attribute value	N/A	Name-based	N/A	Available	Available
CoWSAMI	Infrastructural	Attribute value	N/A	SQL-based	Available	N/A	N/A
CADeComp	Stand-alone	Markup scheme	N/A	XML-based	Available	Available	Available
CAMPH	Infrastructural	Attribute value	Hybrid	SQL-based	Available	N/A	N/A

well as multiple PSG-side query processors, which combine as a whole to enable the distributed query processing in the infrastructure. Historical results of a query can be stored in the context database at the server or at a PSG.

CAMPH uses an attribute-value approach for context modeling and a hybrid approach for context reasoning. The local context reasoner at a PSG focuses on applying different data mining algorithms to recognize the human activities of daily living that occurs in the physical space, such as eating, watching TV, and falling down. Such activity detection is very useful for pervasive elderly homecare applications. A rule-based engine is deployed at the middleware server as the global context reasoner to deduce application-specified, cross-space events.

The service management layer in CAMPH implements a two-tiered global and local framework for the discovery of context-aware services from multiple service providers. A *service provider* can be united with a PSG or be a separate software entity. CAMPH uses different P2P protocols, such as Gnutella [83] and Chord [84], to organize both the physical spaces in the context data management layer and the service providers in the service management layer for context or service lookup.

The application layer contains context-aware applications in different domains. Each application can operate over one or more physical spaces and communicate with the middleware server or PSGs for context and service access.

In summary, Table 8.1 shows a classification of the context-aware middleware systems we have surveyed on the basis of our taxonomy. We summarize our observations from the table as follows:

1. An infrastructural architecture is more widely adopted for context-aware middleware than a standalone architecture.
2. Three common context modeling approaches used in current middleware systems are ontology-based, markup scheme, and attribute-value approaches.
3. The context query languages used in existing systems are quite diversified. Context reasoning has not been investigated in many systems. Context storage is not provided in several systems.
4. Security and privacy issues as well as quality of context have not been well addressed in current context-aware middleware.

8.4.21 Other Context-Aware Systems

There have been a large number of prior research papers on developing context-aware systems for mobile applications and services. In addition to those in Table 8.1, we briefly introduce a few other systems.

PARCTAB [85] is one of the earliest context-aware systems. The Context Toolkit [86,87] provides a number of context widgets as building blocks for context acquisition and processing in applications. CoolTown [88] attaches tags or beacons that store the URLs of Web pages to different geographical areas. These Web pages represent context data related to the areas.

Aura [89] utilizes existing applications as services to facilitate the fulfillment of user tasks. RCSM [90] is a reconfigurable middleware that allows applications to specify their objects through a context-aware interface description language. The PICO middleware [91] enables the pervasive sensor data exchange among multiple types of devices using goal-oriented software communities on the devices.

MobiPADS [92] is a mobile middleware that supports context-aware service deployment and reconfiguration. CASS [93] is a service-oriented middleware that separates rule-based context reasoning from application codes. Wang et al. [94] proposed the semantic space infrastructure for context data management over a smart space using semantic Web technologies. Henricksen and Indulska [95] designed a set of models for context, preference, and programming and used these models to build an infrastructure for software engineering of context-aware applications.

iCAP [96] supports the development of context-aware applications in a visual and coding-free manner. PersonisAD [97] allows the distributed modeling of persons, sensors, devices, and places in the physical world. Context-aware applications can locate these models via a service discovery mechanism. OmniStore [98] performs data management over multiple mobile devices in a personal area network.

CoSAr [99] is a context sharing architecture for mobile Web services. Yang et al. [100] used an ontology model to accelerate the discovery and adaptation of context-aware Web services. Vimoware [101] is middleware that supports the development of mobile Web services and the collaboration of mobile devices. HiCon [102] provides a hierarchical abstraction for real-world context monitoring that consists of three context levels: personal, regional, and global.

8.5 Challenges for Future Research

Following a survey of the literature, we propose in this section a number of outstanding research problems in context-aware middleware systems and discuss potential approaches to address these challenges.

8.5.1 Real-World Deployment

Most current context-aware middleware systems are equipped with real-world case study applications for functionality demonstration and performance evaluation. Although the scenarios of these applications are generally fair, to the best of our knowledge very few of them have been field tested with real-world settings. Many systems are still deployed and evaluated in a simulated or lab-scale environment within a research scope. There is hardly any attempt to standardize context-aware middleware architecture or technologies, except a few related system components such as OWL for ontologies and SOAP/WSDL for Web services.

We believe the real-world deployments of system prototypes with domain-oriented application scenarios are crucial for the future advancement of context-aware middleware technologies. This allows real-world users to experience and evaluate the technologies in practice. Valuable feedback can be obtained from the users for the enhancement of individual systems components. As an example, Bardram et al. [31] have presented a real-world deployment of context-aware infrastructure in a hospital for coordinating the tasks of nurses and clinicians, and have discussed the observations they obtained from the deployment. The deployment illustrates the usefulness and acceptance of context-aware technologies for real-world users with scenario-specific adjustments to address the user concerns.

8.5.2 Security and Privacy

Current security and privacy research in context-aware computing has not been tightly integrated with the work on middleware technologies. More than half of the systems that we surveyed in previous sections do not have a mechanism for security and privacy control. Even for those systems with such a mechanism, most of the solutions have not been tested and evaluated rigorously. There is a need for future research to develop full-fledged context-aware middleware systems that are equipped with practical security and privacy control for mobile users.

8.5.3 Multidomain Support

A generic middleware supporting applications in multiple domains requires substantial amount of system resources and an efficient method of management for storage and processing the knowledge bases, such as ontologies. Approaches to ontology matching [103] may be applied to categorize and integrate local ontologies used by context sources into a set of global, hierarchical ontologies for the whole middleware. Note that issues of ontology matching encompass that of schema matching and are part of the issues of context model integration. More efficient processing techniques for the complex ontologies are also required in future context-aware middleware.

8.5.4 Novel User Interfaces

A context-aware application should have new types of user interfaces specially tailored for the tiny mobile devices and the pervasive computing environment. For example, a user may use a finger to select a service in the application or issue a service search request to the application via speech. An application having different interfaces personalized for individual users or adapted to different contextual situation of a user is also desirable. The agents of a middleware infrastructure on the devices should provide efficient context acquisition, processing, and storage capabilities to enable the provision of novel user interfaces.

8.5.5 Quality of Context

Current middleware systems mostly measure quality of context using a set of quality parameters whose values are explicitly specified by either an application or a context source. An automatic mechanism for quality of context measurement [104] can be another future research topic to be explored in context-aware middleware.

8.5.6 Scalable Context Processing

With numerous context sources distributed in the global network, a context-aware middleware must provide efficient techniques for context lookup and processing over these sources. A client–server architecture is not scalable as the context lookup and processing for all applications are performed at the centralized middleware server. A P2P-based middleware infrastructure is more scalable, whereas an unstructured P2P protocol [83] suffers from a large query routing delay and a structured DHT-based protocol [84] is not suitable to handle dynamic sensory context frequently changed over time from multiorganizational sources. A future context-aware middleware requires the design and implementation of a novel P2P protocol that accelerates the distributed lookup of context sources and the coordinated context processing over these sources.

8.5.7 Human Activity Recognition

An important class of context data is the activities of human users in their daily life. Examples of activities include making a phone call, having a meeting, and taking a shower. Human activity recognition forms a subcategory of context reasoning that detects the activities a user has performed over time and the boundaries of every activity. A context-aware middleware should provide effective human activity recognition schemes for applications [82]. Personal modeling with cultural and regional features can be combined with human activity recognition to deduce the latent implications when a user performs an activity.

8.5.8 Separation of Context and Service Management

Most existing middleware do not separate the management of application-level services from the management of context data into different layers in the architecture. Many systems regard the agents for context sources as system services for context provision. The separation allows different mechanisms to be used for the organization and discovery of context sources or service providers [82]. The services in the service management layer can acquire context data through the context API of the context data management layer. In-depth investigations are required to identify the design and implementation trade-offs of such separation.

8.6 Conclusion

In this chapter, we present state-of-the-art middleware technologies for supporting the development of context-aware mobile applications and services. A generic middleware provides a set of system services to facilitate context data management for applications across various context sources. We have listed a few design requirements and proposed a multilayered architecture for context-aware middleware systems. We define a taxonomy for context-aware middleware and analyze a number of middleware systems in the literature on the basis of the taxonomy. Despite the volumes of work reported, we should also be aware that current context-aware middleware systems are still far from being mature. Indeed, several challenging research issues are yet to be fully understood and addressed before a widespread deployment of context-aware pervasive computing systems can take place.

DEFINITIONS

Context can be any information that is used to characterize the situation of an entity involved in the user–application interaction.

Context attribute is a specific kind of context data. Example context attributes provided by context sources include *name*, *location*, *heartbeat* of a person, and *temperature*, *opening hours* of a shop, *crowd level* of a shop.

Context-aware computing refers to a paradigm of pervasive computing that includes the context-aware applications and services running on the mobile devices of human users, such as laptops, PDAs, and phones, as well as the system technologies to support the development of these applications and services.

Context-aware middleware is a software system lying between the context-aware applications and services on the top and the context sources in the bottom. The middleware consists of several interconnected functional components that can be used as system services to accelerate the cross-source context acquisition, lookup, and processing in the applications.

Context awareness is a key enabling feature in pervasive computing that suggests the applications continuously sense the changing context in their environments and dynamically adapt their operations or behaviors to react to the sensed context with little or no human intervention.

Context modeling is the process of using a generic data model to describe how context data can be represented and accessed.

Context schema is a specific description of the context data available from a context source in terms of the generic context model used by the source.

Context source refers to any person, object, or place in the world that can provide one or more kinds of context data. Example classes of context sources are persons, vehicles, homes, and shops.

Context reasoning is the process of inferring the values of high-level, deduced context attributes from the values of simple, low-level attributes.

Ontology in computer science is a formal representation and shared vocabulary that is used to model a set of concepts as well as their properties and relationships in a specific domain. It is a knowledge base agreed upon by multiple parties in a domain and provides a general, consistent perception of the domain.

Pervasive computing is a next-generation computing paradigm in which various kinds of tiny, inexpensive, and networked computing devices, such as sensors, cameras, and phones, are integrated into physical-world persons, objects, or places so as to interact with human users and software agents seamlessly. This paradigm is also called *ubiquitous computing*.

Quality of context refers to a set of parameters that describes the quality of the sensed context data. Examples of these quality parameters are accuracy, resolution, freshness, and certainty.

Service in a computing system is a software entity having a well-defined access interface that encapsulates and performs certain functionalities.

Taxonomy is a specific, hierarchical, and structural classification of entities.

QUESTIONS AND SAMPLE ANSWERS

Q1: What is the relationship between pervasive computing and mobile computing?

A1: Pervasive computing is not a paradigm of mobile computing. It is more recently proposed than mobile computing and involves many other research areas, such as wireless sensor networks, artificial intelligence, software engineering, human–computer interaction, and social computing.

Q2: What is the relationship between context-aware computing and pervasive computing?

A2: Context-aware computing is a paradigm of pervasive computing. A mobile application or service in pervasive computing is context-aware if the application or service can automatically sense and react to the changing context of its running environments.

Q3: How have context-aware systems evolved in history?

A3: The general evolution trend of previous research on context-aware systems in a chronological order is as follows: (i) location tracking systems, (ii) location-aware information systems, (iii) application-specific systems, (iv) standalone middleware servers, and (v) distributed middleware infrastructures. The first two classes of systems focus on specific kinds of location context. The third class focuses on supporting a specific application in a domain. The last two are generic middleware systems for applications in different domains. An infrastructural architecture is widely known to be more scalable and reliable than a single server for practical real-world deployment.

Q4: Why is a generic middleware system beneficial for context-aware computing?

A4: A context-aware middleware manages numerous context sources in the world that use heterogeneous sensing technologies to acquire diversified context data. The middleware provides an abstract and unified view of context data from all these sources as well as shields the underlying details of context acquisition for the mobile applications and services in pervasive computing. It also provides a user-friendly query interface and/or programming APIs to ease the development of context-aware applications and services. Multiple applications in different domains can be developed in an easy and rapid way with such generic systems support.

Q5: Why is a multilayered system architecture desirable for context-aware middleware?

A5: A multilayered context-aware middleware architecture decouples the dependency of the technologies that are used in individual layers. This architecture allows a better flexibility in each layer for both the technique adaptation by the system administrator as well as the API utilization by the application developers. Each layer in the architecture contains a number of system services each of which provides certain functionalities and can be separately utilized by the mobile applications and services.

Q6: What are the advantages of a context-aware middleware infrastructure over a standalone middleware server?

A6: A standalone middleware server implies a traditional client–server communication model in which multiple clients of context sources or applications communicate with the middleware server individually. The server brings a single point of failure and performance bottleneck. In comparison, an infrastructural architecture improves the scalability, usability, and reliability of context-aware middleware. An infrastructure has a suite of well-defined communication and service protocols that allow the pervasive and public access of context-aware middleware technologies by mobile applications and services throughout the Internet. A successful example of network infrastructure is the Internet itself. More flexible and distributed network topologies, such as P2P, may be used to organize the software entities in an infrastructure.

Q7: Is there any context model that generally outperforms the other alternatives?

A7: We don't think so. A more expressive context model, such as ontology, tends to be complex and verbose, more difficult for the users to understand and specify, as well as requires heavier processing cost than a simple model like the attribute-value model. There is always a tradeoff. Which context model is better depends on the application domains that the middleware aims to support and the design requirements of the system services.

Q8: How do context reasoning and querying relate to context modeling in a middleware?

A8: The context reasoning scheme and context query language in a middleware system are usually chosen with regard to the context model in order to fully utilize the strengths of the model. For example, an ontology-based context modeling approach intuitively leads to the choice of a rule-based reasoning engine and an RDF-based query language. However, this is not absolute. A middleware can surely use machine learning techniques to reason or provide a particular set of APIs to query context data modeled using ontology. The exact selection principles are specific to the operating environments and target application domains of a middleware system.

Q9: How to develop a hybrid context reasoner?

A9: A possible solution is to periodically collect historical data from various context sources back to the middleware server and apply machine learning algorithms at the server to automatically generate system-defined rules from the collected data. These rules can then be distributed and executed onto the source agents. The machine learners are resource-demanding so they are more suitable to run on host-grade computers. In comparison, a rule-based engine is more lightweight and feasible for the mobile devices.

Q10: What is the relationship between the context sources and the operating environment of a context aware-middleware?

A10: Operating environment and context source are two different concepts for a context-aware middleware. A context source is a data provider whereas the operating environment is the running environment of the middleware. The operating environment of a context-aware middleware is geographical area specific, such as home, office, or vehicle, although the middleware design may allow the area to be sufficiently large to cover the whole world. This environment can be abstracted as one or more context sources depending on the requirements of a specific middleware. As an example, for a middleware that operates over a smart home, the home can be a single context source or partitioned into multiple sources including the living room, kitchen, bedrooms, and balcony. Context sources of internet-scale are likely to span over multiple operating environments of different kinds. This raises many new and difficult research issues some of which are discussed in this chapter.

References

1. Weiser, M., The computer for the 21st century, *Scientific American*, 265, 94, 1991.
2. Weiser, M., Some computer science issues in ubiquitous computing, *Communications of the ACM*, 36, 74, 1993.
3. Schilit, B.N., Adams, N., and Want, R., Context-aware computing applications, in *Proceedings of the 1st IEEE Workshop on Mobile Computing Systems and Applications*, Santa Cruz, CA , USA, December 8–9, 1994, IEEE Computer Society, p. 85.
4. Abowd, G.D., Dey, A.K., Brown, P.J., Davies, N., Smith, M., and Steggles P., Towards a better understanding of context and context-awareness, in *Proceedings of the 1st International Symposium on Handheld and Ubiquitous Computing*, Karlsruhe, Germany, September 27–29, 1999, Springer, p. 304.

5. Dey, A.K., Understanding and using context, *Personal and Ubiquitous Computing*, 5, 4, 2001.
6. Chen, G. and Kotz, D., A survey of context-aware mobile computing research, *Report TR2000-381*, Department of Computer Science, Dartmouth College, New Hampshire, USA, 2000.
7. Baldauf, M., Dustdar, S., and Rosenberg, F., A survey on context-aware systems, *International Journal of Ad Hoc and Ubiquitous Computing*, 2, 263, 2007.
8. Kjær, K.E., A survey of context-aware middleware, in *Proceedings of the 25th IASTED International Conference on Software Engineering*, Innsbruck, Austria, February 13–15, 2007, ACTA Press, p. 148.
9. Virtual Worlds Review, http://www.virtualworldsreview.com, November 9, 2011.
10. Strang, T. and Linnhoff-Popien, C., A context modeling survey, in *Proceedings of the 1st International Workshop on Advanced Context Modeling, Reasoning and Management*, Nottingham, UK, September 7, 2004, Springer.
11. Want, R., Hopper, A., Falcão, V., and Gibbons, J., The active badge location system, *ACM Transactions on Information System*, 10, 91, 1992.
12. Priyantha, N.B., Chakraborty, A., and Balakrishnan, H., The cricket location-support system, in *Proceedings of the 6th Annual International Conference on Mobile Computing and Networking*, Boston, MA, USA, August 6–11, 2000, ACM Press, p. 32.
13. Bahl, P. and Padmanabhan, V.N., RADAR: An in-building RF-based user location and tracking system, in *Proceedings of the 19th Annual Joint Conference of the IEEE Computer and Communications Society*, Tel-Aviv, Israel, March 26–30, 2000, IEEE Computer Society, p. 775.
14. Marmasse, N. and Schmandt, C., Location-aware information delivery with comMotion, in *Proceedings of the 2nd International Symposium on Handheld and Ubiquitous Computing*, Bristol, UK, September 25–27, 2000, Springer, p. 157.
15. Espinoza, F., Persson, P., Sandin, A., Nyström, H., Cacciatore, E., and Bylund, M., GeoNotes: Social and navigational aspects of location-based infor-mation systems, in *Proceedings of the 3rd International Conference on Ubiquitous Computing*, Atlanta, GA, USA, September 30-October 2, 2001, Springer, p. 2.
16. Burrell, J. and Gay, G.K., E-graffiti: Evaluating real-world use of a context-aware system, *Interacting with Computers*, 14, 301, 2002.
17. Abowd, G.D., Atkeson, C.G., Hong, J., Long, S., Kopper, R., and Pinkerton, M., Cyberguide: A mobile context-aware tour guide, *Wireless Networks*, 3, 421, 1997.
18. Dey, A.K., Abowd, G.D., and Wood, A., CyberDesk: A framework for providing self-integrating context-aware services, *Knowledge Based Systems*, 11, 3, 1998.
19. Dey, A.K., Salber, D., Abowd, G.D., and Futakawa, M., The conference assistant: Combining context-awareness with wearable computing, in *Proceedings of the 3rd International Symposium on Wearable Computers*, San Francisco, CA, USA, October 18–19, 1999, IEEE Computer Society, p. 21.
20. Dey, A.K. and Abowd, G.D., CybreMinder: A context-aware system for supporting reminders, in *Proceedings of the 2nd International Symposium on Hand-held and Ubiquitous Computing*, Bristol, UK, September 25–27, 2000, Springer, p. 172.
21. Orr, R.J. and Abowd, G.D., The smart floor: A mechanism for natural user identification and tracking, in *Proceedings of CHI 2000 Conference on Human Factors in Computing Systems*, The Hague, The Netherlands, April 1–6, 2000, ACM Press, p. 275.

22. Davies, N., Cheverst, K., Mitchell, K., and Friday, A., "Cache in the air": Disseminating tourist information in the guide system, in *Proceedings of the 2nd IEEE Workshop on Mobile Computing Systems and Applications*, New Orleans, LA, USA, February 25–26, 1999, IEEE Computer Society, p. 11.
23. Cheverst, K., Davies, N., Mitchell, K., and Friday, A., Experiences of developing and deploying a context-aware tourist guide: The GUIDE project, in *Proceedings of the 6th Annual International Conference on Mobile Computing and Networking*, Boston, MA, USA, August 6–11, 2000, ACM Press, p. 20.
24. Cheverst, K., Davies, N., Mitchell, K., Friday, A., and Efstratiou, C., Developing a context-aware electronic tourist guide: Some issues and experiences, in *Proceedings of the CHI 2000 Conference on Human Factors in Computing Systems*, The Hague, The Netherlands, April 1–6, 2000, ACM Press, p. 17.
25. Oppermann, R. and Specht, M., A context-sensitive nomadic information system as an exhibition guide, in *Proceedings of the 2nd International Symposium on Handheld and Ubiquitous Computing*, Bristol, UK, September 25–27, 2000, Springer, p. 127.
26. Fleck, M., Frid, M., Kindberg, T., O'Brien-Strain, E., Rajani, R., and Spasojevic, M., Rememberer: A tool for capturing museum visits, in *Proceedings of the 4th International Conference on Ubiquitous Computing*, Göteborg, Sweden, September 29-October 1, 2002, Springer, p. 48.
27. Pascoe, J., Ryan, N., and Morse, D., Using while moving: HCI issues in fieldwork environments, *ACM Transactions on Computer-Human Interaction*, 7, 417, 2000.
28. Asthana, A., Cravatts, M., and Krzyzanowski, P., An indoor wireless system for personalized shopping assistance, in *Proceedings of the 1st IEEE Workshop on Mobile Computing Systems and Applications*, Santa Cruz, CA, USA, December 8–9, 1994, IEEE Computer Society, p. 69.
29. Yan, H. and Selker, T., Context-aware office assistant, in *Proceedings of the 5th International Conference on Intelligent User Interfaces*, New Orleans, LA, USA, January 9-12, 2000, ACM Press, p. 276.
30. Muñoz, M.A., Rodríguez, M., Favela, J., Martinez-Garcia, A.I., and González, V.M., Context-aware mobile communication in hospitals, *IEEE Computer*, 36, 38, 2003.
31. Bardram, J.E., Hansen, T.R., Mogensen, M., and Soegaard, M., Experiences from real-world deployment of context-aware technologies in a hospital environment, in *Proceedings of the 8th International Conference on Ubiquitous Computing*, Orange County, CA, USA, September 17–21, 2006, Springer, p. 369.
32. Favela, J., Tentori, M., Castro, L.A., Gonzalez, V.M., Moran, E.B., and Martínez-García, A.I., Activity recognition for context-aware hospital applications: Issues and opportunities for the deployment of pervasive networks, *Mobile Networks and Applications*, 12, 155, 2007.
33. Jiang, X. and Landay, J.A., Modeling privacy control in context-aware sys-tems, *IEEE Pervasive Computing*, 1, 59, 2002.
34. Buchholz, T., Küpper, A., and Schiffers, M., Quality of context: What it is and why we need it, in *Proceedings of the 10th International Workshop of the HP OpenView University Association*, Geneva, Switzerland, July 6–9, 2003.
35. Xue, W., Pung, H.K., Palmes, P.P., and Gu, T., Schema matching for context-aware computing, in *Proceedings of the 10th International Conference on Ubiquitous Computing*, Seoul, Korea, September 21–24, 2008, ACM Press, p. 292.
36. Rahm, E. and Bernstein, P.A., A survey of approaches to automatic schema matching, *VLDB Journal*, 10, 334, 2001.

37. Ardissono, L., Furnari, R., Goy, A., Petrone, G., and Segnan, M., Context-aware workflow management, in *Proceedings of the 7th International Conference on Web Engineering*, Como, Italy, July 16–20, 2007, Springer, p. 47.
38. SOAP Specifications, http://www.w3.org/TR/soap, November 9, 2011.
39. Web Service Definition Language, http://www.w3.org/TR/wsdl, November 9, 2011.
40. Hong, J.I. and Landay, J.A., An infrastructure approach to context-aware computing, *Human-Computer Interaction*, 16, 287, 2001.
41. Lua, E.K., Crowcroft, J., Pias, M., Sharma, R., and Lim, S., A survey and comparison of peer-to-peer overlay network schemes, *IEEE Communications Surveys and Tutorials*, 7, 72, 2004.
42. Blanco, R., Ahmed, N., Hadaller, D., Sung, L.G., Li, H., and Soliman, M.A., A survey of data management in peer-to-peer systems, *Report CS-2006-18*, University of Waterloo, Canada, 2006.
43. Ramakrishnan, R. and Gehrke, J., *Database Management Systems*, 3rd ed., McGraw-Hill, 2002.
44. Chandrasekaran, B., Josephson, J.R., and Benjamins, R.V., What are ontolo-gies, and why do we need them? *IEEE Intelligent Systems*, 14, 20, 1999.
45. OWL Web Ontology Language, http://www.w3.org/TR/owl-ref, November 9, 2011.
46. Resource Description Framework, http://www.w3.org/RDF, November 9, 2011.
47. DAML + OIL, http://www.w3.org/TR/daml+oil-reference, November 9, 2011.
48. RDF Schema, http://www.w3.org/TR/rdf-schema, November 9, 2011.
49. Unified Modeling Language, http://www.uml.org, November 9, 2011.
50. Román, M., Hess, C., Cerqueira, R., Ranganathan, A., Campbell, R.H., and Nahrstedt, K., A middleware infrastructure for active spaces, *IEEE Pervasive Computing*, 1, 74, 2002.
51. Gu, T., Pung, H.K., and Zhang, D., A service-oriented middleware for building context-aware services, *Journal of Network and Computer Applications*, 28, 1, 2005.
52. Padovitz, A., Loke, S.W., and Zaslavsky, A., Towards a theory of context spaces, in *Proceedings of the 1st International Workshop on Context Modeling and Reasoning*, Orlando, FL, USA, March 14–17, 2004, IEEE Computer Society, p. 38.
53. Padovitz, A., Zaslavsky, A., and Loke, S.W., A unifying model for representing and reasoning about context under uncertainty, in *Proceedings of the 11th International Conference on Information Processing and Management of Uncertainty in Knowledge-Based Systems*, Paris, France, July 2–7, 2006, p. 1983.
54. Padovitz, A., Loke, S.W., Zaslavsky, A., and Burg, B., Verification of uncertain context based on a theory of context spaces, *International Journal of Pervasive Computing and Communications*, 3, 30, 2007.
55. Jena Semantic Web Framework, http://jena.sourceforge.net, November 9, 2011.
56. Alpaydin, E., *Introduction to Machine Learning*, MIT Press, 2004.
57. Haghighi, P.D., Zaslavsky, A., and Krishnaswamy, S., An evaluation of query languages for context-aware computing, in *Proceedings of the 17th International Conference on Database and Expert Systems Applications*, Krakow, Poland, September 4–8, 2006, Springer, p. 455.
58. SPARQL Query Language for RDF, http://www.w3.org/TR/rdf-sparql-query, November 9, 2011.
59. XQuery, http://www.w3.org/TR/xquery, November 9, 2011.
60. Chen, G., Li, M., and Kotz, D., Data-centric middleware for context-aware pervasive computing, *Pervasive and Mobile Computing*, 4, 216, 2008.

61. Muhtadi, J.A., Ranganathan, A., Campbell, R.H., and Mickunas, M.D., Cerberus: A context-aware security scheme for smart spaces, in *Proceedings of the 1st IEEE International Conference on Pervasive Computing and Communications*, Fort Worth, Texas, USA, March 23–26, 2003, IEEE Computer Society, p. 489.
62. Miraoui, M., Tadj, C., and Amar, C., Architectural survey of context-aware systems in pervasive computing environment, *Ubiquitous Computing and Communications Journal*, 3, 2008.
63. Korpipää, P., Mantyjariv, J., Kela, J., Keranen, H., and Malm, E.J., Managing context information in mobile devices, *IEEE Pervasive Computing*, 2, 42, 2003.
64. Capra, L., Emmerich, W., and Mascolo, C., CARISMA: Context-aware reflective middleware system for mobile applications, *IEEE Transactions on Software Engineering*, 29, 929, 2003.
65. Bellavista, P., Corradi, A., Montanari, R., and Stefanelli, C., Context-aware middleware for resource management in the wireless Internet, *IEEE Transactions on Software Engineering*, 29, 1086, 2003.
66. Corradi, A., Montanari, R., and Tibaldi, D., Context-based access control management in ubiquitous environments, in Proceedings of the 3rd IEEE *International Symposium on Network Computing and Applications*, Cambridge, MA, USA, August 30–September 1, 2004, IEEE Computer Society, p. 253.
67. Davidyuk, O., Riekki, J., Rautio, V.M., and Sun, J., Context-aware middleware for mobile multimedia applications, in *Proceedings of the 3rd International Conference on Mobile and Ubiquitous Multimedia*, College Park, Maryland, USA, October 27–29, 2004, ACM Press, p. 213.
68. Solarski, M., Strick, L., Motonaga, K., Noda, C., and Kellerer, W., Flexible middleware support for future mobile services and their context-aware adaptation, in *Proceedings of the 2004 IFIP International Conference on Intelligence in Communication Systems*, Bangkok, Thailand, November 23–26, 2004, Springer, p. 281.
69. Khedr, M. and Karmouch, A., ACAI: Agent-based context-aware infrastructure for spontaneous applications, *Journal of Network and Computer Applications*, 28, 19, 2005.
70. Bardram, J.E., The Java context awareness framework (JCAF)—A service infrastructure and programming framework for context-aware applications, in *Proceedings of the 3rd International Conference on Pervasive Computing*, Munich, Germany, May 8–13, 2005, Springer, p. 98.
71. Soldatos, J., Pandis, I., Stamatis, K., Polymenakos, L, and Crowley, J.L., Agent based middleware infrastructure for autonomous context-aware ubiquitous computing services, *Computer Communications*, 30, 577, 2007.
72. Wu, Z., Wu, Q., Cheng, H., Pan, G., Zhao, M., and Sun, J., ScudWare: A semantic and adaptive middleware platform for smart vehicle space, *IEEE Transactions on Intelligent Transportation Systems*, 8, 121, 2007.
73. Castelli, G., Rosi, A., Mamei, M., and Zambonelli, F., A simple model and infrastructure for context-aware browsing of the world, in *Proceedings of the 5th IEEE International Conference on Pervasive Computing and Communications*, White Plains, New York, USA, March 19–23, 2007, IEEE Computer Society, p. 229.
74. Kim, E. and Choi, J., A context-awareness middleware based on service-oriented architecture, in *Proceedings of the 4th International Conference on Ubiquitous Intelligence and Computing*, Hong Kong, China, July 11–13, 2007, Springer, p. 953.

75. Santos, L.O.B.S., Wijnen, R.P., and Vink, P., A service-oriented middleware for ¬context-aware applications, in *Proceedings of the 5th International Workshop on Middleware for Pervasive and Ad hoc Computing*, Newport Beach, Orange County, CA, USA, November 26–30, 2007, ACM Press, p. 37.
76. Hu, P., Indulska, J., and Robinson, R., An autonomic context management system for pervasive computing, in *Proceedings of the 6th IEEE International Conference on Pervasive Computing and Communications*, Hong Kong, China, March 17–21, 2008, IEEE Computer Society, p. 213.
77. Padovitz, A., Loke, S.W., and Zaslavsky, A., The ECORA framework: A hy-brid architecture for context-oriented pervasive computing, *Pervasive and Mobile Computing*, 4, 182, 2008.
78. Minami, K. and Kotz, D., Secure context-sensitive authorization, in *Proceedings of the 3rd IEEE International Conference on Pervasive Computing and Communications*, Kauai Island, HI, USA, March 8–12, 2005, IEEE Computer Society, p. 257.
79. Kapadia, A., Henderson, T., Fielding, J.J., and Kotz, D., VirtualWalls: Protecting digital privacy in pervasive environments, in *Proceedings of the 5th International Conference on Pervasive Computing*, Toronto, Canada, May 13–16, 2007, Springer, p. 162.
80. Athanasopoulos, D., Zarras, A.V., Issarny, V., Pitoura, E., and Vassiliadis, P., CoWSAMI: Interface-aware context gathering in ambient intelligence environments, *Pervasive and Mobile Computing*, 4, 360, 2008.
81. Ayed, D., Taconet, C., Bernard, G., and Berbers, Y., CADeComp: Context-aware deployment of component-based applications, *Journal of Network and Computer Applications*, 31, 224, 2008.
82. Pung, H.K., Gu, T., Xue, W., Palmes, P.P., Zhu, J., Ng, W.L., Tang, C.W., and Chung, N.H., Context-aware middleware for pervasive elderly homecare, *IEEE Journal on Selected Areas in Communications*, 27, 510, 2009.
83. Gnutella, http://www.gnutellaforums.com/, November 9, 2011.
84. Stoicay, I., Morris, R., Liben-Nowell, D., Karger, D.R., Kaashoek, M.F., Dabek, F., and Balakrishnan, H., Chord: A scalable peer-to-peer lookup protocol for Internet applications, *IEEE/ACM Transactions on Networking*, 11, 17, 2003.
85. Want, R., Schilit, B.N., Adams, N.I., Gold, R., Petersen, K., Goldberg, D., Ellis, J.R., and Weiser, M., An overview of the PARCTAB ubiquitous computing experiment, *IEEE Personal Communications*, 2, 28, 1995.
86. Salber, D., Dey, A.K., and Abowd, G.D., The context toolkit: Aiding the development of context-enabled applications, in *Proceedings of the CHI 99 Conference on Human Factors in Computing Systems*, Pittsburgh, PA, USA, May 15–20, 1999, ACM Press, p. 434.
87. Dey, A.K., Abowd, G.D., and Salber, D., A conceptual framework and a tool-kit for supporting the rapid prototyping of context-aware applications, *Human-Computer Interaction*, 16, 97, 2001.
88. Kindberg, T. and Barton, J., A web-based nomadic computing system, *Computer Networks*, 35, 443, 2001.
89. Garlan, D., Siewiorek, D., Smailagic, A., and Steenkiste, P., Project Aura: Toward distraction-free pervasive computing, *IEEE Pervasive Computing*, 1, 22, 2002.
90. Yau, S.S., Karim, F., Wang, Y., Wang, B., and Gupta, S.K.S., Reconfigurable context-sensitive middleware for pervasive computing, *IEEE Pervasive Computing*, 1, 33, 2002.
91. Kumar, M., Shirazi, B.A., Das, S.K., Sung, B.Y., Levine, D., and Singhal, M., PICO: A middleware framework for pervasive computing, *IEEE Pervasive Computing*, 2, 72, 2003.

92. Chan, A.T.S. and Chuang, S.N., MobiPADS: A reflective middleware for context-aware mobile computing, *IEEE Transactions on Software Engineering*, 29, 1072, 2003.
93. Fahy, P. and Clarke, S., CASS—Middleware for mobile context-aware applications, in *Proceedings of the 2nd International Conference on Mobile Systems, Applications, and Services, Context-Awareness Workshop*, Boston, MA, USA, June 6, 2004, ACM Press.
94. Wang, X., Dong, J.S., Chin, C., Hettiarachchi, S., and Zhang, D., Semantic space: An infrastructure for smart spaces, *IEEE Pervasive Computing*, 3, 32, 2004.
95. Henricksen, K. and Indulska, J., Developing context-aware pervasive computing applications: Models and approach, *Pervasive and Mobile Computing*, 2, 37, 2006.
96. Dey, A.K., Sohn, T., Streng, S., and Kodama, J., iCAP: Interactive prototyping of context-aware applications, in *Proceedings of the 4th International Conference on Pervasive Computing*, Dublin, Ireland, May 7–10, 2006, Springer, p. 254.
97. Assad, M., Carmichael, D.J., Kay, J., and Kummerfeld, B., PersonisAD: Distributed, active, scrutable model framework for context-aware services, in *Proceedings of the 5th International Conference on Pervasive Computing*, Toronto, Canada, May 13–16, 2007, Springer, p. 55.
98. Karypidis, A. and Lalis, S., OmniStore: Automating data management in a personal system comprising several portable devices, *Pervasive and Mobile Computing*, 3, 512, 2007.
99. Dorn, C. and Dustdar, S., Sharing hierarchical context for mobile web services, *Distributed and Parallel Databases*, 21, 85, 2007.
100. Yang, S.J.H., Zhang, J., and Chen, I.Y.L., Ubiquitous provision of context aware web services, *International Journal of Web Services Research*, 4, 83, 2007.
101. Truong, H.L., Juszczyk, L, Bashir, S., Manzoor, A., and Dustdar, S., Vi-moware—A toolkit for mobile web services and collaborative computing, in *Proceedings of the 34th Euromicro Conference on Software Engineering and Advanced Applications*, Parma, Italy, September 3–5, 2008, IEEE Computer Soci-ety, p. 366.
102. Cho, K., Hwang, I., Kang, S., Kim, B., Lee, J., Lee, S., Park, S., Song, J., and Rhee, Y., HiCon: A hierarchical context monitoring and composition frame-work for next-generation context-aware services, *IEEE Network*, 22, 34, 2008.
103. Ontology Matching, http://www.ontologymatching.org, November 9, 2011.
104. Kim, Y. and Lee, K., A quality measurement method of context information in ubiquitous environments, in *Proceedings of the 2006 International Conference on Hybrid Information Technology*, Jeju Island, Korea, November 9–11, 2006, IEEE Computer Society, p. 576.

Chapter 9

Interoperability across Technologies for Mobile Services

Anup Kumar, M. Iyad Alkhayat, Bin Xie, and Sanjuli Agarwal

Contents

9.1 Introduction

The last two decades have witnessed a revolution in computers and wireless network services, which has led to an explosive growth in mobile applications. Owing to the increased demand for network resources and the popularity of mobile devices, mobility management became a critical issue in wireless networks. This chapter introduces a service virtualizations approach for integrating different wireless networks. This approach provides an open and flexible architecture to include any wireless technology. This work provides an effective framework toward the development of heterogeneous wireless networks (HWNs).

9.1.1 Motivation

The rapid development of mobile stations (MS) and the application's greedy demand for more resources has pushed the development of many different types of networks. Owing to the wide range in application demands and radio frequency (RF) characteristics, we need to rely on more than one wireless technology to achieve the desired level of service. Many wireless networks coexist in the same physical location such as wireless WANs (wide area networks) (WWAN) [1,2], wireless MANs (metropolitan area networks) (WMANs) [3], and wireless LANs (local area networks) (WLANs) [4,5], combined with upcoming wireless standard IEEE 802.22 for regional area networks [6]. Our goal is to integrate different wireless technologies to provide seamless connectivity to mobile users and satisfy the quality-of-service (QoS) requirements of applications including voice, data, multimedia, and Web browsing.

The wireless communication services are available almost everywhere in the form of cellular networks WLANs, WMANs, and WWANs [7], and different technologies offer various wireless services with different limitations. For instance, WiFi is based on IEEE 802.11 protocols and can possibly satisfy user requirements in terms of connection speed and connection cost, but does not support a large area coverage, which limits the user's mobility. The 802.16 protocol offers a larger coverage area with acceptable connection speed [3], whereas the Universal Mobile Telecommunications Service (UMTS) offers the largest coverage area with modest connection speed. Most MSs these days come with multiple wireless network interface cards (NICs), such as cellular interface, IEEE 802.11, IEEE 802.16, and so on. A user needs to subscribe to different providers to be able to access cellular/WiMax base stations (BSs) and WiFi access points (APs) [8]. As mobility is a great challenge even within the same wireless network, existing protocols address mobility issues using mobile Internet protocol (IP) (MIP) or cellular IP (CIP) [9] These solutions solve the macro/micro mobility issue in the same wireless technology using local handoff or horizontal handoff. However, our goal is to provide seamless connectivity for the MS where user mobility may require switching from one network access service to another on the basis of some QoS criteria. Unfortunately, existing protocols such as MIPs are not suitable for this process because of high latency during the registration period; besides, it is not designed for frequent MS movements. CIP is designed for micro mobility and for a limited geographical area, so the range of signaling in a cellular system paging area is reduced. In an HWN, the offered service should be based on the availability of wireless network access services at the call origination time and may vary because of MS mobility. The proposed integration approach combines the available network access services by introducing a new concept called virtual wireless services (VWSs). The VWS allows MS to switch between wireless networks to support the application execution on the mobile devices with the required QoS demands. Several solutions using multiple air interfaces have been proposed either to get better capacity or to extend the coverage footprint of the served area. Most current solutions do not offer the prospect of unifying the available wireless networks into one integrated service. In the proposed approach, an MS with multiple air interfaces will interact with different wireless access services dynamically without user intervention and, at the same time, the application executing on the MS will receive the appropriate network service by hiding the underlying active wireless interface(s).

9.1.2 Scope and Methodology

The overall goal of this chapter is to build a flexible and effective framework for integrating HWNs. The design of an integrated framework is to provide transparent, self-healing, and self-configurable capability that involves issues spanning through several layers of the protocol stack [10,11]. As the Internet is based on a simple end-to-end communication standard and the network carries data without knowledge of the data type, the core networks only forward the packets from the source edge router to the destination edge router. Network performance is under the control of many network devices in the middle.

The integration framework design needs to include network self-discovery and analysis of the state of other alternative networks. As one of the network states change, the design must adopt to a new network state. The design must be able to control traffic flows between available interfaces on the basis of application performance constraints and current network availability. As MS is equipped with several wireless technologies, the design must be able to route the incoming packets from the upper layer and be able to select the appropriate interface to send the data. This transfer causes significant overhead because the change at the lower layer is frequent. The design needs to build an effective mapping mechanism to forward the incoming and outgoing packets without affecting the overall performance. Moreover, the applications running at the MS and the current mobility status play an important role in the network selection to find the best route.

The existing wireless technologies offer different characteristics that need to be addressed during network selection and routing. From the network perspective all carried data have to be moved from one end to another end regardless of whether the selected path is wired or wireless. As the problem addressed is how to deliver the data flow using the optimal available network, the proposed framework will deliver data from the mobile user to the core Internet via the VWS protocol. Then, legacy protocols will handle the flow at the core Internet.

The proposed approach builds a new VWS framework to enable the mobile devices to effectively exploit the heterogeneous technologies available at any location so that the QoS is maintained while minimizing the cost of using the mobile services. The VWS provides on-the-fly transparent network connectivity for the applications running on the mobile devices in a way that the application will not be aware of which wireless technology is in use. In this research, we provide support for the Global System for Mobile Communications (GSM)/UMTS cellular systems, WiMax IEEE 802.16, and WiFi IEEE 802.11. The MS may be equipped with several wireless network interfaces and could be available on different types of mobile PCs, ultramobile PCs (UMPCs), personal digital assistants (PDAs), smart phones, and so on.

The VWS supports the client–server model to unify the different wireless network access services. The framework uses a software agent that runs on a mobile device to obtain the information about the available communication network links and the signal capacity for each link. Furthermore, we use simulation-based performance analysis to evaluate our solution, using OPNET for implementing VWS on HWNs.

The remainder of this chapter is organized as follows. In Section 9.2, we describe the network fundamentals and different HWNs. We also list the most relevant mobility protocols and explore similar integrated architectures for HWNs. In Section 9.3, we discuss a function to perform link-quality (LQ) assessment for the available wireless network links. In Section 9.4, we describe VWS architecture to integrate different wireless networks in a unified framework. Section 9.5 evaluates the simulation results. Section 9.6 presents the conclusion and future research directions.

9.2 HWN Fundamentals

Wireless networks have become the backbone of digital communication, with a growing demand. To satisfy the customer's needs, many wireless networking options allow customers to connect to the Internet. Owing to rapid development of mobile applications, no single wireless technology is able to satisfy the QoS requirements for diverse applications [12]. Each of the existing wireless networks has it is own constraint: the goal is to overcome these constraints by unifying the existing wireless technologies in a way to satisfy application requirements.

9.2.1 HWN Revolution

The Internet revolution was accompanied by the appearance of the mobile PCs. In the wireless communication domain, Internet connectivity over the air is provided by several technologies including WWAN, WMAN, WLAN, and wireless PAN (personal are a network) (WPAN). Each one of these technologies has features that differ mainly in terms of data rate and coverage. It is envisioned that beyond 3G, the system will integrate these heterogeneous networks to offer overlapping coverage to mobile users [13,14]. The HWNs have created a wide research area for the developer and for businesses.

Currently, each wireless operator efficiently allocates the scarcely available radio resources in a given coverage area. In the existing environment, multiple wireless technologies overlap in certain coverage areas. HWN utilizes this feature by developing an integrated framework of multiple wireless technologies to support a variety of applications with different QoS requirements [15]. Wireless technologies such as UMTS, WiFi, and WiMax provide connectivity to mobile devices through different points of attachment (PoAs). In cellular and WiMax networks, the PoAs are called BSs, whereas in a WLANs the PoAs are called APs. These PoAs are different in coverage area size and have different speed constraints. Figure 9.1 shows different wireless network footprints including the overlapped coverage area.

9.2.1.1 Research Challenges

A mobile device with multiple NICs capable of connecting to different wireless networks must decide upon the best network to use and when to switch from one network to another. Mobility poses a great challenge on wireless networks. A mobile device that moves out of reach of a PoA must switch to another PoA that is closer, while the MS is in motion. Usually, the change of PoA in the same wireless technology is called roaming and is handled by local handoff or horizontal handoff. Existing protocols such as WiFi, WiMax, and GSM rely on mobility protocols such as MIP or CIP to provide local handoff [9,16]. MIP and CIP are designed to solve the macro and micromobility issues, respectively. Macromobility refers to global mobility management and micromobility refers to local mobility management. In local mobility

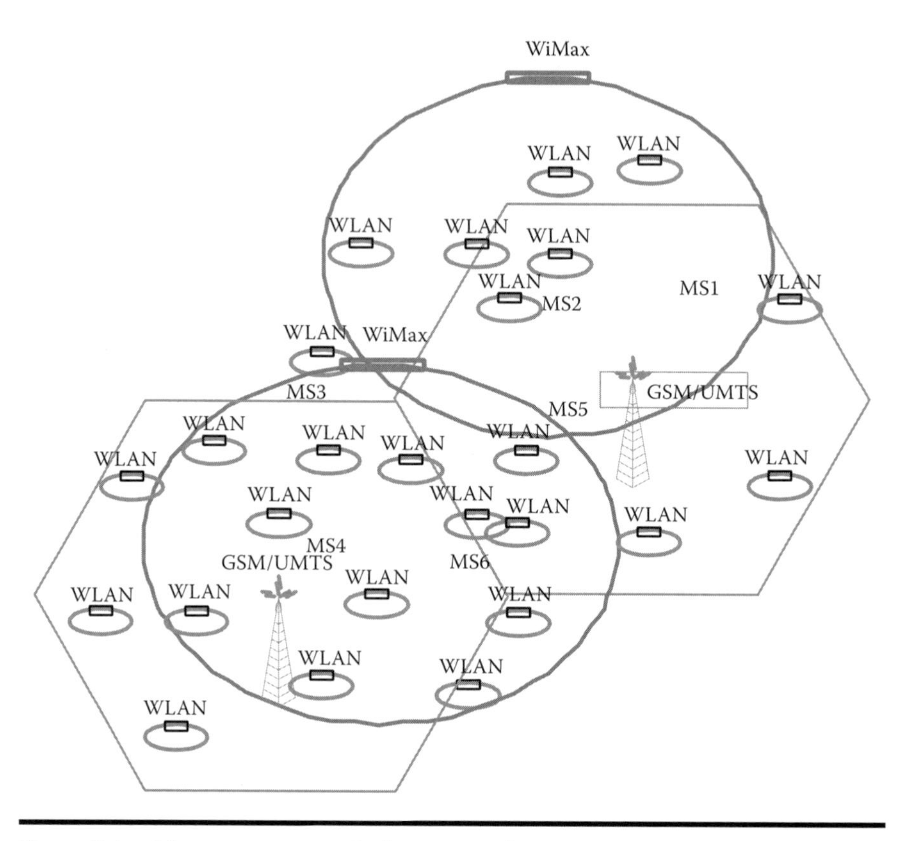

Figure 9.1 A heterogeneous wireless network.

management, the PoA is connected to the same Internet gateway (IGW), whereas in global mobility management the PoA is connected to different IGWs. Our aim is to provide seamless connectivity for an MS (with multiple NICs to connect to different wireless technologies) regardless of types of wireless access services available in the given area. This implies that the MS has a choice to switch from one wireless access to another when some handoff criteria are satisfied. Existing mobility protocols, such as MIP and CIP, are not effective for performing the handoff between different wireless technologies. MIP is not suitable for frequent MS mobility [16] because of the latency posed by the preregistration process at the home network and at the visiting network. CIP is designed to reduce the overhead of the signaling traffic during the mobility management process, by decreasing the management area boundaries [9]. In addition, local mobility is not propagated to the home network. HWN must offer a wireless connectivity service that relies on the availability of one or more wireless technologies, which may vary on the basis of MS location at that particular time. This connectivity must provide required quality of service for the application demands at the MS. The proposed approach combines the available wireless access services by introducing a new concept that supports switching between different wireless network services or using the vertical handoff. The design needs to perform the vertical handoff while the MS is in motion and maintain the connection with the required QoS. To perform the integration between different wireless networks, many related factors need to be investigated. For that purpose, we study in detail the basic network components for each wireless network access service. An MS may vary from a fully functional mobile PC to a smart phone; in our study we assumed that these MSs have cellular air interface, WLAN air interface, and WiMax air interface. Each wireless network access service differs in RF band and allowed transmitted power. So the MS will be able to connect to the Internet via the cellular network, WiMax, and WiFi hotspots with the available air interfaces. All communication schemas operate in infrastructure mode (communicating directly to the PoA). These PoAs represent BSs in cellular and WiMax networks and APs in WiFi networks. PoAs are connected to the IP core network through the IGW. For a better understanding of the network infrastructure in the HWN, we briefly explore the computer network stack.

9.2.1.2 The Network Stack

The OSI reference model divides the network structure into seven layers: application, presentation, session, transport, network, data link, and physical layer. Each of these functional layers provides a service to the above layer and receives services from the layer underneath. This stack concept provides peer-to-peer communication between two stacks at the two nodes. As each layer is served by the layer underneath, the lower layer will send and receive the data in the appropriate representation format to/from the other side across the network. In the end, the application layer can communicate with another application layer at a different computer node across the network. The Internet protocol utilizes the Transmission Control Protocol (TCP)/Internet Protocol (IP) model, which predates the OSI model. The TCP/IP model logically divides network functionality into five layers; each one of them communicates with its peer layers. Each of these layers is responsible for a specific task [17]. Different wireless networks differ only by lower layers—physical layer and data link layer. The physical layer in wireless networks utilizes RF signals, and the data link layer handles the necessary tasks for the physical layer. In Figure 9.2, the Internet TCP/IP stack and the different wireless technologies with three air interfaces—GSM, WiMax, and WiFi—at the physical layer are shown.

The data link layer is split into the medium access control (MAC) sublayer that holds a unique physical address for the network interface over the network. Another data link sublayer is the logical link control that performs appropriate data chunking. The MS has multiple wireless interfaces or network cards: the lower two layers are different and the top three layers are shared in the network stack. In the case of a single interface device, the traffic generated by an application layer heads down to the physical layer, processing several rules at each layer. The Internet infrastructure

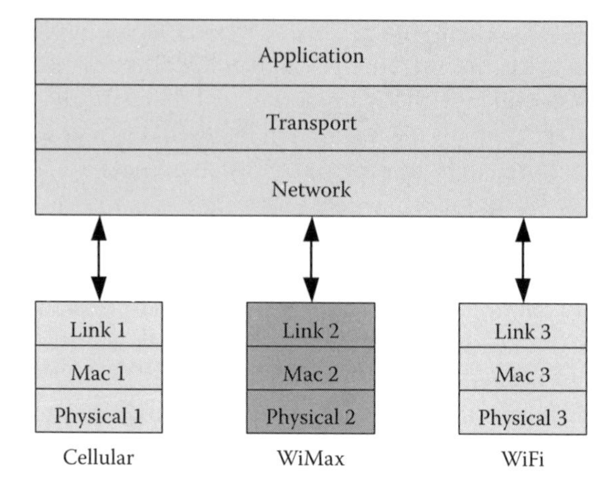

Figure 9.2 Protocol stack for a multiple air interface mobile station.

is built on top of a collection of protocols called TCP/IP protocol suite. The TCP and the IP are the core protocols in this suite [16]. Any device connected to the Internet will be distinguished by its logical IP address. This address is assigned to the devices by the serving host at that physical location. Incoming traffic from upper layers will be tagged by this IP address at the network layer, and this traffic will be navigated in the Internet infrastructure using this address. In the mobility scenario, when MS moves from an old location to a new location, its IP address is still tied to the old location. This means that another IP address has to be assigned to the MS when it changes the PoA during mobility. Unfortunately, the TCP/IP suite does not offer solutions to deal with this mobility and the traffic flows will not be delivered to the destination device when the IP address is changed. So the TCP/IP model will not be able to maintain the communication between two hosts through the network during mobility. Mobility was not considered in the design of the TCP/IP suite; instead it focuses on end-to-end delivery. Also, the IP address is assigned at the third layer of the stack and this address is tied to the physical location or PoA. That increases the challenge in cases in which the IP address needs to be changed when the PoA is changed due to host mobility. Also, in case of a multi-interface MS, the protocol stack will have a different structure at the bottom layers. It is not an easy task to select which lower layers will serve the upper layer, and decide the parameters that will provide the best service. The task to provide seamless connectivity is hard because it needs to maintain the connection during MS mobility in HWN.

9.2.2 Mobility Protocols

The original TCP/IP network did not support mobility and to overcome this issue, *ad hoc* protocols have been added. MIP protocol performs the mobility management for the mobile devices by defining the permanent home address and a visiting address while the MS is not in the home network [16]. The other protocol for mobility management is CIP [18]. CIP is based on the fact that BSs in cellular communication systems are connected in clusters and each cluster is managed by a single gateway. Mobility management in CIP relies on the cluster structures and lets the MS move freely without re-registration if it is under the coverage of the same cluster. The gateway at each cluster periodically tracks the MS by using a paging mechanism to update the last known location of the MS.

These two protocols offer mobility solutions to the TCP/IP network, but they are not sufficient for seamless connectivity and cannot support real-time applications such as VOIP, video conferencing, FTP, and so on. In MIP transition, the time required for registration at the home agent (HA) and the foreign agent (FA) is significant, and it is not feasible for frequent MS mobility or to change its PoA quickly to support many applications. On the other hand, CIP provides mobility in a smaller coverage area on the basis of the network cluster design.

9.2.2.1 Cellular Internet Protocol

The Internet has become the core of communication infrastructure. Mobility of the wireless phone requires continuous connectivity with the BS. As the mobile user moves, the connectivity to the BS may switch to another BS, and that needs to be done without affecting or dropping the call. MIP protocol provides a simple global mobility solution over the Internet [19]. It works for macro mobility, but it is not appropriate to support micromobility because of the overload caused by the registration at each new location. CIP protocol offers a practical model to address the connectivity during the handoff in the micro mobility domain [18]. Figure 9.3 shows three gateways, each controlling one network domain. In each domain, there are several BSs connected to its gateway. These BSs provide the coverage inside this domain or cluster. When a mobile device accesses one domain network, the CIP performs location management to keep track of the mobile device and reroutes packets to it. CIP handles micromobility management in each domain, whereas MIP handles macromobility between different clusters or domains.

Assuming a mobile device at location 1 is moving to location 2 in Figure 9.3, the CIP protocol allows packets to be routed to the mobile device without interruption. The CIP routing protocol ensures that packets are delivered to the mobile device at its new location. Moreover, the transmitted packets by the mobile device are routed toward the gateway and from there to the Internet. The challenge is to keep track of the mobile devices. The cost to locate a mobile device during its mobility is relatively high in MIP because of signaling overheads that keep updating the location of the mobile device at the HA. The CIP reduces these overheads by limiting the geographical control area for each domain, and by keeping the last location of the MS at the gateway. When packets are transmitted by a mobile device, the CIP routing algorithm carries a packet hop by hop through BSs up to

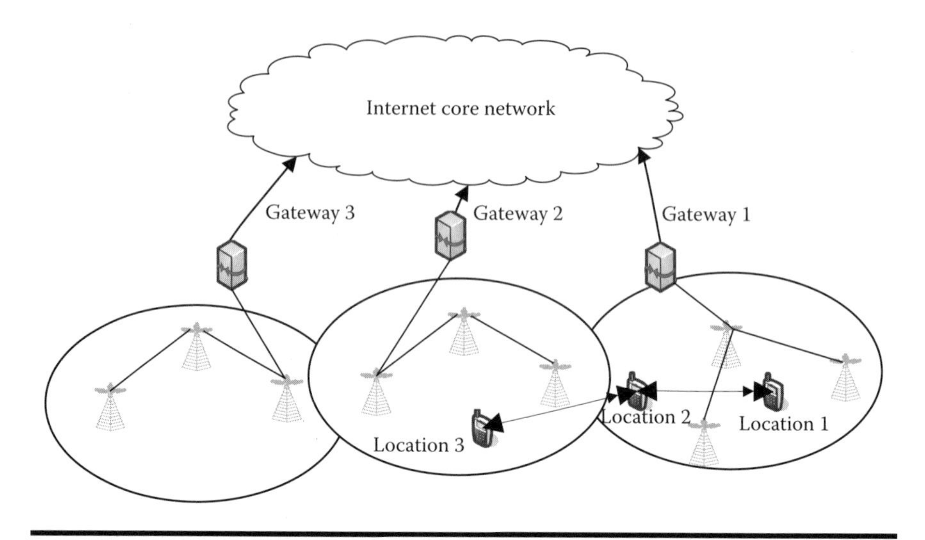

Figure 9.3 Cellular Internet protocol paging area.

the gateway. The path taken by these packets is cached by all intermediate BSs [9]. In a similar manner, when the gateway receives packets that need to be delivered to the MS, the gateway uses the most recently cached path. If there is no traffic to send, the mobile device will keep sending short packets to the gateway periodically to maintain its paths in the cache. For idle mobile devices, a paging mechanism is used to explore the connection route. The cached paths need to be cleared after a certain period of time to remove old paths that are not updated by the mobile devices.

The paging mechanism in the CIP protocol requires the gateway to broadcast beacon packets periodically [9]. These packets are forwarded to all BSs connected to the gateway. When the beacons arrive at the BS, the path where the beacons went through is used to route all packets toward the gateway. All packets transmitted by mobile hosts, regardless of their destination address, are routed toward the gateway using these routes [20]. The trail of the beacon packet from the gateway to the BS is essential to deliver packets in the other direction. The BSs save and update this trail in the routing cache table. This cache table keeps the MS IP address and the path from the gateway to the MS. The intermediate BS will keep the routing path, to make sure it has a valid link from the gateway to the MS. To validate the link route, the MS will send an update route message periodically, even if there is no data to transmit. This message will validate the routing cache mappings as is the case with normal data packets. However, update messages do not leave the CIP access domain [9]. Figure 9.4 shows the update route packet flows in two scenarios. In the first scenario, the MS moves from location 1 to location 2, and there are no delays as the incoming packets partially use the old route. In the second scenario,

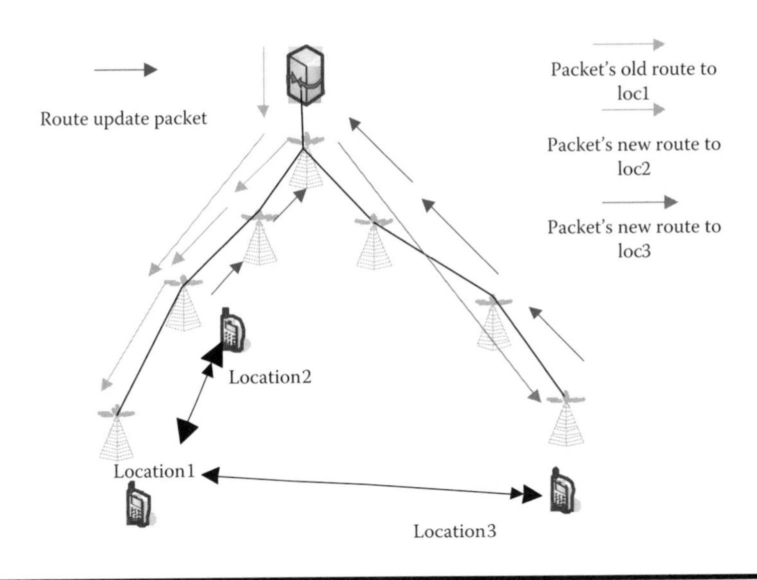

Figure 9.4 Cellular Internet protocol route update packet flow.

the MS moves from location 1 to location 3, and the delay is maximum as the update route has to go all the way to the gateway and then to the BS.

9.2.2.1.1 Handoff in CIP

The handoff process is vital to maintain mobile device connectivity between one PoA and another. CIP protocol introduces two types of handoffs: hard handoff and semisoft handoff [18]. The hard handoff is a straightforward approach in which a connection leaves the current PoA and associates with a new PoA. Some packets will be lost and there is no overhead on the network. In the semisoft approach, the packets are sent to the PoAs, the current one and the new one, during the handoff process. This enhances the QoS over IP networks but requires more control over duplicate packets transmitted. Generally, a PoA advertises its existence by transmitting special beacons periodically. The MS keeps listening to those beacons and decides which advertised PoA to connect to, on the basis of signal strength and other parameters included in the beacon. When an MS is connected to the PoA and moves far away from it, the signal strength gets weaker and the MS searches for another BS with a better signal. In hard handoff, the mobile device just leaves the current PoA and connects to the new one, and it updates the new route by sending the update message to the gateway. In turn, the gateway maps to the new route. During the handoff period, some packets will be lost.

To reduce packet loss during the handoff process, several efforts have been carried out. One approach relies on the interaction between the current and new PoA during handoff. Semisoft handoff minimizes packet loss during the conversion between the two PoAs. The idea is to create a new route before the handoff is initiated. The MS sends an association request to the new PoA before disconnecting from the current one. In this case, the MS keeps accepting packets from the current PoA and the new PoA. This association request specifies the semisoft packet [21], and this request establishes a new routing cache mapping between the crossover and PoA.

9.2.2.1.2 Paging in CIP

The MS has two states: an active state when the MS is exchanging packets with the PoA or an idle state where there is no traffic between the mobile device and the PoA [22]. The location of the MS is important to facilitate delivery of the packets from or to the gateway through the appropriate BSs using the update messages. As a cellular BS is designed to work in clusters, each cluster includes several PoAs and is considered one paging area. When the mobile device is in an idle state, the paging mechanism helps the MS recognize its location. So when a gateway needs to deliver an incoming call, it sends a control message to the last known paging area in which the MS had responded. The PoA keeps transmitting paging packets at regular intervals. The MS responds using an update route message, which is an ICMP packet. The paging update packets are routed toward the gateway on a hop-by-hop basis.

9.2.2.2 Mobile Internet Protocol

The Internet infrastructure guarantees to deliver traffic flows between two nodes connected to the Internet using the TCP/IP protocol suite. Any device on the Internet will be distinguished by a unique IP address. This IP address is assigned to the devices either dynamically or statically by the administration system at the connection location. When the device is moved to another location, a different IP address is assigned, as the IP address is tied to the location administration. Unfortunately, the TCP/IP protocol does not support device mobility over the network, because traffic delivery between devices is based on these IP addresses.

MIP protocols offer a way to solve the issue raised by IP address changes during MS mobility, by assigning a new IP address at the new PoA and keeping the old one without affecting the TCP/IP suite. The MIP design introduces a permanent home address and a temporary visiting address. These are maintained by the MIP mechanism by binding the two addresses [23]. The temporary address, called "care-of address" (CoA), will be changed every time the mobile node changes its PoA. MIP aims to make the mobility transparent to the higher-level protocols and to make minimum changes to the existing Internet infrastructure [24,25]. Figure 9.5 shows the MIP layout [16,23].

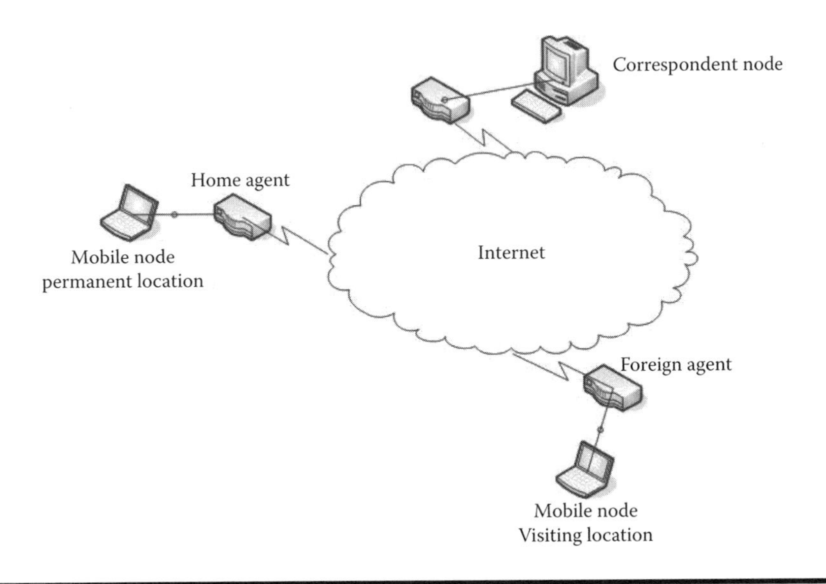

Figure 9.5 Mobile Internet protocol layout.

9.2.2.2.1 MIP Design Principles

MIP introduces two agents, HA and FA. The HA resides in the home network router and the FA resides in the visiting network. Each of these agents periodically sends an advertisement message in their domains. An MS will get this advertised message and, if the MS is in the home domain, no registration is required. If the MS is in a visitor domain, it initiates a registration procedure. When the registration is completed at the visitor network, a visiting CoA is assigned temporarily to the MS during the visiting period, and at the same time it maintains the permanent address at the home network. HA uses CoA to guarantee that incoming packets to the MS at the home address from a correspondent node (CN) will be forwarded to the visiting address CoA [16]. To maintain the relationship between these IP addresses, the permanent IP address and the CoA at the visiting network, HA and FA are used. When the MS moves to a new PoA, the visiting network grants a new CoA IP address that represents the node at the visiting domain after the FA informs the HA about the MS request. In this case, the incoming packets to this MS will be intercepted by the HA at the home network and it will be rerouted to the visiting network using the CoA. This process is carried out using three procedures: agent discovery, registration, and encapsulate/decapsulate.

9.2.2.2.2 Agent Discovery

When the MS turns on, it requests an agent message or it waits until it gets the agent message [16]. The FA/HA will periodically advertise its existence by broadcasting an agent message. The MS will catch this message where it is located and determine whether or not it needs to register with an agent.

9.2.2.2.3 Registration

The MS will locate an FA for the registration if the MS is not in the home network. The MS will send a registration request message containing its permanent IP address [23]. The HA will receive the registration request from the MS in addition to the CoA assigned by FA. The HA will intercept all the packets destined for the mobile node permanent IP address, and forward them using the CoA at the visiting network.

9.2.2.2.4 Encapsulation/Decapsulation Service

In MIP, the MS has a permanent IP address and this address will stay valid for senders over the entire network, and the CoA will be undeclared. That means the CN communicating with the MS will use the permanent IP address [16]. The packets that arrive at the home network of the mobile node will be captured by HA, and it encapsulates it in a new packet with the CoA as destination address. The incoming encapsulated packets from the foreign network will be captured by the FA and decapsulated to get the original packet, and the stripped packet will be delivered to the MS. If the mobile node needs to communicate with the CN, there is no need to go through the HA, it uses the CoA as the source address.

9.2.3 Mobility in Heterogeneous Multihop Networks

As discussed earlier, MIP can support global mobility but is not efficient in a micro mobility management because of the re-registration process. On the other hand, CIP cannot support global mobility because of the overhead of the paging and route update signaling in large clusters. Both MIP and CIP protocols suffer from the fact that they do not support communication between different wireless networks. Also, in single wireless technology, MIP or CIP may not provide services when the MSs are out of coverage or have poor signal quality from a BS [15,26].

Multihop architecture is based on the availability of intermediate stations in the neighborhood; these intermediate stations provide dynamic alternative paths through other wireless interfaces that are supported by the MS. When an MS goes out of reach of the PoA, the MS loses the connection to the PoA. Multihop mechanisms recover the lost connection with the PoA, by relaying through any intermediate station using different wireless technologies. This mechanism assumes that the main connection will go through one wireless technology and the alternative connection will use a different wireless technology. That means the intermediate station must be equipped with two or more air interfaces, the first air interface connected to one device/PoA and the second one connected to another device. The first hop will be based on the infrastructure mode, and other hops will work on the *ad hoc* network mode. This concept requires MSs to accumulate information about all the mobile devices close by to create alternative routes on the fly while maintaining the current connection [27].

Figure 9.6 shows how MS4 has several options to reach to BS using multihop techniques, because it cannot reach it directly as it is out of the coverage area of the BSs. In this case, MS1 and MS3 or MS2 may relay the connection for MS4. The location management requires that single hop or multihop MSs periodically report their locations to the network so that they can be reached when data packets are received from the Internet [27]. The multihop mechanism of the system needs to build a path to the MS to maintain the connectivity. But the connection may be lost owing to the movement of the intermediate MSs. The system needs to be active all the time to find another alternate route every time the connection is broken. The quality of the alternative path varies on the basis of the intermediate MS status. After solving the location management issue the multihop system must deal with the routing mechanism to integrate the connection over multihop communication. Several approaches have been proposed to use the multihop communication to extend the system capacity and the coverage area. The *ad hoc* network is a network structure between peer nodes; the *ad hoc* concept is utilized to provide an alternative path to MS when it faces difficulties in reaching a PoA.

Network architectures for multihop communications are based on 3G cellular link and IEEE 802.11, where the IEEE 802.11 operates in *ad hoc* mode [28]. The 3G BS forwards packets for the destination MS with poor channel quality to the intermediate MS with better channel quality [22,28–30]. The intermediate client is named "proxy." The MS proxy is connected to the destination MS using IEEE 802.11 interface in *ad hoc* mode, which improves the cell throughput. The other

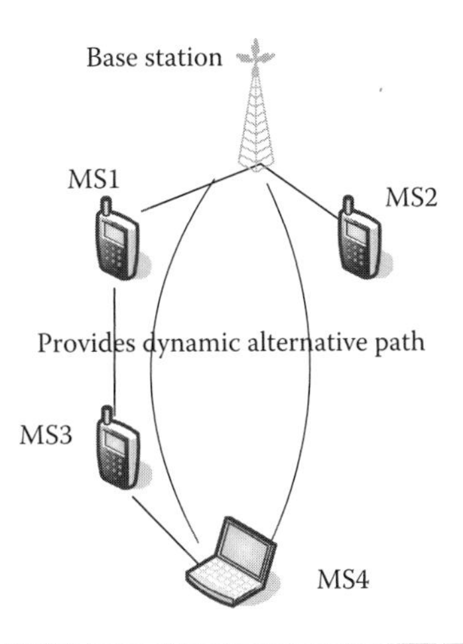

Figure 9.6 MS4 has two alternative connections to reach the BS.

approaches in this group include the two-hop design scheme, two-hop relay, iCAR [31], and so on. The two-hop design performs seamless roaming between 3G WWANs and IEEE 802.11 WLANs. This architecture increases the 3G cellular system capacity and extends the coverage area of IEEE 802.11 terminals [28]. The MSs periodically evaluate the channel status of neighboring MSs and forward this information to the BS/PoA. The MSs with the best channels will serve as intermediate relay gateways. The MSs with a weak link to the BS/PoA will rely on the intermediate to provide better throughput. The two-hop relay scheme substantially boosts the capacity of the cellular networks under a broad range of conditions [28]. The hybrid wireless network architecture presents a simple system where a BS/PoA can work like any MS in the network. The MS uses a special dedicated control channel to keep the BS updated about its local topology information. The BS/PoA directs the MS to reduce or increase the power on the basis of the MS status, to conserve the power, and at the same time maintain required throughput.

Another approach is known as the integrated cellular and *ad hoc* relaying system (iCAR). The iCAR provides relay MS to forward the traffic from one cell to another to balance the load between BSs. It reduces the dropped and blocked calls at congested cells by forwarding them to uncongested cells without increasing the infrastructure cost. Also, in the same manner, mobile-assisted data forwarding (MADF) performs load balancing between congested and uncongested cells. MADF forwards data between cells using unallocated channels [22]. The MS plays the role of a traffic forwarder agent between the two cells. A-GSM is a concept to extend communication in dead locations. A-GSM performs reconfiguration and dynamic control over the topology to instantly setup communications infrastructure [30].

9.2.4 Mobility in Heterogeneous Single-Hop Networks

In previous solutions, despite the use of multimode interfaces only one wireless technology serves as gateway to the Internet. The other wireless interfaces work in an *ad hoc* mode as bridges to other MSs. In a heterogeneous single-hop network, different technologies could be used to serve the MS, and no intermediate MSs or bridging is needed. Several approaches have been proposed to provide the routing to the same MS through multiple wireless technologies.

The generic virtual link-layer (GVLL) mechanism supports the always best-connected QoS level for an application in a heterogeneous network [32]. The heterogeneous network consists of IEEE 802.11e WLAN and IEEE 802.16d WiMax access networks [32]. A user mobile device should support both WLAN and WiMax access capabilities. The GVLL is placed above the WLAN and WMAN MAC layer in the mobile device. The best access network is chosen dynamically on the basis of QoS parameters such as throughput, packet loss, and delay. Media independent handover is used to integrate various wireless technologies through seamless handover at the IP layer [32,33]. Each wireless technology not only independently provides services to a user but also interacts with others in a collaborative manner so as to provide a given QoS guarantee.

The traditional protocol stack, consisting of a physical layer, a data link layer, a network layer, a transport layer, and an application layer, however, cannot meet the design requirements of an HWN because different wireless technologies should coexist in a multimode MS. The vertical and horizontal cross-layer interaction features allow a multimode protocol stack to coordinate multiple wireless interfaces in a flexible manner [26]. Meanwhile, the multimode protocol stack is not expected to require modifications of the underlying protocol stack functionalities so that different wireless technologies can also operate independently [19,31].

9.3 Link-Quality Assessment

To evaluate the connectivity strength of the available links for different wireless technologies, we introduce a new metric called "link quality" (LQ) to perform the required evaluation [27]. As one network parameter may not be sufficient to evaluate the connectivity strength we have combined several parameters to represent LQ. In this study, we describe an LQ estimation mechanism called automatic selection function (ASF) for HWNs that can be used to evaluate connection strength of various links [11]. ASF takes into account all wireless technologies and provides the best affordable service. ASF processes the selected wireless network parameters in addition to MS mobility speed. In other words, on the basis of the network conditions, user profiles, and required QoS, ASF selects the best link to connect without human intervention.

Different wireless parameters with different acceptable limits need to be adjusted and that makes the network convergence decision harder to achieve. Besides, packet routing from one air interface to another needs to address how different gateways will coordinate their function in this convergence process. In addition, the design must be able to work with any existing wireless protocols, and should be flexible enough to adapt new wireless technology in the future. The wireless technologies to be integrated include IEEE 802.16, IEEE 802.11, and GSM/UMTS.

The ASF algorithm interacts with the VWS framework to build a suitable environment to perform the vertical handoff and build the new integration solution. In our simulation-based work we have applied the ASF over IEEE 802.16 and IEEE 802.11, and have addressed the problem of selecting the best connectivity given the application requirement.

9.3.1 MS Type of Services

Providing seamless connectivity to the MS in a heterogeneous wireless environment is great challenge. Application requirements basically are represented as QoS requirements. Different applications such as VOIP, SIP, HTTP, HTTPS, FTP, and

POP3 can be served by heterogeneous networks. QoS requirements will depend on the applications; some may require constant bandwidth, whereas others can tolerate bandwidth fluctuation. For instance, real-time applications such as VOIP and video applications do not tolerate wide changes in bandwidth (jitters). Other applications such as Internet surfing HTTP and HTTPS are more tolerant to the throughput fluctuation. In our work, we consider two QoS levels, one for real-time applications and the other for nonreal-time applications. The network performance will be measured on the basis of this classification.

9.3.2 Mobile Station Mobility Profiles

MS mobility is a crucial factor in HWN conversations. Besides, it is an essential parameter in the network connection decision. The mobility of an MS affects the application performance. ASF needs to provide effective assessment of available links on the basis of MS mobility. For that purpose, we classify MS mobility into three categories:

- No mobility, where the MS is not moving and picks the best advertised PoA to connect.
- Pedestrian mobility, where the MS is moving at a low speed and may change the PoA on the basis of signal availability.
- Vehicular mobility, where the MS is moving at high speed and needs to change the PoA more frequently.

9.3.3 A Connectivity Prospect

A connectivity prospect represents the network status including network access opportunity, type of service, and mobile station mobility profiles (MSMPs). The network to which an MS should connect is based on the availability of wireless services. Several related network parameters need to be analyzed to select the best network service. In another words, the MS needs to find best-advertised services in the heterogeneous network and identify the corresponding PoA automatically without user intervention. In our study, the HWN includes WiFi, WiMax, and UMTS, and the connection decision relies on QoS demands. Also, the MS has three air interfaces to connect to: WiFi, WiMax, and UMTS. Each connected wireless interface receives an IP address from the corresponding service domain gateway. Each air interface has many associated parameters such as link throughput, link delay, link interference, and packet loss rate on the link. The ASF relies on these parameters to take connection decisions. MSs detect the availability of wireless services from individual networks by probing the advertisement frame called the "beacon." An MS can get this beacon in two ways: passive or active scan. In active scan, MS solicits this beacon by sending soliciting request frame to the PoA in the neighborhood. In passive scan, the MS waits until the PoA advertises it voluntarily. These beacon frames contain the necessary information that allows an MS to initialize and/or update the wireless network parameters that defines the estimation LQ. At the beginning, we need to determine the most influential parameters that affect the LQ at each wireless service, and then evaluate the quality of link as shown below:

$$\mathrm{LQ} = f(P1, P2, P3)$$

where $P1$, $P2$, and $P3$ are the network parameters for each wireless service. Each of these selected parameters must have a significant impact on the QoS of the link. The value of LQ is proportional to the QoS, which implies that the higher the QoS the higher the LQ and thus better the quality of the connection. Therefore, we can also write:

$$f(P1,\ P2,\ P3) = \alpha(P1/P1_{\mathrm{Max}}) + \beta(P2/P2_{\mathrm{Max}}) + \gamma(1/P3)/(1/P3_{\mathrm{Min}})$$

where $P1$ and $P2$ affect the quality by proportional increase and $P3$ affects the quality by proportional decrease. The weights α, β, and γ are tuning parameters where

$$\alpha + \beta + \gamma = 1$$

and they represent the weighted factors for $P1$, $P2$, and $P3$, respectively. $P1_{\mathrm{Max}}$, $P2_{\mathrm{Max}}$, and $P3_{\mathrm{Min}}$ represent the maximum and the minimum values based on the parameter influence on the LQ. The weights α, β, and γ can be selected on the basis of user preferences and are adjustable. Following this we must find the boundary condition for this equation to choose the dynamic range for selections.

Now, we introduce the term $\mathrm{LQ}_{\mathrm{max}}$, which is the best value obtained under optimal conditions. The normalized LQ can be represented as

$$\mathrm{LQ}_{\mathrm{fraction}} = \frac{\mathrm{LQ}}{\mathrm{LQ}_{\mathrm{max}}}$$

where $\mathrm{LQ}_{\mathrm{fraction}}$ is the normalized value of LQ and is proportional to the maximum value. So this factor must be processed for every available link such that

$$\mathrm{LQ1}_{\mathrm{fraction}} = \frac{\mathrm{LQ1}}{\mathrm{LQ1}_{\mathrm{max}}}$$

$$\mathrm{LQ2}_{\mathrm{fraction}} = \frac{\mathrm{LQ2}}{\mathrm{LQ2}_{\mathrm{max}}}$$

$$\mathrm{LQ3}_{\mathrm{fraction}} = \frac{\mathrm{LQ3}}{\mathrm{LQ3}_{\mathrm{max}}}$$

As the heterogeneous network deals with WiFi, WiMax, and UMTS, we need to choose the parameters $P1$, $P2$, and $P3$ for each wireless technology. LQ of the links is an important factor in any heterogeneous network that combines technologies with different characteristics in terms of data rate, delay, throughput, and so on. Therefore, the LQ metric must take into account the differences in heterogeneous links as they differ in capacity.

The above framework represents a generic concept to evaluate different links in a heterogeneous environment. Therefore, in our work a user applies the LQ concept to differentiate cellular, WLAN, and WiMax wireless network connectivity.

9.3.4 Automatic Selection Function

To provide seamless connectivity and assure the best-allowed QoS service, ASF needs to select the most appropriate connection. For that purpose, we propose the ASF that takes into account the user's mobility profile, traffic requirements, and the estimate of network condition in the selection procedure. The ASF collects all available connection conditions at the MS along with the respective parameters. These parameters will be processed to pick the best link for connectivity, and the ASF function is defined as

$$\text{ASF}\left(\text{LQ1, LQ2, LQ3, TOS, MSMP}\right)$$

where TOS is the type of service. The MS periodically collects the network parameters at different air interfaces, combines the MS data, and performs the calculations at the ASF to guarantee that best link serves an MS all the time.

9.4 Virtual Wireless Service

The demand for wireless Internet requiring broadband is expected to increase in a dramatic way owing to the popularity of numerous applications including voice, video, and data transmission. Existing wireless technologies such as UMTS, WiFi, WiMax may not satisfy customer requirements for many reasons including availability, coverage, cost, and, most importantly, mobility. Each technology has limitations related to the above characteristics. We present a seamless connectivity scheme for a mobile user by aggregating different wireless technologies to make these networks transparent for the running applications on the mobile devices, and satisfy the QoS requirements over a wide range of applications including voice, data, multimedia, and Web browsing.

This section proposes a new architecture for combining different wireless technologies including cellular networks, GSM/UMTS, WLANs, and WMANs. First, we discuss why we need a new convergence concept. We describe the architecture of the VWS and its components. The VWS mechanism includes network detection, connection selection, and packet forwarding. The VWS aims at providing a generic solution for the different wireless networks by hiding the network selection and connection management schemes from the running application.

In addition, adding a new wireless technology will be as simple as adding a wireless interface to an MS.

9.4.1 Motivation for New Integrated Architecture

Because of the popularity of wireless communication services that can serve users everywhere using many network forms including WLANs, WMANs, and WWANs [1], most advertised MSs in the market are equipped with multiple air interfaces to interact with different wireless technologies. At present, many wireless service providers offer different plans varying in cost, coverage, and offered throughput. For instance, WiFi, which is based on the IEEE 802.11 protocol, can possibly satisfy many user applications and maybe meet their requirements in terms of connection, speed, and cost. However, WiFi does not support large coverage and that will limit user mobility. In addition, WiMax, which is based on the IEEE 802.16 protocol, supports larger coverage areas and may support different mobility speeds, but it does not offer connection speeds like WiFi. Third-generation cellular networks including UMTS/HSDPA (high-speed downlink packet access) have the largest coverage area but the connection speed is modest compared to that of WiFi and WiMax. These facts motivate us to provide one wireless service plan that unifies all these services under one framework. MSs that come with multiple wireless NICs (cellular interfaces, IEEE 802.11, IEEE 802.16, etc.) should be able to handle the integration of heterogeneous services in a transparent manner. In the existing framework, a user needs to subscribe to these different providers to access cellular/WiMax BSs, and WiFi APs [31] to be able to reach any of these services. Moreover, user mobility poses a great challenge when a user moves from one location to another and one needs to decide which MS interface to use. In a seamlessly integrated network, a user does not need to check the best network service to use to execute an application. The existing mobility management protocols use either MIP for macromobility or CIP for micromobility [27]. Unfortunately, these approaches are not sufficient for frequent macromobility because of the latency overhead of the registration process. Moreover, MIP and CIP provide mobility management in a single type of network. As our goal is to provide a seamless connection across HWNs, a new architecture needs to address the handoff between different technologies without service interruptions. The new architecture will offer the integration of old services under one umbrella on the basis of the availability of the wireless connectivity at the MS locations. The proposed approach will combine the available network access services by introducing a new concept to support switching between different wireless services for seamless execution of application on mobile devices with the required QoS demands.

9.4.2 VWS Architecture

This section introduces a novel scheme to aggregate different wireless networks into a VWS. Hence, different radio spectrums with different channels may be colocated under a coverage area. The protocol developed in this chapter uses a new VWS to perform location and connection management dynamically and will provide the best connection service for the applications being executed on an MS. The application at the mobile devices will not be aware which wireless technology is used to provide the service. The VWS can support GSM/UMTS cellular systems, WiMax IEEE 802.16, and WiFi IEEE 802.11. It can handle any future service without additional modifications. MSs represent all types of mobile PCs, UMPCs, PDAs, smart phones, and so on. These MSs need to be equipped with multiple wireless air interfaces or network interfaces to work with different networks seamlessly.

VWS is designed using the client–server model. The client is designated as a virtual wireless agent (VWA), which is a software driver residing at an MS. It controls all the available wireless interfaces on that MS. The server is termed as virtual Internet service provider (VISP), which administers the traffic from/to the MS. Network service infrastructure includes different types of PoAs and wireless gateways (WGWs) corresponding to each available network access service. PoA can be a BS or an AP. WGW will be different for each wireless technology and provides access to the PoA for that network. Figure 9.7 displays the VWS components and shows different wireless networks. The GSM/UMTS cellular systems, WiMax IEEE 802.16, and WiFi IEEE 802.11 are considered in the coverage area (Figure 9.7). An MS is equipped with multiple wireless network interfaces to connect to any of the wireless networks as directed by VWA and VISP. Network infrastructure components, cellular BSs, WiMax BSs, WLAN APs, and WGWs are connected to the VISP via a wired or wireless connection. VWA and VISP combinations work together to provide a new wireless service called VWS that aims to support seamless connectivity for the MS HWNs.

9.4.2.1 VWS Architecture and Design Challenge

In this section, we discuss VWS members and their roles. VWSs perform the vertical handoff to provide continuous connectivity to the MS. During MS mobility it may be served with the same WGW in the same wireless technology or by a different WGW in a different technology. In Figure 9.8, we differentiate between local handoff, which is done within the same technology, and vertical handoff, which is done in different technologies. The VWS lets the local handoff be handled by each wireless technology protocol. For example, in WiFi, the local handoff addressed in IEEE 802.11 utilizes a modified version of CIP. UMTS uses MIP for macro mobility and CIP for micro mobility [20]. In WiMax services, the IEEE 802.16 protocol utilizes WiMax global roaming alliance [3] to perform the local handoff.

Performing vertical handoff is a challenge in heterogeneous networks. The VWS needs to establish multiple connections through different air interfaces that are using different IP addresses. The VWA periodically checks the LQ and when

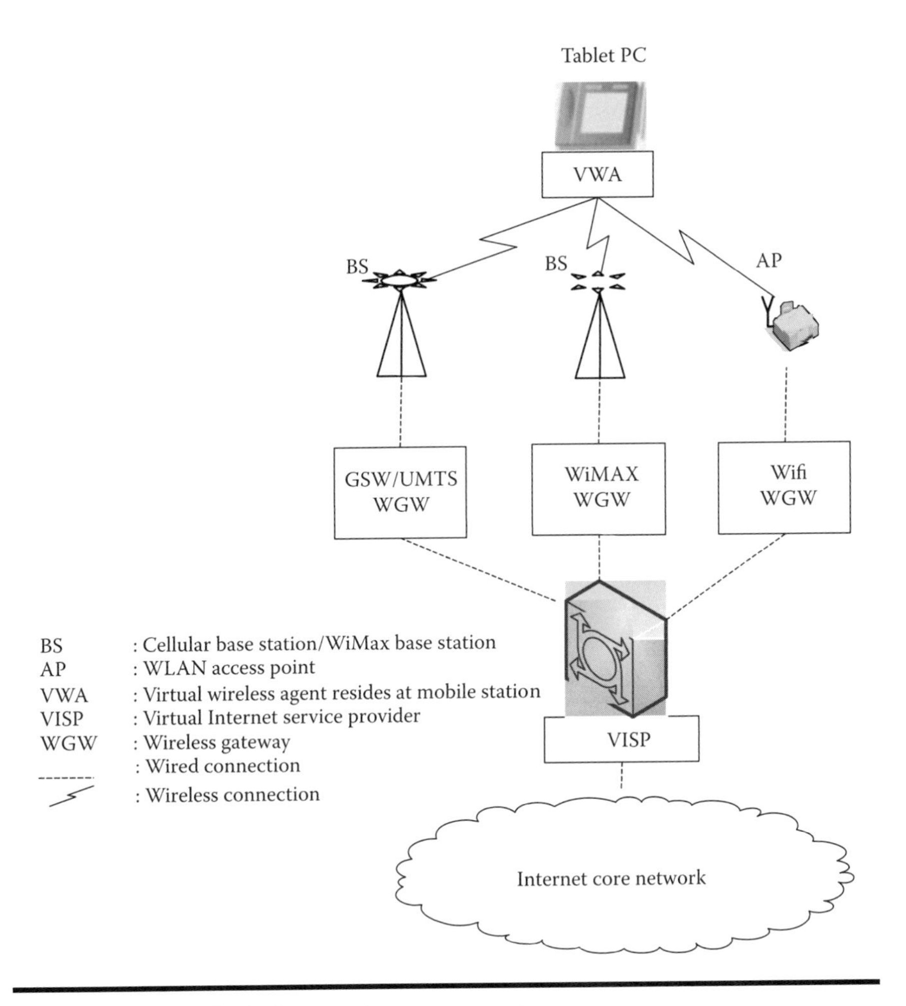

Figure 9.7 Virtual wireless service components.

the quality of used link drops the VWA activates an alternative link from different technologies.

A significant challenge is determining when and how to switch between different wireless technologies. Why do we need to change the current link to another one? The answer is based on the LQ evaluation for the alternative service links. The ASF periodically scans and evaluates the LQ and selects the best link. How do we keep the MS connected and change the link? The answer is based on VWA/VISP as the VWA will periodically send an updated packet to the VISP. Figure 9.9 shows how local handoff inside each wireless technology is carried out and the vertical handoff between the GSM, WiFi, and WiMax is performed. The main VWA challenge is to provide transparent network connectivity to the MS while satisfying application requirements.

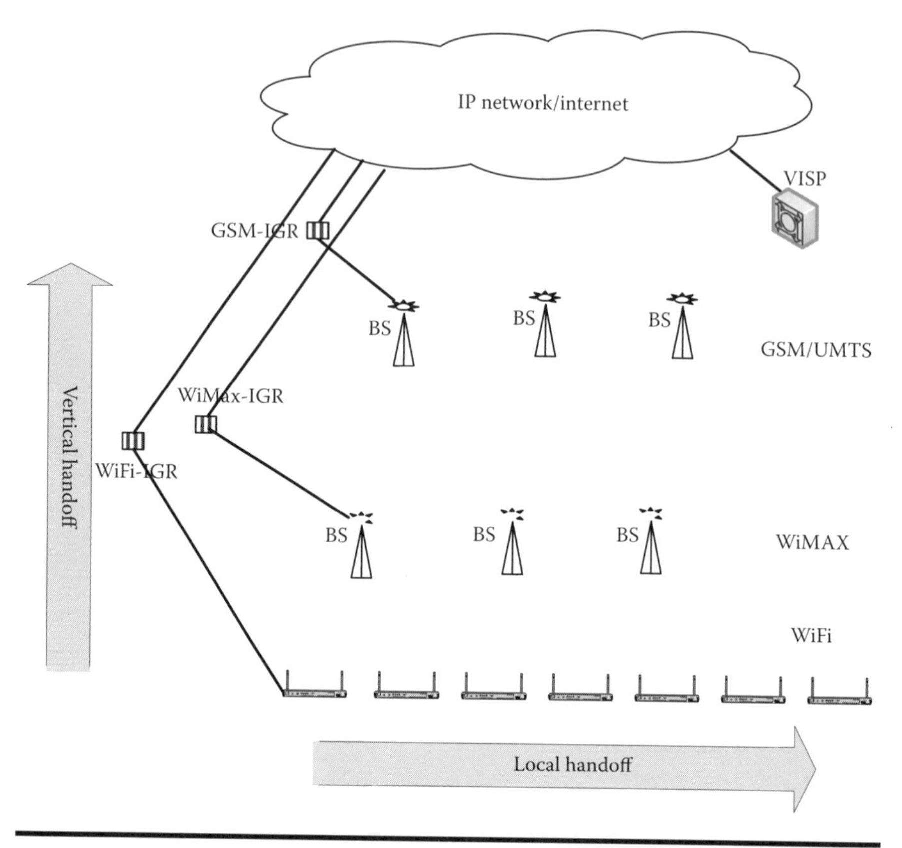

Figure 9.8 Interaction with different technologies in the same physical area.

9.4.2.2 Definitions and Characteristics of VWS

The design objective is to let the MS user stay connected to the Internet with the best network access link without human intervention. As we know, the wireless service signal strength is not consistent in a given coverage area in any technology, due to the influence of many natural phenomena including noise, interference, signal propagation degrading, and so on. For this reason the VWS framework needs to keep evaluating the quality of existing wireless connections and other alternatives periodically. This will help the VWS provide the best network access to an MS. When an MS is in motion, the PoA may change using either horizontal handoff or vertical handoff depending on the quality of the network link available in the coverage area. In summary, VWS aims at creating a new mechanism that helps change a wireless service connection on the basis of application requirements and physical signal properties so that this transfer is smooth and transparent to users.

The VWS assumes the served area is covered with one or more network access services. As the VWA has the privilege to take over all the available wireless interfaces,

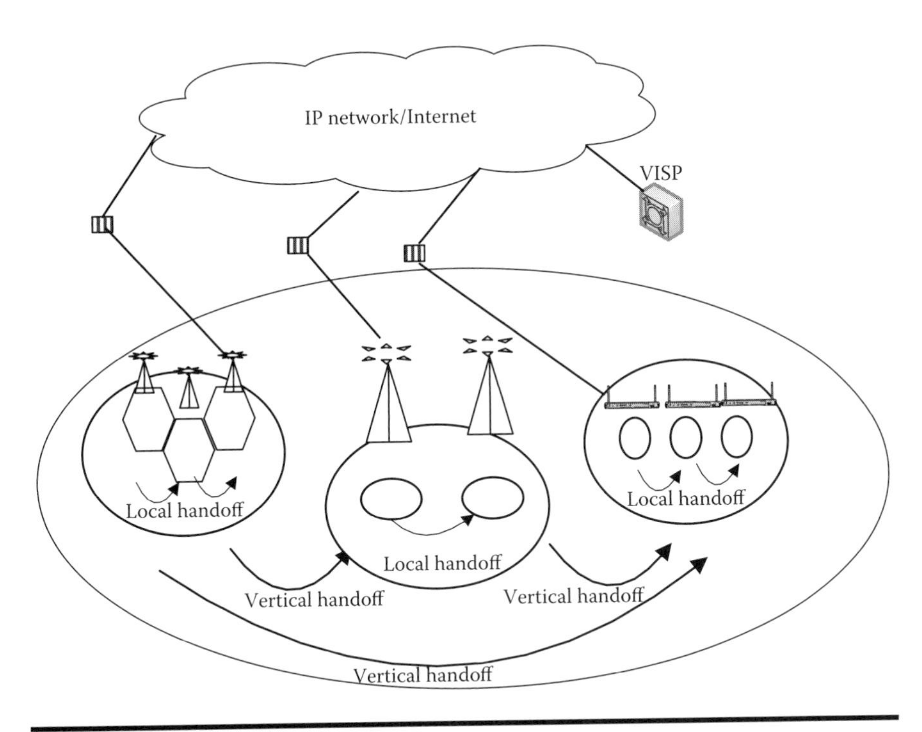

Figure 9.9 Vertical handoff and local handoff or horizontal handoff.

VWA can activate or deactivate any wireless interface on the basis of the connection selection algorithm. The connection selection will provide a mechanism to pick the best affordable link using the ASF, discussed in Section 9.3.

As VWA resides on the mobile device, it has full control over the wireless interfaces to pick the best path to connect over the Internet. The ASF works in cooperation with the VWA. Figure 9.10 shows different MSs with different wireless coverage. The MS needs to pick the best available wireless service. VISPs may be replicated near the gateways on the Internet to provide better performance to various MSs in a coverage area.

9.4.2.3 Server/MS Connection Design

In 4G networks, multiple technologies including UMTS/GSM, WiMax, and WiFi provide services to various MSs. As different wireless networks may coexist in the same coverage area, and these technologies are connected to the Internet with their respective gateways, each gateway assigns an IP address for each air interface at the MS. The MS receives more than one IP address. These IP addresses will allow limited connectivity to deliver the traffic from the air interface to the VISP and vice versa. At the server, the authentication procedure will be performed for the MS regardless of how many air interfaces are available. After authentication, a special

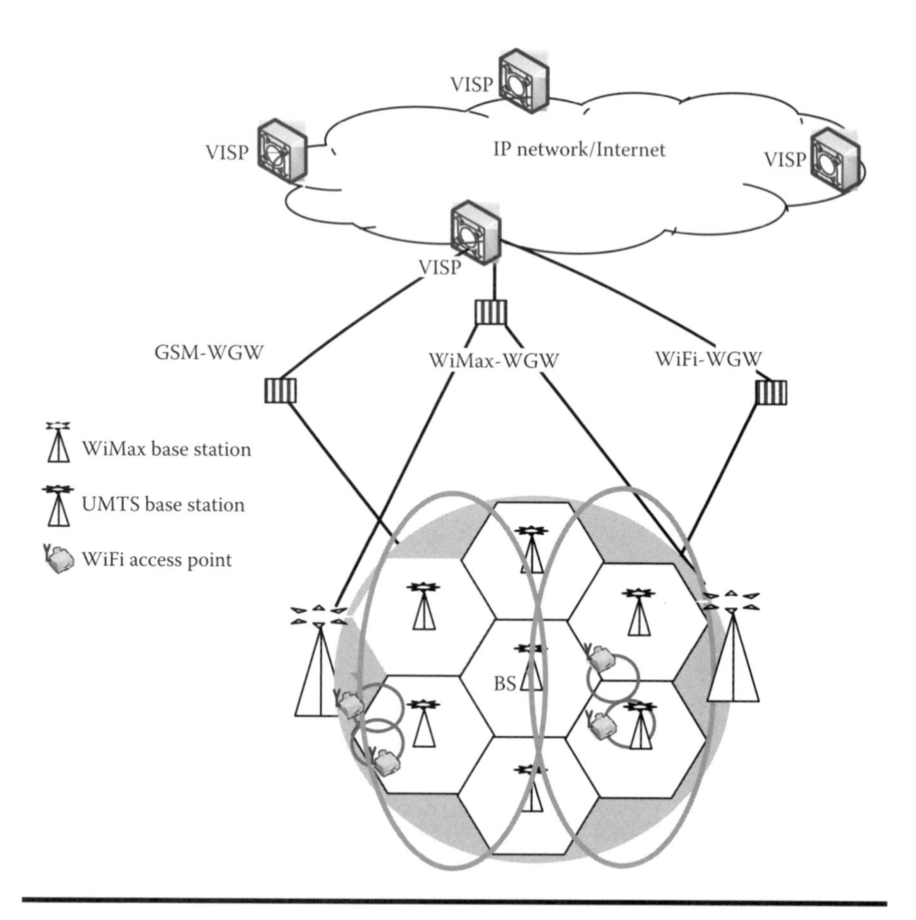

Figure 9.10 Virtual wireless service connectivity in heterogeneous wireless networks.

table will be formed at the VISP for each authenticated MS. We call this the "tracking table" (TT), as it tracks the MS's updated status including the delivery and the MAC address of the different air interfaces. As the MS must have a unique address or one distinguishing IP address during the communication in the network, the VISP will assign a unique address called the master IP address,and the MS will be identified by this master IP address behind the VISP. This means that all the traffic form/to the VISP will be routed via the master IP address, and remaining IP addresses will work as temporary delivery addresses from the MS to the VISP and vice versa. These addresses may change during MS mobility if the gateway is changed or the PoA is changed. The master IP address will not be altered and it will stay active while the MS is connected to the VISP. The VWA at the MS needs to activate or deactivate the air interfaces in response to the ASF algorithm that selects the best available link to reach the VISP. ASF constantly measures different LQs,

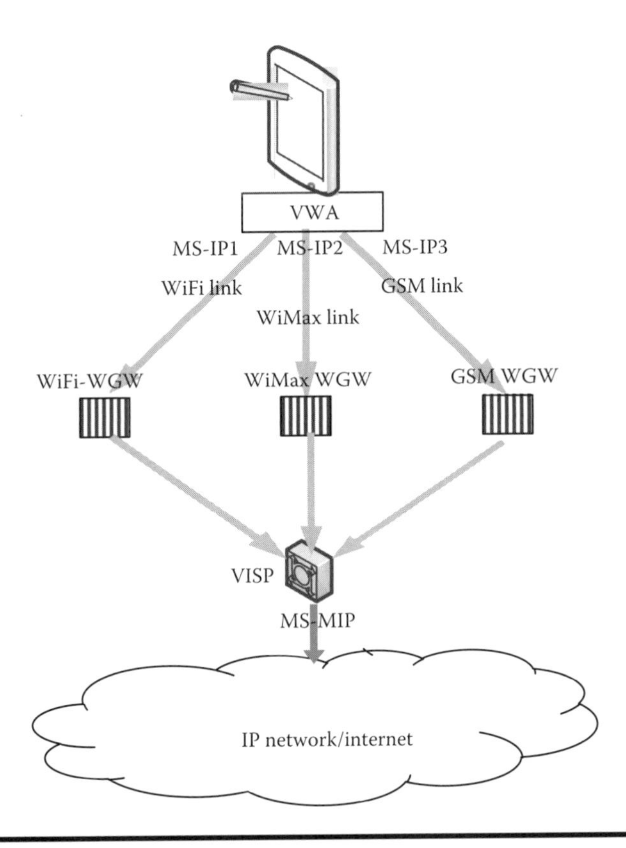

Figure 9.11 Server/MS architecture.

and decides which air interface has better quality to deliver the traffic to the VISP. This estimation will be sent to the VISP periodically. This interaction between the MS and the VISP is required to keep the best connection alive between the VISP and MS. Figure 9.11 shows the network flow between client and the server through different gateways.

9.4.2.4 Mobility Management in VWS

The VWA at the MS proactively monitors the current connection and other alternative connections to evaluate the highest LQ connection all the time. The information collected from different air interfaces is evaluated with ASF and the available links sorted so that the top link is always the best available link. By using the mechanism discussed earlier, MS/VISP, the MS stays connected with the best available link. Besides, the VISP acts like a traffic control center where the traffic from the MS is forwarded to the destination after it encapsulates it with the master IP address. In the same manner, the return data for the MS with a master IP address will be processed by the VISP and forwarded to the MS after striping the master IP address and by using the latest updated top delivery IP on the list. This mechanism represents seamless mobility management through horizontal and vertical handoffs between wireless services by using the client–server module. In the proposed approach, the MS will always be connected to the VISP and the correspondent station on the Internet will stay connected to the MS via the VISP using the master IP address.

9.4.2.5 Hard Handoff/Soft Handoff

The VISP offers two types of handoffs to meet the required QoS for executing an application at the MS. When the MS is in motion from one coverage area to another, the application being executed on an MS should be transferred to the new PoA. Otherwise, the application will go down because the link with the current PoA and MS is either too weak or there is no coverage. The transition time between transfers (disconnection and reconnecting) may affect the running applications. Some applications can tolerate this transition time during the handoff and may run smoothly, whereas other applications may stop and may not tolerate packet loss during transfer.

To overcome this dilemma, two types of handoffs are used: hard handoff and soft handoff. In hard handoff, the link to the prior connection is terminated before the new link is established and the MS is transferred to the new link. This means that the MS only uses one link at a given time. Initiation of the handoff is triggered when ASF senses that the LQ of the alternative link is more suitable than the LQ of the current link. The advantage of the hard handoff is that it is fast and does not involve complex hardware. The drawback is that some packets are lost during the handoff process.

In soft handoff, the VWA at the MS keeps the connections to the current link while attempting to connect to the new link and thus two links exist during the transfer. This means that the MS can start to communicate with the new link without breaking the prior link and it only breaks the old link after all the packets in transit have arrived at the MS. Soft handoff is a more complex technique and requires double the bandwidth during the transition period, but it assures a higher QoS.

In summary, the main purpose of handover is to maintain the connection between the MS and the VISP during transfer between two different PoAs. The goal of the VWS is to keep the application connected and avoid getting disconnected.

9.4.3 Integrated Architecture Workflow

The vast development of wireless network access services allows the Internet to be accessible by large-scale wireless PoAs. A wide range of mobile applications are expected to be a combination of real-time applications and Web applications, and these applications require a diverse QoS level. Integrating different wireless networks that differ in network stack architectures for supporting diverse QoS is a challenging task. The cross-layer approach relies on getting data across the stack layer protocol. The proposed architecture will enable the use of alternate interfaces on the basis of information obtained from nonadjacent layers. We plan to develop a generic architecture identifying basic operations required to establish wireless connections including network discovery, LQ evaluation, and selections of the appropriate PoA. We introduce VWA as the control center at the MS, where the ASF algorithm keeps periodically evaluating each wireless service and then combines these evaluations to select the higher LQ connections on the basis of predefined criteria. The VWS architecture workflow diagram in Figure 9.12 shows the integration of three different wireless services. Each of these services is managed by its own protocols: IEEE 802.11, IEEE 802.16, and UMTS 3GPP, respectively. The VWS does not modify these protocols, which means that each air

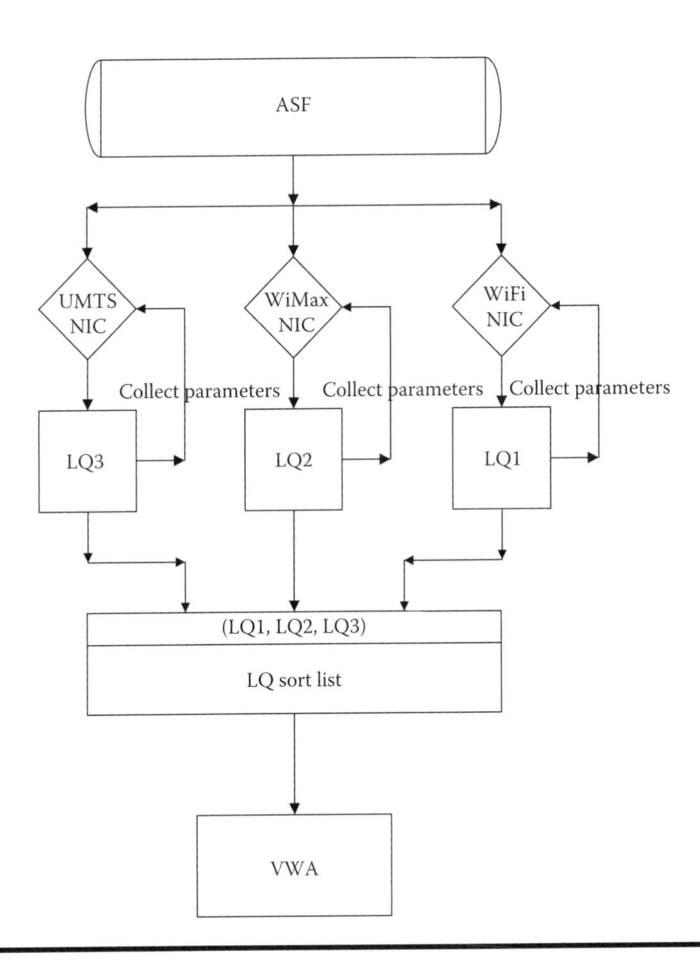

Figure 9.12 ASF algorithm flow work.

interface will be managed by its own protocols and the VWS will provide a VWA at the application layer that manages each air interface. The VWS can activate or deactivate an interface on the basis of the LQ of each service. When the MS turns on, the available air interfaces at the MS will be managed by their respective interface drivers, and they will establish limited link connectivity to the VISP through the appropriate gateway WGW. Assuming that the three network access services are available, the MS is granted three IP addresses representing the connection on the subnet between the air interface and the WGW. So MS obtains IP1 from WGW1, the WiFi gateway; IP2 from WGW2, the WiMax gateway; and IP3 from the WGW3, the GSM/UMTS gateway. These IP addresses have limited connectivity just to deliver the traffic to VISP. The VWA connects the air interfaces with the VISP default address and then different WGWs grant the IPs addresses. The ASF algorithm collects the different wireless parameters to evaluate each individual link as discussed in the previous section.

The LQ obtained from each available link will be used together to select the most appropriate link on the basis of the type of service and MS mobility. ASF will form a routing list containing the best link at the top. This list will be updated by the ASF to always find the best path. Further, this list will be forwarded to the VWA, which has control over the air interfaces and the traffic produced by the applications being executed on the MS.

The ASF must compute LQ dynamically periodically to adjust for changes due to the speed of the MS. Figure 9.12 shows the ASF workflow. The connection management mechanism is handled by the VWA. The VWA provides connection management using the ASF selection to provide seamless vertical handoff across heterogeneous networks. From another perspective, the VISP will keep track of the active link and supports rerouting mechanisms when the VWA performs the switch from one interface to another. Our proposed integrated routing mechanism supports terminal mobility through its route maintenance functions by updating periodically the active routes and the alternative routes between the VWA at the MS and the VISP. However, our integrated routing solution designed to implement the handoff protocol by changing the active routes at two sides does not need to establish or send control messages for authentication, as all the interfaces at the MS are authenticated. This flexibility in handling mobility is reflected not only between the MS and the VISP but also between the VISP and the correspondent MS. The proposed mechanism will handle complete end-to-end connections between the MS and the correspondent MS and guarantees that the correspondent MS will be connected to the VISP. In this chapter, the vertical handoff between heterogeneous networks is performed by the VWA/VISP without interacting with the OSI/ISO network.

9.4.4 VWS Case Studies

We will explain the operation of VISP in detail through two case studies. The first case includes wireless services WiMax/IEEE 802.16e and WiFi/IEEE 802.11.

These services share the same coverage area. Different PoAs such as BSs are used to serve WiMax and WiFi. The MS has two wireless NICs, one runs on the WiMax air interface and the other on the WiFi air interface. The wireless services are connected to different gateways. The WiMax gateway and the WiFi gateway are assigned an IP address for each wireless interface adapter, namely, IP1 for the WiFi NIC and IP2 for the WiMax. IP1and IP2 are delivery IPs from the gateway and to/from the VISP.

9.4.4.1 Case Study 1

In the first scenario, the MS is in stationary mode and the application running at the MS is a Web application. When the MS goes into a discovery mode the ASF will check the available wireless services and calculate the LQ. As the MS is in a hotspot area and is covered with good WiFi signal strength, the VWA at the MS activates the WiFi air interface. The correspondent WiFi gateway will grant the IP1 to the WiFi air interface with restricted connectivity just to deliver and receive traffic from the VISP. When the MS request reaches the VISP, the authentication procedure is launched; the user at the MS needs to provide the username and appropriate password to complete the authentication process. Simultaneously, the MS will be granted an IP2 address for the WiMax interface card from the WiMax gateway. There is no need to repeat the authentication for this service, as it is centrally done at the VISP for the MS with different air interfaces. The WiMax interface at the MS will be idle or in sleep mode as the WiFi air interface is active. The MS is ready to communicate and the applications may send and receive traffic through the WiFi interface. During registration, the VWA will send a packet to the VISP. This packet includes different wireless NICs at the MS, assigned IPs, and MAC addresses. Later, the VWA periodically will send an update packet including the best link to reach the MS. The ASF will continue to evaluate the LQ for all alternative links and sort the best one to be at the top of the list to be used. In our scenario, the WiFi link will be at the top and the WiMax will be underneath it, as the WiFi link has better LQ. The MS is ready to initiate connections, and sends data to the VWA; in turn, the VWA will forward it to the active WiFi interface. The transmitted packets will be stamped with IP1 as the source address. The next hop for these packets will be the VISP. At the VISP, these packets will be encapsulated with the master IP address and forwarded to the destination. The VWA keeps periodically sending the updated packet to promote the best link for the MS. At the VISP, a TT is created when the MS is authenticated. This TT will include the following information: the MS ID, master IP, IP1, MAC1, IP2, and MAC2. The information at the TT will be continuously updated on the basis of the incoming packets from the VWA. In this scenario, we assume that the MS has moved from location 1 to location 2. At location 2, the WiFi signal is still acceptable, but the PoA has been changed from AP1 to AP2 during the MS movement. MS moves again from location 2 to location 3. The ASF identifies that LQ for WiMax became better than LQ1 for WiFi and the VWA sent this information to the VISP through the WiFi connection. The VWA will respond to this change and will send the traffic over the WiMax air interface. The applications continue to send data, as on the basis of the information received from the VWA, the VISP will start forwarding the packets to the MS using the IP2 WiMax connection.

9.4.4.2 Case Study 2

This case setup includes UMTS/GSM and WiFi /802.11 as wireless services. The UMTS/GSM service has a larger coverage spot than WiFi. Usually, UMTS/GSM BSs are located two to three miles apart, and WiFi APs are located in arbitrary hotspot locations in the study area. MS has two wireless NICs, one for the 3G UMTS/GSM and the other for the WiFi air interface. Each wireless service is served by different providers, through different gateways. The UMTS/GSM gateway assigns IP1 for the UMTS/GSM interface at the MS and the WiFi APs assign IP2 for the WiFi interface. IP1and IP2 are delivery IPs from/to the VISP.

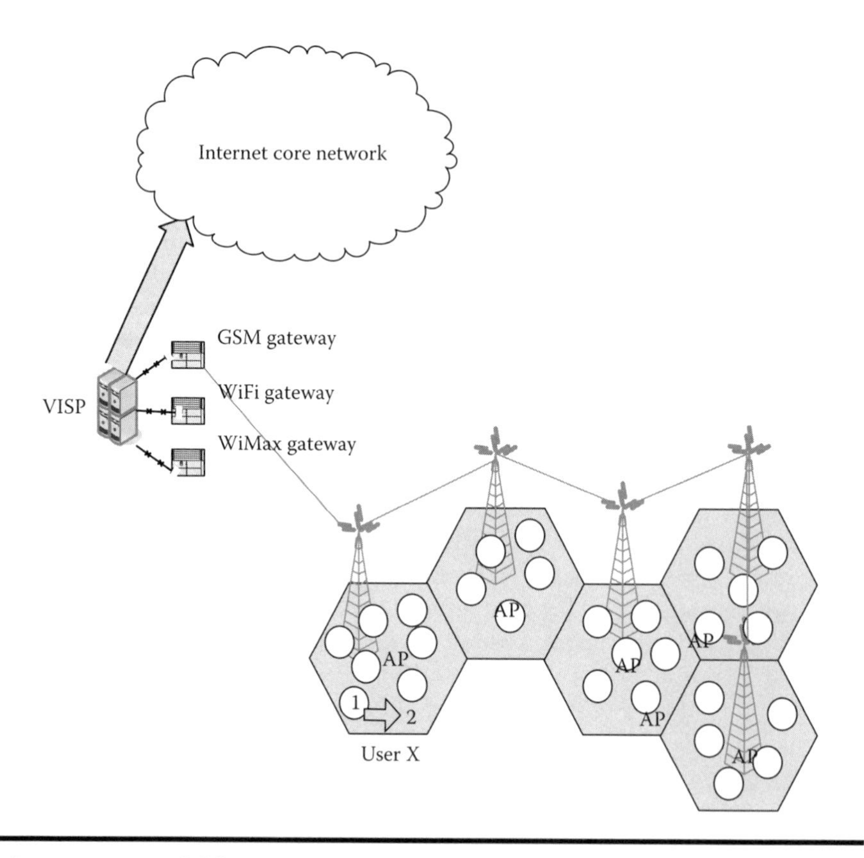

Figure 9.13 Model for user motion from locations 1–2.

Figure 9.13 shows the MS equipped with WiFi and the GSM/UMTS cards. User X at location 1 has coverage from WiFi and GSM. The VWA needs to pick the appropriate technology between WiFi and GSM. On the basis of LQ values, the status of the MS where it was in low speed mobility (of less than 1 m/s), and the LQ2, which indicates better connectivity to the WiFi, the VWA picks WiFi. The IEEE 802.11 interface is activated and the GSM interface is idle. Both interfaces are connected to their gateway and are assigned two delivery IP addresses, IP1 and IP2. These addresses will be registered at the TT in the VISP, after the authentication process is completed.

The VISP will update the TT based on the update packets that arrive from the VWA at the MS. The VISP will bind the delivery addresses and the master IP address together. The VISP will forward the traffic to the appropriate delivery address through the activated wireless interface adapter.

Let us return to our example where user X in position 1 starts moving to position 2 where there is no WiFi coverage. The VWA senses the change in the LQ values and immediately activates the UMTS/GSM adapter and sends the update packet to the TT at the VISP. At the VISP, all packets will be forwarded to the active delivery address—"UMTS/GSM" NIC—and smooth vertical handoff will be carried out. As the main route on the Internet is already available, we just need to activate the right interface at the VISP for delivering client packets.

9.5 Simulation and Performance Evaluation

In this section, we provide a comprehensive performance analysis of our VWS architecture. We have implemented an HWN with the integrated scheme in the OPNET simulator modeler 14.5 with WiFi, WiMax, and UMTS modules. For a better understanding of the OPNET environment, several cases for each individual wireless technology have been implemented to simulate the local handoff under different speeds and for different applications. To study the performance of the VWS concept in our simulation, two different wireless technologies are connected to two different WGWs. These WGWs are in turn connected to one application server. Also, several MSs with multiple air interfaces are found to roam under this heterogeneous network scenario. Application performance was studied under different mobility conditions including no mobility, pedestrian mobility, and high-speed mobility. FTP and voice application were studied in various scenarios.

The planned setup contains HWN with a VWS scheme where the simulated area was 3000×3000 m^2. This area included WiMax BS, wireless AP, customer premise equipment (CPE), core router, MSs with multiple interfaces, and application server. WiMax configuration was as follows: the antenna was set to 15 dbi gain, physical layer modulation was set to WMAN-SCP1, power transmission was set to 0.5 W, and the WiMax advertisement was set every 10 frames. The WiFi AP was configured as follows: BS identifier 2, physical layer modulation in direct sequence, data rate of 11 Mbit/s, transmission power was set to 5 mW, the reception power threshold was set to −95, and the AP beacon interval was set to 20 ms. The CPE/WLAN configuration has two sets of configurations: the first set matched the WiMax parameters mentioned earlier, and the second set matched the WiFi AP parameters. The network router used three ports that support point to point protocols and data rates of up to 10 Mbps. Each of these ports was connected to the WiMax BS and the WiFi AP and then to the application server. The services performed included file transfer using FTP protocol, and voice services were set to simulate GSM quality speech.

As the MS had multiple interfaces, the MS was equipped with IF0, IF1 air interface, and was able to use WiMax or WiFi services. WLAN APs were set to work in infrastructure mode and WiMax BS was set to communicate to CPE. For better reference, another setup of individual wireless networks was created: WiFi, WiMax, and UMTS. These setups included several BSs/APs in each technology, and used local handoff to handle MS mobility. As the local handoff is much simpler to handle than the vertical handoff, we compared both handoffs for the same set of applications being executed at the mobile devices. Note that local handoff depends on the implementation of the data link layer, which provides better performance to upper layers, whereas vertical handoff needs to be done in a network layer. The proposed simulation is suited to validate the effectiveness of the VWS scheme and shows the work in a realistic heterogeneous scenario, exploiting heterogeneous routing to improve the overall WLAN and WiMax coverage.

Three basic metrics that have been used in the performance analysis include throughput, received traffic, and delay.

- *Throughput*: This represents the total data traffic in bits per second received successfully forwarded to the higher layer by the MAC layer.
- *Received traffic*: This refers to the data successfully received by the MAC layer from the physical layer in bits per second. This statistics includes all data traffic received regardless of the destination of the received frames. While computing the size of the received packets for this statistics, the physical layer and MAC headers of the packet are also included.
- *Delay*: This represents the end-to-end delay of all the data packets that are successfully received by the MAC and forwarded to the higher layer. This delay includes queuing and medium access delays at the source MAC and reception of all the fragments individually.

In the VWS scenario, we have implemented a simulation to perform vertical handoff between different wireless networks using the VWS architecture. This scenario contains two different wireless services, WiMax and WiFi, represented by AP and BS as PoAs. The coverage area for each one of the BS/AP is collocated in the same serving area. The MS includes two wireless interfaces and it moves inside a serving area, from one location to another. During MS mobility the VWA will select

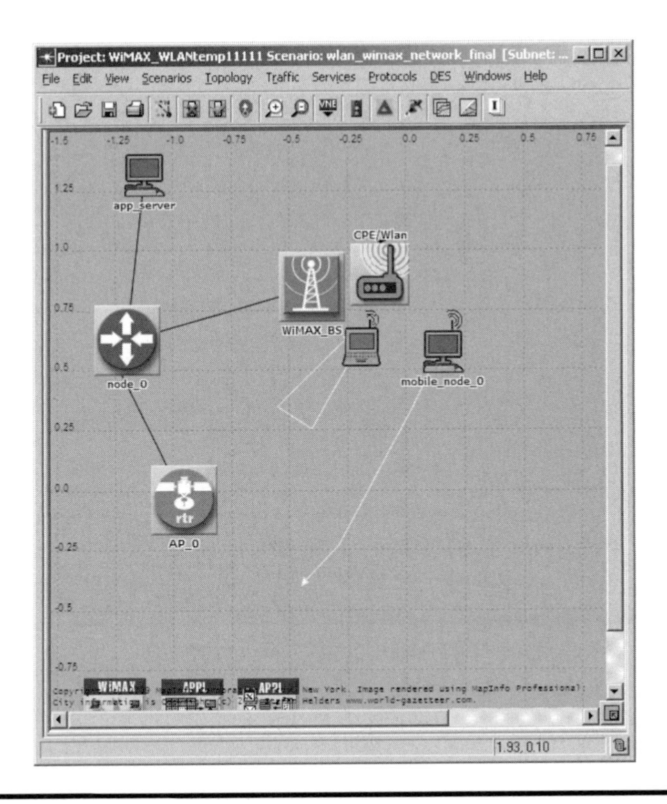

Figure 9.14 The virtual wireless service setup with WiFi and WiMax services.

the best available signal for connection. The WiMax BS will provide the service through the CPE router. The BS, CPE, AP, and MS are shown in the Figure 9.14.

In this scenario, the MS is equipped with two wireless interfaces. Figure 9.15 shows the throughput at the two air interfaces serving the HTTP application. The blue curve represents the throughput at the IF0 interface that was connected to the WiMax BS and the red curve represents the throughput at the IF1 interface that was connected to the WiFi AP.

In Figure 9.16, the blue dots represent the delay for the traffic on the IF0 interface. The red dots represent the delays on the IF1 interface. Figure 9.17 shows the throughput of the two air interfaces serving the FTP application. The blue curve represents the throughput at the IF0 interface that is connected to the WiMax BS and the red curve represents the throughput at the IF1 interface that is connected to the WiFi AP. In Figure 9.18, the blue dots represent the delay in seconds for the traffic on the IF0 interface and the red dots represent the delays on the IF1 interface.

Figure 9.19 shows the throughput at the two air interfaces, serving the voice call with GSM quality application. The blue curve represents the throughput at the IF0

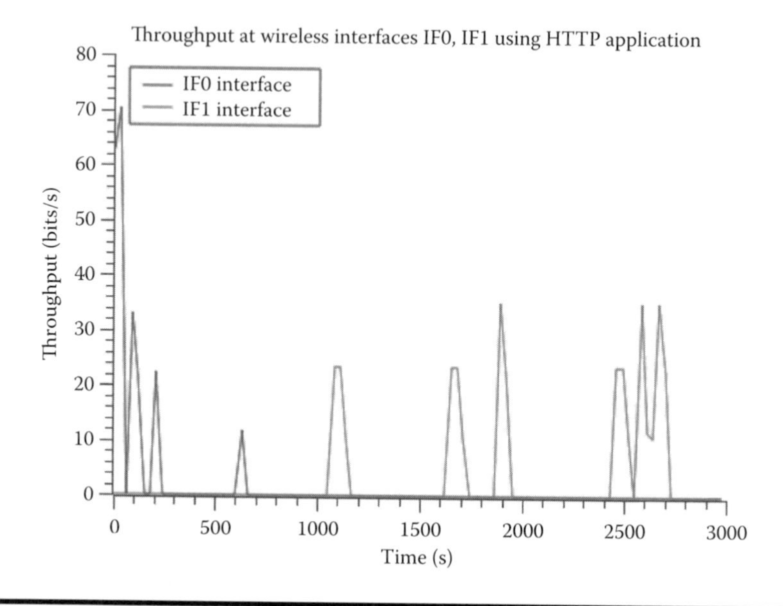

Figure 9.15 Throughput of multimode MS with HTTP.

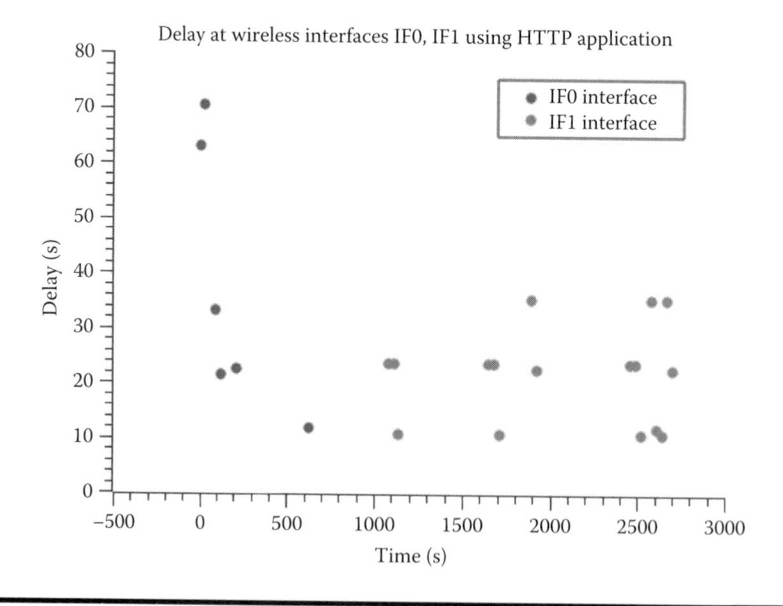

Figure 9.16 Delay of multimode MS with HTTP.

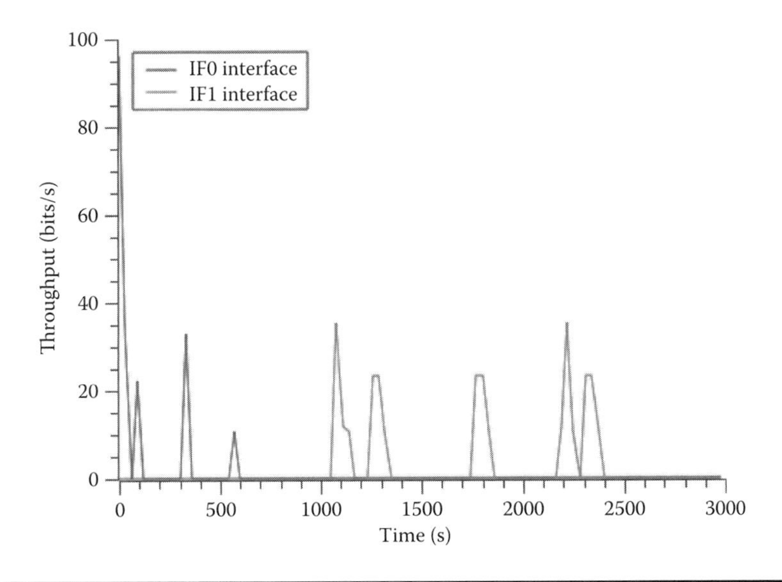

Figure 9.17 Throughput of multimode MS with FTP.

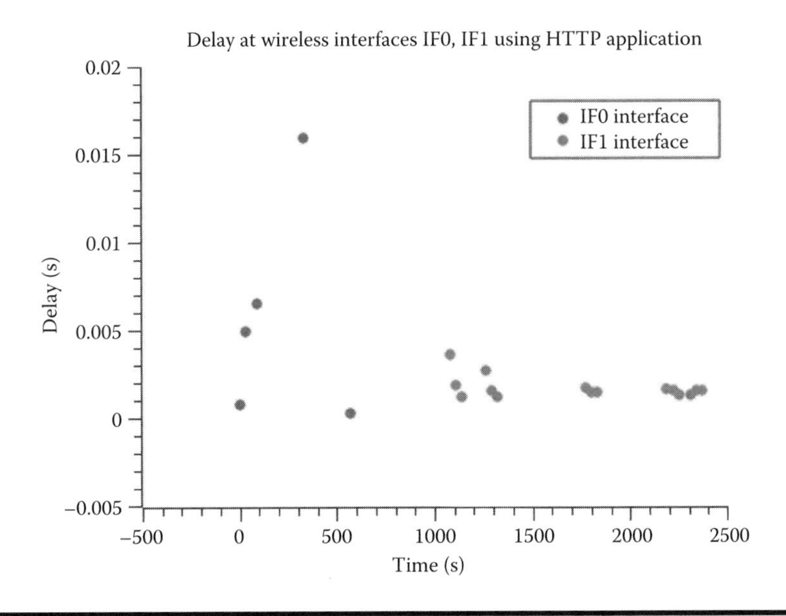

Figure 9.18 Delay of multimode MS with FTP.

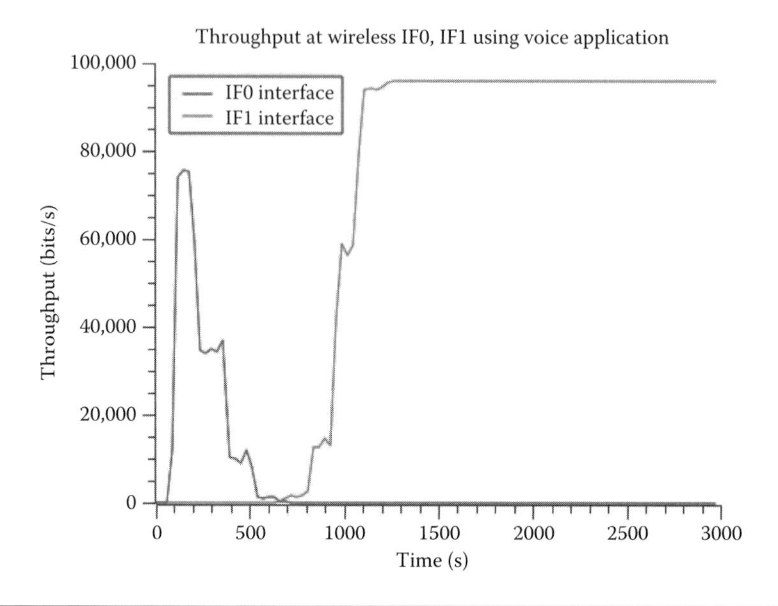

Figure 9.19 Throughput of multimode MS with voice call.

interface that is connected to the WiMax BS and the red curve represents the throughput at the IF1 interface that is connected to the WiFi AP. The throughput at the WiMax connection was around 75 Kbits/s and the throughput at the WiFi connection became better (100 Kbits/s) after the handoff. During the vertical handoff the traffic moved from the IF0 to the IF1 interface and the connections did not break, but the throughput dropped dramatically during handoff on interface IF0. The VWS was able to select the second available service at the IF1 interface and connected the application to that interface.

In Figure 9.20 the black curve represents the delay in seconds for the traffic on the IF0 interface and the gray curve represents the delay on the IF1 interface. We observe that the traffic at the first link increased and the delay at the second link was decreased up to cross point when the vertical handover occurred.

As the VWS offers a mechanism to control the handoff initiation, the delay time can be controlled or reduced to suit the QoS we are looking for. Figures 9.19 and 9.20 show the performance using hard handoff, but the performance can be improved using soft handoff. To illustrate the effect of soft handoff, Figure 9.21 shows the throughput of the same previous setup and the same served application (which is voice call with GSM quality). In this test, we increased the quality during vertical handoff by duplicating the traffic on the two interfaces. The black curve represents the throughput at the IF0 interface that is connected to the WiMax BS and the gray curve represents the throughput at the IF1 interface that is connected to the WiFi AP.

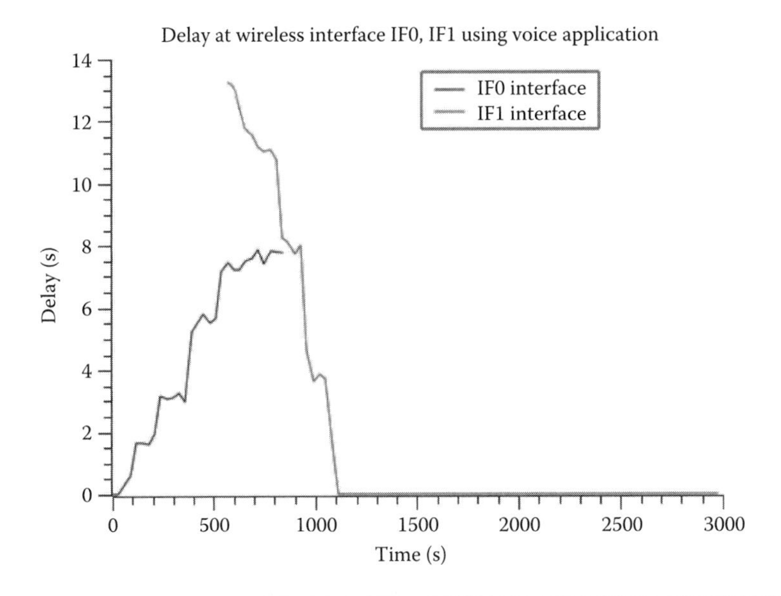

Figure 9.20 Delay of multimode MS with voice call.

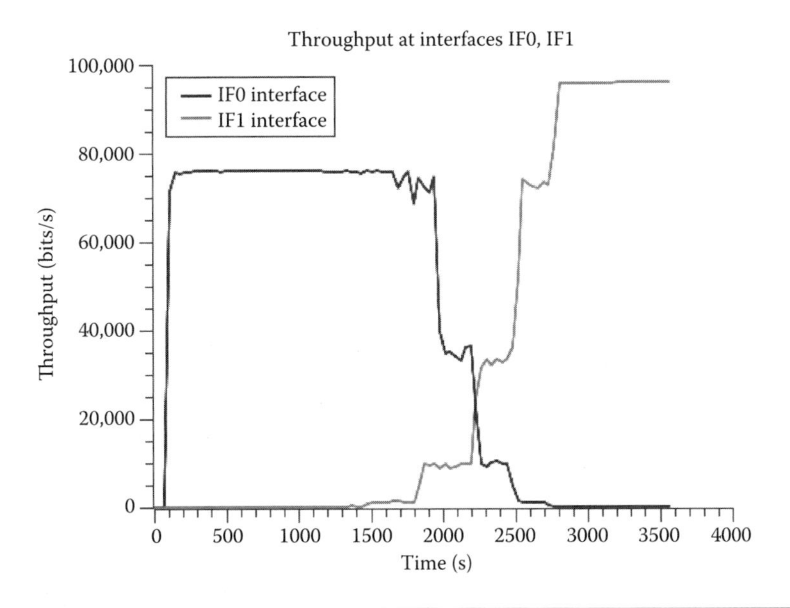

Figure 9.21 Throughput of multimode MS with better voice quality.

During vertical handoff, the traffic was transferred from the IF0 to the IF1 interface and the handoff occurred at a throughput of 20 Kbits/s. This means that soft handoff provides better performance during vertical handoff. The throughput level where the handoff occurs or crosses is well managed by the VWS.

9.6 Conclusions and Future Research Issues

Nowadays, advance mobile computing devices include several air interfaces, but very soon it will be more popular to have multiwireless interfaces embedded in all mobile devices. The challenge is to let these interfaces integrate their functionalities to deliver the best QoS for an application. In this chapter a new architecture is developed to integrate different wireless services without making any changes in their original infrastructure or in their protocols. The VWS offers an integrated environment and at the same time keeps the independence of the current wireless technology operators.

It can be observed from the simulation results obtained in Section 9.5 that VWS improves significantly our ability to handle user mobility in many different business scenarios. In addition, this innovative solution can lead to efficient consolidation of many wireless communication plans by utilizing their infrastructure more efficiently to improve their return on investment and provide advanced wireless services. To achieve this, companies have to update their current business plans and create a new market oriented toward applications that need to be executed over the mobile broadband all the time. The VWS does not require new infrastructure to be added to the existing ones. The VWS requires adding only a piece of software such as the VWA in the existing mobile devices or it may be embedded in the new mobile devices. Hence any wireless communication system may be used along with VWS and thus any new communication system can be added to the current major wireless technologies such as WWANs (2G, 2.5G, 3G, and 4G) WMANs (IEEE 802.16), and WLANs (IEEE 802.11a/b/e/g/n). Integrating theses different wireless technologies with the VWS will provide new opportunities for the users and the service providers.

The VWS framework can be extended in many ways to enhance the vertical handoff performance. Correlating available signal quality with location provides the VWS with a mechanism that helps the MS to select the alternative path. In addition, adding user route behavior predication will provide better VISP resource management and will increase the system robustness. Security can be enhanced by encrypting the traffic between the VISP and the MS, as the traffic is terminated between these two ends.

In summary, VWS provides a rich environment for vertical handoff through the interaction between the distinguishing IP and the master IP addresses during the flow of traffic. Also, VWS gives more flexibility for the connection to be changed on the fly and is based on premeditated parameters.

References

1. D. P. Agrawal and Q.-A. Zeng, *Introduction to Wireless and Mobile Systems*, 2nd edition, Brooks/Cole, April 2005, ISBN No. 0-534-49303-3.
2. The Third Generation Partnership Project (3GPP), http://www.3gpp.org/
3. IEEE 802.16 Standard, *IEEE Standard for Local and Metropolitan Area Networks Part 16: Air Interface for Fixed Broadband Access System*, April 2009, http://standards.ieee.org/getieee802/download/802.16-2009.pdf
4. IEEE 802.11 Standard, *Wireless LAN Medium Access Control (MAC) and Physical Layer (PHY) Specifications*, 2007, http://standards.ieee.org/getieee802/download/802.11-2007.pdf
5. M. Paolini, Wi-FI, WiMAX and 802.20—The disruptive potential of wireless broadband. http://www.eyeforwireless.com/wimaxupdate.pdf
6. IEEE 802.22 Working Group on Wireless Regional Area Networks ("WRANs"), http://www.ieee802.org/22/index.html
7. G. Aggelou and R. Tafazolli, On the relaying capacity of next-generation GSM cellular networks, *IEEE Personal Communications*, 8, Feb. 2001.
8. R. Chakravorty, J. Crowcroft, P. Rodriguez, and S. Banerjee, Performance issues with vertical handovers, Experiences from GPRS cellular and WLAN hot-spots integration, in *Proceedings of the 2nd IEEE Conference on Pervasive Computing and Communications*, March 2004.
9. The Cellular IP Project at Columbia University, http://comet.columbia.edu/cellularip
10. K. Ahmavaara, H. Haverinen, and R. Pichna, Interworking architecture between 3GPP and WLAN systems, *IEEE Communications Magazine* 41(11), 2003.
11. D. Cavalcanti, Integrated architecture and routing protocols for heterogeneous wireless networks, PhD Thesis, University of Cincinnati, Cincinnati, OH, USA, 2006.
12. R. Brannstrom, E.R. Kodikara, C. Ahlund, and A. Zaslavsky, Mobility management for multiple diverse applications in heterogeneous wireless networks, in Proceedings of the *IEEE Consumer Communications and Networking Conference (CCNC '06)*, January 2006.
13. I. AlKhayat, Virtual wireless network service design and evaluation, PhD Dissertation, University of Louisville, Louisville, KY, USA, 2009.
14. P. Bhagwat, C. Perkins, and S. Tripathi, Network layer mobility: an architecture and survey, *IEEE Personal Communications*, 3(3), 1996, 54–64.
15. A. George, A. Kumar, D. Cavalcanti, and D. P. Agrawal, Protocols for mobility management in heterogeneous multi-hop wireless networks, *Pervasive and Mobile Computing* 4(1), 2008, 92–116.
16. D. Ghosh, Mobile IP, *ACM Crossroads Magazine*, 7(2), 2000, 10–17.
17. A. S. Tanenbaum, *Computer Networks*, 4th edition. Prentice Hall, Inc., Upper Saddle River, New Jersey.
18. A. G. Valko, Cellular IP: A new approach to Internet host mobility, *ACM Computer Communication Review*, January 1999.
19. B. Xie, A. Kumar, D. Cavalcanti, D.P. Agrawal, and S. Srinivasan, Mobility and routing management for heterogeneous multi-hop wireless networks, *IEEE Mobile Ad hoc and Sensor Systems Conference*, 2005.
20. A. Valko, J. Gomez, S. Kim, and A. Campbell, Performance of cellular IP access networks, in *Proceedings of the 6th IFIP International Workshop on Protocols for High Speed Networks (PfHSN'99)*, Salem, August 1999.

21. R. Ramjee et al., HAWAII: A domain-based approach for supporting mobility in wide-area wireless networks, in *Proceedings of the IEEE International Conference Network Protocol*, 1999.
22. J. Yang, Z.-y. Zhang, and Z.-z. Tang, On the performance of a novel multi-hop packet relaying ad hoc cellular system, in *Proceedings of the IEEE International Symposium Personal, Indoor and Mobile Radio Communications (PIMRC '04)*, 2004.
23. RFC 2002—IP mobility support, http://www.ietf.org/rfc/rfc2002.txt
24. I. Alkhayat, A. Kumar, and A. Elmaghraby, Seamless connectivity scheme for heterogeneous wireless networks, in *Proceedings of The 8th IEEE International Symposium on Signal Processing and Information Technology (ISSPIT '08)*, 2008.
25. I. Alkhayat, A. Kumar, and S. Hariri, End-to-end mobility solution for vertical handoff between heterogeneous wireless networks, in *Proceedings of the 7th ACS/IEEE International Conference on Computer Systems and Applications (AICCSA-2009).*
26. B. Xie, A. Kumar, and D. P. Agrawal, Enabling multi-service on 3G and beyond: Challenges and future directions, *IEEE Wireless Communication Magazine*, 2008.
27. A.George, I. Alkhayat, and A. Kumar, Connection selection scheme for multi-hop heterogeneous wireless networks, in *Proceedings of the 2006 IEEE International Conference on Mobile Adhoc and Sensor Systems (MASS)*, pp. 747–752, October 2006.
28. H. Luo, R. Ramjee, R. Sicha, L. Li, and S. Lu, UCAN: A unified cellular and ad hoc network architecture, in *Proceedings of the 9th Annual International Conference on Mobile Computing and Networking*, pp. 353–367.
29. H. Shulzrinne and E. Wedlund, Application-layer mobility using SIP, *Mobile Computing and Communications Review*, 4(3), 2000.
30. A. George and A. Kumar, M3HN: An adaptive protocol for mobility management in multi-hop heterogeneous networks, in *Proceedings of the Tenth IEEE Symposium on Computers and Communications (ISCC)*, Spain, June 2005.
31. D. Cavalcanti, C. M. Cordeiro, D. P. Agrawal, B. Xie, and A. Kumar, Issues in integrating cellular networks, WLANs, and MANETs: A futuristic heterogeneous wireless network, *IEEE Wireless Communications Magazine, Special Issue on Toward Seamless Internetworking of Wireless LAN and Cellular Networks*, June 2005.
32. J. Roy, J. Jackson, V. Vaidehi, and S. Srikanth, Always best-connected QoS integration model for the WLAN, WiMAX, in *Proceedings of the First International Conference on Heterogeneous Network Industrial and Information Systems*, pp. 361–366, August 8–11, 2006.
33. L. Mo, K. Sandrasegaran, and T. Tung, A multi-interface proposal for IEEE 802.21 media independent handover management of mobile business, in *Proceedings of the International Conference on Mobile Business*, 2007.

Chapter 10

Mobility and Middleware

Enabling Ubiquitous Quality of Service for Mobile Services

Bing He, Bin Xie, and Sanjuli Agarwal

Contents

10.1 Introduction

As a result of the widespread availability of mobile terminals, mobile service technology has been emerging as a means of providing seamless and ubiquitous mobility and middleware support for various applications. Mobile terminals such as cell phones and laptops are equipped with Internet-capable wireless interfaces such as IEEE 802.11, WiMAX, Bluetooth, and/or universal mobile telecommunications system (UMTS) cellular interfaces to select appropriate wired or wireless connections for various services (e.g., web services). These connections should be quality of service (QoS)-enabled with the ability to exploit network and application contexts to meet the requirements of the users in various aspects. For such a purpose, QoS provisioning and QoS-aware middleware have to take into account many aspects, at very different abstraction layers, from application-specific bandwidth requirements to needs of mobility, from user preferences to energy consumption. For example, Xie et al. [1] advocate wireless network configuration technologies with mobility support for end-to-end QoS (e2eQoS) with the consideration of multiple networks and multiple services. These wireless network configuration technologies aim to promote QoS for mobile services from the angle of available Internet connectivity. In contrast, Bellavista et al. [2] consider the suitability of mobility-aware middleware to relieve the application logic from the burden of determining the most suitable interface and connectivity provider for each client/application at the time of service provisioning. Furthermore, many mobility and middleware protocols have been proposed to achieve accurate QoS evaluation of mobile subscribers and subsequently perform application connection establishment, selection, and service adaptation.

QoS for network services is first studied on the Internet or telecommunication networks. The Internet Engineering Task Force (IETF) terms QoS as an implicated meaning of the achieved service quality. This goal is usually achieved by resource reservation and traffic control in the computer network and packet-switched telecommunication networks. These networks offer better service to selected network traffic and have different priorities for different applications, users, or data flows. For this purpose, QoS metrics are used for evaluating the services provided to users. These metrics include the bandwidth, delay, jitter, and packet-dropping probability (or bit error rate). Furthermore, QoS-aware middleware is then developed to enable data exchange and services in these computer and telecommunication networks. The main purpose is to enforce QoS for application from a specific layer that is referred to as the middleware layer. This reduces the QoS deploying complexity at the application layer. With the expansion of computer networks, QoS-aware middleware further allows interoperability in support of coherent distributed architectures to simplify complex and distributed applications.

However, QoS mechanisms developed for the computer or telecommunication networks are not enough to assure QoS for mobile services as many novel challenges originate from the wireless communication and mobile environments. As shown in Figure 10.1, the QoS Internet Standards Track Protocols, including differentiated services, integrated services, and multiprotocol label switching (MPLS), have no mobility support for mobile users. In other words, wireless network communication and mobility impose new constraints for QoS provisioning to support network services. The QoS Internet Standards Track Protocols should be enhanced with mobility support as shown in Figure 10.1. Owing to insufficient wireless network capacity (e.g., bandwidth) such as in cellular networks, the QoS for users should be carefully reconsidered for real-time streaming multimedia applications such as voice over Internet protocol (VoIP), online games, and IP for mobile TV (IP-TV). These services often require a fixed bit rate and they are delay sensitive. In mobile services, novel QoS mechanisms are needed for controlling the performance from the aspects of reliability and usability of mobile service. QoS control for mobile services is especially critical and complex in the mobile environment. This is because of the variations of the wireless resources over time. To ensure QoS for these real-time applications, QoS control becomes an essential issue to meet the satisfaction of the customers, compared to that on the Internet. In this chapter, we will first illustrate these QoS Internet Standards Track Protocols and then investigate the QoS mobility enhancement protocols that address the mobility challenges.

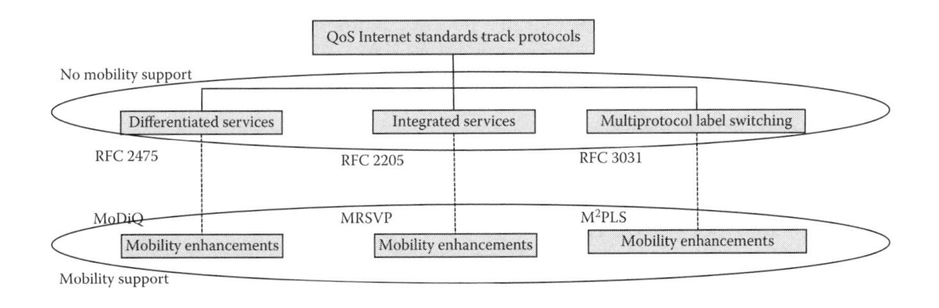

Figure 10.1 Quality of service with mobility support.

QoS-aware middleware achieves service transparency for mobile services. By implementing QoS at the middleware layer, the applications are relieved from the need to monitor the environment and thus leave the QoS adaptation completely to the decision of the middleware. In this application-transparent approach, the middleware is designed to provide best-effort adaptation to general mobile computing, where the context information is hidden from the application. Current research [2,3] shows that the QoS-aware middleware is an effective way of supporting mobile services with real and multimedia applications during application runtime. The QoS-aware middleware is able to probe, instantiate, and adapt applications. These operations are transparent to user and optimize the mobile service performance in accordance with the user behaviors and pervasive mobile environments. The use of QoS-aware middleware for multimedia services in mobility support has the following benefits [3]. At first, during the application instantiation time, the middleware are able to select among available service configurations that are componentized inside multimedia cervices, based on resource availability and user preferences. Secondly, middleware components in multimedia applications may take the responsibility of QoS probing, which could be offline or runtime probing on the performance of the applications' QoS parameters. Thirdly, during the application runtime, the user requirements or the resource availability may change due to various reasons such as statistical multiplexing of concurrent resource usage, user preference and/or environment changes because of user mobility, etc., and the middleware layer may automatically assist the application to adapt to these changes. Finally, the use of mobility-aware middleware in mobile service relieves the application logic from the burden of determining an appropriate network interface and connectivity to the service provider at the time of service provisioning. In this chapter, we will explore the QoS-aware middleware that achieves the above-addressed benefits for mobile services.

The remainder of this chapter is structured as follows. Section 10.2 illustrates the QoS architecture and protocols for Internet services. Then, the design challenges for mobile services in the mobile network and the QoS design requirements are discussed in Section 10.3. Section 10.4 investigates the QoS mobility enhancements of the QoS Internet Standards Track Protocols and Section 10.5 illustrates the QoS-aware middlewares in the literature. Open issues are discussed in Section 10.6. Finally, concluding remarks are included in Section 10.7.

10.2 QoS Architectures and Protocols for Internet Services

In this section, we illustrate QoS mechanisms and corresponding protocols designed for Internet services.

10.2.1 QoS Internet Standards Track Protocols

QoS has been initially addressed with the services on the Internet in the design of communication protocols for both integrated and differentiated service networks. The QoS for Internet services is studied at both the network/transport layer and the session/application layer. At the transport layer, there are three different QoS architectures [4]: differentiated services, integrated services, and Multiprotocol Label Switching (MPLS).

10.2.1.1 Differentiated Services

The differentiated services architecture, proposed by the Differentiated Services Working Group, aims to differentiate the IP traffic so that the traffic is transmitted on the network with certain priority on a per-hop basis. The differentiated service is referred to as DiffServ for simplicity and the corresponding Internet specification is defined in the standard RFC 2475. DiffServ can differentiate the traffic in several ways such as by user, service requirements, or other criteria. The packets of traffic then are marked such that the network nodes can offer different levels of service. The routers are also the network resource and dedicated routers could be used for specific traffic flows. In particular, a policy management system controls service and resource allocation. It uses a mechanism called "per-hop behavior" (PHB) for hop-by-hop resource allocation. To provide differentiated QoS commitments, DiffServ has four commonly defined PHBs: default PHB, expedited forwarding (EF) PHB, assured forwarding (AF) PHB, and class selector PHB. Network resources such as bandwidth and buffer are allocated to traffic flows according to these service differentiations. The DiffServ network has defined a variety of network elements such as classifier, marker, policing, and shaper to facilitate the implementation of the service differentiation. All these elements (classifier, marker, policing, and shaper) are implemented at the network boundaries (i.e., boundary nodes) instead of at the Internet routers. Therefore, these elements and their associated behaviors decouple traffic management and service provisioning functions from the forwarding functions, which are implemented within the core network nodes, that is, core routers. This reduces the complexity on the core network and results in high scalability. The service provisioning policing is an element that determines: (i) how traffic is marked and conditioned on entry to a differentiated services-capable network, and (ii) how that traffic is forwarded within that network. On entering the network, the traffic is first classified by the classifier and possibly conditioned at the boundaries of the network (e.g., edge routers). Once the data packet is classified into a traffic class, the traffic is differentiated by DiffServ-aware routers in the network. To ensure preferential treatment for higher-priority traffic, each traffic class can be managed differently according to their class priority. As a coarse-grained and class-based traffic management mechanism, DiffServ could be used to provide low-latency, guaranteed service to critical network applications such as voice or video while providing best-effort traffic guarantee to noncritical applications such as Web pages browsing or file download.

10.2.1.2 Integrated Services

The integrated services architecture, proposed by the Integrated Services Working Group, aims to provide guaranteed QoS negotiation and reservation for traffic flows. For such a purpose, the Resource Reservation Protocol Working Group proposed a signaling protocol called Resource ReSerVation Protocol (RSVP) for Internet resource reservations. As defined in the Internet standard RFC 2205, RSVP has been specified for receiver-initiated setup of resource reservations for multicast or unicast data flows to achieve scalability and robustness. Specifically, RSVP is first used by the service in a host to request specific qualities of service from the network for particular data streams or flows. RSVP is also used by routers to deliver QoS requests to all nodes along the path(s) of the flows and to establish and maintain the state to provide the requested service. RSVP requests will generally result in resources being reserved in each node along the path from the sender to the receiver. Two types of messages, PATH and RESV, are used in RSVP to set up resource reservation states on the nodes along the path between the sender and the receiver. Initially, the sender learns the IP address of the recipient using some out-of-band mechanism. It then sends a PATH message to the recipient to find a path all the way from the sender to the receiver for a specific flow. When a router receives a PATH message, it will record which upstream router the PATH message was received from and forwards the PATH message to a downstream router. The PATH message is then passed from one to another downstream router and finally received by the receiver. The receiver will respond with a RESV message to make a resource reservation for the specific flow. The RESV message will be transmitted in reverse along the same path as the PATH message was originally transmitted. On receiving an RESV message, each router or host on the path will reserve the required resources for the specific flow if sufficient resources are available. In this way, RSVP can provide specific levels of QoS for application data streams or flows.

10.2.1.3 Multiprotocol Label Switching

The MPLS architecture proposed by the MPLS working group defines the packet forwarding mechanism for the packets of a connectionless network layer protocol. The MPLS architecture and protocol are specified in the RFC 3031 standard. When the packet travels from one router to the next, each router makes an independent forwarding decision for that packet by analyzing the packet's header and runs a network layer routing algorithm to determine the next hop. In this way, each router independently determines the next hop for the packet, according to its analysis of the packet's header and the results of the routing algorithm. MPLS adds a label containing specific routing information to each IP packet. This allows the routers to assign explicit paths to various classes of traffic. By assigning labels to data packets, packet-forwarding decisions are made solely on the contents of this label, instead of examining the packet itself. In each labeled packet, packet headers contain more information than is needed simply to choose the next hop. As each router independently chooses a next hop for the packet on the basis of its analysis of the packet's header and the results of running the routing algorithm, MPLS allows one to create virtual end-to-end circuits across any transport medium types. MPLS is mainly used to forward IP datagrams and Ethernet traffic. Some major practical applications of MPLS include telecommunications traffic engineering and the MPLS virtual private network (VPN).

10.2.2 Comparison of QoS Internet Standards Track Protocols

Table 10.1 compares these three QoS architectures that are used for the Internet or telecommunication networks. IETF defines corresponding specifications for differentiated services, integrated services, and MPLS. The QoS architectures and protocols are described in these specifications. As a result of the limitations and the strengths of each protocol, they are used in different network scenarios. DiffServ can be used for the large-scale Internet whereas RSVP can only be used for small computer networks because of lack of scalability. MPLS is mainly applied for telecommunications traffic engineering and MPLS VPN. As shown in Table 10.1, all these architectures and protocols have no support of mobility which means they do not have the ability to re-establish the application session once the user has moved to a new network attachment. Table 10.1 also gives some examples of mobility enhancements to address the limitation in mobility support, which will be investigated in the following sections.

10.3 Wireless Network Accessibility and QoS Requirements of Mobile Services

Compared to the Internet,[*] which has a relative stable network connection for users, wireless access with user mobility imposes two novel issues for mobile services: mobile handoff and limited network capacity. In this section, we describe these two issue issues supporting mobile service QoS.

10.3.1 Frequent Mobile Handoff

Mobile service offers ubiquitous services for mobile users that may span through the wireless networks that are again connected with the Internet. The wireless network provides Internet accessibility with certain radio coverage while the Internet is where the service providers are located and provides rich services for subscribers. In the last few years, a variety of wireless access technologies have been developed

[*] The user terminal connects to the Internet or telecommunication networks by a wired link.

Table 10.1 Comparison of Internet Quality of Service (QoS) Specifications

	Differentiated Services (DiffServ)	*Integrated Services (ReSerVation Protocol—RSVP)*	*Multiprotocol Label Switching (MPLS)*
Related IEFF specifications	RFC 2474, 2475, 2638, 4594 (an architecture for differentiated services)	RFC 1633, 2210, 2211, 2212, 2215, 2205 (Integrated services in the Internet architecture)	RFC 3031 (multiprotocol label switching architecture)
Protocol layer	Network layer	Network layer	Between the data link layer and the network layer
Description	Implementing scalable service differentiation in the Internet	Resource reservation for services in the Internet	Packet forwarding decision mechanism for the packet of a connectionless network layer protocol
Applications	Large-scale Internet protocl networks	Small-scale Internet protocol networks	Telecommunications traffic engineering and MPLS virtual private network
Strengths	Simplicity and scalability (more scalable than RSVP)	Complete signaling	Scalability
Limitations	Lack of signaling protocol for admission control and error reporting	Less scalable than differentiated services	Sophisticated; each router makes the decision for packet forwarding priority
Traffic classification	Yes	Yes	Yes
Mobility	No	No	No
Applicability for mobile services	Not for wireless network that frequent connection re-establishment is required	Not for wireless network that frequent connection re-establishment is required	Not for wireless network that frequent connection re-establishment is required
Mobility enhancements	Mobile differentiated service QoS model (MoDiQ) [5]	RSVP tunnel [11], mobile RSVP [14,15]	M^2PLS (mobile multiprotocol label switching) [17]

to satisfy the ever-growing demand for enhanced service, requiring support ubiquitous communication. These different wireless technologies, such as wireless wide area networks (WWANs) (e.g., 3G and 4G cellular systems), wireless metropolitan area networks (WMANs) (e.g., IEEE 802.16), wireless local area networks (WLANs) (e.g., IEEE 802.11a/b/g/n), and others, can be located in some areas, offering their own network connection by supporting different communication scenarios. Mobile users access a large variety of services with seamless mobility and QoS guarantee. A WLAN has a small coverage, but offers high-speed communication with the Internet at a low cost. On the other hand, a cellular network has a larger coverage but low data rate, and thus it is more suitable for high-mobility users in support of seamless roaming. Broadband wireless access technologies (such as WiMAX and WiBro) can be used to provide desirable Internet connection in terms of throughput, capacity, latency, and low cost per megabyte. Therefore, QoS-enabled protocols are required with the integration of the Internet protocol and the available wireless networks. With the access from the wireless network, mobile users are mobile and they may change their Internet accessing points during an ongoing communication session. The handoff [1] for mobile services occurs in two different ways, depending on the change of the wireless networks:

- *Horizontal Handoff*: A horizontal handoff migrates from a wireless connection and its application between two homogeneous networks (e.g., from a cellular base station to another base station, or from a WLAN access point to another access point).
- *Vertical Handoff*: A vertical handoff enables a mobile terminal to migrate from a wireless network to another type of network (e.g., from a WiMAX base station to a WLAN access point or from a WLAN access point to a WiMAX base station).

Figure 10.2 shows the horizontal handoff where the mobile terminal migrates from a WLAN access point (WLAN1) to another WLAN access point (WLAN2). It also shows a vertical handoff where the mobile terminal migrates from a WLAN access point (WLAN2) to a WiMAX base station. Both types of handoffs result in re-establishing the Internet connection which involves the operations from the physical layer to the transport layer on the network protocol stack. It also results in redirecting the application from one access point to another access point. With respect to QoS, a handoff can lead to resource shortage and application performance degradation when the earlier negotiated resource for a session is no longer available in the newly migrated networks. During the handoff, the network resource renegotiation is not only required on the migrated wireless network but also needed in the Internet. This is because when the mobile user has changed its access point, a different Internet router will be used for connecting to the service provider (e.g., a multimedia server) on the Internet. When the handoff resource shortage happens, the ongoing communication session may be terminated or the application will

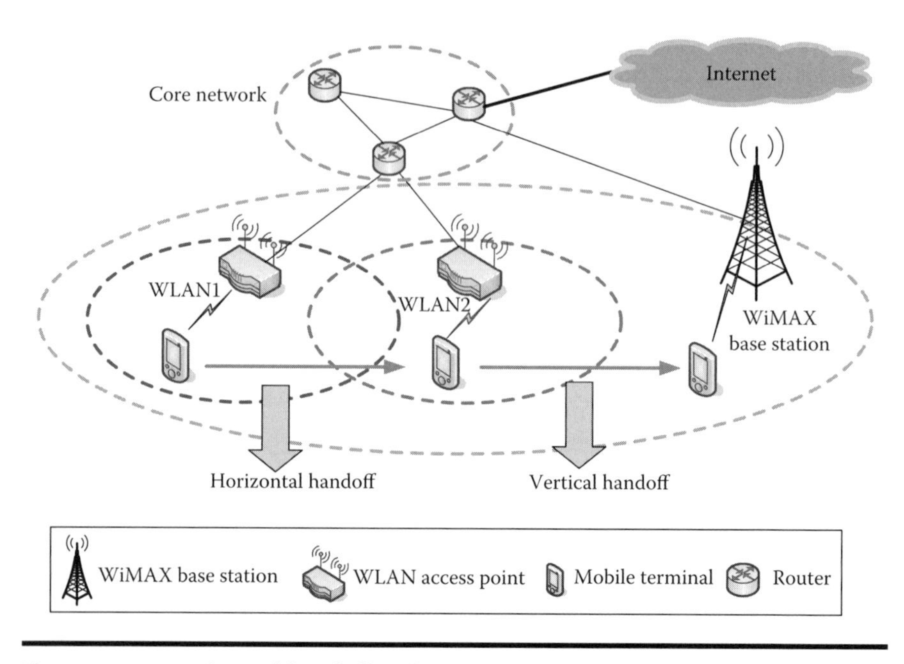

Figure 10.2 Horizontal handoff and vertical handoff.

adapt itself to the resource context on the new networks, which may come with a resource renegotiation and adaptation process to ensure QoS for mobile services. Diederich et al. [5] call this scenario the "handoff-caused resource shortage problem," which may highly degrade the QoS of the mobile service as the users of mobile services expect stable and continuous services without service interruption, especially for real-time services. Therefore, a major QoS requirement of mobile service during handoff is to avoid or mitigate the handoff-caused resource shortage problem. There are two possible ways to address this problem: (i) the mobile terminal adapts to the change of available resource by adjusting the applications; (ii) the network provides seamless mobile communication by reserving the resource for the application through a handoff resource pre-reservation. We describe these two approaches in detail later.

10.3.2 Limited Wireless Network Capacity

Compared to the Internet, a wireless network has limited throughput in serving a number of mobile users. As the user traffic demand increases, the limited wireless link capacity at the access points (e.g., base station and WLAN access point) will result in congestion. As popularity of the wireless network grows, it has been used for applications that require high-bandwidth communications as well and wireless link capacity is a major problem in satisfying the required throughput. Despite recent advances in wireless physical layer technologies, a wireless link still cannot offer the same level of reliable bandwidth as a wired link. Owing to significant medium access control layer overhead, the available data rate of an IEEE 802.11 channel is significantly less than their maximum physical speed (e.g., IEEE 802.11b at 11 Mbit/s, IEEE 802.11a/g at 54 Mbit/s, even IEEE 802.11n at 100 Mbit/s). On the other hand, different applications of mobile services impose different bandwidth requirements on the QoS. Usually, the QoS is mainly evaluated from the customer's point of view. Stringent real-time end-to-end packet delivery is the most basic QoS guarantee required by the multimedia applications in mobile services. In mobile multimedia services, quantitative communication parameters such as network throughput, delay, jitter, and packet loss are the major concerns, which indeed affect the QoS perceived at the application level. For the applications sensitive to the perceived latency, including real-time applications such as IP telephone applications and video chat, the response time dominates the application performance. On the other hand, complex distributed applications such as Web access to multimedia data may not need to meet the stringent real-time requirements even when they exhibit specific latency requirements.

In general, QoS parameters describe the speed and reliability of data transmission, for example, throughput, transmission delay, and error rate. The above two issues (i.e., handoff and network capacity) for mobile services have motivated many researchers to evaluate the QoS in mobile networks. In addition to commonly used QoS metrics in the Internet, there are two novel QoS-related parameters to measure the QoS and wireless network performance, which are services accessing success probability and handoff success probability. The service accessing success probability says that a service request from a wireless network may be blocked because of a contingent network resource. The handoff success probability indicates that a handoff may fail if the new application connection could not be successfully established on the new network access point. Table 10.2 illustrates the QoS parameters and compares them with respect to the traditional Internet services and mobile services. As seen in Table 10.2, more constraints are imposed on mobile services as a result of wireless communication. The constraints, for example, include throughput bottleneck of the wireless networks, high wireless network delay, high wireless transmission error rate, high jitter on the wireless transmission, limited coverage of wireless network, accessibility of multiple networks, and others. Owing to these constraints, novel QoS protocols are essential for addressing the limitations in wireless networks.

10.4 QoS Mobility Enhancements for Mobile Services

The three QoS architectures discussed earlier are developed to support QoS for real time, multimedia, and other traditional Internet services in which wireless networks are not a part of the network for packet delivery. Recently, these QoS architectures have been enhanced to support QoS for mobile services in mobile networks.

Table 10.2 Comparison of Quality of Service Parameters

Quality of Service Parameters	*Internet for Network Services*	*Wireless Networks for Mobile Services*
Bandwidth for a service	Throughput on the Internet for a specific service that has a much higher throughput than the wireless network.	Wireless and Internet throughput allocated for a specific service and the wireless throughput is the bottleneck.
End-to-end transmission delay	The transmission delay between the host and the service provider on the Internet.	The transmission delay includes the wireless network and the Internet. The wireless network may dominate the delay, depending on the traffic at the access point.
Error rate	Information loss that can be evaluated by frame error rate to reflect the proportion of transmitted frames that contains bit errors.	The same definition as in the Internet. The wireless network has a high error rate.
Jitter	Delay variation that is the variability in the arrival time of each individual packet.	The same definition as in the Internet. The wireless network has a high jitter. Jitter on the wireless network depends on the traffic load at the access point.
Accessing success probability	None. All the connections are assured by wired links.	A user may not access the network owing to congestion at the wireless network. The request for accessing the network for a service may be dropped at the accessing wireless network.
Handoff success probability	None. During the service period, the user has no mobility on the wired connection.	A service may handoff from one accessing point to another, and the handoff may fail because of congestion at the migrating accessing point.

10.4.1 Differentiated Services with Mobile Support

The DiffServ design has been improved for enabling QoS in wireless networks such as UMTS/3G. In the wireless network, a mobile terminal may suffer from the shortage of network bandwidth during handoff or while requesting a new service. This problem is further described in Figure 10.3 for UMTS/3G networks.

Assume a telephony session with a bandwidth (e.g., 4.75 kb/s) in the connection from the mobile terminal. The application initially resides in the lightly loaded cell A as shown in Figure 10.3, and the user moves to the neighboring cell B. The application handoff therefore occurs which migrates the application from cell A to cell B. However, the handoff may not be processed because of resource shortage in cell B where the network resource (e.g., frequency band) has been utilized by the mobile terminals in cell B. To address the problem of handoff resource shortages in wireless networks, a mobility-specific QoS parameter has to be considered: the handoff success probability. This parameter represents the confidence of successfully transferring a service to the targeted network.

On the basis of the differentiated services QoS architecture, Diederich et al. [5,6] have proposed an extended QoS framework for wireless mobile networks, which is called the mobile differentiated service QoS (MoDiQ) model. MoDiQ integrates the UMTS network with the IP to provide efficient support for mobile applications with burst traffic characteristics such as www browsers. MoDiQ consists of three parts, a data forwarding plane based on the differentiated services approach, a control plane providing assurances on the handoff success probability and the necessary signaling protocols, and a service model especially suited for usage in wireless mobile networks. MoDiQ enhances the legacy DiffServ service model with QoS-enabled handoffs and with separate services for loss-sensitive and loss-tolerant applications. The three service types in the legacy DiffServ architecture, i.e., premium service, Olympic service, and

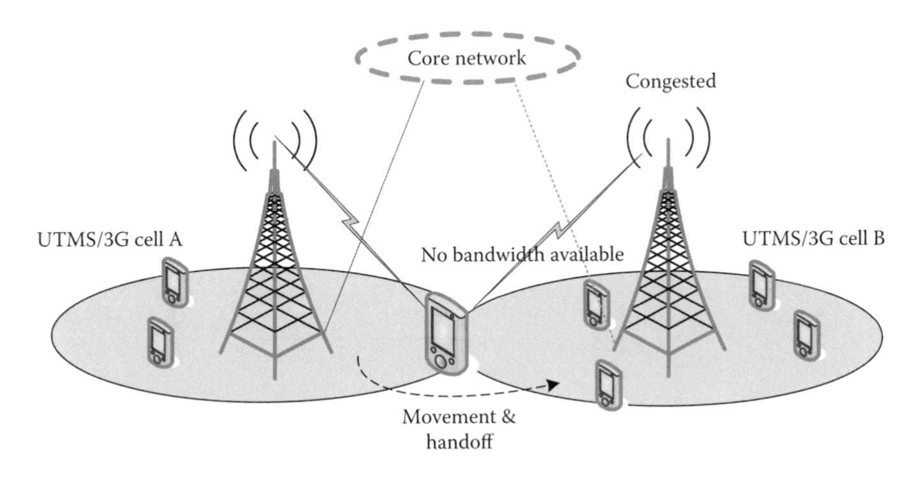

Figure 10.3 Handoff in the UMTS/3G wireless networks.

the traditional best-effort service, are extended to six service classes as shown in Figure 10.4, which are mobile premium service, portable premium service, best-effort low-delay service, mobile Olympic service, portable Olympic service, and best-effort service. The QoS support capabilities of these service types are

- *Mobile Premium Service*: It provides a low-delay packet delivery with support for assurances on the handoff success probability.
- *Portable Premium Service*: This is a low-delay service without the support of handoff.
- *Mobile Olympic Service*: It provides a high percentage of packet delivery with support for assurances on the handoff success probability.
- *Portable Olympic Service*: It provides a high percentage of packet delivery without the support of handoff.

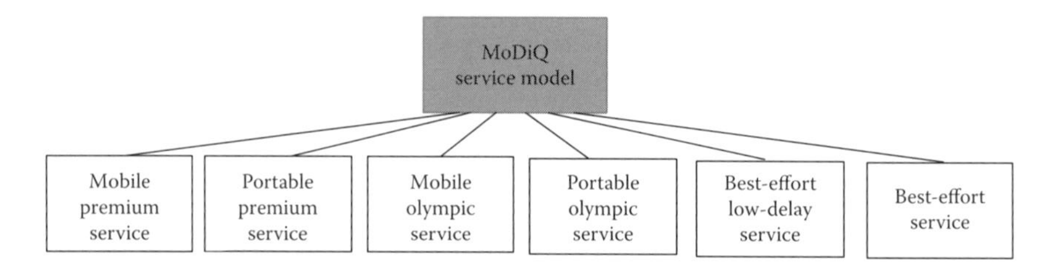

Service models		Delay sensitivity	QoS sensitivity	QoS-enabled handoff support
DiffServ	Premium service	High	High	No
	Olympic service	Low	High	No
	Best-effort service	Elastic	Low	No
MoQiQ	Mobile premium service	High	High	Yes
	Portable premium service	High	High	No
	Mobile olympic service	Low	High	Yes
	Portable olympic service	Low	High	No
	Best-effort low-delay service	High	Low	No
	Best-effort service	Elastic	Low	No

Figure 10.4 Legacy DiffServ and MoDiQ service model. (Adapted from J. Diederich, L. Wolf, and M. Zitterbart, *Computer Communications*, 27(11), 1106–1114, 2004.)

- *Best-Effort Service*: It provides QoS guarantee for the best-effort traffic that tolerates a relatively high packet delay.
- *Best-Effort Low-Delay Service*: It provides QoS guarantee for the best-effort traffic with a low packet delay.

The QoS support capabilities are defined in terms of the requirement of delay, packet loss sensitivity, and mobility, as shown in Figure 10.4. A premium service (i.e., a mobile premium service or a portable premium service) is high delay-sensitive compared to an Olympic service (e.g., a mobile Olympic service or a portable Olympic service). On the other hand, both the premium service and the Olympic service have a stringent requirement on the packet delivery ratio, compared to the best-effort services. MoDiQ separates the mobile terminal from the no-mobile terminals to enhance the efficiency of wireless network resource utilization as not all mobile terminals are actually moving. The QoS-enabled handoffs are only enabled in the mobile premium service and mobile Olympic service. During the handoff from a UMTS cell to a neighboring cell, the amount of bandwidth assigned in the old cell will be available in the new cell with a certain probability. These services will be given the assurances on the handoff success probability for delay-sensitive and less delay-sensitive applications. In addition, best-effort low-delay services are separated from the best-effort service that can tolerate a relatively high delay. To evaluate the performance of the mobile QoS architecture, a new mobility-specific QoS parameter, the handoff success probability, is defined by Diederich et al. [5]: they claim that providing assurances on the handoff success probability is crucial, especially for future cellular mobile networks where the cell size is decreased to accommodate more mobile terminals in a given area. In this case, the handoff will be more frequent and the probability of handoff resource shortage will be high.

On the basis of the above-proposed DiffServ-enabled wireless network, the QoS for mobile services could be further incorporated with the price where the allocation of the network resource is mapped to the user payment. Ozianyi et al. [7] proposed a pricing model to differentiate the user flows with different QoS levels by user payment. The available network resources get strained with the increased usage levels, which results in poor service to all the users. On the contrary, all users prefer receiving the high-quality service at an affordable cost. This requires the provision of QoS guarantees for network services at a low cost. In a real business scenario, this relationship is hard to achieve. The revenue sources for network operators have been shifting from the provision of network access to that of rich services, for example, multimedia services. To attain a functional compromise, Ozianyi et al. [7] proposed a pricing model that relies on service profiles to manage resource utilization. The service profiles define the QoS achieved for accessing services through a common resource pool, in which resource sharing is used to maximize network resource utilization, user satisfaction, and profits for the network operators. Users would select pricing profiles according to their budgets, and the network will map these profiles to a set of QoS options that may translate to the choice of an access network for service access.

10.4.2 RSVP with Mobile Support

RSVP is a signaling protocol designed to provide QoS guarantee for integrated services in the Internet. RSVP is designed without the consideration of the user mobility and, therefore, it cannot be used directly for mobile services [8]. According to the original RSVP signaling protocol, the resource reservation path cannot be dynamically adapted along with the movement of the mobile terminals. Once a mobile terminal roams to a new area, its previous reserved resources are no longer maintained and the QoS would be degraded significantly owing to the lack of necessary resource reservation in the new region. In the popular mobile IP scheme [9], the IP-in-IP encapsulation technique [10] is used to route IP packets to a mobile terminal that is moving away from its home network. If RSVP protocol is used, RSVP messages will be invisible to the intermediate routers of the IP tunnel, and thus the routers could not provide the required QoS on the new path. Recently, many schemes have been proposed to enhance the RSVP for the mobile environment and two of them are the RSVP tunnel and mobile RSVP.

10.4.2.1 RSVP Tunnel

The RSVP tunnel is proposed by Terzis et al. [11] to resolve the RSVP message invisibility problem. The RSVP tunnel is combined with mobile IP in their QoS signaling protocol [12] to provide QoS provisioning for wireless and mobile networks. Mobile IP [9] is a mobility management protocol standardized by the IETF to offer continuous Internet connectivity for mobile users. Continuous Internet connectivity means that a mobile terminal is able to maintain its Internet accessibility when it moves from one network to another. Mobile IP defines two key entities to support user mobility: home agent (HA) and foreign agent (FA). The home agent is the server on the mobile terminal's home network that maintains the information about the terminal's current location, identified as care-of address (CoA), and security credentials. On the other hand, the foreign agent is the server on the visiting network providing the CoA and security administration of the visiting network. When the mobile terminal changes its Internet attachment, the mobile terminal performs a registration process at the visiting network to create a mobility binding at the home network.

The Internet packets from a sender, which is referred as the correspondent node, can be accordingly redirected from the home agent to the foreign agent by using the binding of the home and the visiting IP address (e.g., the mapping between the home address and the CoA). The foreign agent then forwards the packets to the mobile terminal on the visiting network. In other words, the packets are first delivered from the sender to the home agent by using the home address of the mobile terminal. On receiving the packets, the home agent then forwards these packets to the foreign agent by using the CoA. Finally, the foreign agent transmits the packets to the mobile terminal. This process is called tunneling that performs the IP-in-IP encapsulation between two different networks. This tunneling process can be optimized in such a way that the senders can directly deliver the packets to the foreign agent if the CoA is updated at the sender. This avoids forwarding at the home agent and thus reduces packet delay.

On the basis of the mobile IP tunneling mechanism, the RSVP tunnel is used to establish an RSVP session between the corresponding tunnel end-points, that is, the entry points and exit points. In the RSVP tunnel, an extra pair of tunnel PATH and RESV messages without encapsulating IP headers are sent to establish a QoS-guaranteed communication path between the entry point and the exit point. The encapsulated end-to-end RESV PATH messages are exchanged between the sender and the receiver, and all routers on the tunnel path may reserve the desirable resources for the receiver if there are sufficient resources available. In the QoS provisioning protocol proposed by Terzis et al. [12], two scenarios are considered for a unicast RSVP session, that is, the mobile terminal as the receiver and as the sender. As illustrated in Figure 10.5, when the mobile terminal is the receiver, it receives PATH messages from a correspondent node CN (i.e., the sender). When a mobile terminal moves to cell B, it informs its home agent of its new location. Then the home agent will set up a tunnel RSVP session between itself and the foreign agent

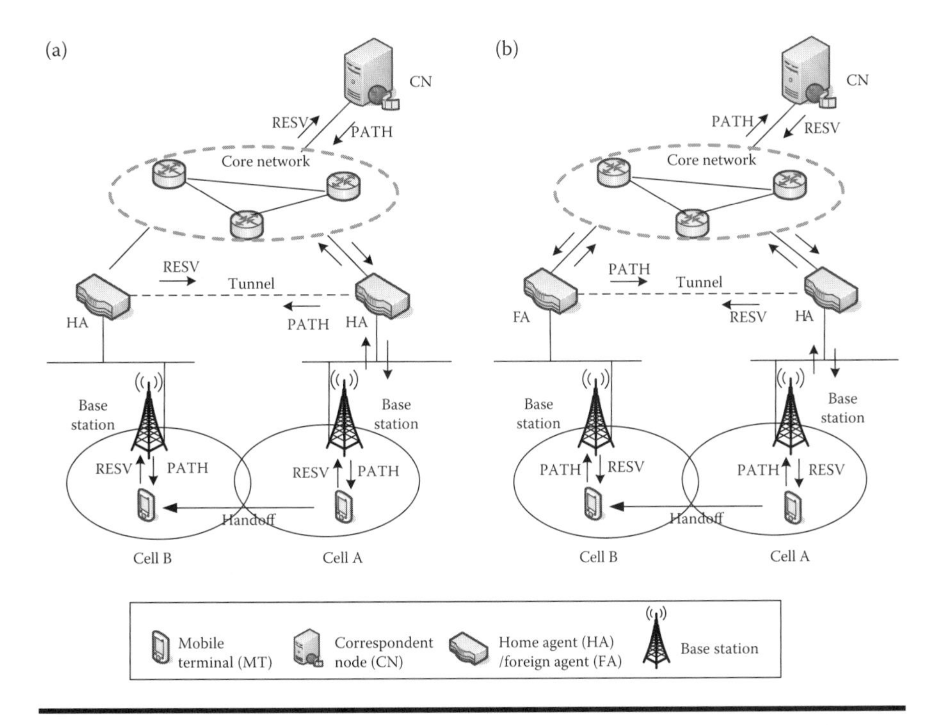

Figure 10.5 ReSerVation Protocol (RSVP) tunneling-based quality of service provisioning protocol. (a) Mobile terminal (MT) as a receiver and (b) mobile terminal as a sender. (Adapted from A. Terzis, M. Srivastava, and L. Zhang, In *Proceedings of IEEE INFOCOM,* 1999.)

and encapsulates PATH messages from the sender through the tunnel to the mobile terminal's new location. When the foreign agent receives an RESV message from the visiting mobile terminal, it will send a RESV message through the tunnel session between itself and the home agent and wait for the home agent to complete the reservation over the tunnel. After receiving the confirmation for the tunnel reservation from the home agent, the foreign agent encapsulates the end-to-end RESV message and sends it to the home agent, which will decapsulate and forward the message to the sender. When the mobile terminal is the sender, as shown in Figure 10.5b, the mobile terminal will send the PATH messages while the correspondent node sends the RESV messages with the reservation request. When the mobile terminal moves to cell B, the path set up using regular mobile IP tunnels will not necessarily go through the home agent, and may follow a different path, up to some merging point toward the correspondent node.

By making modifications to the mobile IP and RSVP, nested RSVP sessions are established between the tunnel end-points, and the RSVP signaling invisibility problem is resolved by the RSVP tunnel. However, the RSVP tunnel is still limited in supporting the mobility of the mobile terminals. When a mobile terminal moves to a new foreign region, the mobile terminal's reserved resource and service may be terminated because of the lack of reserved resources in the new region [8]. To provide QoS support for handoff, Lai et al. [13] present a handover RSVP (HO-RSVP) that is a resource reservation protocol for seamless handover in wireless networks. HO-RSVP maintains a continuous QoS guarantee between the sender and the receiver on the basis of the mobile IPv4. In this protocol, the resource reservation remains unaffected in the unchanged segments of the signal path. When the mobile terminal is moving, new reservations are made in the changed paths as a result of mobility, thus ensuring end-to-end resource reservation. HO-RSVP also has a resource pre-reservation function to prevent the mobile terminal from moving into a performance-degraded access network. On the basis of the proposed HO-RSVP scheme, if the receiver is mobile, resource reservation can be refreshed in the data path for the sender after handover. When the sender is mobile, after the sender moves to new subnet, a procedure is initiated to tear down the old reservation after handover. As a better QoS signaling mechanism in the mobility environment, the HO-RSVP approach provides higher throughput for multimedia applications and less packet delay for real-time services.

10.4.2.2 Mobile ReSource reserVation Protocol

The Mobile ReSource reserVation Protocol (MRSVP) was proposed by Talukdar et al. [14,15] to achieve the desired mobility-independent QoS guarantees in integrated services packet networks for real-time multimedia applications. The MRSVP makes advance resource reservations at multiple locations where a mobile terminal may possibly visit them during service time. The mobile terminal can thus achieve the required service quality when it moves to a new location where resources are reserved in advance. Through the exchange of a pair of PATH and RESV messages between the sender and the receiver, the mobile terminal will make an active reservation to its current location but it only makes passive reservations to each of its locations in its mobility specification, which is the set of locations that the mobile host will visit when it is participating in the flow. The difference between an active and a passive reservation defined in MRSVP is that an active reservation is the path on which packets are actually transmitted, whereas passive reservation paths are only reserved in advance without any actual packet flows over the link. When the mobile terminal moves to a new location, MRSVP changes the passive reservation of the new visited location into an active state and the original active reservation is altered into a passive state at the same time. Active and passive reservations for the same session are merged in the same way that usual RSVP requests get merged in routers. In this way, the needed resources for the mobile terminal in the new region can be retrieved rapidly because the resources were already preserved in the original passive reservation path. That is, a seamless handoff for QoS guarantees can be retained using the MRSVP protocol.

However, MRSVP demands too much bandwidth in making advance resource reservations by predictions. This excessive resource waste may degrade the system performance significantly, especially when the mobile terminals are moving randomly. Also, it requires a high accuracy in the mobility prediction. Like MRSVP, fast handoff RSVP (FH-RSVP) [16] is an integration of the RSVP and mobile IP interworking scheme to support QoS for real-time applications in mobile environments. FH-RSVP uses unicast sender-oriented RSVPs with HMIPv6 protocol and makes anticipated reactive resource reservations on the basis of fast handover principles when an intrasite handoff can happen to the WLAN access radio medium. FH-RSVP is shown to achieve better QoS guarantees than MRSVP in terms of setup time for the resource reservation path, throughput, and packet losses. Furthermore, FH-RSVP is more efficient than MRSVP in terms of reservation blocking, forced termination, and session completion probabilities.

10.4.2.3 Hierarchical Mobile RSVP

To overcome the disadvantages of RSVP in supporting the mobility and to improve the efficiency of MRSVP, Tseng et al. [8] proposed a hierarchical mobile RSVP protocol (HMRSVP) to achieve mobility-independent QoS-guaranteed services. The HMRSVP integrates RSVP with mobile IP registration and makes advance resource reservations only when inter-region movement may occur with a high probability, that is, when a mobile terminal moves into the overlapped area of the boundary cells between two regions.

The HMRSVP integrates RSVP with a mobile IP regional registration protocol and makes advance resource reservations only when the handoff delay tends to be long. In the standard mobile IP protocol, each time a mobile terminal moves, it must register with its home agent via the visiting network (e.g., foreign agent) to create an

IP address binding between the home address and the visiting address. In cases when the home agent is far from the visiting network, this registration process may become too expensive across multiple Internet domains, resulting in high service delay. The mobile IP regional registration protocol localizes the registration process within a region when a mobile terminal makes an intraregion movement. A region refers to a cluster of routers or subnets encompassed by an enterprise or campus network. Mobility agents in a region are organized hierarchically according to its topology. Owing to the hierarchical nature and IP routing properties of the Internet, the foreign agent can perform the registration process with some level of independence from the home agent. The registration for mobile terminal intraregion movements can thus be isolated within the region. Thus, the setup time for the resource reservation path for an intraregion handoff is significantly reduced. Therefore, HMRSVP adopts the hierarchical concept of mobile IP regional registration and makes advance resource reservations for a mobile terminal only when the mobile terminal visits the overlapped area of the boundary cells between two regions. It is shown with numerical results from simulation that HMRSVP not only could achieve the same QoS guarantees as MRSVP but also could outperform MRSVP in terms of reservation blocking, forced termination, and session completion probabilities.

10.4.3 MPLS with Mobile Support

As illustrated in Section 10.2.2, MPLS uses a label-based switching technique that is based on the integration of layer 2 switching and layer 3 routing for wired networks. Ashraf et al. [17] developed an enhanced version of MPLS, the mobile multiprotocol label switching (M^2PLS), to provide QoS provisioning for real-time traffic in mobile networks. It is designed to provide an efficient QoS mechanism for resource reservation and adaptation. M^2PLS integrates label forwarding functionality with resource reservation to guarantee QoS. The resource reservation admission and decision is performed along the path after per-hop computation of resources, for example, the available bandwidth, the delay of the link, and the jitter experienced by the packets.

To enable label distribution in the mobile network, the mobile label distribution protocol (MLDP) is proposed in M^2PLS to establish a label switch path (LSP). MLDP is implemented in two phases: control flow and data flow. Control flow includes on-demand LSP establishment between the source and destination, which is used for resource reservation, maintenance, and hop-by-hop routing. Once the control flow phase has been successfully performed, the source can initiate a data flow where data are forwarded in a hop-by-hop manner through pre-established LSP. The forwarding decision is made entirely on the basis of the label carried by the data packet that uniquely identifies an LSP and the data flow packets are forwarded below the IP layer. MLDP also addresses QoS violations caused either by end-to-end delay violation as a result of channel deterioration and congestion along the existing LSP or by LSP breaks as a result of node mobility and power failure.

MLDP handles e2eQoS violations by maintaining LSP delay and LSP jitter information at the source and the destination. To facilitate instant violation detection, some packets called the QoS violation packets are set as the highest priority in the network. Once QoS violation is detected, an MLDP recovery message will be initiated. With this adaptation mechanism, the M^2PLS is able to provide instant QoS violation detection and recovery mechanism for multimedia applications.

Sa-Ngiamsak et al. [18] further discussed a QoS recovery scheme for integration of mobile IP and MPLS networks for QoS guarantees over a dynamic mobile environment. The recovery scheme uses traffic management to support QoS-guaranteed tunnels, according to link/node failure or topology changes in mobile network. It addresses the limitation of the current schemes that focus only on wire network and are limited for multiple failure recovery in such dynamic topologies. The proposed modified flexible MPLS signaling for mobile IP over the MPLS network utilizes the merging point of recovery optimization to evaluate the nearest active upstream node to be the merging point of recovery node. This can address the multiple failures problem and gain faster rerouting response time to improve performance.

10.5 QoS-Aware Middleware in Mobile Services

The network QoS mechanisms discussed above target the satisfactory QoS for applications. However, it is still tedious and error-prone for software developers to develop applications that interact directly with these low-level network QoS mechanisms, which are developed by application programming interfaces written in third-generation languages such as C++ or Java. To mitigate the nontrivial implementation problem for QoS-enabled applications, QoS-aware middleware emerges as a way for QoS implementation while hiding the implementing complexity from the applications. It enforces the QoS adaptation on the mobile network by implementing various QoS mechanisms in such a way to allow applications to specify their coordinates (source and destination IP and port addresses) and per-flow network QoS requirements via higher-level frameworks. The middleware frameworks—rather than the applications—are responsible for converting the higher-level QoS specifications into the lower-level network QoS mechanism application programming interfaces. In this section, we first discuss the needs and the corresponding challenges of QoS-aware middleware and then present the current solutions for mobile services.

10.5.1 Needs of QoS-Aware Middleware

The mobile services provided by QoS-aware middleware should be able to dynamically reconfigure themselves via internal QoS adaptation procedures that allow them to react timely to the changes in the execution environment. The adaptation procedures may be implemented in the mobile applications or by the operation systems. However, adaptation mechanisms by mobile applications usually suffer from the problem of unfairness to other applications. Adaptation by the operation system focuses more on the overall system performance and may neglect the needs of individual applications. Hence, the adaptation task is best coordinated by a middleware that is able to cater to an individual application's need on a fair ground while maintaining the optimal system performance. This is achieved, for example, by a context-aware mobile middleware that sits in between the mobile application and the operating systems [19].

When a mobile terminal moves into the region of weak signal or migrates to a new access point (e.g., handoff), it may violate the QoS requirement of ongoing communication sessions. This can cause the termination of current sessions, and requires a new QoS adaptation and even a renegotiation process. However, the extension of the QoS paradigm to the end-user would be a hard and complex task, which requires the applications to be adaptive in the sense that applications can directly and actively react to varying resource availability within the network and the end systems. To simplify the programming of mobile broadband applications and to allow support of dynamic QoS changes, these active adaptation mechanisms should be hidden from application developers. Middleware provides a smooth transition from the protocol implementation in the distributed objects to suitable abstractions of the underlying infrastructure. The idea of shifting adaptation mechanisms from the application level to a flexible middleware featuring QoS functions will thereby result in simplified application development for mobile environments. Figure 10.6 illustrates the middleware for facilitating the QoS adaptation over a network.

The QoS-aware middleware for mobile services in a dynamically changing environment faces the following challenges [20,21]:

- *QoS-Awareness for Limited Resource and Multiple Applications*: The resources such as the network bandwidth, transmission queue space, and machine computational resources are limited in the mobile terminal and network (e.g., base stations, routers, and paths). Therefore, QoS awareness has to be considered in composing multiple services while each service may have multiple instances in the networked environment for resource-demanding applications. The QoS-aware middleware should be aware of the current state of resources and the current network load with which to know the application requirements and the available resources. From the perspective of a single application, all service instances related to the application must run on top of a physical infrastructure with abundant resources, satisfying the specific QoS requirements. From the perspective of the networks as a whole, wisely allocating limited resources such that the network can sustain more requests simultaneously is fundamental to achieving global resource optimization. These two goals may be in conflict because of limited resources.
- *Efficiency in QoS Diagnosis and Reaction*: The QoS-aware middleware should be able to react efficiently to the change of the state of resources. For this, the QoS-aware middleware should be able to diagnose the possible causes of violations and

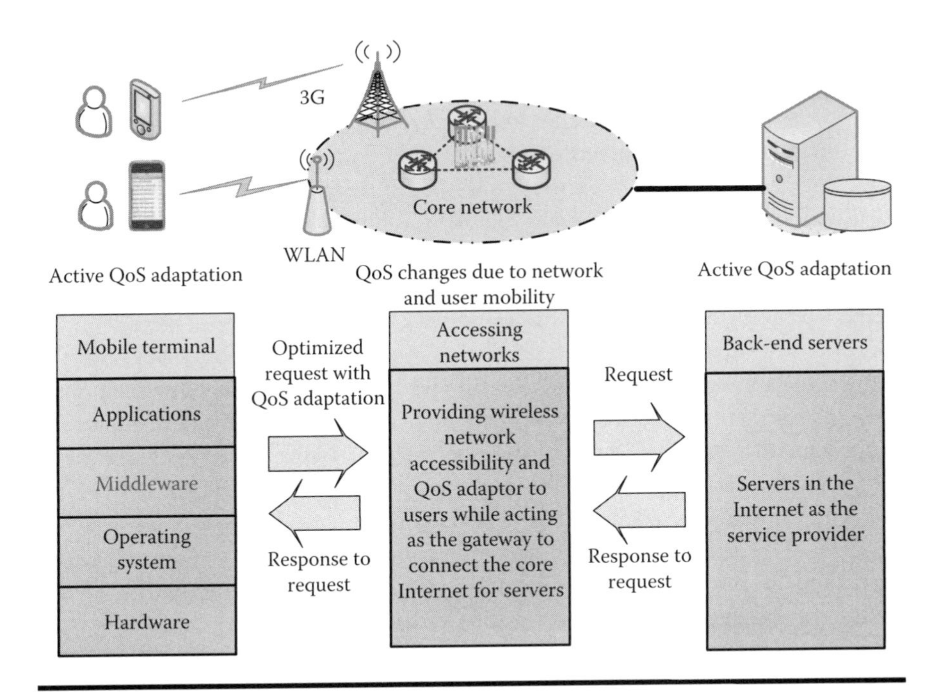

Figure 10.6 Mobile middleware for optimizing quality of service over a network.

make proper reactions in managing and allocating the resources. As a result, the ability to predict QoS violations would be helpful in doing so.

- *Scalability for Large-Scale Networks and Applications*: The QoS-aware middleware should be able to effectively perform QoS resource management not only in a small-scale network and on an end-to-end basis but also in large-scale networks and in large group-based applications. The scalability could be perceived in two dimensions: network size and application's client population size. When the number of applications in a large-scale network increases, the QoS-aware middleware operations do not significantly increase such that the efficiency is not highly degraded.
- *Adaptability and Reconfiguration*: As a result of the dynamic properties of the network, the resources that essentially comprise a path of component services may vary during the lifetime of an application session. The QoS-aware middleware should be able to perceive the fluctuations of these resources and be adaptively reconfigured in the existing paths so that the QoS of the application is still satisfied or optimally configured.
- *Robustness and Self-Healing*: During the application runtime, services and resources may fail, but such failures should be hidden from the applications. The QoS-aware middleware could immediately reconfigure the application. The self-healing operations to the failure achieve fast recovery from any failures and the recovery should be unperceivable by the users.

- *Transparency*: The QoS provision should be transparent to end-users and the complexity of QoS control should be hidden from applications. The QoS-aware middleware should be independent from the network architecture in the heterogeneous wireless network and the Internet.

The development of QoS-aware middleware for mobile services has to address these challenges from different aspects. The service discovery enables the user to locate the available service provided on the network. After the service is negotiated, the network should be able to offer the required QoS and this QoS is enforced on the network. The QoS requirement translation provides the mapping between the domain-specific requirements and the middleware-specific configuration options. The QoS reservation for necessary resources may need to be conducted [22] in accordance with the QoS requirement. The reservation is implemented on the basis of complex resource reservation, admission control, and resource negotiation mechanisms. In the end, QoS adaptation is carried out to periodically monitor the available resources of the execution environment and accordingly make adaptation and reconfigurations. In the mobile environment, user mobility has to be specifically considered in the design of the QoS-aware middleware for mobile services. In the following sections, we investigate these aspects in implementing the QoS-aware middleware for mobile services.

10.5.2 Mobile Service Discovery

Discovering local services offered by different service providers is necessary because of the mobility of mobile users. The underlying software provided to the mobile users or the servers usually comes with different discovery protocols implemented over different middleware platforms. When a user uses a specific discovery protocol or middleware platform, s/he may not find all the available provided services as s/he moves from one location to another. To solve this problem, Jarir and Erradi [23] proposed an adaptive reflective middleware MAQAME, which is a meta-level architecture for QoS adaptation in mobile environments. MAQAME allows the dynamic adaptation of both discovery and interaction protocols for mobile users while hiding the heterogeneity of these protocols. It also provides a QoS management component to select the service with better QoS among available local services. By implementing a uniform interface for the services provided by heterogeneous protocols and middleware, MAQAME guarantees discovery and middleware transparency to mobile applications, thus providing great flexibility, adaptation, and customization of client applications.

10.5.3 QoS Requirement Translation

In QoS-aware middleware, a problem is how to translate the QoS requirement to the network configurable options. The QoS requirement translation is to map the domain-specific requirements into the middleware-specific configuration options, that is, the domain-specific QoS requirements should be correctly translated into configuration options of the underlying middleware. The translation mapping can be established by defining translating rules. This problem becomes complicated when mobile services are deployed on heterogeneous middleware platforms that are based on their own configuration mechanism and options. Kavimandan and Gokhale [24] presented a middleware QoS configuration approach to address this challenge. This approach allows the parameterized transformation to meet the desired QoS requirements in the distributed real-time and embedded systems. In the proposed scheme, the domain-specific, platform-independent model QoS requirements are transformed iteratively to more refined and detailed middleware platform-specific model QoS requirements. There are three steps in the proposed transformation parameterization and specialization approach:

- *Identifying configuration variability points*: In this step, developers develop the transformation rules in a search-and-replace pattern using source and target modeling languages of transformation.
- *Specifying configuration variabilities*: A modeling language is automatically created from the transformation project and the platform-specific configuration variabilities are specified.
- *Accommodating configuration variabilities*: Configuration variabilities are accommodated into transformation algorithms that are specialized for a specific middleware platform.

On the basis of the proposed middleware configuration approach, platform-specific configuration variabilities are automatically extracted from transformations, which allow reusability and rapid software development.

10.5.4 Resource Reservation

In general, resource reservation aims to provide suitable QoS guarantee for services by performing network resource reservation based on the prediction of service resource variation, such as resource utilization, user movement/handoff, and possible resource degradation. A layered QoS middleware is proposed by Nikolaou et al. [25], which has a two-layer umbrella style in a reservation-based QoS middleware architecture. The layered QoS middleware enables the applications to specify and request network resources while the application would be independent from the underlying network reservation protocols. Figure 10.7 shows the layered QoS middleware structure that contains two layers: a generic QoS module at the upper layer and a specialized QoS module at the bottom layer. The upper generic QoS module works as an intermediate layer between the application and the lower specialized QoS module. The specialized QoS module interacts with the specific reservation mechanism or protocols, such as RSVP or Q.2931 in the ATM network.

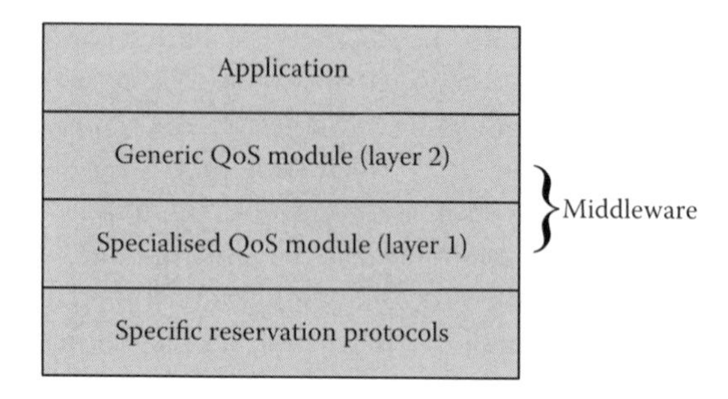

Figure 10.7 A layered quality of service middleware: Two-layer umbrella-based quality of service middleware.

To enable the QoS-enabled reservation, the generic QoS module [25] supports a minimum set of operations that include SetupReservation, ModifyReservation, and ReleaseReservation:

- *SetupResearvation*: It has the functionality to initiate the establishment of a new reservation for an application.
- *ModifyReservation*: This operation is to update and modify the reservation, if required, in a dynamic environment.
- *ReleaseReservation*: This is used to initiate the release of a previously established reservation from both directions.

The actual operations are implemented at the specialized QoS module. The parameters defined in the generic QoS module are generic and will be mapped to the reservation parameters of the specific reservation mechanism/protocol through the mapping mechanism of the specialized QoS module. In the proposed scheme, the support for the RSVP protocol and G.2931 signaling are described. This two-layer middleware scheme could also be applied to the environment that supports different reservation schemes.

Like the layered QoS middleware, the transparent QoS mechanism (TQM) [26] is also motivated to reduce the dependence of the applications on the QoS mechanisms. Such a design provides the transparent feature in QoS implementation. TQM considers the fact that the complexity of communication between diverse applications and underlying QoS architectures leads to the deployment problem of decreasing the utility of QoS provisioning. Without modifying legacy applications, TQM presents a transparent QoS mechanism to communicate with underlying QoS architectures and also provide swappable modules to support different QoS setups for diverse applications. Compared to the layered QoS middleware, TQM can be adapted to both IntServ and DiffServ networks, which has been examined by authors in the UNIX platform.

The QoS reservations for diverse applications are not static in a dynamic network environment. Therefore, it is a problem to determine when and how to make resource reservation, especially in the mobile environment where network resources are fluctuating with time, user movement, and so on. To appreciate resource reservation, Gu and Nahrstedt [27] presented a QoS-aware middleware framework that conducts QoS reservation based on prediction/pre-reservation. This approach maintains user preference by a user request model and a user preference model. The maintenance of user preference would be then useful to predict and manage the QoS reservation on the network. QoS management is responsible for advance resource reservation, prefetching, and advance service discovery. The QoS reservation operations are triggered not only by the fluctuation of network resources but also by the user's behavior, life habit, and movement. With the consideration of user preference, the QoS-aware middleware focuses more on the user level instead of only on mobile terminals or applications. This can be illustrated by a simple example. Suppose a user has a habit of watching CNN news at 8:00 a.m. from his mobile smart phone. Ideally, the QoS pre-reservation middleware would pre-reserve the network resource for the user before 8:00 a.m. on the network wherever the user is. As a result, the user experience of the mobile service is improved and the reservation will be more efficient and effective. In the architecture proposed by Gu and Nahrstedt [27], user preference is described using asynchronous messages encoded in XML format, which enhances the applicability of the solution.

10.5.5 QoS Adaptation

Adaptation-based QoS middleware dynamically reacts to the changes of the operating context. They are usually developed in between the mobile applications and the operating environment, having the knowledge of the mobile application, the operating system, and the operating environment. The adaptation-based QoS middleware should be able to react to the following types of change [3]:

- *Variations in Resource Availability*: The available resources for an application vary over time and their variations can be generally quantified by specific resource-level QoS parameters. At the user side, these parameters could be the CPU computational speed and its availability, the communication bandwidth at the interface for the application (e.g., multimedia streaming), and the usable buffer size of the mobile terminal. At the service side, similar parameters as at the user side can be developed for modeling the capability of the application server. Moreover, the parameters, which represent the network performances between the service and the client, are the end-to-end packet delay, jitter, and packet loss ratio. All these parameters change over time according to the applications and environmental conditions.
- *Variations in User Preferences*: There are two categories of user preferences that a user may specify and change at the application running time. The first one

is the level of user's satisfaction at different application QoS levels. The second could be the trade-off policy among QoS parameters of conflicting interests. For example, for video streaming, one user may require the best image quality with lower frame rate. On the other hand, another user may prefer the highest frame rate with relative lower image quality.

- *User Mobility*: In a distributed network environment, the user mobility should be addressed in the QoS-aware middleware design to offer smooth application even if the user changes Internet attachments. When the user moves to a new location with a different network access point, the application should be accordingly reconfigured to the new access point. This is called application-level handoff. The handoff is caused by the loss of link or network at the previous network attachment and the application will be disrupted in this period. After the new connection is established at the new location, the application with the QoS should be adaptively migrated to this newly established connection. If the location of an application component is considered as one of its core parameters, user mobility may be regarded as a spontaneous reconfiguration in a new location for the application component graph or parameters.

The adaptation-based QoS middleware allows the provision of a highly reconfigurable and adaptive execution environment to dynamically react in response to the changes in the above operating contexts. To achieve this goal, the adaptation-based QoS middleware is able to organize and implement its system components as a collection of services that are highly configurable and robust enough to enable the system itself to respond to the varying conditions in the environment. Therefore, the adaptation-based QoS middleware should be capable of managing the end-to-end QoS (e2eQoS), dynamically adapting QoS, and cross-layer mapping for the above purposes [28]. The e2eQoS represents the desirable QoS requirement of the user and can be parameterized at the application layers. The corresponding QoS parameters are delivered by the infrastructure within the specific time constraints over the network. Then, the middleware performs QoS adaptation on the basis of the time-varying application requirement, operating conditions, and available resources on the network. The cross-layer mapping constructs the real-time link between the higher mission-layer concept and the lower resource-layer for provisioning the QoS and resource control. On the basis of this linage, the middleware allocates resources among the participants and provides system-wide management of all participants, available resources, and QoS and application requirements. Local management for real-time adaptation could be enforced with the constraints of end-to-end operation. It provides local mechanisms for controlling the system resources and shaping data streams to meet resource constraints.

A typical adaptation-based QoS middleware model [21] includes four basic components: monitor/detect/predict, diagnoses, analysis, and QoS action. As shown in Figure 10.8, these four components perform their functions in four stages, respectively. In phase I, the monitor/detect/predict component evaluates the availability of the resource of the execution environment, the application profile, communication QoS requirements, and systems state information to detect or predict QoS violations. Once the violation is detected, the causes of violation are diagnosed in phase II and possible solutions are proposed in phase III. In phase III, different QoS solutions obtained from phase II are analyzed and compared to determine the appropriate QoS actions that need to be implemented. In phase IV, the selected QoS actions are implemented and corresponding QoS adaptation actions are carried out to meet the QoS requirements.

As a result of the difference of the relationship between applications and middleware in the adaptation phase, the adaptation QoS middleware could be classified into two types: inactive middleware and active middleware [22]. Middleware that adapts itself and provides the application with a transparent and stable execution environment is called "inactive" middleware. On the contrary, active middleware masters and controls the adaptive behavior of the application and the application can reconfigure and adapt itself under such control. The inactive middleware implements the adaptation and configuration in itself while the active approach allows the separation of application-specific adaptation strategies from the parameters of the control middleware.

There are several works on the adaptation-based QoS middleware. Li et al. [3] proposed an integrated middleware architecture for runtime application support, specializing in runtime probing (monitoring) of application states, runtime instantiation of a specific service configuration, and runtime adaptation to application state variations. Shriram and Sugumaran [29] proposed a middleware consisting of

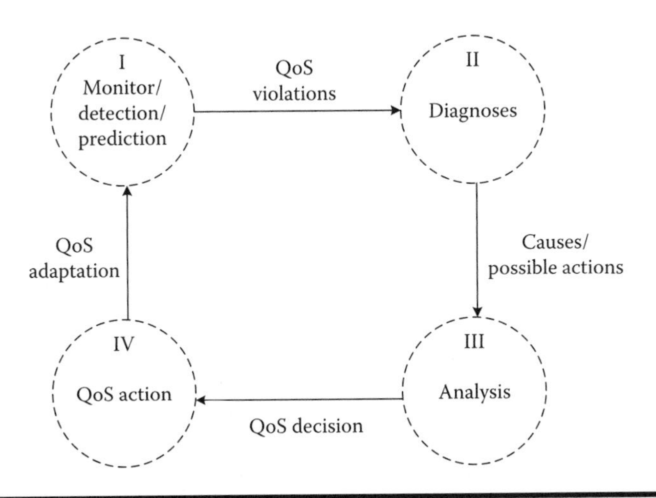

Figure 10.8 Quality of service adaptation model for the adaptation-based quality of service middleware. (Adapted from B. Shirazi, M. Kumar, and B. Y. Sung, in *Proceedings of the 37th Annual Hawaii International Conference on System Sciences*, Jan 2004.)

five managers on the mobile peer as well as the community network. These managers are user interface manager, context manager, communication manager, service manager, and memory manager. The overall process flow [29] can be illustrated with an interactive application as the shown below. The user sends an application request to the middleware and the request contains device profile that is further passed on to the user interface manager. The middleware determines whether or not the request can be executed in the system by querying the service manager. The service manager initiates a process of service discovery for processing the request. Additionally, service discovery specifies a unique identification of the process in the community network. It then sends out the request to the communication manager for delivering to the community network. The processing result may be transcoded and transformed by the middleware to accommodate the resource constraints of the mobile phone. This is to ensure that the results are displayed at the mobile phone properly according to the hardware device. In the end, the response is sent to the user and presented in the proper form.

Chan and Chuang [30] proposed an adaptation-based QoS middleware, named Mobile Platform for Actively Deployable Service (MobiPADS), which is designed to support context-aware processing by providing an executing platform to enable active service deployment and reconfiguration of the service composition in response to environments of varying contexts. Unlike most mobile middleware, MobiPADS supports dynamic adaptation at both the middleware and application layers to provide flexible configuration of resources to optimize the operations of mobile applications, which enlarges the amount of adaptation space available to the middleware to optimize processing in the event of adverse conditions, at the cost of application transparency. Thus, MobiPADS is an active adaptation-based middleware. Panzieri et al. [22] proposed another active adaptation-based middleware, which has two major components: a resource periodic monitoring (RPM) service and a QoS control (QSC) service. The scheme is an extension of the DCOM software commercial platform. In this scheme, the latency is regarded as a major QoS index for Internet telephony. On the basis of latency, a QSC service invokes an adaptation procedure that allows the application to modify those internal parameter values (e.g., audio playout buffer size) in order to meet its own QoS requirements. The QSC service operates periodically at the same rate as the RPM service updates the latency value. As a consequence, the application parameters can adapt dynamically to possible variations of both the network and the processing loads. The estimation of the upper bound for the packet transmission delays experienced during an audio communication is obtained at periodic intervals.

Chuang et al. [31] proposed a Web Proxy for Actively Deployable Services (WebPADS) framework to exploit the desired features of the Web service architecture operating over a wireless environment. WebPADS includes the components for active deployment, dynamic reconfiguration, and flexible service migration. Acting as a Web proxy system to a Web client, WebPADS has a proxy client and a proxy server. By intercepting Web traffic on both ends, WebPADS can transform and optimize the traffic for maximum performance. To overcome the shortcoming of the static agent-proxy model, WebPADS extended the agent-proxy model and proposed an active service model that supports dynamic service reconfiguration [32]. The proxy in this active service model is composed of a chain of service objects called mobilets, which could be dynamically reconfigured to adapt to the vigorous characteristic changes of the wireless environment. Mobilets usually exist in pairs and cooperate to provide a specific service; that is, a master mobilet resides at the client and a slave mobilet resides at the server. The slave mobilet shares the major portion of processing burden, whereas the master mobilet instructs the slave mobilet to take actions. Further, mobilets can migrate to a new proxy server when the mobile terminal moves to a different network domain, which allows the system to track the mobility of the mobile terminal and serve it continuously. By flexible adding, deleting, and reconfiguring of the mobilets, the proposed service model is able to adapt to the changes in heterogeneous wireless environments.

10.5.6 Always-Best Connected

With the rapid development of the semiconductor and wireless communication technologies, many portable devices are equipped with multiple wireless interfaces (e.g., radios), such as WiFi, WiMAX, Bluetooth, GMS, and/or UMTS. Multiple interfaces allow the device to connect to different networks that has distinguished network operating features such as coverage, throughput, delay, services available, and cost for applications. Thus, the probability of the middleware refers to the ability of the middleware to work with multiple interfaces and to simultaneously exploit several heterogeneous interfaces with different connectivity providers to facilitate the QoS provision. The selection of the service providers and wireless interfaces is highly context-dependent and several aspects should be taken into account. It is preferable that the QoS-aware middleware performs the connectivity management (e.g., interface and connectivity provider selection) and provides a better QoS during the movement of devices. With the support of middleware, the burden of determining the most suitable interface and connectivity provider at the service provisioning time would be relaxed from the applications.

Bellavista et al. [2] proposed the mobility-aware connectivity (MAC) middleware solution that can dynamically exploit the potential of any available connector in a mixed infrastructure-based (such as WiFi APs and UMTS base stations) or ad hoc manner and select the best one for the services. This process is performed without affecting the normal operations of final users and service providers. The proposed MAC middleware separately considers the requirements affecting the behavior of the whole user (e.g., power consumption and security concerns), and the ones related to each application (e.g., channel bandwidth and jitter needs). Also, by explicitly gathering and modeling the mobility indicators as crucial metrics, the MAC middleware is able to perform better in making decisions on the channel selection and connection establishment.

To fulfill the QoS requirements of the nomadic mobile service (NMS), where the mobile devices are the data producers, providing data to the clients located in the Internet, Pawar et al. [33] proposed a context-aware middleware architecture to support vertical handover for the NMS providers in mobile devices. The QoS requirements are considered in terms of its required uplink throughput and delay, which are represented by the e2eQoS parameters. A prediction-based interface selection scheme is applied during the process of vertical handover. The handover decision is made not only on the basis of the theoretical throughput of the available networks but also on the basis of the relevant context information from the mobile device and the fixed network. The proposed context information for the mobile network selection includes service requirements, user preferences, device capabilities, interface power consumption, user mobility, service criticality, and e2eQoS prediction information. On the basis of the collected context information sets, a context reasoning component applies a utility function to assign a score to each available network and makes the handover decision by using the score. To improve the estimation accuracy of the e2eQoS and minimize the mobile device power consumption in searching for available wireless networks, the QoS prediction context source in the fixed network also provides predictions on the availability of mobile networks and their estimated e2eQoS. Thus, the proposed middleware architecture provides satisfactory QoS performance with optimized battery usage.

10.6 Future Research

QoS-aware middleware for mobile services allows the application to focus on the functionalities without the need to directly configure network resource in a mobile environment. There are still limitations in the most current QoS middleware protocols. Firstly, these protocols are usually designed to meet QoS requirements for mobile terminals rather than for end-users (e.g., application layer). Secondly, the implementation of these QoS protocols requires sophisticated explicit coordination of various resources that spans through the mobile terminal, the accessing wireless network, and the Internet. However, the sophisticated explicit coordination is unacceptable from the viewpoint of mobile users and thus it needs to provide transparent QoS provisioning that hides the complexity of QoS control from the users. Furthermore, the QoS architecture should be independent from the different network architectures to support transparent mobility. For example, a mobile user who is receiving a service has no need to know whether s/he is moving from one network domain to another or from one cellular cell to another.

The QoS resource reservation and reconfiguration over the wireless networks render many constraints such as limited link capacity, multihop communication delay, and unpredicted network interference. Compared to the Internet that has relative network stability in resource control, QoS has to be automatically enforced to support the application needs with dynamic network conditions. Rather than a static network access point, the user mobility over wireless networks also causes many frequent handoffs in serving mobile service, and the QoS provision becomes more complicated for bandwidth-intensive applications such as multimedia. Therefore, the QoS-aware middleware has to be smoothly integrated with the service-oriented architecture framework to offer high flexibility for not only network resource configuration but also the service discovery and configuration. With increase of movement speed, it becomes a problem for the QoS-aware middleware to precisely evaluate application QoS performance and its tendency on the time-varying network conditions. In other words, the robustness and efficiency of the QoS-aware middleware have to be improved in response to dynamic user and network conditions. On the basis of the precise QoS and resource evaluation, the QoS-aware middleware makes decisions to effectively reconfigure the network and user resource to maximize resource utilization. At the same time, the configuration for an application could not degrade QoS performance of other applications that may be provided for other users. When the reconfiguration spans heterogeneous platforms and networks, a unified network QoS specification language is required across heterogeneous environments such that the network components can understand the required operations in resource allocation.

When a user moves from one Internet attachment to another, the application handoff is still a problem in designing the QoS-aware middleware. The handoff considers not only the capacities of accessing wireless networks but also the QoS requirements of the application. When multiple accessing points are available, QoS-aware middleware needs to evaluate the radio interfaces to select the appropriate wireless networks. For example, the energy consumptions should be considered for saving the energy of the mobile terminal due to power constraint at the user. The QoS-aware middleware for application handoff should determine when and how the application migrates from one network to another. The application handoff should be smoothly performed such that the user cannot experience the performance degradation during network switching.

10.7 Conclusion

QoS mechanisms and IETF standards for network services are initially developed for the Internet and telecommunication networks to ensure user-desirable QoS performance such as delay, jitter, and packet loss ratio. However, these approaches are limited to supporting mobile services when the mobile user moves from one domain to another or from one network to another. The main problems arise from the changes of network attachments (e.g., application handoff) and the limited wireless network capacity. To ensure QoS for mobile users, Internet standards have been improved to adaptively reconfigure the required user and network resource according to the QoS variation and user preference. Furthermore, these QoS mechanisms are designed as the QoS-aware middleware to facilitate the QoS operations over the network. These operations adaptively react to the network context by the service discovery, QoS diagnosis, parameter adaptation, user and network resource reservation, network selection, and other QoS reconfiguration actions.

References

1. B. Xie, A. Kumar, and D. P. Agrawal, Enabling multi-service on 3G and beyond: Challenges and future directions, *IEEE Wireless Communication Magazine*, June 2008.
2. P. Bellavista, A. Corradi, and C. Giannelli, Mobility-aware management of Internet connectivity in always best served wireless scenarios, *Mobile Network Application*, 14, 18–34, 2009.
3. B. Li, D. Xu, and K. Nahrstedt, Towards integrated runtime solutions in QoS-aware middleware, In *Proceedings of the International Workshop on Multimedia Middleware*, 2001.
4. G. Neureiter, L. Burness, A. Kassler, et al., The BRAIN quality of service architecture for adaptable services with mobility support, In *Proceeding of the 11th IEEE International Symposium on Personal, Indoor and Mobile Radio Communication (PIMRC)*, 2000.
5. J. Diederich, L. Wolf, and M. Zitterbart, A mobile differentiated services QoS model, *Computer Communications*, 27(11), 1106–1114, 2004; [Extended version of the paper presented at the *Workshop on Applications and Services in Wireless Networks*, 2003].
6. J. Diederich, T. Lohmar, M. Zitterbart, and R. Keller, A QoS model for differentiated services in mobile wireless networks, In *Proceeding of the 11th IEEE Workshop on Local and Metropolitan Area Networks*, 2001.
7. V. G. Ozianyi, N. Ventura, and E. Golovins, A novel pricing approach to support QoS in 3G networks, *Computer Networks*, 52(7), 1433–1450, 2008.
8. C. Tseng, G. Lee, R. Liu, and T. Wang, HMRSVP: A hierarchical mobile RSVP protocol, *Wireless Network*, 9(2) 95–102, 2003.
9. C.E. Perkins, *Mobile IP Design Principles and Practices*, Addison-Wesley, Reading, Massachusetts, 1998.
10. C.E. Perkins, IP encapsulation within IP, RFC 2003; Internet Engineering Task Force, 1996.
11. A. Terzis, J. Krawczyk, J. Wroclawski, and L. Zhang, RSVP operation over IP tunnels, RFC 2746, 2000.
12. A. Terzis, M. Srivastava, and L. Zhang, A simple QoS signaling protocol for mobile hosts in the integrated services Internet, In *Proceedings of IEEE INFOCOM*, 1999.
13. S. Lai, Y. Ma, and H. Deng, HO-RSVP: A protocol providing QoS support for seamless handover between wireless networks, In *Proceedings of the 2nd ACM International Workshop on Quality of Service & Security for Wireless and Mobile Networks*, 2006.
14. A.K. Talukdar, B.R. Badrinath and A. Acharya, Integrated services packet networks with mobile hosts: Architecture and performance, *Wireless Networks*, 5(2), 111–124, 1999.
15. A. K. Talukdar, B. R. Badrinath and A. Acharya, MRSVP: A resource reservation protocol for an integrated services network with mobile hosts, *Wireless Networks*, 7(1), 5–19, 2001.
16. S. Elleingand and S. Pierre, FH-RSVP scheme for intra-site handover in hierarchical mobile IPv6 networks, *Computer Communication*, 2007.

17. S. Ashraf, S. A. Khan, N. S. Khattak, and A. Rehman, M2PLS: Mobile multiprotocol label switching, In *Proceedings of the 4th Innovations in Information Technology*, 2007.
18. W. Sa-Ngiamsak, P. Krachodnok, and R. Varakulsiripunth, A recovery scheme for QoS guaranteed mobile IP Over MPLS network," In *Proceedings of the 1st International Symposium on Wireless Pervasive Computing*, 2006.
19. R. Cheung, An adaptive middleware infrastructure for mobile computing, In *Proceedings of the 14th International Conference on World Wide Web*, 2005.
20. J. Jin, and K. Nahrstedt, QoS-Aware service management for component-based distributed applications, *ACM Transactions on Internet Technology*, 2008.
21. B. Shirazi, M. Kumar, and B. Y. Sung, QoS middleware support for pervasive computing applications, In *Proceedings of the 37th Annual Hawaii International Conference on System Sciences*, 2004.
22. F. Panzieri, M. Roccetti, and V. Ghini, The implementation of middleware services for QoS-aware distributed multimedia applications, In *Proceedings of the International Workshop on Multimedia Middleware*, 2001.
23. Z. Jarir and M. Erradi, A meta-level architecture for QoS awareness in a mobile environment, In *Proceedings of the 8th International Conference on New Technologies in Distributed Systems*, 2008.
24. A. Kavimandan, and A. Gokhale, A parameterized model transformations approach for automating middleware QoS configurations in distributed real-time and embedded systems, In *Proceedings of the Workshop on Automating Service Quality, at the International Conference on Automated Software Engineering*, 2007.
25. N. A. Nikolaou, C. A. Tsetsekas, and I. S. Venieris, A QoS middleware for network adaptive applications, In *Proceedings of the IEEE International Conference on Multimedia Computing and Systems*, 1999.
26. C.-H. Shih, C.-C. Liao, C.-K. Shieh, and W.-S. Huang, A transparent QoS mechanism to support IntServ/DiffServ networks, In *Proceedings of the IEEE Consumer Communications and Networking Conference*, 2004.
27. X. Gu, and K. Nahrstedt, An event-driven, user-centric, QoS-aware middleware framework for ubiquitous multimedia applications, In *Proceedings of the 2001 International Workshop on Multimedia Middleware*, 2001.
28. P. K. Sharma, J. Loyall, R. E. Schantz, J. Ye, P. Manghwani, M. Gillen, and G. T. Heineman, Managing end-to-end QoS in distributed embedded applications, In *Proceedings of the IEEE Internet Computing*, 2006.
29. R. Shriram, and V. Sugumaran, Adaptive middleware architecture for information sharing on mobile phones, In *Proceedings of the 2007 ACM Symposium on Applied Computing*, 2007.
30. A. T. S. Chan and S.-N. Chuang, MobiPADS: A reflective middleware for context-aware mobile computing, *IEEE Transactions on Software Engineering*, 2003.
31. S. N. Chuang, A. T. S. Chan, J. Cao, and R. Cheung, Actively deployable mobile services for adaptive web access, *IEEE Internet Computing*, 2004.
32. S. N. Chuang, A. T. S. Chan, J. Cao, and R. Cheung, Dynamic service reconfiguration for wireless web access, In *Proceedings of the 12th International Conference on the World Wide Web*, 2003.
33. P. Pawar, K. Wac, B. Beijnum, P. Maret, A. Halteren, and H. Hermens, Context-aware middleware architecture for vertical handover support to multi-homed Nomadic mobile services, In *Proceedings of ACM Symposium on Applied Computing*, 2008.

Chapter 11

Middleware Systems Architecture for Distributed Web Service Workflow Coordination

Janaka Balasooriya, Sushil K. Prasad, and Shamkant B. Navathe

Contents

11.1 Introduction

Services-oriented computing (SOA) has emerged as the leading cross-disciplinary distributed computing paradigm [1]. "According to Forrester Research, the SOA service and market had grown by $4.9 billion in 2005, and it is expected to have an interesting growth rate until 2010, with a compound annual growth rate of 88 percent between 2004 and 2009" [2:1]. SOA-based applications span domains as diverse as enterprise e-commerce applications [3], personal applications [4–7], and scientific applications [8,9]. Therefore, the users and developers of these applications are usually noncomputer scientists. In addition, the proliferation of hardware and communication technologies have made handheld wirelessly connected devices more powerful. Software technologies that we develop today should be able to be deployed on both wired and wireless devices. Therefore, efficient technologies are required to rapidly develop and deploy robust collaborative applications leveraging off the existing Web services [10].

11.1.1 Web Service Workflow Composition: Current State of the Art

Web service composition is the process of combining autonomous (Web) services to form another value-added service encapsulating predefined application logic. The composed service encompasses the workflow coordination logic. The constituent services in the workflow can be from different organizations providing ways to develop interorganizational collaborative applications. For example, Figure 11.1a illustrates the purchase order workflow presented in the WS-BPEL specification (Web Services Business Process Execution Language) [3]. In the purchase order workflow, once the purchase order is received by the "Receive Purchase Order" Web service, it triggers "Price Calculation," "Shipping," and "Product Scheduling" Web services simultaneously (AND-Split logic). Then, the "Shipping" Web service sends shipping information to "Price Calculation" and "Product Scheduling" Web services to complete their execution (dataflow). Once the "Invoice Processing" Web service receives control and data from all three Web services, it starts processing the invoice (AND-Join logic). Figure 11.1b illustrates the software architecture of the WS-BPEL-based implementation of the workflow.

As shown in Figure 11.1b, the composite workflow process is modeled as a separate state-preserving Web process encapsulating all the dataflow and control-flow dependencies such as AND-Join and AND-Split. Modeling and enforcing workflow control flow require complex constructs and runtime support [11–13]. Therefore, the current workflow software architecture typically results in complex and centralized logic for workflow coordination. Owing to its centralized nature and the inability of participant Web services to share the burden of enforcing composition and coordination constraints, the composed Web process has to encapsulate numerous functionalities ranging from application logic to transaction management. There are several drawbacks in this model.

11.1.1.1 Specific Problems in Centralized Workflow Coordination

1. *Centralized Coordination*: There are both pros and cons in centralized coordination. A positive aspect is total control over the behavior of the Web process. However, distributed coordination has two possible advantages over centralized coordination: (i) owing to security, privacy, or licensing imperatives, some Web-based objects will only allow direct pair-wise interactions without any coordinating third-party entity; and (ii) centralized coordination/workflows suffer from issues such as scalability, performance, and fault-tolerance [14]. For example, data transfer and message passing among participant Web services need to go through the central Web process generating more network traffic and making the composed Web process more complex.

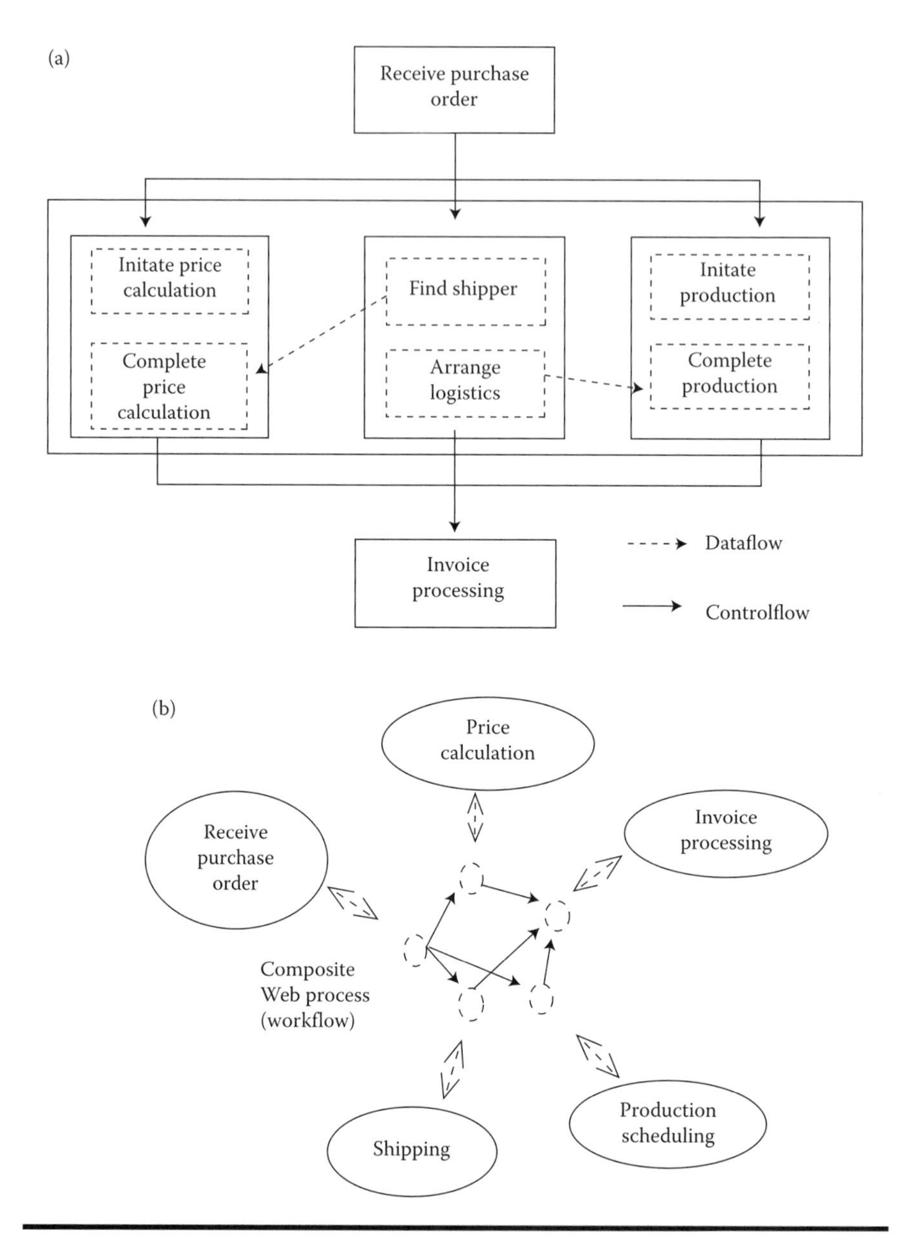

Figure 11.1 Web service workflow development: A bigger picture. (a) Purchase order workflow. (Adapted from Workflow Patterns Initiative, Queensland University of Technology, Australia, http://www.workflowpatterns.com/evaluations/standard/index.php.) (b) Software architecture of traditional WS-BPEL implementation of the purchase order workflow.

2. *Deployment Platforms*: Being able to execute workflows over a diverse set of devices including wireless devices has significant benefits [4,7,15–17]. Portions of long-running workflows can reside on handheld devices providing monitoring and controlling capabilities. The current platforms consume significant amount of resources and are difficult to deploy on limited resource devices such as wireless devices.

One of the viable solutions for these drawbacks is to distribute the workflow coordination among (possibly) participating Web services.

11.1.2 Distributed Web Service Workflow Coordination

In a Web service-based workflow application, each participating Web service is an activity in the workflow. Each activity has its own pre- and post-execution dependencies. Execution of an activity means a method invocation in Web service-based workflows. Pre-execution dependencies have to be met before executing the Web service method whereas post-execution dependencies are enforced after executing the method. In true distributed workflow coordination, each participating Web service has to be aware of its own workflow dependencies and enforces them as methods are executed. The following section briefly explains our Web service coordination management middleware (WSCMM) approach. Section 11.2 also presents other major approaches and architectures in distributed workflow coordination.

11.1.2.1 WSCMM Approach

The main objective of the WSCMM is to distribute the complexity of the centralized workflow coordination logic among participating Web services. The WSCMM system adopts a layered workflow software architecture that clearly differentiates the high-level workflow definition and runtime artifacts. The stateless Web services are empowered by the "Coordinator Proxy Objects (CPOs)" into self-coordinating stateful entities. Next, the high-level "Web bond" primitives are used to interlink the CPOs, capturing the workflow logic. Previously, we have proposed the ideas of Web bonds as a set of primitives for Web service coordination/choreography and proved those Web bonds are capable of modeling all the workflow control flow dependencies [11]. Each CPO maintains and enforces its own dependencies during the execution of the workflow. Thus, the workflows we create are inherently distributed. Once workflows are developed, the WSCMM system distributes the workflow coordination responsibilities among participating Web services. Web service method invocations go through the CPO, which enforces pre- and post-Web service method invocation dependencies. The development of a WSCMM is analogous to the development of a DBMS (database management system) to coordinate the execution of queries and transactions in the Web services domain. Sections 11.4 and 11.5 explain the WSCMM system in detail.

The remainder of this chapter is organized as follows. Section 11.2 presents related work. Section 11.3 outlines the proposed solution toward our WSCMM system. Section 11.4 illustrates the two-layered workflow software architecture that facilitated distributed workflow coordination. Section 11.5 presents the WSCMM architecture and simulation and verification of the middleware architecture. Section 11.6 describes the implementation of the BondFlow system, which is the prototype implementation based on WSCMM. It focuses on the handheld-based execution of workflows. Future research challenges are presented in Section 11.7. Definitions of commonly used words chapter review questions and answers have also been given at the end of the chapter.

11.2 Related Work

In this section, we explain the main approaches in distributed workflow coordination, comparing them with the WSCMM approach.

11.2.1 IBM Symphony [14]

IBM symphony decentralizes the coordination by partitioning centralized workflow specification into separate modules so that each module can be executed in a distributed setting. The motivation behind the IBM symphony approach is to achieve better performance in terms of throughput, scalability, and response time. In this approach, first, the workflow is developed using the BPEL4WS [3] as a centralized workflow. Then, a module in the IBM symphony analyses the workflow logic (dependencies) and partitions the workflow so that each workflow node (participating Web service in this case) has its own logic. Code partition is based on both control flow and dataflow dependencies [14]. In addition, each partition has its own Web Service Description Language (WSDL) interface and acts as a Web service of its own. Thus, each partition can be deployed in the same server where the participating Web service is deployed as workflow is being executed. IBM symphony uses the BPWS4J engine as the workflow runtime environment [14].

11.2.2 OSIRIS [18]

In the OSIRIS system, the composite Web service is seen as a collection of ordered set of activities. Each activity defines a Web service call. In this architecture, each service sits behind the OSIRIS middleware layer called the hyperdatabase layer that handholds message routing and service invocation dependencies. In addition to having a middleware layer that facilitates distributed execution of the workflow, the OSIRIS system runs several metadata repositories that defines the workflow process. These metarepositories push the workflow definition and perhaps change to each node in the workflow as the workflow gets executed enabling dynamic changes.

OSIRIS used the O'Grape [18] visual tool to develop workflows as a graph. Each node in the graph represents participating Web services in the workflow. At the time of decomposing the workflow into distinct activities, it adds fork and join nodes to represent (AND-Split and AND-Join) workflow coordination logic.

11.2.3 Self-Serv [10]

The Self-Serv project presented in Reference [10] proposes more fundamental changes to the way workflows are composed and executed. Self-Serv differentiates the high-level workflow definition and runtime artifacts by modeling the workflow using statechart [10]. A statechart is a collection of states and transactions. Transactions can fire if an event occurs that stratifies the condition of firing the transaction. As transactions are fired states are changed in the statechart. In Web service workflows, states and transactions represents Web services and invocation of Web service methods. Workflow dependencies are attached as transaction firing conditions. At runtime, Self-Serv uses the concept of a "coordinator," which can act as a scheduler for participating Web services. Each coordinator has transactions and conditions that correspond to workflow dependencies. Thus, the coordinator in the execution environment corresponds to a state and transactions associated with the state in the statechart represents the workflow. Each participating Web service can run the Self-Serv middleware and each coordinator representing the participating Web service can be deployed in the middleware by allowing coordinators to control the execution of the workflow in a peer-to-peer fashion.

11.2.4 Silver: A BPEL Workflow Process Engine for Mobile Devices [19]

The main objective of the Silver system is to modify the BPEL workflow execution engine so that the mini version BPEL engine can be deployed on handheld devices. Silver uses only a subset of features supported by BPEL. Especially, it does not support complex workflow control flow patters such as multiple instances and multimerge [19]. It also, uses lightweight components based on kSOAP and kXML [19], making it a system with minimum functionality. Hackmann et al. [19] claim that the code base of Silver is 114 KB.

With regard to other handheld-based workflow execution systems, Jørstad et al. [16] describe issues related to service composition in mobile environments and evaluate criteria for judging protocols that enable such composition. The composition protocols are based on distributed brokerage mechanisms and utilize a distributed service discovery process over *ad hoc* network connectivity. Papazoglou et al. [5] present architecture for mobile device collaboration using Web services. Mnaouer et al. [4] present a rapid application development environment for mobile Web services. However, a key limitation in most of these technologies is that they treat handheld devices only as clients limiting true peer-to-peer architecture.

11.2.5 Other Systems

Schmidt [20] proposes a system to distribute the execution of business applications using Web services by adding business rules into the Simple Object Access Protocol (SOAP) messages. Business rules encoded in the SOAP header specify the order of execution. Messages are decoded and processed by special processing units called SOAP intermediaries. Rosenberg and Dustdar [21] propose a service-oriented distributed business rules system and its implementation on the basis of WS-Coordination. The Web Service Resource (WSRF) framework is another proposal toward stateful Web services. It provides standardization representation to stateful resources and the Web service interface provides functionalities to access (read, update, and query) state information. This state information is used to process Web service messages [22]. Comparative study of various implementations of WSRF is presented by Humphrey et al. [23].

11.2.6 Discussion: Limitations of Current Approaches

Even though IBM symphony distributes the workflow execution, there are several limitations. First, it is necessary to develop the centralized BPEL code and then partition and distribute it among participant entities. Second, usually, there are problems partitioning the code in complex application scenarios such as long-running transactional applications without proper infrastructure support. Also, the workflow development still needs a significant amount of effort. Finally, the workflow execution environment consumes lot of resources and is difficult to deploy in resource-constrained devices such as handheld devices.

In the OSIRIS approach, though metainformation enables dynamic changes to the workflow, it needs some sort of a centralized coordinator. Also, to enforce fork/join dependencies, the OSIRIS system introduces new nodes to the workflow by making it more complex. In addition, the fork/join nodes are coordinated by the metarepository. Self-Serv is one of the most comprehensive distributed workflow coordination approaches.

Self-Serv uses statecharts to model workflow dependencies. Statechart has limitations in modeling complex workflow control flow patterns such as advanced AND-Split, AND-Join, and patterns involving multiple instances [24].

The WSRF-based approaches require the Web service interface to be changed to store state information, thus making Web services more complex entities. Although the above approaches solve some issues in centralized workflow coordination, we believe more fundamental enhancements are needed to the Web services infrastructure facilitating more fine-grained decentralization of the coordination. The following section describes our approach in detail.

11.3 WSCMM Solution, the Core Idea

To overcome the above limitations, it is necessary to differentiate the high-level workflow definition and runtime artifacts and distribute workflow coordination responsibilities among constituent Web services and Web entities [25–30]. We envision Web services to actively participate in workflows enforcing their own dependencies. A good motivating analogy would be to consider the evolution of database application development platforms.

11.3.1 Evolution of Database Application Development Analogy

Figure 11.2 illustrates the evolution of database technologies from simple file systems to a three-tier system, equipped with layers to manage the database, user interface, and workflows, progressively reducing the burden of application development. In the early 1960s, the application developer had the burden of capturing all the logic of data manipulation, constraint checking, and concurrency control (Figure 11.2a).

With the introduction of DBMSs, most of the data handling functionalities were transferred to these systems. In Figure 11.2c, the two layers of the architecture are shown as DBMS and application layer with the user interface management system on the top for user interaction. The application layer has typically been implemented with a set of clients and hence this popularly became known as the client–server architecture that has been in operation since the mid-1980s. Since the advent of the worldwide Web around 1994, an additional layer for Web servers got added to this architecture (Figure 11.2d).

Development of various middleware technologies and workflow management systems further reduced the burden of the application developer. The current Web server-

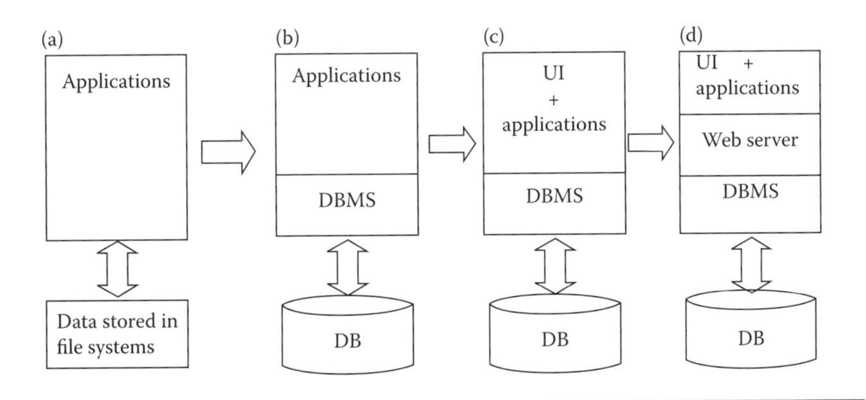

Figure 11.2 Evolution of database application infrastructure. DB, database; DBMS, database management; UI, user interface. (Adapted from W.M.P. van der Aals, *Journal of Circuits, Systems and Computers*, 8(1), 1998, 21–66.)

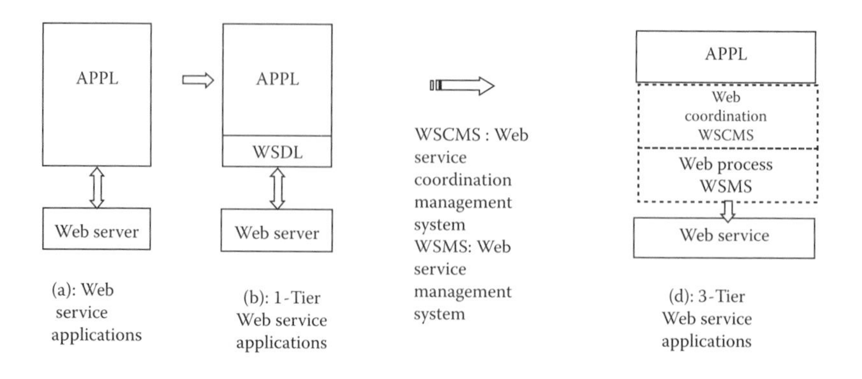

Figure 11.3 Proposed enhancements to the Web service infrastructure. APPL, Application Logic; WSDL, Web Service Description Language.

based database applications have similar characteristics. The application programmer has the burden of capturing all the application logic as well as housekeeping tasks.

The individual Web services, encapsulating information and data stores, with its access methods described using WSDL, lack even the basic management system (Figure 11.3a), not to mention any support for transactions, composition, or workflows. The application programmer has the burden of capturing almost all of the coordination logic. From this perspective, Web services infrastructure is still in its early developmental stage. Therefore, we propose to (a) enhance the Web services infrastructure so that it has a management system for Web services to manage methods and method invocations more effectively, akin to DBMS in databases; (b) evaluate coordination and composition techniques for Web services and transfer generic functional layers to the Web service so that they become capable entities to enforce distributed coordination, akin to the workflow management system (WFMS) in databases. We call these systems the Web service management system (WSMS) and the Web service coordination management system (WSCMS), respectively (Figure 11.3d).

11.3.2 Our Approach and Contributions

Here, we present a layered workflow software architecture for distributed coordination of workflows over Web services and propose fundamental enhancements to the Web services infrastructure that will facilitate the layered workflow software architecture. Then, we verify our layered approach using a detailed simulation and a proof-of-concept prototype implementation [11,31].

1. *Two-Layered Workflow Software Architecture*: we propose a layered workflow software architecture that will greatly simplifies the workflow development task by distributing the complexity of the centralized workflow coordination logic over stateless Web services of traditional systems such as BPEL.

The stateless Web services are empowered by CPOs into self-coordinating stateful entities. Next, the high-level Web service coordination constructs can be used to interlink the CPOs, capturing the workflow logic. Previously, we proposed the ideas of Web coordination bonds as a set of primitives for Web service coordination/choreography and proved those Web bonds to be capable of modeling all the workflow control flow dependencies [11]. We have used Web coordination bonds to enforce workflow control flow dependencies. Each CPO maintains and enforces its own dependencies during the execution of the workflow. Thus, the workflows we create are inherently distributed. Details of our layered architecture are presented in Section 11.4.

2. *Web Service Coordination Management Middleware*: The primary objective of the WSCMM system is to distribute the workflow coordination responsibilities among participating Web services. Subsequently, it simplifies the workflow development process. WSCMM consists of two components: WSMS and-WSCMS. WSCMS maintains and enforces workflow dependencies while WSMS transforms the stateless Web service into a stateful entity through the CPO. Web service method invocations go through this object, which enforces pre- and post-Web service method invocation dependencies using the functionality of WSCMS. The development of a WSCMM is analogous to the development of a DBMS to coordinate the execution of queries and transactions in the Web services domain. We have carried out a detailed simulation to identify and key components and design issues of our middleware. Proof-of-concept experiments demonstrate that we can develop both centralized and distributed workflows over the architecturally enhanced Web services with relative simplicity.
3. *Prototype Implementation*: We have prototyped our BondFlow system using Java as a platform to configure and execute workflows over Web services [31]. The footprint of the BondFlow runtime system is 24 KB and the additional third-party software packages, such as those of the SOAP client and XML parser, account for 115 KB. The execution time workspace used by the BondFlow system is 5.4 MB including JVM (Jeode 1.2 handheld Java version). Therefore, we have been able to test the BondFlow system on both wired and wireless infrastructures. For communication among the coordinator objects, we used SOAP in wired devices and our System on Mobile Devices (SyD) middleware in wireless devices. SyD is our recently prototyped middleware platform to develop and execute distributed applications over handheld devices [6]. Lightweight SyDListener enables handheld devices to host server objects.

11.4 Two-Layered Workflow Software Architecture

As mentioned in Section 11.3, Web coordination bonds have been applied as coordination primitives to model and enforce workflow dependencies. First, we briefly present the idea of Web coordination bonds. Details can be found in References [11,31].

Web bonds have been proposed as a set of primitives for Web service coordination/choreography. There are two types of Web bonds: subscription bonds and negotiation bonds. The subscription bonds allow automatic flow of information and control from a source entity to other entities that subscribe to it. This can be used for synchronization as well as for more complex changes, needing data or event flows. The negotiation bonds enforce dependencies and constraints among entities and trigger changes on the basis of constraint satisfaction. A negotiation bond from *A* to *B* has two interpretations: pre-execution and post-execution. In the case of pre-execution, to start the activity *A*, *B* needs to complete its execution. In case of post-execution, to start the activity *A*, *A* needs to make sure that *B* can be completed afterwards. Without loss of generality, both pre- and post-execution interpretations of negotiation bonds enforce atomicity. In this chapter, unless specified, we have used the pre-execution type of negotiation bonds implicitly. We have established that Web bonds have the modeling power of extended Petri nets. They can express all the benchmark patterns for workflow and for interprocess communication, a feat that almost all previously proposed artifacts and languages are not capable of doing comprehensively [11]. Section 11.4.2 illustrates the modeling of purchase order workflow using Web coordination bonds.

11.4.1 Two-Layered Workflow Software Architecture

As shown in Figure 11.4a, the architecture of the traditional workflow code is "single layer" where the developer needs to program the workflow from scratch (ensure communication, workflow coordination, and intermediate data processing) (Figure 11.4b).

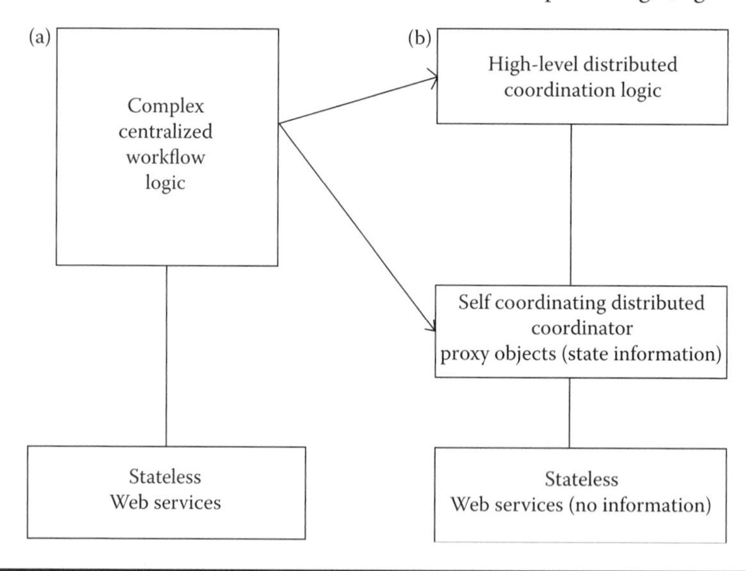

Figure 11.4 Two-layer workflow software architecture. (a) Architecture of the single-layer workflow using traditional systems. (b) Architecture of the proposed two-layer workflow in software architecture.

In contrast, in our proposed workflow software architecture, workflow coordination has been encapsulated as a separate layer (using Web coordination bonds in our implementation). In addition, the coordinator objects represent participating Web services in the workflow. The coordinator object encompasses all the coordination capabilities of Web bond artifacts (Figure 11.4b). CPO communicates with the Web service from method invocations and is state preserving. Capabilities of Web coordination bonds including modeling workflow dependencies have been encapsulated in the upper layer (Figure 11.4b). The developer's responsibility is to configure the workflow using high-level constructs by linking Web service appropriately and specifying constraints (high-level configurability).

11.4.2 Web Service CPO

Figure 11.5 illustrates the CPO. The coordinator object provides the same interface as the Web service provides to the outer world. Web service method invocations of the workflow take place through the coordinator object and the Web bond coordination layer ensures that pre- and postmethod invocation dependencies are satisfied. As shown in Figure 11.5, each coordinator object has a bond repository, a set of user-defined constraints (if any), and runtime information associated with it. The bond repository consists of all the workflow dependences related to the coordinator object (participating Web service). This indirection allows us to bring transparency to the system and hide the necessary coordination and communication logic behind it. It also maintains the status of method invocations such as intermediate date and partial results.

User-defined constraints represent the additional dependency conditions (dependencies not defined using Web bonds) that need to be satisfied while enforcing workflow dependencies. As shown in Figure 11.6, each Web service method call

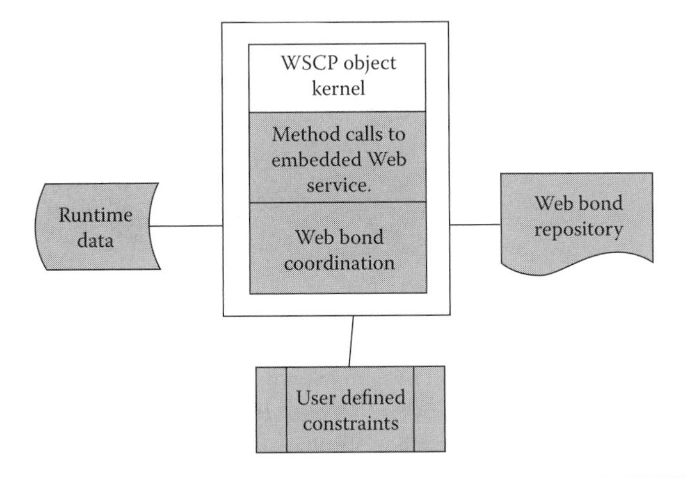

Figure 11.5 Web service coordinator proxy object.

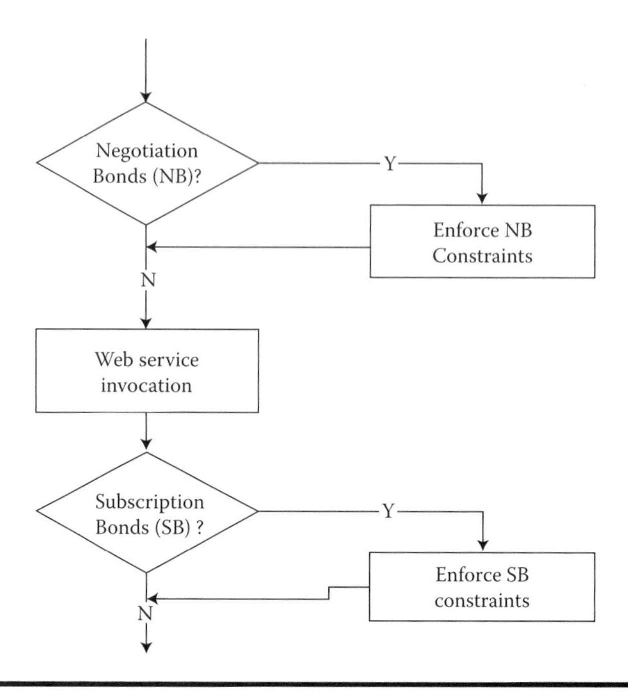

Figure 11.6 Flow within a proxy object.

is encapsulated by a negotiation and a subscription bond check. The negotiation bonds enforce pre-method invocation dependencies whereas the subscription bonds enforce postmethod invocation dependencies.

This logic ensures that workflow dependencies are satisfied with associated Web service method invocation. The idea of Web service CPO together with underlying Web bond primitives encapsulates the workflow coordination layer. This simple yet powerful idea empowers Web services and makes workflow configuration less programming intensive. We believe this concept has enough potential to lead to a fundamental shift in workflow development over Web services.

The workflow configuration process starts by creating bonds among methods of selected Web services to reflect dependencies (negotiation and subscription bonds). Bond constraints are specified during the bond creation time and the bond configuration is stored in a persistent storage in XML format.

11.4.3 Coordination Logic Layer Using Web Bonds

First, we illustrate the workflow configuration using high-level Web coordination bond constructs using the purchase order case study workflow. Figure 11.7 illustrates the modeling of a purchase order workflow using a network of Web coordination bonds. Five Web services are involved in the workflow. The system generates CPOs for each Web service. Then, a network of Web bonds is created among methods of these

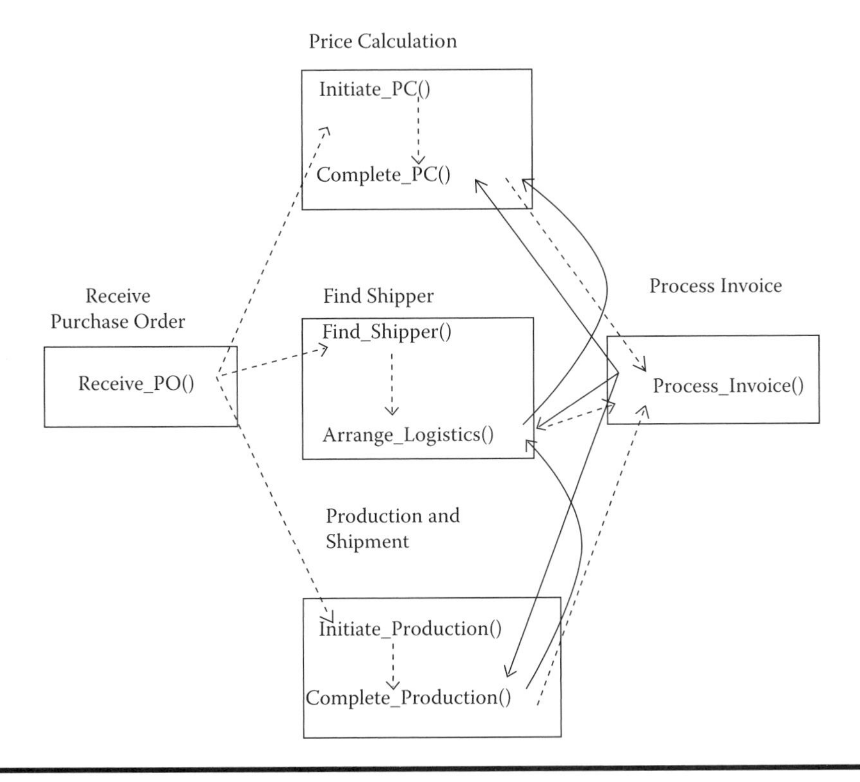

Figure 11.7 Purchase order workflow using Web coordination bonds

coordinator objects to enforce the workflow constraints. For example, the "Receive Purchase Order" Web service needs to pass control to "Price Calculation," "Find Shipper," and "Production and Shipment" Web services once it is completed. To model this split dependency, *Receive_PO()*method has three subscription bonds to each of *Initiate_PC()*, *Find_Shipper()*, and *Initiate Production()* methods. Similarly, the rest of the dependencies have been modeled using other negotiation and subscription bonds. The configured workflow consists of five coordinator objects representing each Web service with bond repositories associated with them.

11.5 Web Coordination Management Middleware

On the basis of our two-layered workflow software architecture, we extract three key layers of functionality encapsulated by the composite Web process (Figure 11.8). The top layer encapsulates the abstract workflow process defined using high-level constructs. The middle layer represents the code that enforces workflow dependencies (implemented on the basis of underlined language constructs). The last layer implements actual communication with individual Web services that are participants of the workflow. For each workflow, all three layers need to be implemented from

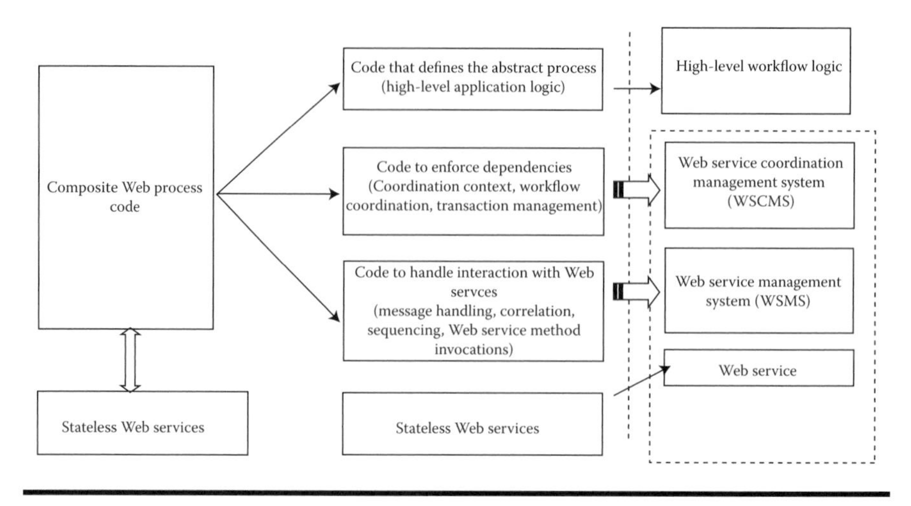

Figure 11.8 Functional decomposition of composite Web process.

scratch. However, the second and third layers represent significant amounts of generic functionalities such as enforcing basic workflow coordination logic, Web service invocations, message handling, and storing corresponding state information. Therefore, generic functionalities of these two layers can be extracted and provided as a middleware layer for distributed workflow coordination. We identify three categories of functionalities for a middleware system for distributed workflow coordination over Web services.

11.5.1 Implications on Functionality

1. *Enforce dependencies*: Workflow activities need to satisfy various kinds of constraints to accomplish the task successfully. For example, before initiating the activity, the workflow management system may need to satisfy application-specific data, control, and resource dependencies and once an activity is completed the activity may need to inform results and pass control to other activities on the basis of various conditions. In a distributed coordination environment, each Web service needs to maintain its own dependencies and enforce them locally.
2. *Preserve state information*: Long-lived workflow applications require state of method invocations (success or failure) and intermediate results to be stored and make global decisions. Such state information needs to be maintained and correlated with proper application context.
3. *Process messages:* Web services communicate exchanging messages. Therefore, to become live participants in distributed applications, Web services should bear enough capabilities to process messages and make decisions accordingly. This entails maintaining proper communication context for each application, message correlation and sequencing, and reliable messaging.

In our middleware, functionalities pertaining to workflow dependency are carried out by the WSCMS layer. Processing messages and maintaining state information is handled by the WSMS. The next section discusses these components in detail.

11.5.2 WSCMM Architecture: An Overview

This section starts with a generic description of our Web WSCMM architecture, its components, and related issues. Then we discuss each component in detail. The Web coordination middleware consists of two main components: WSMS and WSCMS (Figure 11.9).

11.5.2.1 Web Service Management System

WSMS handles two functionalities: (i) preserving state information for long-lived interactions and (ii) processing messages locally and initiating appropriate actions. The core functionality of the WSMS is to transform the stateless Web service into a state-preserving self-coordinating entity. WSMS performs this transformation by generating a coordinator object to represent the Web service, which encompasses all the coordination capabilities of WSCMS implementation. Sections 11.4.2 and 11.4.3 discussion in more detail the coordinator object and enforcing workflow control flow logic.

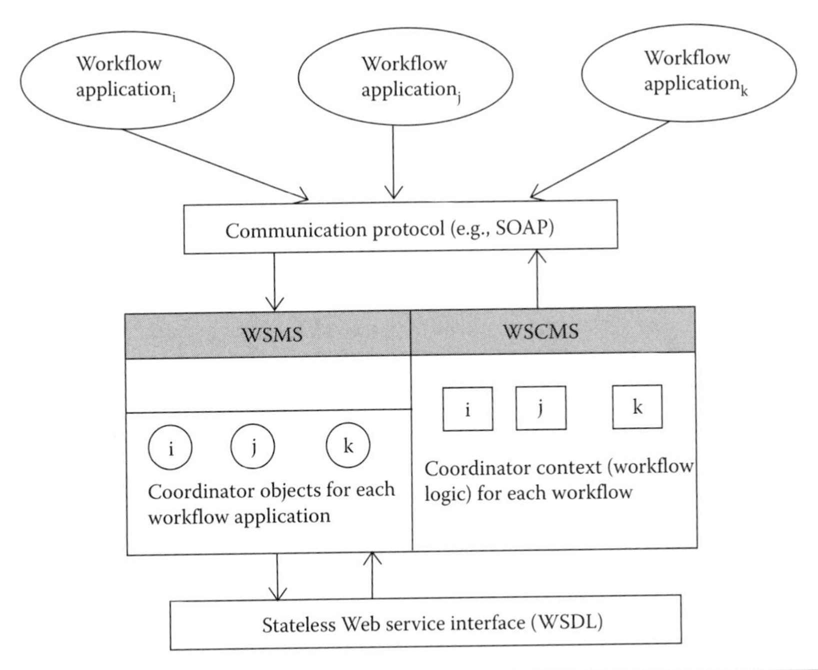

Figure 11.9 Web service coordination middleware architecture.

- *Stateful view*: The state/instance handler instantiates a coordinator object on the basis of the WSDL description for each such application. The coordinator object has a binding to the original Web service method calls. Moreover, each coordinator object has a corresponding status context stored in the persistent storage. Web service method invocations go through the coordinator object. Each method invocation has a series of steps including enforcing dependencies and updating state information.
- *Message handling*: The message handler of the WSMS handles the inter-Web service communication and keeps the state information of interactions. On arrival of a message, the communication server (SOAP server) passes it to the message handler. The message header conates a unique identification for each message (ConvID). ConvID consists of a reference to the application, the method being invoked, parameter set, status of the tags of the invocation such as "Ready," "Commit" in transaction processing. On the basis of this information, the message handler resolves the message and takes appropriate actions.

11.5.2.2 Web Service Coordination Management System

The WSCMS keeps the coordination (dependency) information (coordination context) for each application and enforces dependencies. As coordination and dependency enforcement is local to each participating Web service, the WSCMS maintains a coordination context for each application locally to reflect dependencies. The WSCMS supports two types of dependencies: premethod execution dependencies and postmethod execution dependencies.

In Web service-based workflow applications, individual Web services represent a particular workflow activity performed by invoking Web service method calls. Workflow dependencies need to be associated with Web service method invocations. Typically, workflow activities enforce two types of dependencies. Before initiating the activity (triggered by the workflow engine), it needs to make sure that all the dependencies (including data, control, and resource) have been satisfied. If not, the activity waits until it receives all the control and data items or it can start fulfilling these requirements. These kinds of dependencies can be characterized as "pre-execution" dependencies. Other types of dependencies arise once workflow activity is completed. On completion of the activity, control/data may need to be passed to other entities in the workflow on the basis of workflow-specific constraints. These kinds of dependencies are characterized as "post–execution" dependencies.

- *Pre-Execution Dependencies* (*Join Dependency*): Pre-execution dependency for workflow j, defined over the method m_i of Web service w_i with parameter set k can be represented as $J_j{\cdot}w_i{\cdot}m_i(\text{parameter}_k) = \{D, \text{constraints}\}$, where D is the set of destination methods and constraints are workflow constraints such as AND-join and Sync-Merge.

  WSCMS ensures that join dependencies are met before making the Web service method call. Series of events take place in both the local and the destination WSCMSs while enforcing join dependencies. Figure 11.10 illustrates the interaction among WSCMM components while enforcing join-dependency constraints.

Message handlers maintain an inbox and outbox for each workflow application. Both the inbox and outbox have entries for each join-dependency point. When it receives control/data from destination entities, the message handler directs them to the appropriate inbox. Once the activity receives trigger (control) to perform the method call (step 0), it sends a message to the WSCMS for dependency check (steps 1, 2, and 3). If all the dependencies are met, the Web service methods get invoked and state information is updated (step 5). Otherwise, the WSCMS sends messages to all the remaining destination entities for dependency check (step 4). Dependency check performs two operations. First, it requests state information from the state handler of the destination Web service related to this particular application join-point. If status information is available, a response is sent. Otherwise, it tries to invoke the remote method and send the response to the requester Web service. This invocation requires a similar dependency check.

- *Post-Execution Dependencies* (*Split Dependency*): Split dependency for workflow j, method m_i of Web service w_i with parameter set k can be represented as $S_j{\cdot}w_i{\cdot}m_i$ (parameter$_k$) = {D, constraints}, where D is the set of destination methods and constraints are workflow constraints such as AND-Split and XOR-Split.

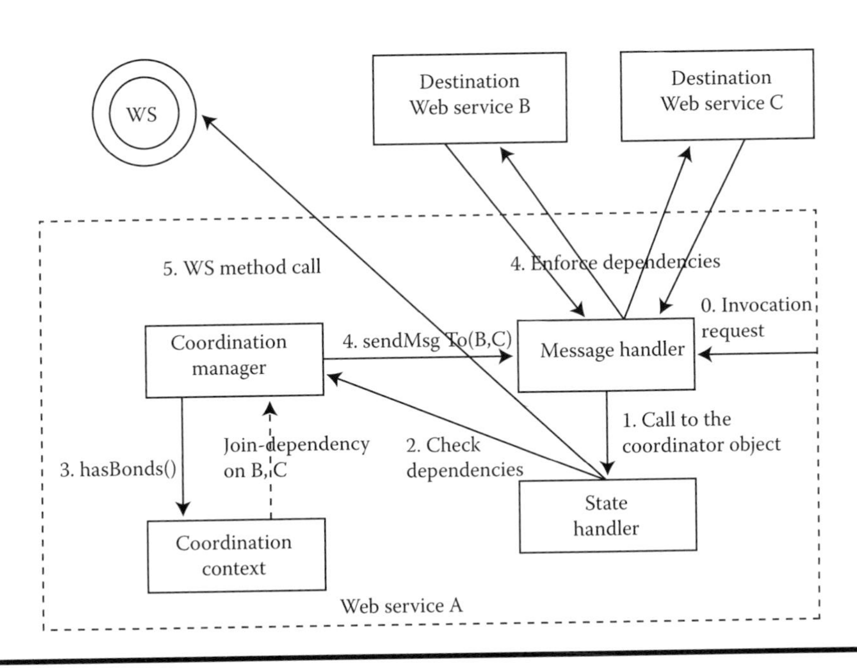

Figure 11.10 Enforcing pre-execution dependencies.

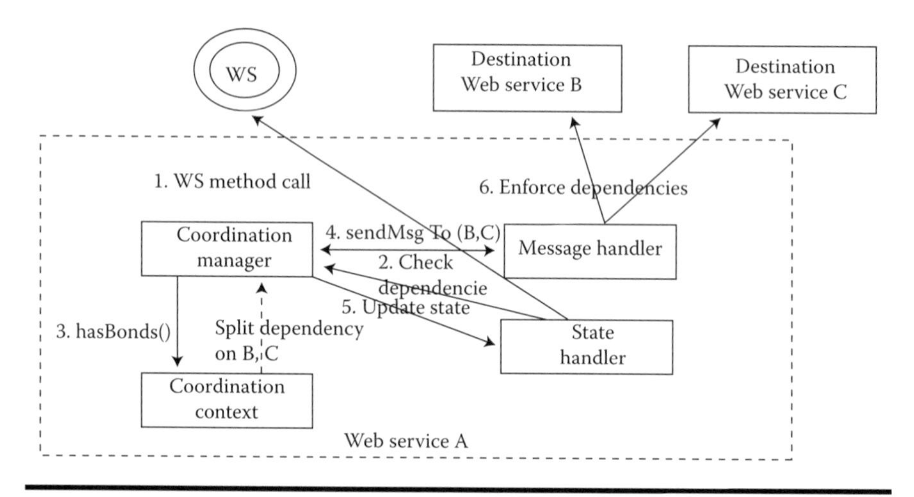

Figure 11.11 Enforcing post-execution dependencies.

Enforcing split dependencies requires a Web service to trigger a set of remote Web services depending on the workflow constraints specified for the split-point. Figure 11.11 illustrates the interaction among WSCMM components while enforcing split-dependency constraints. As shown in Figure 11.11, the WSCMS requests the message handler to send data/control to remote Web services according to the workflow split criteria. The message handler places remote invocations to the outbox (dispatcher) and triggers remote Web service methods (step 6). At the same time, the WSCMS updates state information (step 5).

11.5.3 Simulation and Verification of WSCMM Architecture

We have used the Discrete Event System Specification (DEVS) modeling tool to simulate our WSCMM and verify various dependency modeling scenarios. DEVS provides a formal framework that facilitates simulation and verification of distributed systems. DEVS is derived from a mathematical dynamical system theory [32].

It supports hierarchical modular composition and object-oriented implementation. There are two primary modules: atomic model and coupled model. One can combine these models to specify complex simulations. Atomic models have input events, output events, state variables, state transition functions, external transition, internal transition, time advance function, computing function, and transitions. The current state can be specified using state variables and input and output functions are computed on the basis of the current state and the computing function. The coupled model has components, interconnections, internal couplings, external input couplings, and external output couplings. The main purpose of the simulation is to verify the correctness of our middleware and identify design issues. In order to do that, we simulate the interactions among components of the middleware for different incoming messages including pre- and postmethod invocation dependencies. Figure 11.12 shows our simulation model for the middleware. It consists of several modules: message handler (msgHandler), Web service coordination management system (wsCoMys), Web service management system (wsMgtSys), and a Web service component. Here, we briefly describe the simulation of the message handler module. Simulation details of other modules and sample workflows can be found in Reference [33].

11.5.3.1 Message Handler Simulation

The message handler consists of two components: two incoming ports to receive messages and three outgoing ports to send messages. The message receiver (mrec) receives messages from remote services (Figure 11.12).

On receiving the message, it places the message in a FIFO queue. Then, mrec passes messages to the message resolver (mres). The message resolver's job is to identify the type of message (Table 11.1). On the basis of the message type, the message

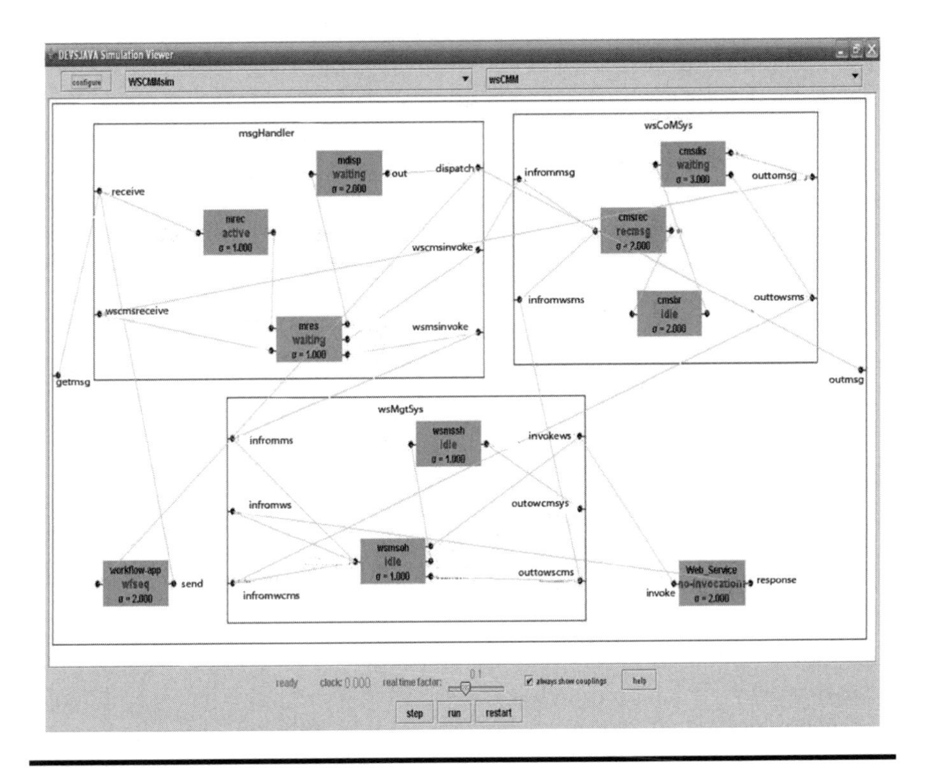

Figure 11.12 Simulation model.

Table 11.1 Message Tag and the Outgoing Message Ports at the Message Handler

Tag	*Outgoing Port*
0—Method invocation	Send the message to WSMS through "outwsms" port
1—Data/control from subscription bonds	Send the message to WSCMS through "outwscms" port
2—Enforce dependency (method invocation), negotiation bond	Send the message to WSMS through "outwsms" port
6—Enforce post-method execution dependencies (data/control through outgoing subscription bonds)	Send the message to dispatcher through "outdispatcher" port

is directed to the appropriate component. For our simulation, we have used the following message format:

Workflowid:fromwebservice:method:parameterset:tag

In this message format, the first portion is to identify the workflow because any Web service can participate in different workflows at a given time. The second portion is to identify the message sender. The third and fourth portions contain method details and parameters. Finally, the tag is to identify the type of message. For example, suppose Web service wl receives the message, wf1:ws2:m2:p2:0. This means that the message belongs to workflow 1. The sender is Web service 2 and the tag is 0. Tag 0 means the message is a method invocation. In this case, invocation of method m2 with a parameter set p2. Once, the resolver receives this type of message, it resolves the message using the tag and directs it to the appropriate output port. Table 11.1 shows the relationship between a tag and the outgoing message port. Our simulation results show that the Web bond-based realization of the WSCMM behaves correctly while enforcing workflow control flow dependencies.

11.6 Prototype Implementation and System Evaluation: The BondFlow System [31]

On the basis of the idea of WSCMM, the BondFlow system has been developed as a platform to configure and execute workflows over Web services (Figure 11.13). The workflow configuration module consists of a Web service interface module, CPO generator module, and workflow configuration module. The workflow execution module consists of a Web bond manager, communication layer, and JVM runtime. Implementation details can be found in Reference [31].

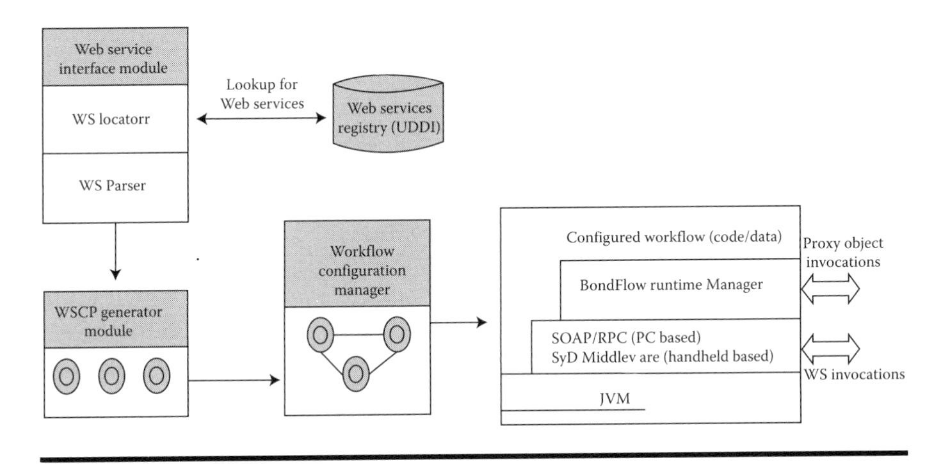

Figure 11.13 BondFlow system.

11.6.1 BondFlow System

The BondFlow system can be used to deploy workflows in both wired and wireless devices. Handheld devices can act as service providers hosting Web services or clients requesting services deployed in other wired/wireless devices. The main challenge here is to provide a framework that hides communication and device heterogeneity of the ubiquitous environment. Previously, we have designed and developed the SyD middleware platform that addresses the key problems of heterogeneity of device, data format and network, and mobility [34]. The BondFlow system uses SyD middleware as the underlying software infrastructure for handheld-based communication and coordination.

Workflow applications have been executed on HP's iPAQ models 3600 and 3700, with 32 and 64 MB storage, respectively, running Windows CE. There are two possible deployment strategies. First, the entire workflow can reside in a single wireless device. In this case, communication among coordinator objects is via local in-memory calls. Actual Web service call is made using SOAP (kSOAP). Second, the workflow can be distributed among several iPAQs (Figure 11.14). This scenario is important in cases where some portions of the workflow can be monitored and executed by a selected set of users on specific devices and/or with specific security settings.

In this case, coordinator objects need to communicate using a remote messaging system to enforce dependences. The SyD middleware has a listener component that facilitates device-to-device communication. The SyDListener is a lightweight module in our SyD middleware framework for enabling mobile devices to host server objects. It can be considered as a tiny SOAP server for handheld devices. The SyDListener enables handheld devices to communicate among applications deployed on other peer devices (Figure 11.14).

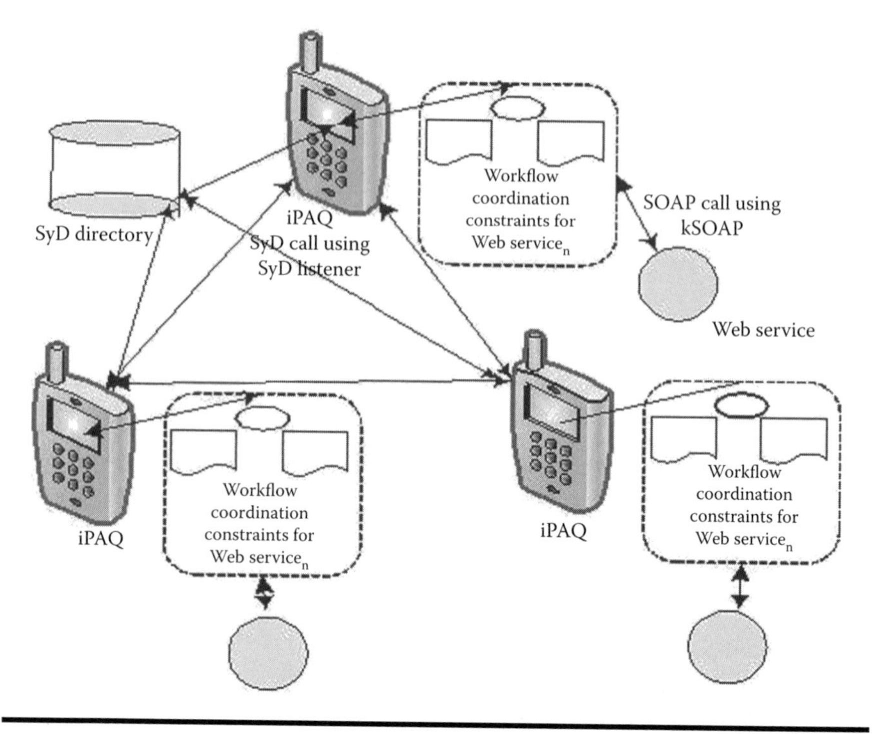

Figure 11.14 Workflow distributed among several iPAQs.

11.6.2 System Evaluation

11.6.2.1 Hardware Software Setup

We ran our experiments on a high-performance SunOS 5.8 server. We built wrappers using JDK 1.4.2. The WSDL parser was built using WSDL4J API. WSLD4J API is an IBM reference implementation of the JSR-110 specification (JavaAPI's for WSDL). NanoXML 2.2.1 was used as the XMLparser for JAVA. Various publicly available Web services including Xmethod's SOAP-based Web services (http://www.xmethods.net/) were used for our experiments. For the wireless device experiments, we used HP's iPAQ models 3600 and 3700 with 32 and 64 MB storage, respectively, running Windows CE/Pocket PC OS interconnected through IEEE 802.11 adapter cards and a 11 MB/s wireless LAN. Jeode EVM personal Java 1.2 compatible was used as the Java Virtual Machine.

11.6.2.2 System Performance Details

We deployed and executed case study workflows including the purchase order workflow on both wired and wireless infrastructure. Table 11.2 shows the workflow execution timings for the two case study workflows for both wired and wireless set-

Table 11.2 Workflow Execution Timings

Workflow	*Total Execution Time (ms)*	*BondFlow Related Timings (ms)*	*BondFlow Related (%) Computation*
Purchase order:	7820	1048	13.4
No. NB = 4, no. SB = 9			
Online book purchase:			
No. NB = 5, no. SB = 6	2483	102	4.1

tings. Bond-related times for both workflows are approximately ~10% of the time without the BondFlow system. The bond-related time accounts for times taken to check workflow dependencies in bond repository and initiate appropriate method calls on remote Web services (coordinator objects). Table 11.3 shows the footprints of two workflows. The coordinator objects and corresponding bond repositories account for ~25% and ~75%, respectively. The footprint of the proxy object is small (~10 KB) and typically increases by 0.3 KB for was additional operation (method) of the Web service. Intermediate system-generated files are less than 100 KB for a sufficiently large workflow. Typically, the footprint of the bond repository increases 0.3 KB for each additional bond. Thus, within a very small amount of additional storage for the proxy objects, we were able to get substantial gains in the speed of the workflow.

11.7 Discussion and Challenges for Future Research

11.7.1 Quality of Service

An important aspect of distributed coordination is enforcing quality-of-service (QoS) requirements. It is imperative to investigate how to enforce QoS requirements such as service guarantees, response time, and security levels. One can investigate how Web bonds, mentioned in this chapter, can be extended to enforce such QoS requirements [35].

Table 11.3 Footprint of the Workflow

Workflow	*Bond Repository (KB)*	*Proxy Objects (KB)*	*Total Workflow (KB)*
Purchase order	7.10	25.4	32.5
Online book purchase	5.82	19.8	25.62

11.7.2 Disconnections (a Proxy Architecture)

Disconnections and bandwidth issues are still hindrances to handheld-based computing. One of the research challenges is to provide solutions that mitigate the effects of frequent disconnections and limited bandwidth availability. Frequent disconnections and limited bandwidth in the wireless network have made the proxy an important concept to enhance the fault tolerance for mobile devices. The whole purpose of the proxy is to carry out operations even if the mobile device gets disconnected in the middle of the transaction. Much work needs to be done in this area [36,37].

11.7.3 Service Composition in the Cloud

On-demand computing is currently the buzz in the world of distributed computing. The concept of cloud computing further strengthens the feasibility of such on-demand computing. Cloud computing provides resource and service needed to perform computation, and it provides a framework that facilitates, self-healing, self-monitoring, resource registration and discovery, and dynamic reconfiguration [38].

Figure 11.15 shows a generic cloud computing infrastructure. Here, virtualization and cloud resource management is the middleware layer that presents cloud resources to users. A resource can be anything from network computing resources to structured/unstructured data. Resources can be added, removed, or modified in the cloud. Thus, compared to the traditional software architecture, data integration in the cloud architectural design is agile and highly configurable for two reasons: (i) new services can be added into the system without stopping the application, and (ii) new applications can be composed from the existing services by altering the application specification without changing the basic underlying framework. One of the major challenges

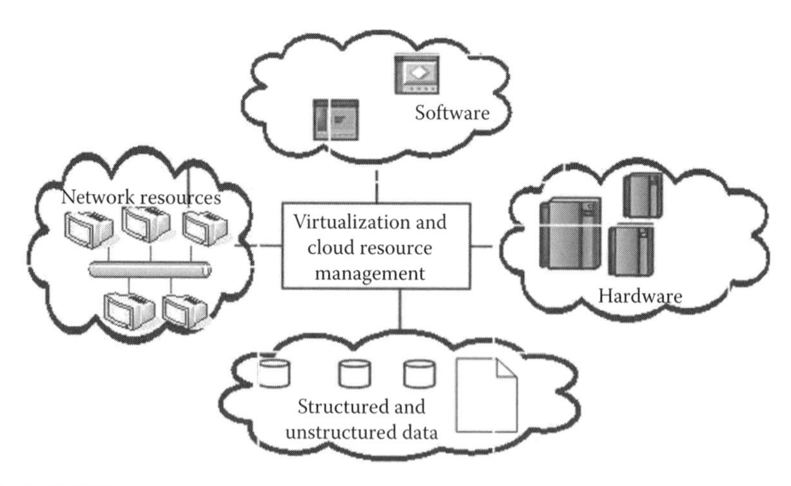

Figure 11.15 Cloud computing: A bigger picture.

in cloud computing is to provide versatile middleware components that present resources to the cloud including a dependable and fully functional service broker that provides reliable and efficient service hosting and discovery [38,39].

11.7.4 Web Service-Based Data and Tool Integration in Bioinformatics

Biological data and tool integration is one of the emerging research areas where Web services will play a significant role. These data sources are heterogeneous (data, technology) in nature and Web service infrastructure is an ideal platform to hide this heterogeneity. Among different approaches of data and tool integration, Web services provide better interoperability and necessary scalability. Many efforts are already underway to convert these tools and data sources into Web services. Typically, noncomputer scientists would prefer to compose their workflows (for any application in that matter) easily. Thus, the Web is a very attractive environment for them. Extending the BondFlow system as a Web-based tool to configure and execute biological (scientific) workflows is a worthwhile research effort [40].

DEFINITIONS

Distributed coordination is the process in which the coordination context is managed by more than one participating entity.

Middleware is a software component that provides specified services such as transaction management and data and process integration that facilitates the interaction between multiple applications on one hand and multiple databases/software platforms on the other.

Peer-to-peer communication is the exchange of messages among different applications (parties) where each party is treated equal and any party can initiate a message.

Web service choreography is the process of exchanging "observed sequence of messages" by Web services as peers when performing a unit of work.

Web service composition is aggregating two or more Web services to accomplish a particular task. The composite application itself can be another Web service.

Web service coordination is the process of exchanging messages among Web services to accomplish a particular task such as a workflow with "shared context." This context can be managed by a third-party application other than participating Web services.

Workflow refers to "the automation of a business process, in whole or part, during which documents, information or tasks are passed from one participant* to another for action, according to a set of procedural rules. *participant = resource (human or machine)" (http://www.e-workflow.org/).

Workflow management system (WFMS) is a software system that automates workflows by enforcing dependencies among workflow participants.

QUESTIONS AND SAMPLE ANSWERS

Q1: Name and describe three Web service-based application domains.

A1: (i) *E-commerce applications* that include supply chain, workflows, and virtual organizations; (ii) *biomedical applications* that include biomedical data and tool integration; and (iii) *personal applications* that include travel, calendar, and scheduling.

Q2: What is a workflow application?

A2: In a workflow application several entities interact with each other in a desired order to accomplish a specific task such as supply chain management. Order of execution of activities is determined by workflow control flow logic where sequential order being the simplest. In a sequential workflow, workflow activities are executed in a liner fashion.

Q3: Name and describe two major issues in centralized workflow coordination.

A3: (i) *Less secure*: All message communications go through a single point of coordination. It is difficult to enforce different security levels for each participating Web service messages as it may require different levels of security. For example, the Department of Defense Web service may require a higher level of security attached to incoming messages but, with a centralized coordination, it is not easy to enforce such security requirements. (ii) *Fault-tolerance*, *performance*, and *scalability* issues.

Q4: Compare and contrast the idea of Web service coordination management middleware (WSCMM) and evolution in database management application development.

A4: As we have discussed in Section 11.4, WSCMM proposes an infrastructure change in Web service implementation where Web services become context-aware computing entities. The Web service management system enhances the Web services infrastructure to manage methods and method invocations more effectively, akin to database management systems in databases, whereas the Web services coordination management system transfers generic functional layers to the Web service so that they become entities capable of enforcing distributed coordination, akin to the workflow management system in databases.

Q5: In your words, explain the functionality of the coordinator proxy object described in this chapter. How does the coordinator proxy object help to handle disconnections in the wireless environment?

A5: According to the discussion in this chapter, the coordinator proxy object helps to enforce workflow dependencies and perform actual Web service method invocation. However, it does not store any other application-related data. The coordinator object can be extended to store state information such as data in a transaction for long-lived interactions (duration of such interactions can be several minutes to few weeks). Tentative commitments can be made using intermediate data available locally while disconnected.

Q6: Briefly explain the functionality of Web service coordination management system (WSCMS) and Web service management system (WSMS) components.

A6: *WSMS*: The core functionality of the WSMS is to transform the stateless Web service into a state-preserving self-coordinating entity. WSMS performs this transformation by generating a coordinator object to represent the Web service, which encompasses all the coordination capabilities of the WSCMS implementation.

WSCMS: WSCMS keeps the coordination (dependency) information (coordination context) for each application as particular Web service participates and enforces dependencies. As coordination and dependency enforcement is local to each participating Web service, WSCMS maintains coordination context for each application locally to reflect dependencies.

Q7: Read Reference [6] and discuss how SyD middleware platform hides device heterogeneity.

A7: In the SyD middleware architecture, each device is managed by a SyD deviceware that encapsulates it to present a uniform and persistent object view of the device by hiding system and communication heterogeneity [6].

Q8: Explain the two-layered workflow software architecture presented in this chapter. How does the two-layered workflow software architecture lead to lightweight workflow application?

A8: In the two-layered workflow software architecture, the stateless Web services are empowered by coordinator proxy objects into self-coordinating stateful entities. Next, a workflow dependency modeling technique can be used to interlink the coordinator proxy objects, capturing the workflow logic. Thus, coordination-related information can be separated from the actual workflow process and the workflow application is lightweight. As the coordinator proxy object and its dependencies can be separated the lightweight workflow application can be developed.

References

1. Papazoglou, M.P., Traverso, P., Dustdar, S., Leymann, F., Service-oriented computing: State of the art and research challenges. *IEEE Computer*, 40(11), 2007, 38–45.
2. Di Nitto, E., Sassen, A.-M., Traverso, P., and Zwegers, A., *At Your Service, Service-Oriented Computing from an EU Perspective*, MIT Press, Cambridge, Massachusetts, 2009, ISBN-10: 0-262-04253-3.
3. Ko, I.-Y. and Neches, R., Composing web services for large-scale tasks, *IEEE Internet Computing*, 7(5), 2003, 52–59.
4. Mnaouer, A. B., Shekhar, A., and Yi-Liang, Z., A generic framework for rapid application development of mobile web services with dynamic workflow management, In *Proceedings of IEEE SCC 2004*, pp. 165–171.
5. Ranganathan, A. and McFaddin, S., Using workflows to coordinate web services in pervasive computing environments, In *Proceedings of the IEEE International Conference*

on Web Services (ICWS'04), San Diego, California, USA June 6–9, 2004. IEEE Computer Society, 2004, pp. 288–295.

6. Prasad, S. K., Madisetti, V., Navathe, S. B., Sunderraman R., Dogdu, E., Bourgeois, A., Weeks, M. et al. System on Mobile Devices (SyD): A middleware testbed for collaborative applications over small heterogeneous devices and data stores, In *Proceedings of ACM/IFIP/USENIX 5th International Middleware Conference*, Toronto, Ontario, Canada, October 18–22, 2004, pp. 352–371.
7. R. Steele. A Web services-based system for ad hoc mobile application integration, *IEEE International conference on Information Technology: Coding and Computing [Computers and Communications], 2003. Proceedings. ITCC 2003.* pp. 248–252, April 28–30, 2003.
8. Sinderson, E., Magapu, V., and Mak, R., Portal of NASAs Mars exploration Rovers mission-middleware and Web services for the collaborative information, Invited paper, In *Proceedings of ACM/IFIP/USENIX 5th International Middleware Conference*, Toronto, Ontario, Canada, October 18–22, 2004, pp. 1–17.
9. Indrakanti, S., Varadharajan, V., and Hitchens, M., Authorization service for web services and its application in a health care domain, *International Journal of Web Services Research*, 2(4), 2005, 94–119.
10. Benatallah, B., Dumas, M., and Sheng, Q. Z., Facilitating the rapid development and scalable orchestration of composite web services, *Distributed and Parallel Databases* 17(1), 2005, 5–37.
11. Prasad, S. K. and Balasoorya, J., Fundamental capabilities of Web coordination bonds: Modeling petri nets and expressing workflow and communication patterns over Web services, In *Proceedings of Hawaii International Conference on System Sciences (HICSS-38)*, Big Island, Hawaii, January 4–8, 2005.
12. van der Aals, W. M. P., The application of petri nets to workflow management, *Journal of Circuits, Systems and Computers*, 8(1), 1998, 21–66.
13. Workflow Patterns Initiative, Queensland University of Technology, Australia, http://www.workflowpatterns.com/evaluations/standard/index.php, Last accessed on November 13, 2011.
14. Chafle, G., Chandra, S., Mann, V., and Nanda, M. G., Decentralized orchestration of composite Web services, In *Proceedings of the Alternate Track on Web Services at the 13th International World Wide Web Conference (WWW 2004)*, pp. 134–143, New York, May 2004.
15. Hawryszkiewycz, I. and Steele, R., Extending collaboration to mobile environments, In *Proceedings of the International Conference on Web Technologies, Applications and Services*, pp. 77–82, Calgary, Canada, July 4–6, 2005.
16. Jørstad, I., Dustdar, S., and van Do, T., Service-oriented architectures and mobile services. In *Proceedings of 3rd International Workshop on Ubiquitous Mobile Information and Collaboration Systems (UMICS), co-located with CAiSE 2005*, pp. 617–631, Portugal, June 13–14, 2005.
17. Dustdar, S., Gall, H., and Schmidt, R., Web services for groupware in distributed and mobile collaboration, In *Proceedings 12th Euromicro Conference*, pp. 241–247, Feb. 11–13, 2004.
18. Schuler, C., Weber, R., Schuldt H., and Schek, H.-J., Scalable peer-to-peer process management the OSIRIS approach, In *Proceedings of IEEE International Conference*, pp. 26–34, July 6–9, 2004.

19. Hackmann, G., Haitjema, M., Gill, C. D., and Roman, G.-C., Sliver: A BPEL workflow process execution engine for mobile devices. In A. Dan and W. Lamersdorf, editors, *Proceedings of ICSOC, Lecture Notes in Computer Science*, Springer, Berlin/Heidelberg, pp. 503–508, 2006.
20. Schmidt, R., Web services based execution of business rules. In *Proceedings of the International Workshop on Rule Markup Languages for Business Rules on the Semantic Web*, Sardinia (Italy) pp. 46–50, June 4, 2002.
21. Rosenberg, F. and Dustdar, S., Towards a distributed service-oriented business rules system, In *Proceedings of the IEEE European Conference on Web Services (ECOWS)*, 11pp., November 14–16, 2005.
22. Czajkowski, K., Ferguson, D. F., Foster, L., Frey, J., Graham, S., Sedukhin, I., Snelling, D., Tuecke, S., and Vambenepe, W., The WS-resource framework, http://www-106.ibm.com/developerworks /library/ws-resource/ws-wsrf.pdf, 2004.
23. Humphrey, M. et al. State and events for Web services: A comparison of five WS-resource framework and WS-notification implementations. In *Proceedings of the 14th IEEE International Symposium on High Performance Distributed Computing (HPDC-14)*, Research Triangle Park, North Carolina, July 24–27, 2005.
24. Tran, D. T., Hoang, N. H., and Choi, E., The WORKGLOW system in P2P-based Web service orchestration, In *Proceedings of International Conference on Convergence Information Technology (ICCIT 2007)*, pp. 2312–2317, November 21–23, 2007.
25. Barros, A., Dumas, M., and Oaks, P., Standards for Web service choreography and orchestration: Status and perspectives. In *Proceedings of the Workshop on Web Services Choreography and Orchestration for Business Process Management*, LNCS, pp. 61–74, Nancy, France, September 2005.
26. Benatallah, B., Casati, F., Grigori, D., MotahariNezhad, H., and Toumani, F., Developing adapters for Web services integration. In *Proceedings of CAiSE 2005*, pp. 415–429, Porto, Portugal, June 2005.
27. Wan, F. and Singh, M., Enabling persistent Web services via commitments, *Information Technology and Management*, 6(1), 2005, 41–60.
28. Jørstad, I., Dustdar, S., and van Do, T., A service oriented architecture framework for collaborative services, In *Proceedings of 3rd International Workshop on Distributed and Mobile Collaboration (DMC), IEEE WETICE*, Sweden, June 13–15, 2005.
29. Schmit, B. A. and Dustdar, S., Towards transactional Web services. In *Proceeding of the 1st IEEE International Workshop on Service-oriented Solutions for Cooperative Organizations (SoS4CO '05), co-located with the 7th International IEEE Conference on E-Commerce Technology (CEC 2005)*, Munich, Germany, July 19, 2005.
30. Tai, S., Khalaf, R., and Mikalsen, T., Composition of coordinated Web services, In *Proceedings of the ACM/IFIP/USENIX International Conference on Distributed Systems Platforms (Middleware 2004)*, Vol. 78, no. 5, pp. 294–310, Toronto, Canada, October 2004.
31. Balasooriya, J., Padhye, M., Prasad, S. K., and Navathe, S. B., BondFlow: A system for distributed coordination of workflows over Web services, In *Proceedings of the 14th HCW, in conjunction with IPDPS'05*, pp. 121–128, Denver, Colorado, April 4–8, 2005.
32. DEVS simulator, http://www.acims.arizona.edu/SOFTWARE. Last accessed November 13, 2011.
33. Balasooriya, J., Prasad, S. K., and Navathe, S. B., A middleware architecture for enhancing Web services infrastructure for distributed coordination of workflows, In

Proceedings of 2008 IEEE International Conference on Services Computing (SCC 2008), Honolulu, Hawaii, July 8–11, 2008.
34. Balasooriya, J., Joshi, J., Prasad, S. K., and Navathe, S. Distributed coordination of workflows over Web services and their handheld-based execution, In *Proceedings of ICDCN 2008*, Kolkata, India, January 5–8, 2008; *Lecture Notes in Computer Science*, Vol. 4904, Springer, Berlin/Heidelberg, 2008, pp. 39–53.
35. Ben Halima, R., Drira, K., and Jmaie, M., A QoS-oriented reconfigurable middleware for self-healing web services, In *Proceedings of ICWS 2008*, pp. 104–111, September 23–26, 2008.
36. Lee, C., Helal, S., and Nordstedt, D., The Jini proxy architecture for impromptu mobile services, In *Proceedings of the International Symposium on Applications and the Internet Workshops (SAINT 2006 Workshops)*, pp. 113–117, January 23–27, 2006.
37. Aijaz, F., Adel, S. M., and Walke, B., Middleware for communication and deployment of time independent mobile Web services, In *Proceedings of ICWS 2008*, pp.797–800, September 23–26, 2008.
38. Ramakrishnan, R., An overview of cloud computing at Yahoo!, http://us.apachecon.com/page_attachments/0000/0194/ApacheCon-09cloud.ppt. Last accessed July 2008.
39. IBM whitepaper, Seeding the clouds: Key infrastructure elements for cloud computing, February 2009, ftp://ftp.software.ibm.com/common/ssi/sa/wh/n/oiw03022usen/OIW03022USEN.PDF. Last accessed November 13, 2011.
40. Gonzalez, G. and Balasooriya, J., Web service orchestration for bioinformatics systems: Challenges and current workflow definition approaches. In *Proceedings of the 2007 IEEE International Conference on Web Services (ICWS'07)*, pp. 1226–1227, July 2007.

Chapter 12

Development and Implementation of Mobile Services in Mobile Platforms

Yingbing Yu, Bin Xie, and Sanjuli Agarwal

Contents

12.1 Introduction

Mobile devices such as personal digital assistants (PDAs) and smart phones allow people to access a broad range of applications for business and personal use in a mobile environment. People can use mobile devices to access the Internet for sending and receiving emails and text messages, browsing the Web pages, searching the information of interest, and obtaining semantic Web services anywhere and anytime. A person travelling on business can be expected to instantly initiate a conference call by using a mobile phone and present his/her products by sequential slides in a synchronous way to several customers. The mobile phone is limited in physical storage such that the slides can be stored in an enterprise server. By accessing an online social network from the mobile Safari Web browser, a person can know where his/her friends are by searching the location-tagged content on the social network for possible meetings with them. Recently, microblogging has become a popular way for people to share thoughts, activities, and other information through user-generated content on the social network. In these applications, the content presentation of Web applications could be optimized for the mobile Safari Web browser that addresses the mobile device hardware constraints (e.g., small screens, dialog windows). The mobile device operates with specific operating systems to run the Web application and other mobile services. These operating systems are generally different from those used by laptops or desktops. Compared to laptops and desktops, mobile devices such as BlackBerry and iPhone are small handheld devices that have limited power, limited computational capacity, small storage, small screen, and small keyboard for text input. As a result, the operating system for these devices should be optimized to address these constraints. In addition, the wireless network interface only has limited network throughput for Internet connectivity. The messages from mobile devices to network access points should be made compact for saving the network throughput and consuming less power. Owing to mobility, the connection from mobile devices to the Internet should be smoothly maintained with the ability to migrate from one access point to another. For each operating system for mobile devices, the corresponding development tools are fundamental to providing the framework for the development of all kinds of mobile service applications. This also enables the development of the third-part mobile applications based on the operating systems.

There are many operating systems and development tools in the market for developing mobile applications. The Symbian operating system [1] from Nokia currently is a dominant mobile platform in terms of its market share. This can be attributed to the popularity of Nokia cell phones and smart phones in both the 2.5G and 3G networks. The Windows Mobile [2] platform has a familiar integrated development environment (IDE) for Windows developers and this has proven to attract especially new developers in the field. The Palm operating system [3] was introduced early in 1996, with the first mobile device from Palm, Inc. The Palm platform provided mobile devices with essential business tools as well as the ability to access the Internet via a wireless connection in the early age of mobile devices. Unfortunately, at present, Palm has lost its position as one of the leading companies for mobile devices and thus few people develop applications with the Palm platform. The iPhone operating system [4] is the platform for perhaps the most popular smart phones and PDAs (iPhone and iPod Touch) from Apple, Inc. The iPhone operating system is the only available development platform for developing applications running in their products. The BlackBerry operating system [5] is also a successful platform since the company Research in Motion (RIM) introduced the popular smart phone to support the features of sending and receiving emails over the 2.5G and now the 3G wireless cellular networks. Similar to the iPhone operating system, it is the only option to develop Internet applications running in BlackBerry phones. With the support of the open-source community and strong promotion from several influential companies in the IT industry, embedded Linux is proven as an alternative for mobile platform development even for higher-end devices with powerful processors and large memory.

Java ME [6] aims to allow developers to write programs and run them in different CPUs without further modifications. Unfortunately, this is still a big challenge for Java ME even though a set of specifications and standards have been developed for the purpose. Google announced the open-source platform Android [7] in 2008 and several companies are already beginning to develop mobile devices using the Linux-based operating system. BREW (binary runtime environment for wireless) [8] is the mobile application development platform launched by Qualcomm in 2001. It provides a highly efficient, low-cost, scalable application execution environment to develop applications on all Qualcomm devices.

In this chapter, we investigate these mobile development operating systems as well as their development tools. We provide an understanding of the popular mobile platforms and illustrate some example mobile applications. The remainder of this chapter is structured as follows. Section 12.2 illustrates the Symbian platforms. Subsequently, we discuss other platforms from Sections 12.3 through 12.9. Finally, concluding remarks are included in Section 12.10.

12.2 Symbian Operating System

The Symbian operating system is the mobile operating system jointly developed by Nokia, Sony Ericsson, Panasonic, Samsung, Motorola, and Siemens. Nokia acquired the full software in 2008 and now the highest version is Symbian 9.2. In a high level, a Symbian mobile phone consists of the following hardware: 32-bit processor CPU, ROM, RAM, input/output and power equipments. ROM is non-erasable storage and comes with the various functions of Symbian systems. RAM is used to store the data for those active procedures, as well as temporary swap files, or cache files. Input/output control equipments are devices such as keyboards, touch screen, memory card expansion, Bluetooth, and so on. Symbian has the advantage of the largest market share with the support of the majority of handheld equipment vendors. It has become an open, easy-to-use, and professional development platform for smart phone applications.

With the promotion from Nokia, Symbian has become the most popular mobile operating system with the largest number of developers. A large number of third-party applications and tools are also available. Symbian wireless communication devices not only provide voice communication but also have other types of input such as stylus, keyboard, and so on. The Symbian operating system kernel is an objected-oriented system that supports enterprise communications using standard transfer protocols. The software has a number of functions, including sharing information, browsing the Web, email system, fax and personal information management. The Symbian operating system can support many types of file formats such as Microsoft Word, Excel, and PowerPoint documents and emails. Symbian office and multimedia features are constantly improved, from the initial mono output support to Bluetooth stereo support.

12.2.1 Symbian Operating System Versions

The Symbian operating system itself has three versions, namely, Pearl, Quartz, and Crystal. These versions correspond to the applications for cell phones, smart phones, and handheld PC. Nokia offers three different types of user interfaces (UIs) (i.e., Series 60/80/90) and these allow mobile phone manufacturers to have more choices in developing the applications for Symbian mobile phones. Series 60 is almost dedicated for mobile phones with a digital keyboard. Series 80 is designed for the phone with a full keyboard support. Series 90 is designed to support the operations of the stylus. In addition to these three mainstream platforms, Series 20/30 is mainly designed for low-end mobile phones and Series 40 mostly for mobile phones to support the Java extension. Series 60/80/90 is designed for high-end smart phones and those mobile phones designed for business. Series 60 (S60) is the favorite among many manufacturers and it has further evolved with five versions over the years. Its main functions include (i) personal information management such as calendar, phone book, photo album, directory, document management application for the synchronization with PC, and so on, (ii) emails and short messaging service (SMS), (iii) multimedia such as camera, Image Viewer, RealOne Player, multimedia applications, and so on, (iv) document management that allows users to create, delete, and move files/directories, (v) PC connectivity with USB and Bluetooth, and (vi) various applications such as HTTP advanced MMS services, SMIL application, sound recording, Macromedia Flash, music player, and others.

12.2.2 Symbian Software Development Kit

To develop mobile applications with the use of the Symbian operating system, the first step is to get the Software Development Kits (SDKs). The Symbian operating system SDKs support development on the basis of Java and C++. The SDK provides a set of binary files and tools to build and develop applications for the Symbian operating system. Further, the SDK cell phone emulator contains sample documentation and applications for developers. As the Symbian operating system does not include the UI, a number of partners for the Symbian operating system vendors offer specific UI platforms for development. Thus, SDKs can be integrated on a specific UI platform to build applications. A UI platform provides a unique UI, a series of variety tasks related to system applications. These common tasks usually include sending and receiving messages, browsing, telephony features, multimedia capabilities, and contact/calendar management. These applications typically use a number of applications provided by the Symbian operating system. Some Symbian SDKs are: UIQ, Nokia Series 90, Nokia Series 80, Nokia Series 60, and Nokia Series 40. SDK can be generally divided into two categories: Nokia Series and UIQ. Each platform corresponds to a series of specific development tools and the developer needs to choose the tools according to their own circumstances. For the Nokia Series, the most popular one is the Series 60 platform that is based on Symbian operating system technology and structure, including UI, a variety of applications, and development tools.

The S60 platform includes some commonly used applications such as the personal information management (PIM) program (e.g., phone book, calendar, and photo album), email program, the message sending and receiving procedures, and so on. S60 UI is designed for easy one-handed use. From the user's point of view, perhaps the most important feature is its UI. The UI includes a large-screen color display (176 × 208 pixels, 256 color display) and a variety of input buttons (two soft keys, five-button navigation, send/receive button, etc). The software developer should understand the associated UI platform to develop an application for a particular handset. Hereafter, the developer needs to know the cell phone used by the specific version of the Symbian operating system. With these two aspects of information, the developers decide what version of SDK should be used for development. In most cases, the developer can develop a single version of the application, and then be able to test and run the application in all devices based on the same UI platform and the same version of the Symbian operating system.

The Symbian operating system has evolved over the years and now it is not just designed to support traditional telephony services; the platform itself has a set of rich application programming interfaces (APIs) that can be used to develop some interesting mobile service applications. Oleinicov et al. [9] have proposed a high-level software architecture to develop a Voice over IP (VoIP) client applications on the Symbian S60 platform. Traditionally, VoIP service can only be supported within the existing telephony infrastructure where the public switched telephone network can be reached. This approach combines the essential benefits of IP and mobile telephony in portable mobile devices. It makes it possible to make VoIP calls without depending on current locations. The Symbian S60 provides almost the whole set of APIs that can be used to port the VoIP client application to the platform.

Huebscher et al. [10] discussed the possibility of utilizing the smart phone as a computing platform to develop ubiquitous computing applications over the Symbian S60 platform. The implementation of many ubiquitous computing applications is only become possible in recent years, as now smart phones have a much larger memory and a more powerful processor. The mobile operating system was not originally designed to support the ubiquitous computing applications and this also renders the challenge to create prototypes even for the simplest applications. Huebscher et al. [10] discussed some of the key issues with regard to developing ubiquitous computing applications. At first, the computational capability of the mobile device limits the efficiency of computing algorithm of the application. The second issue is that the radio coverage may cause intermittent network connection in a mobile environment for applications. Thirdly, the application may function incorrectly when a mobile device roams across different network service providers. Furthermore, smooth data transfer for an application may be the question when mobile devices move between different access point networks.

Instead of the discussion on the concept level, Zaykovskiy and Schmitt [11] studied the implementation of distributed speech recognition (DSR) by presenting the DSR software on the Symbian operating system and Java ME platforms and comparing the results. The DSR technology can be deployed in mobile and smart phones owing to the increasing memory and processing power. Although there has been some research on the theory of DSR, the technology has never been implemented in real mobile devices. DSR is a new technology that facilitates the fast and effective processing of speech recognition. The process includes two parts: a front-end on the client side and a back-end on the server side. Figure 12.1 shows the system architecture for DSR on the client and server sides. The client captures the speech signal using a microphone and extracts features out of the signal that is

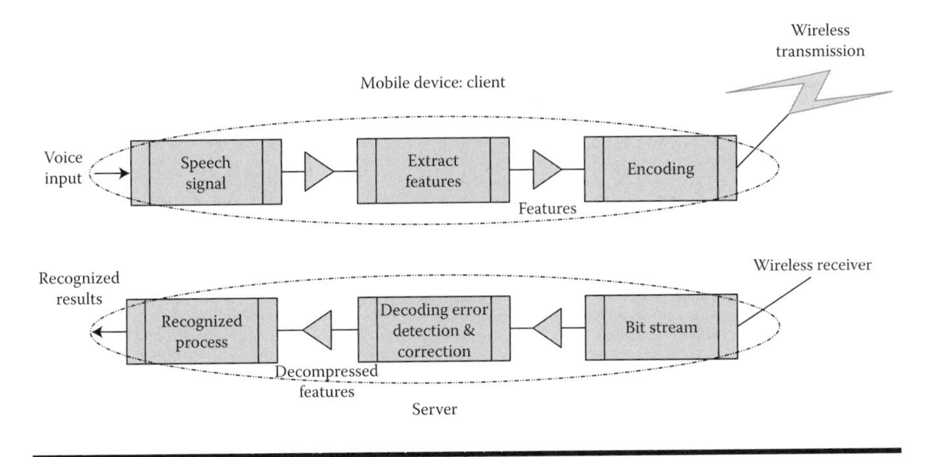

Figure 12.1 Distributed speech recognition architecture.

further compressed for transmission. At the server back-end, the features are decompressed and subjected to the actual recognition process. The implementation is based on the ETSI ES 201 108, the DSR standard that specified a feature extraction scheme based on the widely used Mel Frequency Cepstral Coefficients (MFCC) for both Java ME and Symbian platforms. MFCC has the DSR algorithms for feature compression, error protection and bit-stream generation, and decoding the bit-stream on the server side. Real-time efficiency for the implementation on the two platforms as well as the execution times is the basis of the performance comparative analysis in different platforms and mobile devices.

12.2.3 Advantages and Disadvantages of the Symbian Operating System

The Symbian operating system aims to provide a development platform for wireless communication services to combine computing and telephony technology with low development costs. For such a purpose, it offers the characteristics of an open development environment, low power consumption, and high processing performance. In addition, the security and reliability of the operating system itself, multithreaded operation mode, and a variety of UI alternatives achieve the flexibility, simplicity, and ease of operation. Furthermore, Symbian can be used in low-end mobile phones where many other proprietary-embedded operating systems developed by manufactures also compete with each other. The Symbian operating system has the advantage of being standardized and programs can share the information using the dynamic link library to achieve interoperability among different versions. Mobile phone manufacturers and network operators prefer the flexibility of Symbian customizability that at the same time makes integration more difficult for Symbian mobile phones.

Symbian is designed for small devices, providing a more powerful and more efficient memory management and a more flexible system. Recently, Nokia aimed to combine different versions of Symbian into a unified platform that will make it highly attractive for developers. Symbian has the advantage of being the leader in the market share among the majority of handheld communication equipment vendors. It has become an open, easy-to-use, professional development platform and supports the development of both C++ and Java. Symbian supports various wireless communication modes for multitasking, object-oriented component-based 2G, 2.5G, and 3G systems and application development, GSM, GPRS, HSCSD (high-speed circuit-switched data), EDGE (enhanced data rates for GSM evolution), CDMA (IS-95), and CDMA 2000. Symbian achieves highly efficient battery management, data synchronization, data encryption, certificate management, and software installation management.

Despite the above benefits, the Symbian operating system is weak in terms of supporting the mainstream media formats. Another issue is that Symbian is still limited in its compatibility between different versions of the Symbian operating system. For example, a number of versions are designed for a variety of needs; this draws a barrier for software developers in the form of incompatibility issues as software developers mostly focus on one platform. Whenever a new version of the Symbian operating system is released, developers and third-party software have to update their application software packages because of incompatibility. In addition, the Symbian operating system only provides a core and UI, and this allows manufactures to add other features. The problem is that, although many manufacturers mainly focus on the comprehensive functionality of the product, they often overlook some of the basic functions of the platform instead. All these cause inconvenience to developers and even to users as they need to constantly update applications.

The system capacity and ease of use of the Symbian operating system have been recognized by the industry and manufactures. As a result, the Symbian system has become the leader in the market share, competing with Windows Mobile, the Palm operating system, the embedded Linux operating system, and others. The Palm operating system has been gradually losing market share. However, Windows Mobile and Linux are the two main competitors along with the new Android platform from Google. The Windows Mobile system is from the Windows desktop system; both the interface and function are very similar to that in PCs, which is one major reason for having attracted a large number of users. Most experts believe that Windows Mobile has inevitably eroded the territory of the Symbian operating system. As the most open platform, the Linux system has given developers great flexibility, but it is still not widely used by ordinary consumers. The emergence of Android from Google is the latest high-profile event that is beyond the speculation of the industry. It is not a smart phone operating system but instead an open mobile development platform by collaborating with more than 30 leading companies in the wireless application technology. As a mobile operating system, Symbian provides a totally different platform from that of Microsoft. From the developer's point of view, Microsoft's success lies in constantly providing rich features and lower the difficulty level for beginners. Microsoft has been constantly going through the development of new tools and resources. This is one reason that Microsoft has attracted a larger number of developers. Symbian is learning to adopt a similar strategy, including launching various forms of interaction and enriching the knowledge base to lower difficulty, especially for entry-level developers.

12.3 The BlackBerry Platform

BlackBerry is a mobile email system terminal developed by RIM. It has the largest market share of smart phones in North America. RIM entered the mobile market earlier and developed the mobile email system that became very successful especially in the U.S. market. The product now has subscribers worldwide and is adopted by many governmental organizations and enterprises. The system supports push-style e-mail, mobile phones, text messaging, Internet fax, Web browsing, and many other wireless information services. BlackBerry has a wide-screen design and mainly the QWERTY keyboard input. It also has a number of built-in communication software including BlackBerry Messenger, Google Talk, and Yahoo! Messenger. Unfortunately, the lack of some entertaining features is a weakness of the product.

12.3.1 BlackBerry Enterprise Services

BlackBerry provides the wireless extension for business information platforms. Its enterprise solution offers various enterprise services for business users with integrated email, SMS, and phone functions without leaving the site. BlackBerry has a two-way paging mode mobile email system, compatible with existing wireless data links to allow users to wirelessly access the Internet and Intranet. The synchronization with the IT infrastructure of the company by default is perfect and can automatically forward your Outlook emails to your BlackBerry device. The encrypted connection from BlackBerry to the IT infrastructure provides excellent security features that enable mobile business users to access their corporate email accounts in a secure way, even if they are not in office. Users can send, receive, archive, and delete messages, and read email attachments in a variety of file formats. The BlackBerry "always online" technology can automatically transfer emails and users do not need to perform any operation for the communication. In addition to secure email services, the BlackBerry wireless handheld device has a built-in phone for voice services, call waiting, call forwarding, and call conferencing. BlackBerry also allows a user to wirelessly access a wide range of applications of personal assistants. With the use of BlackBerry, mobile users can access the latest all-day calendar, address book, and task information. Users can update the calendar, address book, and task information on the road to fully enhance work efficiency. BlackBerry supports SMS, and users can enable the text messages and telephone communications equipment. All these functions allow clients and colleagues to collaborate seamlessly to improve customer service and responsiveness.

12.3.2 BlackBerry Software Development Kits

The BlackBerry platform enables the development of advanced enterprise solutions with a wireless connection. It has the necessary components as a major and successful mobile application development platform: innovative software, advanced wireless handheld devices, wireless network service, and integrated applications. The BlackBerry development platform is divided into three categories: BlackBerry browser development (BBD), rapid application development (RAD), and Java application development (JAD).

BBD is to develop Web applications. It allows users to develop standard static and dynamic Web pages as the carrier running in the BlackBerry built-in browser runtime environment. It utilizes the typical client–server architecture to implement business functions. The BlackBerry browser supports different industry standards including HTML, AJAX, data-push technique, and offline queue processing. The BBD development process is smooth for programmers as it mainly needs to develop Web-based programs in the first place. BBD programming knowledge requirement is low and the development is fast and efficient, which makes it suitable for novice programmers. It does not need to develop the client programs and also involves less maintenance work.

RAD (based on the previous mobile data system—MDS) is a rich-client development method to extend Web services to BlackBerry smart phones. RAD can make full use of existing Web services and their resources. Through the WYSIWYG (What You See Is What You Get) interface, developers can use the drag-and-drop mode quickly to develop a graphical UI. The client program can be installed manually by the user or by downloading over-the-air and pushing install to the phone. Phones must install the MDS runtime environment to support the MDS runtime. To develop applications using RAD, programmers need to master the knowledge of Web services.

JAD can be deployed to develop the standard Java ME programs or the dedicated BlackBerry Java programs. JAD can maximize the use of the BlackBerry smart phone operating system, hardware, and built-in procedures. Programmers have much control of the applications to allow more personalized and customized development. Java programs can not only bring their own procedures for the use of the system but also for the use of the functionality of Bluetooth, global positioning system (GPS), and multimedia. JAD also supports the development under the integrated environment including the Eclipse and NetBeans. For example, to develop games or other highly customized applications, JAD is undoubtedly the first choice. In general, to develop complex procedures and applications, the JAD mode is recommended for programming using Java.

The BlackBerry platform has been used to develop many enterprise applications, m-commerce, and other applications. RIM offers the capability to access Web services. The client programs in BlackBerry devices can be developed for remotely assessing Web servers, database servers, or other application servers for different enterprise applications. Figure 12.2 shows such an example architecture that uses BlackBerry MDS Studio. The BlackBerry PIM specifically designed for business application provides the functions to run Java and other programs on the BlackBerry so-called MIDlet. As shown in Figure 12.2, the BlackBerry device is connected with wireless networks that could be the wireless local area network or any cellular networks. To ensure a reliable connection, the middleware could be used between the application and BlackBerry enterprise server. The middleware allows asynchronous data transmission. If the wireless connection from the BlackBerry to the server is not available, the data packets will be temporarily cached for later delivery. Retransmission will be initiated immediately after a new connection is established. The BlackBerry can work virtually as a terminal of the inner enterprise network and the communication is achieved by a simple and compact protocol. The security of the communication between the BlackBerry and the BlackBerry enterprise server is enforced anywhere and anytime, wherever the BlackBerry accesses a network. At the same time, the BlackBerry could be managed by the enterprise network through installation, application maintenance, security policy, and so on.

A number of applications such as Web services can be developed by using Web service development tools. The kSOAP library enables one to create applications in the mobile Java that communicate directly with the Web service and uses SOAP protocol. J2ME Web service APIs can also be used for connecting the BlackBerry devices to Web services. The visual studio of Microsoft also provides the tool for creating mobile applications. Behind the BlackBerry enterprise server, application and database servers are integrated for mobile applications. On the basis of the architecture in Figure 12.2, Kozel and Slaby [12] study a mobile application for a traveling salesman who can directly enter the enterprise orders at the customer's site. The salesman can access the subset of the enterprise's systems for contact browsing,

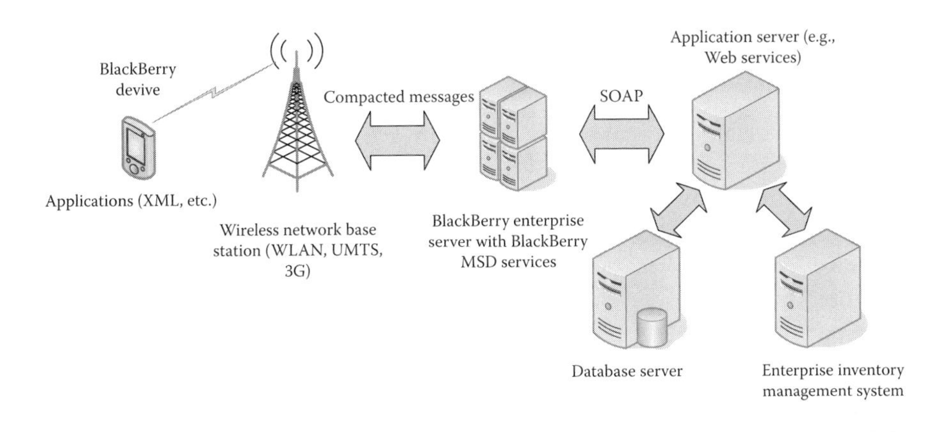

Figure 12.2 Example architecture of BlackBerry MDS Studio.

overview of the contracts, remarks, and so on. The salesman is also able to access various customers' data displaying specific customers and the dealer prices. The salesman can further process the order such as selecting items, dispatching the order, immediate confirmation of items and orders, and so on.

Using a similar architecture as in Figure 12.2, Massoth and Paulus [13] develop the application that BlackBerry uses for mobile acquisition of sale and return operations for shop-in-shop systems. The shop-in-shop refers to the small store inside a large department store, where the small store does not have enough space to accommodate its own cash register, for example, a small jewelry store. In such a case, the acquisition of sales and returns can be performed using a BlackBerry device that is connected to the inventory and accounting systems. The client program in the BlackBerry has all the operations to support product sale, billing, and accounting. Rather than the enterprise applications, many other mobile applications have been developed and mobile healthcare services are examples. The mobile healthcare services [14] are the functional combination of telemedicine with the mobile pervasive services. In the mobile healthcare services, BlackBerry can be used for health monitoring, tracking, and planning, and others for compelling heath applications.

12.4 The iPhone Operating System

The iPhone operating system is Apple's operating system developed for the use of the iPhone and iPod Touch. It is based on the Mac OSX operating system. The iPhone operating system includes the following built-in applications: SMS (text message), calendar, photos, camera, YouTube, stocks, maps, weather, time, contact, memos, system settings, iTunes, and App Store. The iPod Touch retains most of the iPhone applications except the telephone, SMS, and camera.

12.4.1 The iPhone Operating System Software Development Kits

The iPhone operating system architecture is divided into four levels: core operating system layer, core services layer, media layer, and cocoa touch layer. The iPhone and iPod Touch use ARM (advanced RISC machine) architecture-based CPU, rather than Apple's Macintosh x86-based processors. Therefore, Mac OSX applications cannot directly be ported into the iPhone operating system and they need to be rewritten in a way to support the ARM processors. The iPhone operating system 2.0 supports third-party applications through the review process to be downloaded from Apple's App Store. The SDK for the iPhone operating system was released in March 2008. It allows users to develop iPhone and iPod Touch applications and test them in an iPhone simulator. Users can upload an application onto a device only paying the iPhone Developer Program fee. Since the release of Xcode 3.1 [15], it has become the standard iPhone SDK development environment. The iPhone SDK is divided into the following functional sets:

- *Cocoa Touch*: It includes multitouch events and controls, accelerometer support, view hierarchy, localization, and camera support.
- *Media*: It includes OpenAL, audio mixing and recording, video player, image file formats, Quartz, core animation, and OpenGL ES.
- *Core Services*: These include network, SQLite embedded database, GeoLocation, and Threads.
- *Operating System X Core*: This includes TCP/IP protocol, sockets, power management, file system, and security.

The SDK requires an Intel processor running the Mac OSX Leopard system. Other operating systems, including Microsoft Windows and the old versions of Mac OSX are not supported. SDK itself is free of charge, but to release the software, developers must join the iPhone Developer Program, which requires paying a fee to get approval. After joining the program, the developer will receive a license that they can use to release the developed software to Apple's App Store. Apple still has not announced any iPhone to run Java. However, Sun Microsystems has announced that it will release a Java Virtual Machine (JVM) for the iPhone operating system, which is based on the micro edition version of Java (Java ME). This will allow Java applications to run in the iPhone and iPod Touch.

The iPhone operating system uses a variant of the same XNU kernel that is found in Mac OSX. Developing iPhone operating system applications requires Mac OSX to run the Xcode [15] development tools. Xcode is the major tool for iPhone application development which has rich features. In addition to the general IDE, such as code editing and project management, Xcode supports the iPhone application using the iPhone simulator and debugs directly in iPhone and iPod Touch devices. However, Xcode is not the only development tool and there are several other tools for developing iPhone applications. We further describe Xcode and other tools as below.

- *Xcode*: Xcode is an IDE that contains all the necessary tools for the management of iPhone application projects and source files. Xcode has a powerful text editor of supporting code completion, syntax highlighting, code folding (temporary hide code block), and displaying errors and warnings. The Xcode build system provides appropriate default configuration and it allows the programmer to set the environment according to their own preferences. Developers can choose for the iPhone or iPhone device simulator to build applications in Xcode. The simulator provides a local test environment for the procedures. When the program has the basic functionality, one can use Xcode to build it and then connect it to run on iPhone or iPod Touch devices.

- *Interface Builder*: The Interface Builder [16] is the interface design for iPhone operating system programs and it supports direct drag-and-drop programming. Using Interface Builder, developers can visually assemble application UI. The functionality is similar to Visual Studio, which makes it easy to design and arrange interface and save a lot of time. Because it is a visual editor, one can accurately see how the interface looks at runtime.
- *Instruments*: Instruments [17] is a suite of performance analysis tools. With Instruments, the developer can get simulators and actual equipment operating efficiency analysis. Instruments can collect data in the runtime, including memory usage, disk usage, network usage, as well as the graphical interface efficiency and display in chronological order for the view. This tool can analyze all the data of the program for performance optimization and thus gives users the best experience. In addition to providing a timeline view, Instruments also provides a tool for behavior analysis of the running programs. For example, the developer can store the data generated from multiple runs to check if the program performance has been improved or needs further revision. All these data can be saved as files in Instruments and are ready to be opened for viewing.

12.4.2 The iPhone and Web-Based Applications

A number of iPhone and iPod Touch applications such as Web-based applications (e.g., wireless social network for healthcare [18]) have been recently developed. The iPhone has the mobile Safari browser—a Web browser offering the same features as the desktop version. The camera on the iPhone can be used for tracking and mapping in a mobile environment. Klein and Murray [19] proposed a key-frame-based SLAMsystem on a camera phone by using the 3G iPhone. Parallel tracking and mapping systems have been developed to mitigate the impact of the device's imaging deficiencies. The SLAMsystem is capable of generating and augmenting small maps, though with reduced accuracy and robustness compared to SLAM on a PC.

Stelte and Hochstatter [20] developed a monitoring tool for network services on the iPhone and this tool is called Nagios. Nagios on the iPhone can be used to monitor the elementary activity in the network management as the gathered data are the basis for further action for troubleshooting problems to plan network extension. Status and alert messages can be visualized with a Web interface on the iPhone. Compared to other plugins for the Nagios system available for visualization of data, the Nagios developed by Stelte and Hochstatter [20] allows the users of the iPhone to effectively interact with Nagios.

Figure 12.3 shows the Nagios design and depicts how information is transferred between the Nagios application on the iPhone and the iNagMon that is a monitoring system on the Internet. The iPhone in Figure 12.3 shows the start screen of the Nagios application that gives the status information for each component in the

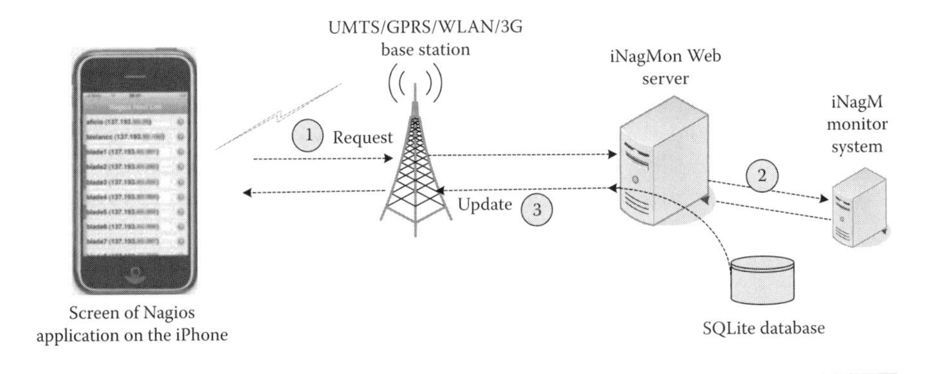

Figure 12.3 Nagios design and information exchange.

monitored network. The visualization of Nagios on the iPhone has to be adapted to the small iPhone screen. It visualizes the Nagios data such as display of the IP process (e.g., a mail system) or a network infrastructure in the enterprise network. By fetching the data on the iNagMon monitoring system, as shown in Figure 12.3, the status of network elements is updated in a certain interval to reflect their current status. The process can be implemented by the request and update operations from the iPhone to the monitoring system, as shown in Figure 12.3. The connection from the iPhone to the iNagMon Web server, as shown in Figure 12.3, can be using UMTS and other wireless networks. The Nagios on the iPhone has strategies to address the intermittent connectivity from the iPhone to the Nagios system and to different network types, and takes into account that data transfer is still quite costly in some networks.

Owing to mobility of the user and network availability, the communication to the iNagMon Web server can be problematic due to intermittent connectivity. To address this problem, SQLite [21] is designed to cope with the intermittent connectivity of the iPhone by replicating Nagios data with a resource-aware strategy. SQLite is a small database on the server and can store files on a Web server (e.g., the Nagios Web server). The application on the iPhone can now update the small database to download the file, depending on the current network connectivity. The server has no resource problem in extracting the data from the database and transforming it into an SQLite file. However, it poses resource problems on the iPhone. Therefore, the information preparation is conducted on the server side. The use of the downloaded SQLite database on the iPhone offers the flexibility. Even if the network connection is not available for the iPhone, the local database copy is still available. The application can use the information available on the iPhone itself and count the number of unsuccessful tries to download the new database file from the Nagios Web server. If the number of tries exceeds a certain threshold, the user is warned that the data may not be up-to-date anymore. The application will still try to update the database in the background, transparently for the user.

Many other applications have been developed for the iPhone. EarthWinds is a Web application for the iPhone that shows one to see the world from the current satellite images of the Earth. The application shows the movement of both clouds and hurricanes. It has a series of controls to move around the earth and present the visualization options by each continent. These satellite images are usually shown four times per day and they are updated from a German weather service called "deutscher Wetter Dienst". WorldWatchr is an application that enables one to monitor webcams that are continuously updated as Web camera pictures. By watching the webcam on your iPhone, you can monitor your home, watch a famous place of the world, see the current traffic maps in a city, and more. It can cycle through all one's cameras and updates the images periodically. Florafolio is also a Web application exclusively for the iPhone designed by Holimolimedia. Florafolio is an easy-to-use, interactive field guide to native plants of North America. This catalog is shown as a list of flowers with data as its vulgar and scientific name, its image, and a series of icons to show the features of their habitat. It covers the trees, shrubs, perennials, ferns, vines, and grasses that are indigenous to Eastern Canada and North Eastern United States. This Web application can be visualized in the Safari browser for the iPhone and is the perfect guide for anyone who wants to identify species in the wild or garden with native plants. Florafolio 2 has been recently released with an awesome new search engine and other handy enhancements.

12.5 Windows Mobile

Windows Mobile is the software platform that Microsoft designed for handheld devices including Pocket PC, smart phones, PDAs, portable music players, and so on. Windows Mobile is the "mobile version of Windows" and it provides the familiar Windows interface extended from desktop PCs to personal devices. The latest version of Windows Mobile is 6.5 though many developers are still using version 6.1. Windows Mobile has released a series of versions including Windows Mobile 2002, 2003, Windows Mobile 5.0. In September 2005, Microsoft launched Windows Mobile 5.0 based on Windows CE 5.0. It contains some main updates: .Net framework 2.0 support, support for Power Point software, virtual GPS port, simplified Bluetooth and WiFi settings, and Windows Media Player version 10.0. In February 2007, Microsoft launched Windows Mobile 6.0 based on the kernel for Windows CE 5.2. The UI is similar to Windows Vista in three versions: Windows Mobile Classic for the traditional Pocket PC and PDA, Windows Mobile Standard for the smart phone, and Windows Mobile Professional for Pocket PC-based smart phones. It has some important updates: support for IP phones, improved online search function through Exchange 2007, support for HTML mail, preinstalled .Net framework 2.0 SP2, and preinstalled Windows Live.

In April 2008, Microsoft launched Windows Mobile 6.1 and the kernel is still Windows CE 5.2. The main new features include improved Internet Explorer, adding scaling, Adobe Flash video, Silverlight, H.264 video, increased text input options, and greatly simplified Bluetooth and WiFi settings. In February 2009, Microsoft introduced the new Microsoft-based generation mobile phone system—Windows Mobile 6.5. The launch of Microsoft's mobile phones has a new UI and richer browsing experience. It also launched two new services: "My Phone" and "Windows Market." The former can be used for the network synchronization of user messages, pictures, video, contacts, and so on. The latter is to provide mobile phone and network applications through online stores. My Phone is a free service that allows mobile phone users to manage personal information and take a backup in a password-protected network without worrying about losing important information when upgrading. It supports automatic synchronization and backup features even if the phone is lost or upgraded. In addition, users can also automatically upload pictures and videos to the My Phone service, which is more convenient to keep the contents secure.

The Windows Mobile Pocket PC version is designed to store and retrieve emails, contact and appointment information, to play multimedia files and games, to exchange text messages with MSN Messenger, and to browse Web content. It is also able to exchange information and synchronize with a desktop computer. With the built-in WiFi capability, it can access the Internet and enterprise network through a wireless access point.

The Windows Mobile smart phone version has similar PDA features integrated with existing mobile devices. Pocket PC and smart phone software is not compatible though the majority of the code is the same. Both the smart phone and the Pocket PC are equipped with ActiveSync software that is used to manage the connection between two devices. Users can configure the ActiveSync software to synchronize email, calendar, appointments, and contact information with the information stored in the desktop or laptop.

12.5.1 Windows Mobile Operating System

Initially, Windows Mobile only supported synchronization with Microsoft Outlook software; the later newly released version can support seamless connection with a desktop. Windows Mobile also adds support for Microsoft Office documents including Word and Excel files, so there is no need to convert the documents into the format only recognized by handheld devices. Windows Mobile enables high-performance email systems by using the direct-push technology. Windows Mobile has almost the best support for multimedia with the preloaded Windows Media Player which supports audio and video formats including MP3, WMA, WMV, ASF, MPG, and so on. Windows Mobile has attracted a large number of allies, and each year a large number of third-party software is developed. The platform is well-known among a variety of applications, which contributes partially to further success. As a version of the operating system for Microsoft's handheld devices, it has the advantage of being compatible with the current desktop PC and Office documents. At the same time, it comes with strong multimedia and entertainment performance. These have made Windows Mobile a mobile development platform with perhaps the most potential.

At present, the Microsoft Windows Mobile system is widely used in smart phones and PDAs. Although Symbian still has the largest market share, Windows Mobile is catching up with the production. An advantage of the Windows Mobile interface is that it has a similar interface to desktop Windows which facilitates those familiar with computer operation. It also has many preinstalled software including Microsoft Office, Word, Excel, Power Point, Internet Explorer, Media Player, and so on. The synchronization feature is very convenient as it is fully compatible with Outlook, Office Word, Excel, and so on. It also supports powerful multimedia functions to support almost any mainstream broadcast audio and video file formats with the support of third-party software. The touch operation is also very good in comparison to the iPhone series. There is rich third-party software including the dictionary and satellite navigation software.

There are also some disadvantages of the Windows Mobile system. In general, it has high requirements for hardware. It is more complicated than other mobile operating systems, especially for those who are not familiar with computer operations. Table 12.1 describes the main functions of the Windows Mobile operating system for the Pocket PC phone series and the smart phone series. These functional features show the benefits of using Windows Mobile and the potential to build applications on the basis of these functions.

12.5.2 Windows Mobile Applications

Wang and Shi [22] present a handheld PDA wireless medical information system that is based on the Windows Mobile operating system. The purpose of this system is to improve the efficiency and thus enable the medical staff to quickly access patient medical information without the limitations of location and time. The PDAs based on the Windows Mobile operating system connect to the back-end Windows server and database from the wireless local area network or other available networks. The medical staff can use the system to download the data from a background database to a PDA. The database is implemented on the platform of the SQL Server 2005 and SQL Mobile 2005. Users only log into the system after authentication and they are given different privileges in data access and operations, according to their predefined roles. The system, in general, can be divided into four modules to deal with the information for patients, doctor's advice, diagnosis and treatment information, and cost.

The traditional video monitoring system transmits the data between the PC and the camera via the wired cable. With the fast deployment of 3G networks in the past several years, much attention has been focused on wireless multimedia monitoring and streaming. Zhang et al. [23] present the design and implementation of a wireless remote monitoring system based on the Windows Mobile platform, as shown in Figure 12.4. The system can transfer the image data between Pocket PCs and remote

Table 12.1 Functions of Pocket PC Phone Series and Smart Phone Series

Pocket PC Phone Series		*Smart Phone Series*	
Functions	*Description*	*Functions*	*Description*
PIM	Personal information management system	Start Menu	This is for smart phone users to run a variety of applications. Similar to the desktop versions, the Start Menu is also the shortcut icon for main programs and the figures for the icon are assigned serial numbers to allow fast running.
Internet Explorer	Internet Explorer as the PC version	Title bar	Similar to the role of the desktop Windows system tray, the title bar is where the smart phone can show all kinds of information including the titles of current running procedures and a variety of tray icons, such as the battery power icon, the phone signal icon, input method, and application icons.
Inbox	Message Center, integrated with Outlook email and newsletter features	Phone features	It is extended to integrate with Microsoft Outlook to provide dial-up, contacts, history, and so on.
Windows Media Player	Windows Media Player as the PC version	Outlook	Windows Mobile has built-in Outlook software including tasks, calendar, contacts, and inbox. The outlook of Windows Mobile can synchronize with the desktop computer and Exchange Server if provided with an Internet connection.

(continued)

Table 12.1 (continued) Functions of Pocket PC Phone Series and Smart Phone Series

Pocket PC Phone Series		*Smart Phone Series*	
Functions	*Description*	*Functions*	*Description*
MSN Messenger/ Windows Live	MSN Messenger/ Windows Live as the PC version	Windows Media Player (WMP)	It is bundled with the Windows Mobile software. Its initial version is 9, but most of the new devices use version 10. For existing equipment, the user can upgrade to WMP10. WMP supports the file formats of WMA, WMV, MP,3 and AVI files.
Office Mobile	Similar to the PC version of Microsoft Office		
ActiveSync	PC connection for the exchange of information		

camera devices via Bluetooth. The system consists of a smart phone based on the Windows Mobile platform, a monitoring unit, and a camera and the Bluetooth module, which is referred to as BTCAM. The monitoring system sends the commands to the BTCAM and then the BTCAM will start to capture the images. The captured images are compressed and transmitted to the Pocket PC. The portable

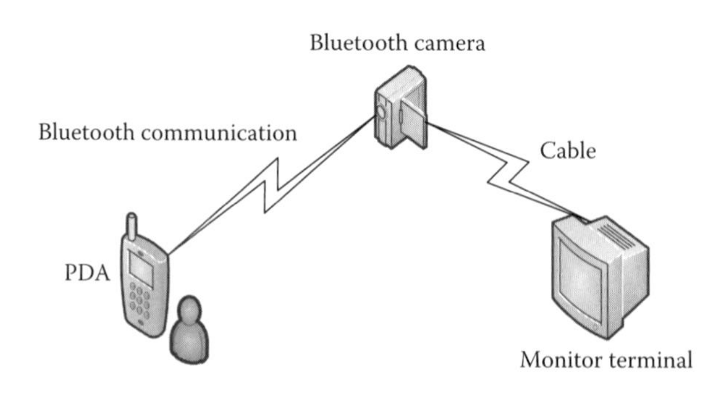

Figure 12.4 Wireless monitoring system architecture.

control terminal is also able to send inquiry commands to the BTCAM. The images are in JPEG format which is suitable for most applications.

In this implementation, the application running on the Pocket PC has two threads. The main thread is responsible for sending control commands to the monitor terminal. A second thread is in charge of data collections, data assembly and disassembly, and image display. The portable terminal is to send control commands to the server and a timer is initiated at the same time. If no response is received in a given time period, the terminal will resend the command and report the failure after multiple trials. The portable terminal waits until it receives the image data packet from the server after sending the transmission request. The data are divided into packets and a sequence number is used for indexing. If the index is out of the ordered sequence, packet loss occurs and it will wait for the server to resend the packet. This is done by sending a request to the server to resend the data. When all the data have been received correctly, the packets will be reassembled as a frame of image and be displayed on the PDAs.

12.6 Embedded Linux

Linux was introduced in 1991, and after more than 10 years it has become a powerful, well-designed operating system. Linux not only competes with a wide range of commercial operating systems, it also has been customized as the platform for mobile handheld devices in the emerging embedded operating systems. Linux, as a free and open-source operating system, has a market share in the mobile phone market. Embedded Linux is converted from the standard version of Linux into a much smaller size burned into the memory chips for dedicated applications.

12.6.1 Embedded Linux Operating System

The embedded Linux operating system has many benefits such as being copyright free, high system performance, high security, reliability, and the availability of open source of many applications. As a platform, the advantage of Linux is being free and there are a large number of developers and existing applications. Therefore, the product development cycle usually can be reduced and the product can go to the market quickly. It is possible for independent development groups or individuals to more efficiently utilize the hardware capability with the availability of source code. It is also possible for the industry to develop more secured software modules with their own security or authentication modules. The Linux operating system supports multimedia applications including MP3, WMA, RM, and AVI audio and video formats. The open source of applications allows rapid development of new applications.

In addition, Linux has a wide range of hardware support including x86, ARM, MIPS, ALPHA, PowerPC architecture, and so on. It has been successfully transplanted to dozens of hardware platforms and is able to run on almost all popular CPUs. Linux has a wealth of driver resources to support a variety of mainstream hardware devices. The Linux kernel has high performance and stability. Linux has a very compact core design of five components: process scheduling, memory management, interprocess communication, virtual file system, and network interface. The most unique mechanism of the Linux system is that the modules can be added or removed from the kernel depending on needs. These features make the Linux kernel very small and very suitable for embedded systems. Linux provides users with the maximum freedom and often requires application-specific modifications and optimization, so the access of source code has become essential. In the development of Linux applications, in general, we do not have to start from scratch, but can choose a similar application as a prototype to start. Linux supports all the standard Internet network protocols that can be easily migrated to embedded systems. In addition, Linux also supports ext2, fat16, fat32, romfs file system, and so on. The key to the development of Linux systems is the availability of a set of comprehensive development and debug tools. Currently, embedded Linux provides developers with a complete tool chain that uses the GNU compiler gcc, together with gdb, kgdb, xgdb for debugging, which makes it easy for all levels of debugging.

12.6.2 Application Development with Embedded Linux

Embedded Linux system development is booming and has a large market share. In addition to traditional Linux companies engaged in the development, IBM, Intel, Motorola, and many other companies have begun to study embedded Linux. There are three challenges for developing a mature embedded Linux system:

- *Real-Time Performance*: The existing Linux system is in essence not an embedded real-time operating system. Linux kernel scheduling strategy is based on the UNIX system that has defects when applied to an embedded real-time environment, such as the interrupt being closed when running kernel threads, the uncertain time slot of time-sharing scheduling strategy, the lack of high-precision timer, and so on. For this reason, the use of Linux as the underlying operating system and then constructing real-time processing capabilities on top of it is a possible solution.
- *Core Structure*: The Linux kernel is a single large program that enables direct communication between various components to reduce the switching time between tasks. Embedded systems with limited resources do not conform to the characteristics. An embedded system usually has microkernel architecture, that is, the kernel itself provides only the most basic operating system functions, such as task scheduling, memory management, and interrupt handling. Other functions such as file system and network protocols and additional features run in the user space, and can be added according to the needs. Though the microkernel shows a poorer performance than the monolithic architecture,

it can greatly reduce the size of the kernel, which makes it easy to maintain and transplant and to better meet the requirements of embedded systems.

- *Integrated Development Environment*: A good IDE for Linux is to provide a complete simulation capability and to avoid the problem of application development relying heavily on the availability of embedded hardware. An integrated development platform for the Linux system should include a compiler, linker, debugger, tracer, optimization, and integrated UI. Most of the current Linux-based graphical interface is tailored to specific system platforms and there is a gap compared to Windows or other commercial embedded operating systems. The IDE is an integral part of designing a more mature Linux system.

Kim et al. [24] describe the hardware and software architecture of a Linux-based smart phone platform. The hardware has a main application processor and a coprocessor for multimedia acceleration. The Linux-based software architecture includes the modification in Linux kernel and application middleware. It implements all the key features in battery-powered mobile devices including power management and communication support. As described, two sets of APIs are developed for power management which is the core component. One set of APIs is used to develop the device driver for the smart phone applications. Another set of APIs is provided for the user-level power management such as getting the status of power consumption. The software layer supports the traditional phone communication applications including voice call, messaging, and also the data-centric applications. It has developed the software layer called telephony API (TAPI) that provides all the necessary APIs to support these services.

The platform supports seamlessly the traditional mobile services including voice call. TAPI, a communication application support software layer, has been designed to support these in addition to some data service applications. TAPI translates the API calls from communication applications using TAPI to the appropriate AT commands, which is implemented in the CDMA module. The TAPI has client–server architecture to provide the transmission transparency of CDMA data. A TAPI server has the full control of the CDMA module and also manages all the potential exceptional scenarios in the communications.

Figure 12.5 shows the software architecture of the Linux-based smart phone platform. The application launcher is the manager to control the overall application and communications management and Windows display management. The information that the application launcher collects includes the status of batteries, CDMA signal strength, and so on [24].

12.7 Android

Android is Google's Linux-based open-source mobile operating system platform. The original meaning of the term refers to the android "robot." Google and the

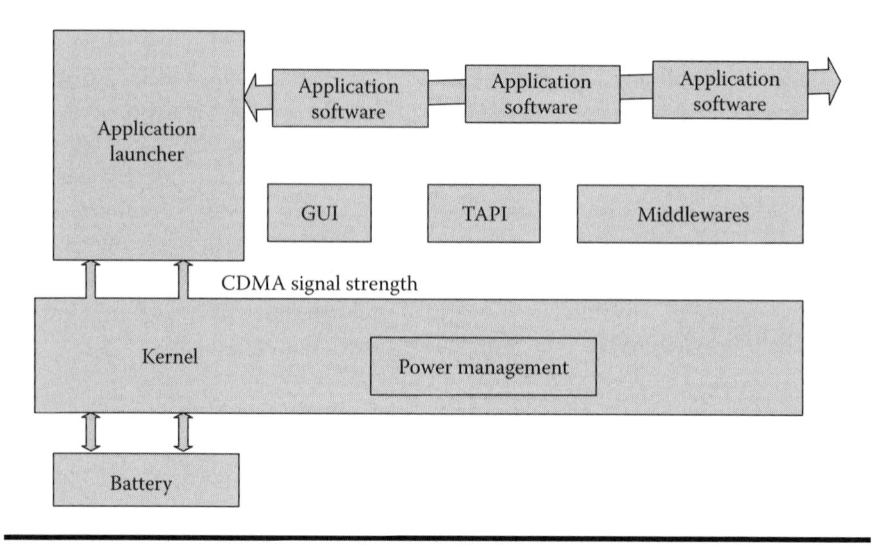

Figure 12.5 Linux-based smart phone software layers.

Open Handset Alliance have cooperated to develop Android. As an important component of corporate strategy, Google will further promote the business goal "anytime, anywhere information for everyone." The Android platform for mobile devices is to promote innovation and allow users to experience the best of mobile services. At the same time, developers will also be more open to joint cooperation and thus greatly reduce the development of new mobile equipment cost. Android includes operating systems, UI, and applications—all the software for a mobile phone. Google worked with telecommunication operators, device manufacturers, developers, and other interested parties to form deep-level partnerships, hoping to establish a standardized, open software platform for mobile phones in the mobile industry. The Open Handset Alliance includes more than 30 technology and wireless leaders such as Motorola, Qualcomm, HTC, and T-Mobile. The companies do not include the global mobile phone giant Nokia, or the iPhone from Apple, or the U.S. operator AT&T and Verizon. Microsoft also did not join and Canada RIM and their BlackBerry were blocked at the door.

For consumers, Google phones have emerged as a common and powerful mobile phone product. In September 2008, the U.S. operator T-Mobile USA officially launched the first Google phone—T-Mobile G1. Taiwan phone HTC OEM is the world's first mobile phone using the Android operating system to support the WCDMA/HSPA network with the theoretical download rates of 7.2 Mbps, and supports WiFi. Google G1 supports a variety of services, including Gmail, Google Maps, YouTube, Google Calendar, Google Talk, and the Chrome Lite browser. Google released the cell phone using Qualcomm MSM7201A processor, support for 7.2 Mbps download speed, 3G network operators in the United States.

MSM7201A is for the single-chip, dual-core, integrated hardware to accelerate multimedia capabilities, support for 3D graphics, 300-megapixel camera that can scan the bar code function, and GPS functionality. Android uses the WebKit browser engine, a touch screen, high-level graphical display features, and access to the Internet. Mobile phone users will be able to view email, search websites and watch video programs over mobile phones that emphasize search capabilities and a powerful interface.

12.7.1 Android Software Development Kits

Developers are given a greater degree of freedom in the development of procedures. Compared with Windows Mobile, Symbian, and other platforms, The Android operating system is free of charge to developers, so costs can be saved. The Android SDK provides all the tools to develop application using the Java. Every application runs in its own Linux process and also has its own JVM. Applications run in separate spaces from all other processes. The SDK has the following features, as specified in Reference [25]:

- Application framework enabling reuse and replacement of components.
- Dalvik Virtual Machine optimized for mobile devices.
- Integrated browser based on the open-source WebKit engine.
- Optimized graphics powered by a custom two-dimensional graphics library; three-dimensional graphics based on the OpenGL ES 1.0 specification (hardware acceleration optional).
- SQLite for structured data storage.
- Media support for common audio, video, and still-image formats (MPEG4, H.264, MP3, AAC, AMR, JPG, PNG, GIF).
- GSM telephony (hardware dependent).
- Bluetooth, EDGE, 3G, and WiFi (hardware dependent).
- Camera, GPS, compass, and accelerometer (hardware dependent).
- Rich development environment including a device emulator, tools for debugging, memory and performance profiling, and a plugin for the Eclipse IDE.

12.7.2 Android-Based Applications

Android SDK comes with a set of common applications written in Java. Like the iPhone, many mobile applications can be developed using Android SDK. Hu et al. [26] present a semantic context management framework in Android SDK. The framework is called ContextTorrent that can make various types of context information be semantically searchable and sharable among local and remote context-aware applications. Context awareness is the key component to support the ability to detect and react to the environmental variable for context-awareness applications. Hu et al. [26] implement the ContextTorrent for a context-awareness application

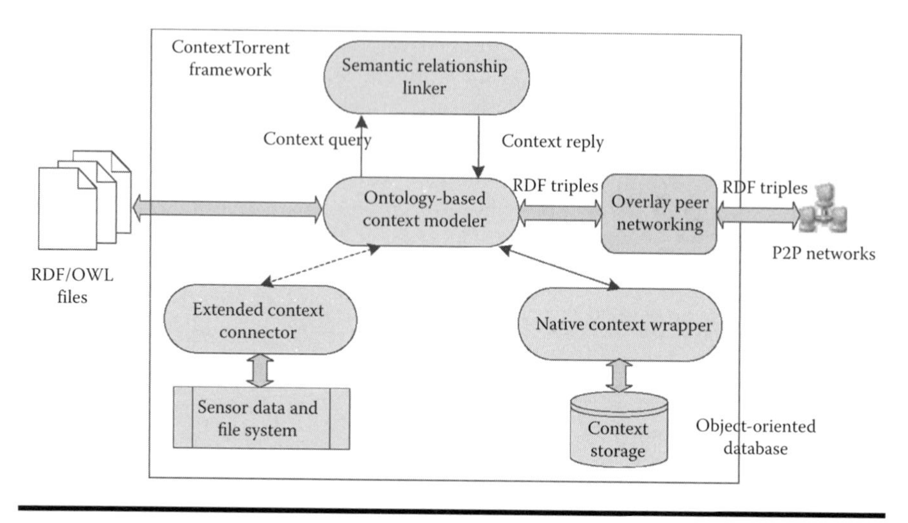

Figure 12.6 ContextTorrent framework for semantic context management.

that could enable us to issue a query for finding a person's phone number with his name as the keyword. Another example is that the location-based social network allows the users to see where their friends are, to search location-tagged content within their social graph, and to meet others nearby.

The main function of the ContextTorrent is that it is able to respond to the context resource (e.g., phone number and address) to a user who initiates a query (e.g., name) from the Google phone or other mobile device. Figure 12.6 depicts the ContextTorrent framework for semantic context management. As shown in Figure 12.6, the framework has five main components: ontology-based context modeler, semantic relation linker, extended context connector, native context wrapper, and overlay peer networking. Their functionalities are illustrated below.

- *Ontology-Based Context Modeler*: It models static and dynamic contexts as semantic resource using the standard semantic Web RDF/OWL (Resource Description Framework/Web Ontology Language) parser that parses the external RDF files or RDF triples. The static context is the information such as user profiles in RDF/OWL files can be imported in the beginning. The dynamic contexts are the spatial–temporal information (e.g., location of moving user) which is updated on the changes. OWL allows processing the content of information instead of just presenting information to humans. RDF represents the information in the Web in a minimally constraining, flexible way. As shown in Figure 12.6, these external RDF files or RDF triples are retrieved from the external networks via the overlay peer networking. In addition to parsing the OWL/RDF schema, the ontology-based context modeler also performs insertion, deletion, and modification of RDF triples either trained from the network or imposed by the local semantic relation linker to dynamically update semantic relationships.

- *Semantic Relationship Linker*: The semantic relation linker manages the semantic relationships among context entities. On the reception of a context query issued by a context-awareness application, it returns the context resource uniform resource identifiers (URIs) ranked by relevancy (e.g., by distance of the location). The context resource URI could also be used to locate all its associated properties to get the property values.
- *Native Context Wrapper and Extended Context Connector*: The native context wrapper and extended context connector are designed to provide a unified interface for accessing the context data. For this purpose, the native context wrapper wraps up the context data using ontology schemas such that these data can be accessed by applications via standard URIs. On the other hand, the extended context connector creates links to external context sources (e.g., file resources on a remote device) and performs the format adaptation of various sensor data (e.g., GPS receiver). Unique URIs are created for all context sources. The separation of access interface and data storage management makes it possible for the upper layer context modeler to build customized and extensible storage independently. As shown in Figure 12.6, ContextTorrent builds the embedded context storage based on an embedded object-oriented database to store the ontological relationships among context entities.
- *Overlay Peer Networking*: This component is implemented to connect the external network for transferring the RDF triples from nearby mobile devices, their desktop counterparts, or other devices in the smart space. The peer-to-peer overlay network is constructed in a way that all these devices are connected to the network. The use of the overlay network could facilitate context searching, provisioning, and delegation efficiently in a large-scale network environment.

The implementation of this framework allows context searching, sharing, and transferring on mobile devices for mobile applications. ContextTorrent considers Android's intent interface to enable context sharing across applications. This feature enables the runtime binding of program code of concerned applications to certain shared context data. Thus, efficient multitasking can be achieved as a key advantage over other Java-based mobile platforms (e.g., J2ME). This again offers better user experience. Furthermore, ContextTorrent implements a notification mechanism of Android to implement context-aware notifications. The use of Android SDK for development facilitates the reuse of various components. That is, any application can publish its capabilities and any other application may then make use of those capabilities. All applications are a set of services and systems, including a rich and extensible set of views, content providers, resource manager, notification manager, and activity manager. Android comes with a set of C/C++ libraries including the

System C library, media libraries, surface manager, LibWebCore, SGL, 3D libraries, FreeType, and SQLite.

12.8 BREW

BREW [8] is a mobile phone value-added services application development platform launched by Qualcomm, Inc. in 2001. It was originally developed for CDMA handsets and later ported to other air interfaces including GSM/GPRS. BREW provides a highly efficient, low-cost, scalable application execution environment. It is to develop applications that can be ported to all Qualcomm devices. Manufacturers and developers can expand on top of the environment to provide various additional modules, such as multimedia, multiconnectivity, location services, UI, and other functions. Through the BREW interface, the developers can provide a complete set of information, business and entertainment features such as Bluetooth, GPS, and data services. As a mobile application platform, BREW can support high-speed Internet access, downloading games, online shopping, and other wireless data services.

12.8.1 BREW Software Development Kits

The BREW runtime environment is a layer of software interface located between the chip system software and applications on wireless devices. BREW supports a variety of programming languages including C/C++, Java, Extensible Markup Language (XML), Flash, and so on. A third party can expand the BREW platform to provide additional functionality. The BREW platform is based on C/C++ that has a huge user base. The BREW Development Kit is a free download. Java programmers can also benefit from the chip-level integration of BREW. IBM has developed a separate JVM as a BREW extension. Java developers can take full advantage of the Java application without the need to consider the different manufacturers. BREW solution includes a comprehensive billing and payment infrastructure—the BREW distribution system and the system developer can guarantee that their applications can make an appropriate profit.

For testing applications during the development process, the SDK includes a BREW emulator or the BREW simulator. The BREW emulator (currently called the BREW simulator) does not emulate the handset's hardware. Instead, the BREW application is compiled to a native code and linked with a compatible BREW runtime library. BREW works for any of the existing 2G networks and even if there is no 3G network, BREW applications can provide powerful services. At the same time, BREW can continue to evolve with the upgrade of network and device. In terms of equipment manufacturers, the ideal situation is that the application should be perfectly applicable to all types of different mobile phones. The BREW mobile platform is a bridge between the low-level features and high-level third-party applications. The platform is independent of air interface and can be transplanted to

CDMA IS-95A, IS-95B, 1x, 1xEV-DO, and GSM/GPRS mobile phones. BREW SDK will enable developers to use standard tools in a familiar environment. Users may choose from a number of applications in the following areas:

- *Communication:* instant messaging, email, ring tones, and interactive messaging.
- *Location:* mapping, navigation, traffic, and other location-specific content.
- *Game:* single-player and multiplayer interactive games.
- *Mobile commerce:* account balance, trading, and other financial transactions.
- *Entertainment:* music, video, humor, and so on.
- *Information:* flight tracking, news, weather, sports, and so on.
- *Expansion:* Developers can use powerful third-party extension.

12.8.2 BREW-Based Mobile Applications

Oda et al. [27] discuss the implementation of a BREW application on mobile phones to monitor women's healthcare information. A temperature sensor device periodically sends the accumulated data to a back-end sever via a mobile phone. The BREW application deployed in the mobile phone has a connection to the server over the 2.5G/3G wireless networks to retrieve the data. The application has an interactive graphical interface to allow mobile users to easily monitor the health information on mobile phones. Figure 12.7 shows the architecture of the BREW development environment. In general, BREW comes in a standard application layer with nonchangeable applications such as instant message, GPS, email, browser, and so on. The developers write the application in C/C++ or other high-level languages for BREW-enabled applications.

Dongre et al. [28] proposed a client–server architecture in which BREW is used in the application development platform on the client side. BREW is selected because it has enhanced UI, the ability of downloading application offline, and so on. The server-side environment is WAMP, the acronym of Apache/MySQL/PHP for Windows. WAMP is the popular combination of an open-source group of software.

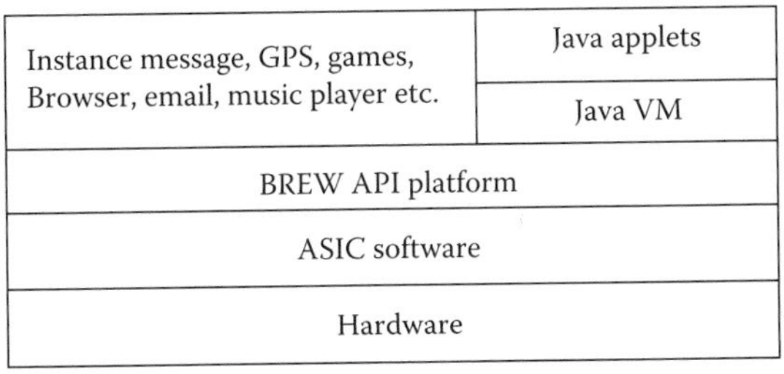

Instance message, GPS, games, Browser, email, music player etc.
Java applets
Java VM
BREW API platform
ASIC software
Hardware

Figure 12.7 Development environment architecture of BREW.

PHP is selected as the server-side scripting language. One advantage of PHP is its platform independence and it can run in multiple platforms including Windows, Mac operating system, UNIX, and Linux. The client application is developed using BREW SDK 3.0 that contains an emulator to test during the development process. The interface is designed via the BREW UI toolkits including IMenu control, IText control, and IStatic control to draw screen contents. WAMP provides the popular combination of Apache with the PHP server-side language and the MySQL database in one easy-to-install package. Figure 12.8 shows the software architecture that is used to build mobile phone applications.

Zhuo et al. [29] proposed an end-to-end wireless media streaming system over the existing CDMA networks. The system is a client–server architecture with the media server and mobile client. The client side is developed on BREW. In the proposed system, H.264 is selected for the video coding compression method because of its high compression efficiency and error detection. The video data is stored in the backend database to be retrieved by clients. PureVoice coding technology from the BREW platform is used for the voice compression method that is a highly efficient coding technique developed by Qualcomm and comes with the BREW platform. The system implemented four basic VCR functions, that is, play, stop, pause, and fast-forward. The system is tested over the CDMA network and shows that it can support the continuous and smooth multimedia contents in the lower network bandwidth around 79 kps.

Figure 12.9 shows the architecture of the wireless media stream system. With regard to the data transmission, real-time protocol is adopted to provide end-to-end service delivery. The protocol is to send the data in TCP format only because the UDP port is blocked in the testing environment. In addition to the video data, the protocol also sends other information including the signal to start and end transmission, data synchronization, frame type, and so on, which is application dependent and not supported by the protocol itself. Real-time streaming protocol at the application layer is to provide a reliable connection for streaming media and support the four VCR functions and other interactive functionalities. An event-driver scheme is adopted in the platform to trigger the

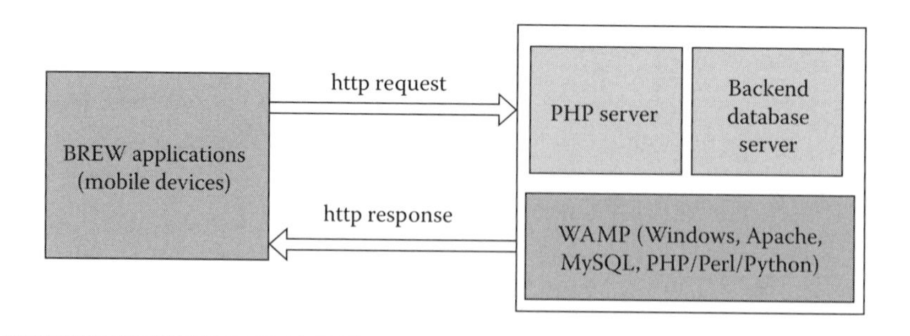

Figure 12.8 Client–server architecture for BREW mobile applications.

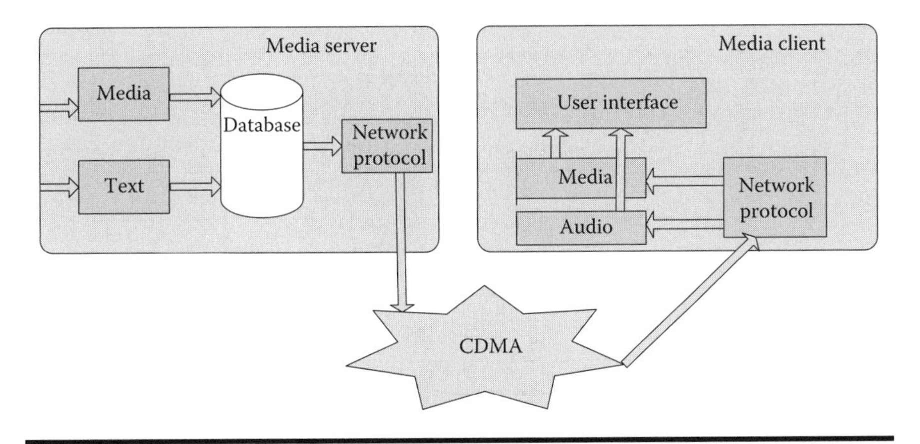

Figure 12.9 Architecture of the wireless media stream system.

modules to work together in terms of packet arrival and receiving, timer, and other messages.

12.9 Java ME

The Java Platform, Micro Edition (Java ME) [6] provides a robust, flexible environment for applications running on mobile and other embedded devices including mobile phones, PDAs, TV set-top boxes, and printers. Java ME includes flexible UIs, robust security, built-in network protocols, and support for networked and offline applications that can be downloaded dynamically. Applications based on Java ME are portable across many devices, yet leverage each device's native capabilities. Java ME is a highly optimized Java runtime environment, mainly for consumer electronics devices such as cellular phones and video phones, digital set-top boxes, car navigation systems, and so on. In the design of Java ME specifications, Sun Microsystems think that it does not make sense to create a single development system for a variety of electronic devices. Accordingly, all embedded Java ME devices generally are divided into two types: one with limited computing functions and power supply embedded device and the other with better computing power and power supply embedded devices.

12.9.1 Java ME Software Development Kits

Java ME development does not need special tools and developers usually just need to install Java SDK and the free Sun Java Wireless Toolkit to start development, compiling, and testing. In addition, some existing Java IDE (Eclipse and NetBeans) also support the development of Java ME. Figure 12.10 gives an overview of the components of Java ME technology and its relationship to other Java technologies.

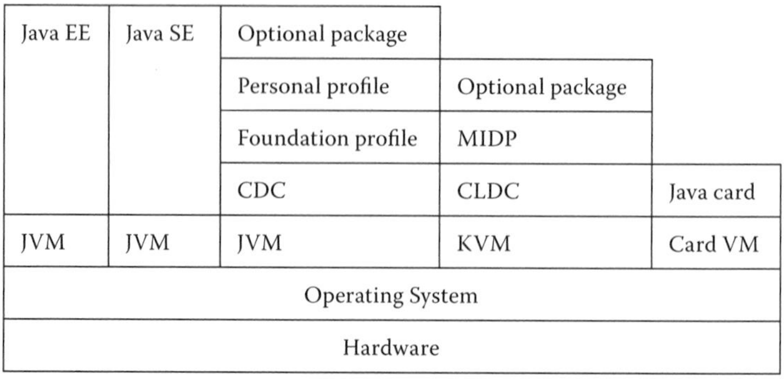

Java EE	Java SE	Optional package		
		Personal profile	Optional package	
		Foundation profile	MIDP	
		CDC	CLDC	Java card
JVM	JVM	JVM	KVM	Card VM
Operating System				
Hardware				

Figure 12.10 Java platform: Micro Edition (Java ME).

Compared to Java SE and Java EE, the operation environment in Java ME is more diversified though the functionality of each device is more specific with limited resource constraints. In order to achieve the standardization and compatibility to meet the needs of different aspects, Java ME is divided into the structure of configuration, profile, and optional packages. A configuration is to provide the most basic set of libraries and virtual machine capabilities for mobile devices of a broad range. A profile is a set of APIs that support a narrower range of devices, and an optional package is a set of technology-specific APIs.

In addition, to differentiate between the two types of embedded devices, Java introduces the concept of "configuration." The configuration for the limited power embedded devices is defined by the Connected Limited Device Configuration (CLDC) [30] specification, and the configuration for other embedded devices is defined by the Connected Device Configuration (CDC) [31] specification. CLDC is specifically designed for mobile devices with limited memory, processing power, and graphical capabilities. Configuration is mainly a vertical classification of devices based on the storage and handling capacity, which defines the characteristics of a virtual machine and basic class libraries. Java further introduces the concept of a "profile." Profile is a framework on top of the specifications in a configuration. Profile is to more clearly differentiate a wide variety of embedded devices in the Java program, as well as to help develop specific features. Profile is built on top of a configuration basis, together constituting a complete running environment. It mainly is a horizontal classification to include specific-purpose libraries and APIs.

The standardized profiles based on CLDC have mobile information device profile (MIDP) [32] and the information module profile (IMP). In general, the combinations of CLDC and the MIDP will provide a complete Java application environment for mobile devices such as cell phones and mainstream PDAs. A Java ME application called MIDlet will be created. Similar to the regular Java programs, the MIDlets can run on any mobile devices following the specification.

With regard to the development tools, the Sun Java Wireless Toolkit can be used to develop applications in combination with suitable IDEs. In addition, the NetBeans mobility pack is a Java IDE that is used to develop applications on mobile devices.

Optional packages are used to provide additional, modular, and more diverse functions. At present, the standardization of optional packages includes database access, multimedia, Bluetooth, and so on. These are extensions to the basic profiles defined by the Java specifications, but are not mandatory. Some devices support them though others may not. To give some examples, the WMA (wireless messaging API) provides support for wireless SMS and the FCA (file access API) profile provides facilities to read and write files on devices.

The Java ME Platform SDK 3.0 is the newest toolbox for developing mobile applications. The SDK provides device emulation, a standalone development environment, and a set of utilities for rapid development of Java ME applications. The SDK itself includes the advanced tools found in the Java Wireless Toolkit 2.5.2 for CLDC. It provides the following features, which are described in detail in the following section [33,34]:

- Integration with third-party emulators and Windows Mobile devices
- On-device deployment and on-device debugging
- CLDC/MIDP, CDC/FP/PBP/AGUI, and BD-J integrated into one SDK
- New CLDC HotSpot Virtual Machine
- Optimized MSA 1.1 stack with extensions
- Profiling support
- BD-J support
- New development environment based on the NetBeans platform
- Lightweight UI Toolkit (LWUIT) integration
- Device search database integrated in SDK
- JavaFX Mobile emulator included

12.9.2 Java ME-Based Applications

Isuru et al. [35] present a software project developed to increase students' interactive participation and engage students in learning specific subjects. The QuizFun project is a mobile-enabled game platform that consists of two applications: a Web application and a mobile application. The Web application is deployed on a server and students download and install the mobile application to mobile devices. Instructors can edit questions in the server and create games using the Web interface. The mobile application from students' mobile devices connects to the server over the Internet to retrieve questions. Figure 12.11 shows the basic idea of the system architecture and the mobile application software architecture.

The Web application was developed on the platform Eclipse using ICEfaces, which is a technology based on Java server faces for building graphical UIs for the

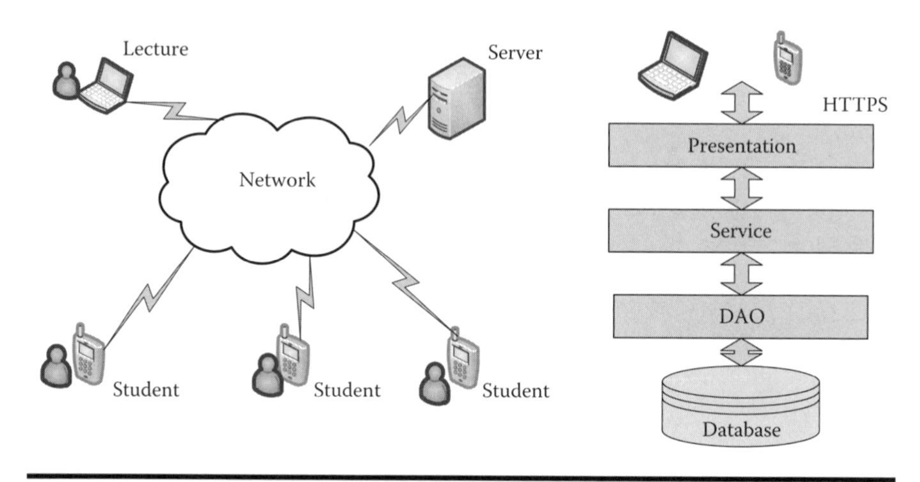

Figure 12.11 System architecture and mobile software architecture of the QuizFun project.

Java server application. The mobile application was built using Java ME and NetBeans was used as the development environment. HTTPS is used between the mobile and Web applications to ensure secure communication over the Internet. For data interchange, XML is used for the standard formatting of information.

The application has a scoring system that will give students immediate responses and feedback when a game is over. Users must register in the server before playing the games. Questions are labeled with three different difficulty levels and are classified as "modules." QuizFun has two game modes, a multiplayer game mode and a single-player game mode, and there are some slight differences between the two modes. In the multiplayer game mode, instructors create games by adding questions and answers and publish it online so students can register with the server to play with it. In the single-player mode, the students get some random questions for each difficulty level and also only the player receives the scores.

12.10 Conclusions

For the mobile platform, the developers can choose from Windows Mobile, Android, BlackBerry, iPhone operating system, Linux, and so on. As the applications for these platforms are not compatible, the choice will limit the available mobile devices, tools, and programming languages that are available to developers. Even for Java ME, the goal of "write once, run anywhere" is still not achievable in reality. To develop applications with a specific platform, we need to consider the variety of devices that are available for the platform and also the features of the platform itself. The choice of the platform will mostly determine the tools that we can use to develop the applications. Most of the platforms are proprietary and run specific programming languages. It also will determine the available IDEs for the users to develop the applications. For example, in Android, Eclipse is mostly used for the Java applications whereas Windows Mobile uses Visual Studio and iPhone uses Xcode for developing applications.

Each of the several mobile platforms in the market has its own advantages and disadvantages. The Windows Mobile platform has the advantages of compatibility with the dominant Windows operating system. It also provides a familiar development environment for Windows developers to lower the difficulty of application development especially for beginners. One of the problems with the Windows Mobile is that it usually consumes more system resources. This makes it difficult, especially for lower-end mobile devices. The cost of purchasing the software itself is also high compared to the options of several open-source-based freeware such as Android and embedded Linux.

Google Android is the new platform that has attracted much attention in the last two years and many service providers are developing products based on it. The main advantage of the platform is that it is open source. It has an application framework that focuses on reusable components to speed the process from development to manufacture. The platform is yet to be proven in the market, including the enterprise application. Embedded Linux is similar in terms of the free and open source features, which has attracted many developers in the past. It has rich resources and applications. The problem with Linux is that its debugging method is complicated and no industrial standards have been fully established to ensure the easy development, deployment, and porting to different mobile devices and products.

The Symbian operating system only requires limited system resources and few memories compared to other platforms. It provides a standard and open platform with many existing applications. It also provides the very powerful communication capability in the SDK. Unfortunately, owing to multiple versions, compatibility is a problem for the applications to run on different versions of the operating system. Also, it has a lack of multimedia functionality which has become more and more important nowadays.

The iPhone SDK has the advantage of the couple of million devices that are already in use. The iphone SDK is open source and has good extension capability. With the SDK, it is able to create the application with the best user experience. Unfortunately, the SDK can only be used to develop applications running on Apple products and it is also the only available platform for Apple iPhone and iPod Touch devices. Similar to the iPhone operating system, the BlackBerry operating system can also be only used to develop applications running on the most popular BlackBerry devices that support wireless emails. As both platforms are proprietary, it will be a big disadvantage for both to become one of the dominant application frameworks in the future. It will also be possible to become obsolete once the product loses the market share, as has happened to the Palm operating system.

REVIEW QUESTIONS

Q1: Explore the new features of the Symbian platform. Compared with another mobile development platform, what are its advantages and disadvantages?

Q2: Download the newest version of Symbian SDK and follow the instructions in the document to write a simple application. Test the application in a mobile device or run it in a Symbian emulator.

Q3: Different tools can be used for BlackBerry Java application development. In this case, use the Eclipse Plugin as the integrated development environment for BlackBerry to create an application. Deploy the project to the simulator to display the simple "Hello World" message.

Q4: Write a BlackBerry application that will support features including contacts, calendar, sending emails, and sending and receiving SMS messages.

Q5: Use the iPhone SDK to create a simple application with some common features. Examine how one can port the iPhone solution to another popular mobile platform such as BlackBerry, Windows Mobile, and so on.

Q6: Create a Windows Mobile application with Visual Studio 2005/2008. Use multiple user interface controls in the application and determine how to connect a backend SQL server database. Finally, make use of the Device Emulator to test your first Windows Mobile application.

Q7: Develop an Android application to use the Android's search framework to perform the search from a local dictionary. Try to expand the application to connect a backend database that can support a large dataset.

Q8: In the past, mobile platforms have been dominated by non-Linux-based proprietary operating systems such as Symbian, BlackBerry, and Windows Mobile. Android has become the fastest growing mobile operating system only recently. Discuss the features and advantages that it can provide and why it can surpass other Linux-based phones to compete in the market.

Q9: If you are assigned to develop a mobile application, from the development point of view, discuss which platform you would choose and why you would choose it.

Q10: Applications developed under different mobile platforms are usually not compatible. Select several mobile platforms in this chapter and discuss the porting process, if possible.

References

1. Symbian operating system, http://symbian.nokia.com/
2. Windows Mobile, http://www.microsoft.com/windowsmobile/en-us/default.mspx
3. Palm operating system, http://developer.palm.com/
4. iPhone operating system, http://www.apple.com/iphone/ios/
5. BlackBerry platform, http://na.blackberry.com/eng/developers/
6. Java ME, http://www.oracle.com/technetwork/java/javame/index.htm
7. Android platform, http://www.android.com/

8. BREW platform, http://www.brewmp.com/platform/platform
9. O. Oleinicov, M. Hassinen, K. Haataja, and P. Toivanen, Designing and implementing a novel VoIP-application for Symbian based devices, in *Proceedings of 2009 Fifth International Conference on Wireless and Mobile Communications*, Cannes/La Bocca, French Riviera, France, August 23–29, 2009, pp. 251–260.
10. M. Huebscher, N. Pryce, N. Dulay, and P. Thompson, Issues in developing Ubicomp applications on Symbian phones, in *Proceedings of the International Workshop on System Support for Future Mobile Computing Applications*, Orange County, California, September 2006, pp. 51–56.
11. D. Zaykovskiy and A. Schmitt, Java vs. Symbian: A comparison of software-based DSR implementations on mobile phones, in *Proceedings of 2008 IET 4th International Conference on Intelligent Environments*, Seattle, Washington, July 21–22, 2008, pp 1–6.
12. T. Kozel and A. Slaby, Mobile access into information systems, in *Proceedings of International Conference on Information Technology Interfaces*, Dubrovnik, June 23–26, 2008, pp. 851–856.
13. M. Massoth and D. Paulus, Mobile acquisition of sales operations based on a BlackBerry infrastructure with connection to an inventory and ERP management system, in *Proceedings of International Conference on Mobile Ubiquitous Computing, System, Services, and Technologies*, Valencia, September 29-October 4, 2008, pp. 413–418.
14. B. Falchuk, Visual and interaction design themes in mobile healthcare, in *Proceedings of MobiQuitous 2009*, 2009.
15. Xcode, http://developer.apple.com/technologies/tools/
16. Interface builder, http://developer.apple.com/technologies/tools/
17. Instruments, http://developer.apple.com/mac/library/DOCUMENTATION/DeveloperTools/Conceptual/InstrumentsUserGuide/Introduction/Introduction.html.
18. W. D. Yu and A. Siddiqui, Towards a wireless mobile social network system design in healthcare, in *Proceedings of International Conference on Multimedia and Ubiquitous Engineering*, Qingdao, China, June 4–6, 2009, pp. 429–436.
19. G. Klein and D. Murray, Parallel tracking and mapping on a camera phone, in *Proceedings of the IEEE International Symposium on Mixed and Augmented Reality*, Orlando, FL, Oct. 19–22 2009, pp. 83–86.
20. B. Stelte and I. Hochstatter, iNagMon—Network monitoring on the iPhone, in *Proceedings of the International Conference on Next Generation Mobile Applications, Services and Technologies*, Cardiff, Wales, September 15–18, 2009, pp. 534–538.
21. SQLite, http://www.sqlite.org/.
22. Z. Wang and L. Shi, A handheld wireless medical information system, in *Proceedings of the 5th International Conference on Information Technology and Application in Biomedicine, in conjunction with the 2nd International Symposium & Summer School on Biomedical and Health Engineering*, Shenzhen, China, May 30–31, 2008, pp. 315–318.
23. Y. Zhang, X. Duan, J. Wang, and L. Zhang, Design and implementation of wireless monitoring system based on Windows Mobile, in *Proceedings of 4th International Conference on Wireless Communications, Networking and Mobile Computing (WiCOM '08)*, October 12–14, 2008, pp. 1–4.
24. D.-H. Kim, M.-H. Yun, S.-J. Kim, and C.-H. Lee, Design and implementation of smartphone edition based on embedded Linux, in *Proceedings of 10th International Conference on Advanced Communication Technology (ICACT 2008)*, Vol. 1, February 17–20, 2008, pp. 328–331.

25. Android features, http://developer.android.com/guide/basics/what-is-android.html
26. D. H. Hu, F. Dong, and C.-L. Wang, A semantic context management framework on mobile device, in *Proceedings of International Conference on Embedded Software and Systems*, Zhejiang, May 25–27, 2009, pp. 331–338.
27. S. A. A. Oda, W. Chen, and M. Kitazawa, BREW implementation of a mobile phone-based monitor for women's healthcare, in *Proceedings of 6th International Special Topic Conference on Information Technology Applications in Biomedicine (ITAB 2007)*, November 8–11, 2007 pp. 288–291.
28. E. Dongre, D. Dongre, and A. Parakh, Database application for mobile phone using BREW, in *Proceedings of Third UKSim European Symposium on Computer Modeling and Simulation (EMS '09)*, November 25–27, 2009, pp. 523–528.
29. L. Zhuo, J. Wang, D. D. Feng, and L. Shen, Wireless media streaming system over CDMA networks, in *Proceedings of the 9th International Conference on Advanced Communication Technology*, Vol. 3, February 12–14, 2007, pp. 2226–2230.
30. Connected Limited Device Configuration (CLDC), http://java.sun.com/products/cldc/.
31. Connected Device Configuration (CDC), http://www.oracle.com/technetwork/java/index-jsp-138820.html
32. Mobile Information Device Profile (MIDP), http://www.oracle.com/technetwork/java/index-jsp-138820.html
33. Java Platform, Micro Edition, Software Development Kit 3.0, http://java.sun.com/javame/sdk/
34. Java ME components, http://www.oracle.com/technetwork/java/javame/index.html
35. M. Isuru, T. C. Perera, K. Lokuge, H. Mudunkotuwa, N. Premarathne, and M. Kularathna, QuizFun: Mobile based quiz game for learning, in *Proceedings of International Workshop on Technology for Education (T4E '09)*, August 4–6, 2009, pp. 95–98.

SECURITY AND APPLICATIONS OF MOBILE SERVICES

Chapter 13

Security, Privacy, and Authorization for Mobile Services

Robert Kelley, Anup Kumar, and Bin Xie

Contents

13.1 Introduction

With the proliferation of mobile services and the enormous amount of personal data stored on mobile devices and transmitted across networks, security has become a central issue. Numerous attacks threaten mobile services including, but not limited to, the compromise of personal data on the mobile device, the compromise of personal data transmitted to and from the mobile device over wireless networks, the compromise of user location privacy, and denial of service. Without security, mobile services will enjoy only limited success as many mobile users will not use them for fear of loss of personal data or infringement of privacy. It is no surprise then that there has been much research on securing mobile services to address security concerns and to encourage users to embrace mobile services.

Attacks against mobile services are multifarious. For instance, a stolen mobile device can be used to access services and data to which the thief would ordinarily not have access. Moreover, private information (e.g., credit card numbers, user location, passwords, etc.) transmitted with a mobile device can be captured by a third party to gain access to restricted systems or for financial gain. Mobile devices can even be used in denial-of-service attacks that are designed to prevent legitimate users of the network from using it. Because the attack vectors for mobile services are multidimensional, providing adequate defense for them is a substantial challenge.

This chapter focuses on the issues associated with mobile security and methods that have been devised to secure all layers of the mobile service stack. First, we describe typical network/device architectures for mobile services and present basic security definitions. We then outline common vulnerabilities that threaten mobile services and present several security solutions found in the literature. Finally, we propose a development model for writing and implementing secure mobile services and provide future directions for research in this critical area.

13.2 Background

The architectures of mobile systems vary widely and are difficult to strictly classify. Generically, the mobile service stack is divided into three layers (see Figure 13.1):

1. *Mobile Device Layer*: various devices used to connect to proprietary and public networks/services, for example, personal digital assistants, smart phones, netbooks, laptops, and tablet devices.
2. *Network Layer*: the network infrastructure used to connect mobile devices to servers. This infrastructure typically involves the public Internet and the public phone network, but may also consist of proprietary networks depending on the mobile service involved.
3. *Back-End Infrastructure Layer*: the servers and applications to which mobile devices can connect and use. These servers may host framework-based applications, Web-based applications, or Web services applications.

13.2.1 Web-Based Architectures

Two dominant approaches comprise the vast majority of mobile service applications: Web-based architecture and framework-based architecture [1]. Web-based architectures use standard Web technologies to provide applications and services to end-users. These architectures typically use a Web browser hosted on the mobile device to interact with applications that are written in xHTML, CSS, and standard server-side dynamic languages such as ASP.NET, PHP, or Perl. Alternatively, a device may use the Wireless Application Protocol (WAP), a proxy-based system, for translating standard Web data into a format that can be consumed by a mobile

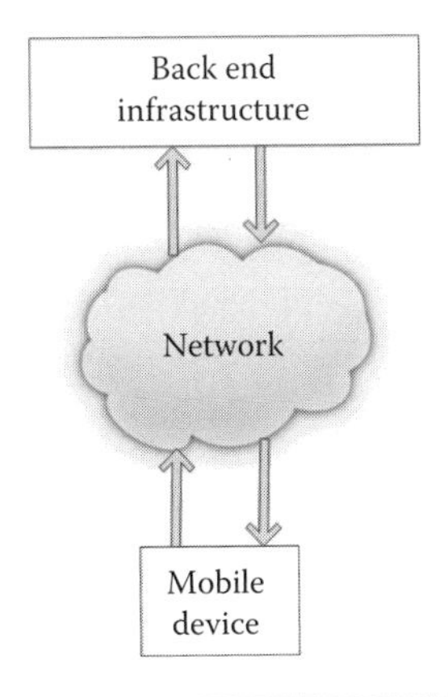

Figure 13.1 Generic mobile services stack.

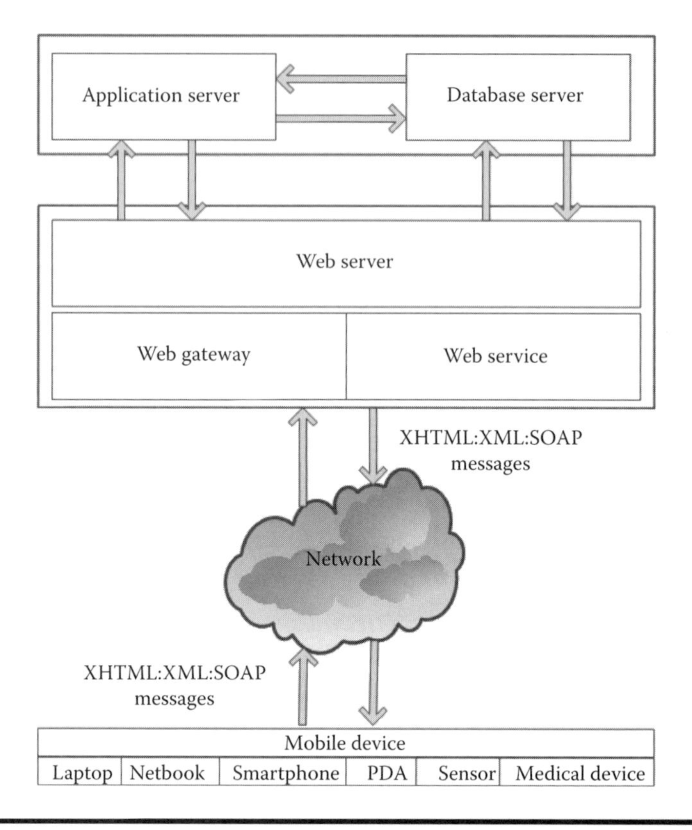

Figure 13.2 Web-based architecture for mobile services.

device.* In Web-based architecture, most of the processing for applications actually occurs on the servers providing them; the device simply provides input/output features. Figure 13.2 shows a typical layout of such architecture.†

In recent years, many Web-based architecture systems have been implemented using the concept of service-oriented architecture (SOA) in which software services can be assembled at runtime, as well as design time, and can be provided to consumers over a network. Originally designed as an architectural concept for promoting software reuse for Web-based applications running on wired networks, SOA has been extended to provide services to mobile devices as well. In fact, SOA has improved the application development landscape for mobile devices, because leveraging Web services on the network enlarges the possible application design space.

* The use of WAP is diminishing with the move toward adapting standard Web technologies to mobile devices directly, removing the need for proxy servers.

† We use the term "network" in our figures generically to refer to any data transmission network from 802.11x or WiMax-based systems to cellular telephone-based systems such as GSM, 3G, 4G, or EDGE.

With the traditional development approach, applications for mobile devices were limited by computational, storage, and power constraints. Now, it is possible to use services on the Web to "outsource" resource-intensive operations. For example, suppose a bioinformatics researcher would like to align DNA sequences as a result of a conversation with a colleague at a conference. Ordinarily, s/he would have to wait until s/he has access to at least a laptop computer to do this. However, a smart phone, with an alignment application that passes the sequences to a Web service that aligns them and returns the alignment information, could provide that information in minutes during the conversation.

The structure of SOA at its core is simple; software developers write software components that are dedicated to a particular task such as credit card authorization. These components (referred to as services), are registered in a service registry that is publicly available. An application needing software components can query the registry to find services that provide the functionality it needs. The application chooses an appropriate service on the basis of some parameter(s), for example, price or performance, and then subscribes to a service that matches its criteria. It will then use that service to provide a particular function by passing messages back and forth to it. Figure 13.3 illustrates the basic architecture of SOA.

In addition to standard Web technologies, SOA uses other standard protocols for providing services. For example, a popular format for trading messages between providers and consumers is the Simple Object Access Protocol (SOAP). The Web Services Description Language (WSDL) is a standard designed specifically to provide structure for describing Web services in the service registries. Universal Description Discovery and Integration (UDDI) and ebXML registries are the realization of the registry concept that provide potential service consumers with

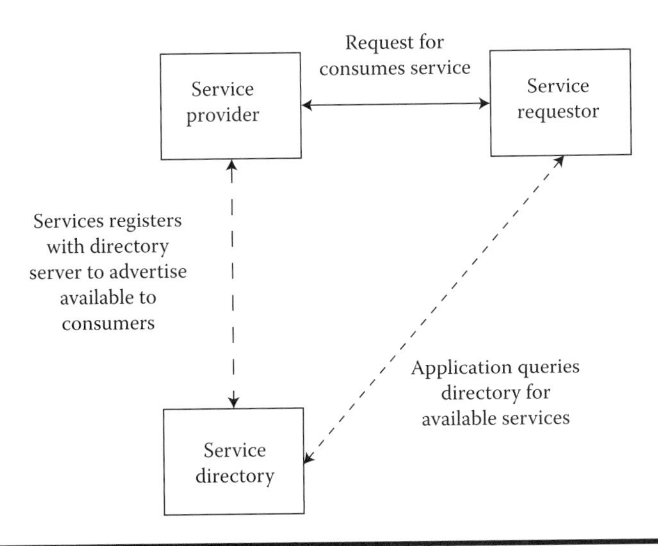

Figure 13.3 Service-oriented architecture.

information about providers and the WSDL documents that describe the services they provide. Further background information regarding SOA as well as SOAP, UDDI, and WSDL can be found in References [2–6].

13.2.2 Framework-Based Architectures

Framework-based architectures typically use non-Web specific technologies such as the Java 2 Micro Edition (J2ME) framework, the .NET Compact Framework, or the Apple iPhone's Cocoa Touch layer to provide applications. Network connections are made with standard Transmission Control Protocol (TCP)/Internet Protocol (IP) sockets using an appropriate application programming interface (API). Common applications written in this manner include global positioning system (GPS) current location services, movie ticket purchase applications, e-book readers, and so on. Figure 13.4 shows a possible architecture for this type of service. Like Web-based architectures, framework-based architectures may use Web services, but are typically proprietary in nature.

13.2.3 Next-Generation Heterogeneous Multihop Architectures

Web- and framework-based architectures are convenient models in which to view the logical associations between mobile devices and the services to which they connect. However, they do not adequately convey the physical structure of the networks. Two

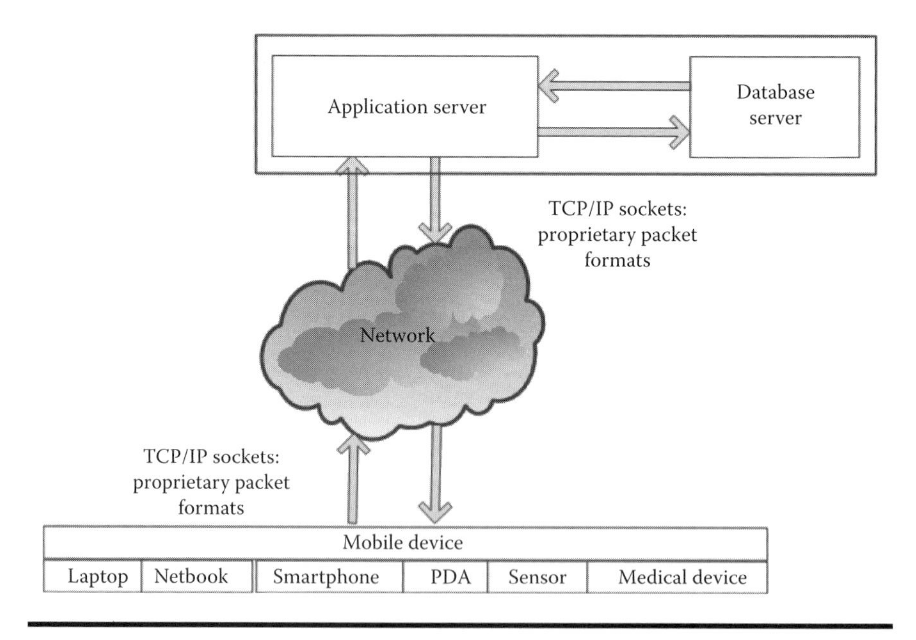

Figure 13.4 Framework-based architecture for mobile services.

architectures dominate the physical structure of most mobile networks, single hop and multihop.

Until recently, wireless devices usually contained only a single network interface (e.g., cellular, IEEE 802.15.4, IEEE 802.11.x). Internet connectivity for these types of devices occurs through a single-hop wireless connection to a base station that is further connected to the Internet through a conventional wired infrastructure (see Figure 13.5). This architecture describes a typical wireless local area network (WLAN) environment often used in corporate offices as well as cellular networks that provide Internet access.

The main limitations of the single-hop architecture includes: restrictive bandwidth access, lower quality of service (QOS) for applications, and no direct interaction with neighboring nodes. Mobile devices may come in contact with several different types of networks in the course of several hours (e.g., cellular network, WLANS, Bluetooth). In fact, in many environments, the mobile device will be within range of more than one of these services. Providing devices with multiple interfaces so they may switch from one network architecture to another depending on QOS, availability, and cost parameters is an attractive feature for mobile users. In addition, multihop connectivity where a mobile device can use a neighboring mobile unit as a router to forward its data allows enhanced coverage, providing a solution to dead-spot locations due to shadowing of buildings [7].

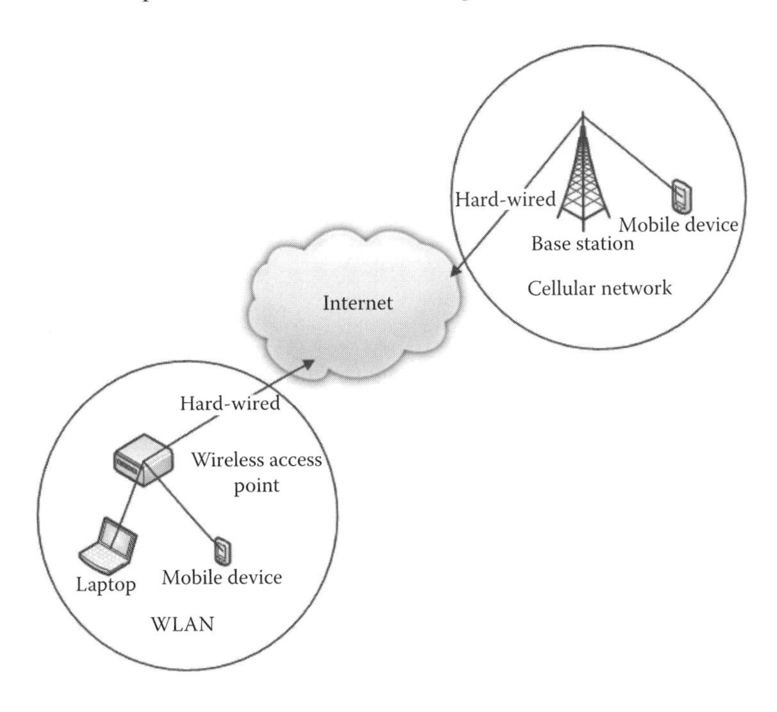

Figure 13.5 Architecture of single-hop networks.

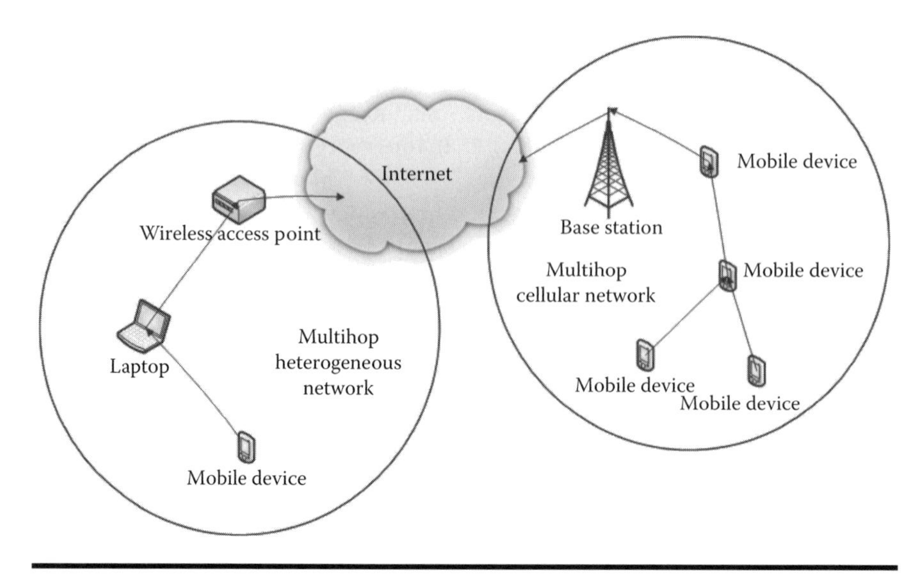

Figure 13.6 Architecture of multihop heterogeneous wireless networks.

The heterogeneous wireless network (HWN) is an emerging networking infrastructure that will eventually become the standard model for wireless interconnectivity. The HWN is an integration of a cellular network, WLAN, and a mobile *ad hoc* network. It presents an exceptional opportunity to provide ubiquitous communication where a given mobile device transmits data packets hop by hop using intermediate devices before they reach a base station. An HWN gives mobile devices with multiple interfaces (e.g., Bluetooth and cellular or 802.11 and cellular, etc.) the ability to switch interfaces dynamically when one interface can provide a better connection than another. HWN also affords the advantage of improving QOS for various applications. In this architecture a mobile device that is outside the coverage range can reach a cellular base station or a wireless access point through intermediate devices. These intermediate devices act as a router (see Figure 13.6).* A more detailed discussion on HWN can be found in Reference [8].

Mobility in wireless networks is provided by the mobile IP protocol [8]. The mobile IP performs its mobility management with two entities: home agent (HA) and foreign agent (FA), using the following steps (see Figure 13.7):

1. Mobility agents (HA and FA) advertise their presence (agent advertisement message).
2. A mobile host receives an agent advertisement and determines whether it is on its home network or a foreign network.

* Similar to wireless sensor networks in which not all nodes can reach the base station in a single hop.

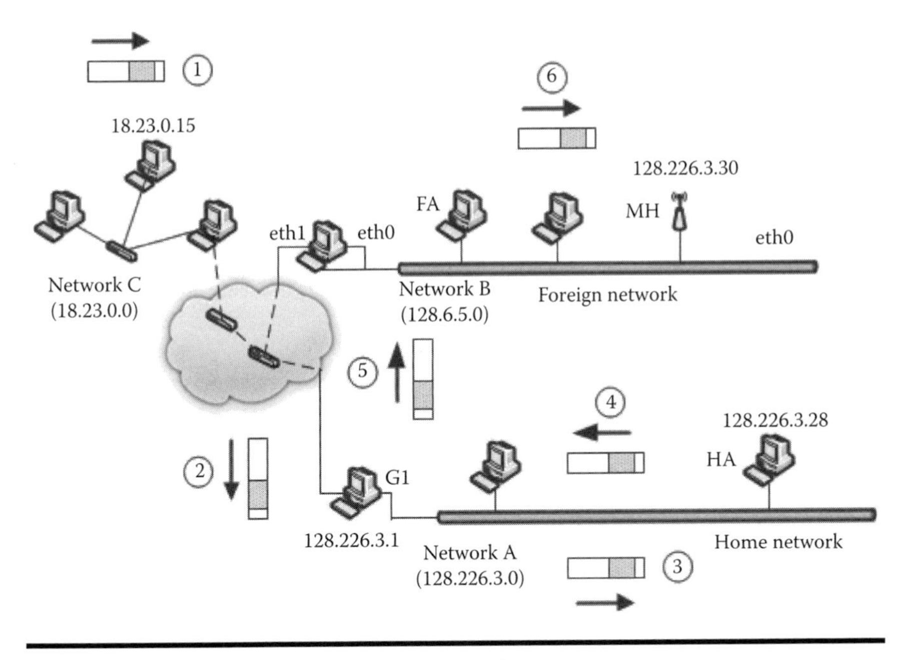

Figure 13.7 Mobile Internet protocol in action.

3. When the mobile host detects that the agent is located on its home network,
   - It operates without mobility services.
   - If it has just returned to the home network, it deregisters itself.
4. When the mobile host detects that the agent has moved to a foreign network, it obtains a care-of address on the foreign network.
5. The mobile host, operating away from home, then registers its new care-of address with its HA (via FA).
6. Datagram sent to the mobile host's home address are intercepted by the HA and tunneled to the mobile host's care-of address.
7. In the reverse direction, datagram sent by the mobile host may be delivered to their destination using standard IP routing mechanisms.

This registration process during the mobility of a device poses several security challenges as outlined in subsequent sections.

13.3 Vulnerabilities for Mobile Services

Mobile services are subject to numerous vulnerabilities typical of any network-based applications. We classify vulnerabilities into three categories: core vulnerabilities, network vulnerabilities, and application vulnerabilities. Each category engenders specific risks (see Figure 13.8) that are introduced and discussed in this section.

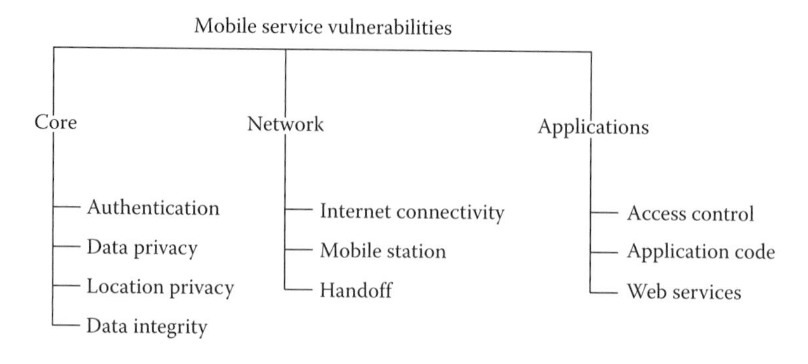

Figure 13.8 Summary of attacks against mobile services.

13.3.1 Core Vulnerabilities

Core vulnerabilities include authentication, privacy, and integrity attacks. These attacks are considered core because they are specifically designed to compromise the data consumed or generated by mobile devices. Also considered are attacks designed to compromise the identity of the user of the mobile device so the attacker may use it to gain access to a larger system.

13.3.1.1 Authentication Attacks

Authentication constitutes a mobile user proving s/he is who s/he claims to be. Authentication vulnerabilities range in complexity from simple to complex. On the simple end, malicious users simply try to guess a mobile user's password; this attack often succeeds because people use easy-to-guess passwords. More sophisticated is the dictionary attack. In this attack, the malicious user captures a mobile user's password (which is encrypted) and using the same encryption scheme as the mobile device, encrypts a list of common words and compares their encrypted form to the captured password; when they match, the adversary has discovered the password. Authentication vulnerabilities often occur in tandem with other vulnerabilities. For instance, eavesdropping which is a confidentiality vulnerability can be used to capture passwords. Compromising message integrity with spoofing or altering the contents of messages can be used to convince a user or mobile device to give up a password or other authentication token so an unauthorized entity may use it.

In many mobile environments, access to network resources involves authenticating with some authority before access to resources is granted. One vulnerability in this area includes the situation in which a legitimate user is authenticated with the mobile device and then loses it (or it is stolen) while its authentication credentials are still valid. In this case, an unauthorized user of the device will be able to access files and data that are restricted.

13.3.1.2 Data/Location Privacy Attacks

Protecting private data from unauthorized viewing and use is a central concern in any computing environment. There are two central issues regarding confidentiality for mobile services: protecting private data of users (e.g., passwords, names, phone numbers, files, etc.) and a relatively new concern that has recently been studied—protecting user location.

13.3.1.2.1 Location Privacy

A unique, but significant, problem for mobile services is location privacy. Many mobile services now use GPS to provide customized services based on a mobile user's geographic location (e.g., finding restaurants or grocery stores, etc.). GPS coordinate data are transmitted to a server that uses it as input for a query to find locations close to those coordinates. A compromised server can be used to locate and track mobile users without their knowledge, which is a severe threat to personal privacy [9]. A thorough overview of location privacy attacks is presented in Reference [10].

13.3.1.2.2 Data Privacy

Data privacy is a significant issue for mobile services. Mobile devices typically transmit data on wireless links in which the radio signal propagates in all directions; any device that is listening on the same channel can receive the traffic leading to possible message compromise. Moreover, mobile devices are designed to be portable; because of their typically small size, they are subject to theft, loss, and accidental breakage.

Of particular security concern is the protection of private data that may reside on the device. For the general consumer, this may include personal information such as passwords to other systems, social security numbers, private phone numbers, text messages, and so on. For the corporate user, the mobile device might contain internal corporate financial data and product data as well as user identification codes and passwords to corporate systems. Other mobile devices such as medical equipment or networked sensors may contain personal health data or local environmental data that should not be accessible to unauthorized parties who may have physical access to the device.

13.3.1.3 Data Integrity Attacks

Messages that are transmitted over a network can be captured by an adversary and changed before being forwarded to another mobile device or server. Not only private data may be changed but also location or device address information can be spoofed. This type of compromise might cause a mobile user to divulge personal information to an unauthorized entity unintentionally. For example, suppose a mobile user receives a text message from a trusted friend that has been forged by an adversary. Posing as the friend, the adversary may be able to get the user to disclose something personal (e.g., home address, personal whereabouts, etc.).

13.3.2 Network Vulnerabilities

Vulnerabilities in this category are attacks that focus on the network infrastructure that connects mobile devices together. Many of the attacks are similar in nature to traditional wired networks. However, device mobility and wireless connectivity change the nature and structure of many attacks as well as opens many more points of possible failure. This section introduces several vulnerabilities that are unique to mobile services.

13.3.2.1 HWN Attacks

The implementation of HWNs, as discussed in the previous section, further complicates securing mobiles services because there are multiple entry points that an adversary can exploit. This section presents several attacks to which HWNs are vulnerable.

13.3.2.1.1 Internet Denial of Service

Service providers who sell mobile services have two primary security problems: attackers using their service to access the Internet for free and attackers using their service to disrupt the network for others. Accessing the network for free can occur if the challenge–response system provided by the Internet service provider is inadequate or by MAC (media access control) address spoofing in which an unauthorized mobile device uses the MAC address of an authorized device. Disrupting the network can occur in several ways. For example, a distributed denial of service attack can be initiated in which the attacker generates bulk paging messages in a cellular system (e.g., CDMA 2000) to increase the delay of regular paging messages, including call setup requests [11]. Or an attacker may feed a wireless system with bulk messages by compromising nodes on the Internet and coordinating an attack to send the messages toward a base station that becomes too overwhelmed to process all the messages [12]. This type of attack can exhaust the radio resources of the base station, thus preventing mobile stations connected to it from transmitting. In addition, this attack works on multihop environments, so the intermediate nodes used to route data also exhaust resources.

Wireless devices on an HWN can also be used to initiate distributed denial-of-service attacks. For instance, if several wireless devices contact the base station at one time, it is possible to prevent others from using it because the radio spectrum is limited. Moreover, a wireless device may be used to remotely control "agents" on the Internet that are programmed to send a large amount of traffic to a particular device. It is difficult to trace this type of attack because the wireless device is mobile and may use a temporary identification to access the network.

13.3.2.1.2 Mobile Station Mobility Attacks

Mobile IP has several potential weaknesses in the registration procedure. First, it is subject to registration poisoning in which an attacker advertises a high-quality route to the base station. Registration messages bound for the base station are received; the attacker may drop or alter them preventing the mobile station from receiving service. An adversary may also send fake registration information to the base station with a spoofed IP address. This allows the attacker to redirect all traffic bound to the IP address to create a denial-of-service attack. Registration requests can also be captured and later replayed to give the malicious node access to the network.

The base station also has vulnerabilities. As HWNs typically involve multihop routing they can be affected by base station cache poisoning. In this scenario, an adversary provides false routing information to the base station with which it builds predetermined routing tables for reaching all mobile stations on the network. Alternatively, an adversary may respond for a legitimate node with incorrect routing information to create a denial-of-service for that node. By poisoning the cache, the adversary can direct traffic anywhere it wants.

A malicious node masquerading as a base station is another problem. The attacker may advertise itself as a genuine base station using forged messages or duplicate beacons captured by eavesdropping. Legitimate nodes see these beacons and assume the attacker is a genuine node and initiates the registration process. Eventually, multiple nodes may begin to use this base station to connect to the Internet. The attacker can then capture all of the data transmitted through it to steal passwords and other sensitive information [7].

13.3.2.1.3 Handoff Attack

Because HWNs include multiple types of networks, nodes often have to perform vertical handoff (moving from one network interface to another). Typically, handoff is initiated when a mobile device detects a base station on a different interface than what it is currently using that can offer better service. In this scenario, the mobile device initiates changing to the new network and sets up the connection. Then, the data flow is redirected from the old interface to the new interface. The handoff attack takes advantage of this process. For instance, an adversary might: initiate a false handoff by claiming a better route or better performance, drop or modify registration packets, or advertise false beacons to entice legitimate nodes to connect to it.

13.3.3 Application Vulnerabilities

Mobile devices involved in mobile services typically have applications and data that reside on the device. This condition leads to the possibility that data and applications can be subverted by a criminal to capture data or control the device. This section discusses application vulnerabilities and how they affect device operation.

13.3.3.1 Access Control

Controlling access to data that resides on a device is a significant security concern. As devices can easily be lost or stolen, personal data on the device is at considerable risk. In particular, identity theft, which has risen substantially in recent years, can be initiated on the basis of information found on a device. Moreover, many devices support removable media such as SD cards. With these devices a criminal could copy data from it and return it before the owner realizes it was ever missing.

13.3.3.2 Application Code

Mobile devices are designed to run arbitrary application codes. End-users often have no way of ensuring the applications they run are safe from viruses and malicious code that will either steal data from the device or subvert the network connection for illegitimate purposes. For instance, buffer overflows, in which a program can put more data into a memory location than it can hold, may cause the mobile operation system (OS) to elevate privileges to the attacker.

Viruses can be delivered to the mobile device without the mobile user's knowledge through downloading a legitimate program in which a virus has been attached. Viruses may be designed to corrupt data or to direct the device to make unauthorized network connections that can incur significant fees for the mobile user (e.g., calling 900 phone numbers). For example, a recently discovered Trojan, called Viver, sends SMS messages to premium-rate phone numbers, incurring a fee for the infected user [13].

Application security is particularly difficult because there are a multitude of OSs and software application programming interfaces used by the mobile device development community. As a result, there are few one-size-fits-all solutions for application security.

13.3.3.3 Web Services Vulnerabilities

Web services are the realization of the SOA. At the basic level, Web services are simply applications that mobile users can use over the network. The most significant vulnerability for a Web service is the XML document that is used to transfer data from the server to the client. As these are plaintext documents, an adversary can capture the traffic stream in which they are transmitted and read the document to identify the information contained in them. Further, securing a Web service so only authorized users can use it is a challenge. Web services, by their very nature, are built on open standards; they are typically listed in public registries for the purpose of making them available to everyone. A balance between restricting access to the service and publishing its existence has to be maintained.

13.4 Security Solutions for Mobile Services

13.4.1 Solutions for Core Vulnerabilities

13.4.1.1 Authentication

Authentication refers to the process of proving the identity of a user with one or more of the three basic qualities [14]:

1. Something the user knows: password, pin numbers, and so on.
2. Something the user has: identity card, badges, and physical keys.
3. Something the user is: fingerprint patterns, retina scans, voice patterns.

Two primary models of authentication for mobile devices are prevalent in the literature. The first model uses Kerberos or Kerberos-like verification systems that use tickets for identity. The second involves using multifactor authorization with items other than tickets. In this section, we will present several different schemes for authentication that have been suggested in the literature.

13.4.1.1.1 Ticket-Based Authentication

Perhaps the best-known solution for network authentication is Kerberos [15]. Originally designed at MIT for the Athena Project, Kerberos has been codified as an official RFC of the Internet Engineering Task Force (RFC 4120). Kerberos uses a central authentication server and a ticketing server that clients use for multiple servers/services authentication. The process to authenticate a user is only performed once at the beginning of a network access session. The client (in this case a mobile device) requests authentication from the authentication server that authenticates the user on the basis of a user-supplied password. The authentication server then returns a session key and a ticket to the client as well as the ticket server.

Once a mobile device user has been authenticated and has obtained a ticket to the ticket server, it can use that ticket to request further tickets to other networks servers. For each resource the mobile device wants to access, it must request a ticket from the ticket server that will verify the user and the access permissions for the desired resource and then provide a session key for accessing that resource (see Figure 13.9).

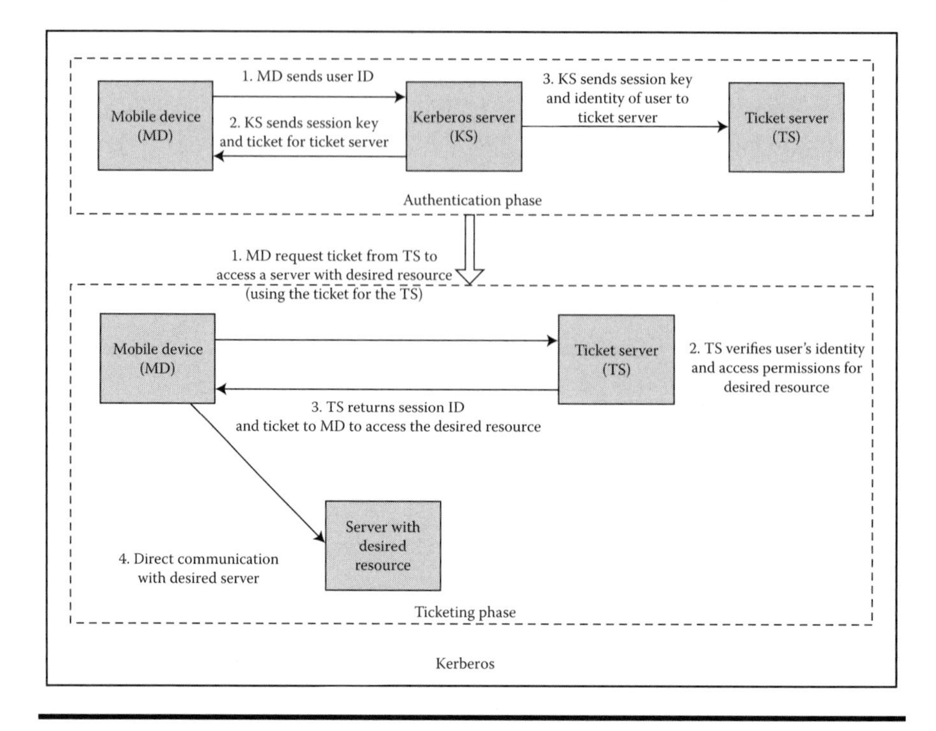

Figure 13.9 Kerberos authentication and ticketing phases.

Kerberos provides many advantages for authentication. For example, passwords are not communicated on the network because they are stored on the server, spoofing is difficult because the ticket server is involved in every request for server access; tickets are only viable for a short amount of time reducing the window for replay attacks. The interested reader is directed to References [14,16] for a more in-depth explanation of Kerberos.

Several implementations of Kerberos (or variants of Kerberos) have been proposed. Wang et al. [17] suggest a ticket-based authentication system for mobile commerce that uses a Kerberos infrastructure but with a Web services approach. This scheme has a Trusted Authentication and Registration Center (TARC) that authenticates users and a Trusted Credential Center (TCC) that serves as the ticket server. Services are described with WSDL, listed in the TCC using UDDI and messages are transported in SOAP format.

Pirzada and McDonald [18] argue the ticket granting server in Kerberos does not scale well for mobile services and instead propose a modification in which authentication and ticketing functions are aggregated into a single server they call a Kaman server. When a mobile device wants to create a trusted connection with another host, it requests a ticket from the Kaman server that provides a session key for the hosts to use. The requesting device then sends the ticket to the destination device which uses that ticket to set up a secure channel on which to communicate. Several Kaman servers can be used for authentication so the load can be distributed. For the system to work, it is assumed that all users have a secret key (password) known only to them, all servers know the hashed passwords of each user, and all servers share a secret key with each other so user information can be replicated across the Kaman network. Like traditional Kerberos, tickets are only valid for a certain amount of time. Should the mobile device be stolen and used by an adversary, the tickets the device contains will eventually expire and reauthentication will be necessary for further access.

Providing direct access to Kerberos infrastructures can be problematic as mobile devices are often resource constrained (e.g., limited computational ability, storage, and power). Proxies that can mediate between mobile devices and Kerberos infrastructures can help overcome this limitation. Fox and Gribble [19] propose such a system with Charon. In this scheme, much of the processing for authentication occurs on a proxy server rather than the mobile device. The proxy server and the mobile device create a secure channel in which to communicate. When the client wants to authenticate to a network service, using the secure channel, the mobile device contacts the proxy device which then contacts the Kerberos authentication and ticketing servers. As far as the Kerberos infrastructure is concerned, it is not aware of the mobile device directly—all interaction occurs between the proxy and the authentication or ticketing server. Fox and Gribble argue this type of implementation of Kerberos is necessary for mobile devices because existing Kerberos programs are too large to install on mobile devices. Moreover, many mobile devices use stripped down versions of OSs that do not support the full API. The API, which provides class libraries that support input/output functions, memory management, file and directory management, network sockets, and so on, is usually smaller for mobile devices because of their limited resources. For example, in the Microsoft Windows OS space, the .Net compact framework for the Windows Mobile OS environment contains only about 30% of the full .Net framework used for Windows XP or Windows 7 environments [20].

Although Kerberos solves many authentication problems, it is not a perfect solution. It suffers from some potential drawbacks that have been introduced in Reference [14] and are briefly presented here:

1. Kerberos requires continuous availability of the trusted ticket granting server as well as a trusted relationship between the ticket granting server and every server users want to access.
2. Kerberos requires timely transactions. Tickets are designed so they are only valid during a certain time frame. Once a ticket has been granted, it has to be used within that time frame or it expires.
3. A subverted device can save and later replay user passwords; strong passwords are not enforced so users can use relatively weak passwords.

4. Kerberos does not scale well as it is envisioned as an end-to-end solution; all servers and devices must use it to be effective and as the network grows, so does the overhead of managing the authentication/ticketing infrastructure.

It can be argued that not all of these issues are that serious. Although it would be convenient and much easier to design systems without the Kerberos infrastructure because designers have more flexibility, it is unreasonable to believe robust solutions can really be built without an infrastructure of some type, whether it is Kerberos or another system. There are many environments in which supporting a Kerberos infrastructure is not unreasonable at all (e.g., corporate environments).

Of the deficiencies listed here, the most critical are Kerberos' vulnerability to password guessing attacks and the fact that a subverted device can replay user passwords under certain circumstances. Less critical, but still a concern is availability of the ticket server. If a device does not have a connection to the ticket server when a ticket expires, it could be prevented from accessing even local data on the device.

Sun and Yeh [21] address the issue of password insecurity (password guessing and replay) with their perfect forward secrecy protocol. In their system, security is divided into seven classes with Class 1 being the least secure (with the least overhead) and Class 7 being the most secure (with the most overhead). Depending on environmental needs, the level of security can be tuned. The perfect forward secrecy system essentially uses various ticketing strategies. Most involve the exchange of a password that has been concatenated to random numbers chosen by the client. Because this random number changes with every authentication, guessing the password is difficult, even if the user has a weak password. The addition of the random number also prevents replay attacks as the number changes for each authentication session. The limitation of this system is that it uses public-key cryptography that is unsuitable for resource-constrained devices. To address this deficiency, Chang et al. [22] suggest replacing the public key infrastructure (PKI) component with much simpler one-way hashing functions and XOR operations. This approach uses much less power than Sun and Yeh's [21] proposed scheme.

13.4.1.1.2 Multifactor Authentication

Besides various Kerberos incarnations, numerous solutions have been proposed for authentication for mobile devices. Many of them focus on the issue that using only passwords as a single authentication factor reduces the networks resilience against compromise because only one piece of information needs to be captured to allow unauthorized users to access the network. Two or more factor authentication in which the user has to provide a password as well as other information to be authenticated (e.g., fingerprint, eyes scan, security badge, etc.) can provide much better security.

One novel solution for two-factor authentication is presented by Soleymani and Maheswaran [23] who suggest using social authentication. In this scheme, mobile device users (specifically cell phones users) declare a list of friends that will "vouch" for them. The list is stored on a central server. Each time the device participates in a call of a certain minimum duration with a friend on the list or comes in contact with a friend's device over Bluetooth, the phone is issued a token. The token essentially indicates that phone has recently had contact with a friend device that is listed on the friend list. To authenticate, users must use a personal identification number and have a recent set of tokens to be able to use the device. The idea behind this scheme is that a malicious user will not be able to collect enough fresh tokens to use the device (or access restricted areas) without alerting the owner of the device because tokens are only issued when a call is of a minimum duration or the devices are physically close enough to connect over Bluetooth.

Other methods for authentication include using device location and images rather than passwords for verifying user identity. Jansen and Korolev [24] suggest using device location to implement authentication. With this scheme a mobile device authenticates a beacon it receives from a fixed device in the environment (e.g., office or warehouse) and downloads the access policy it provides for that location. The mobile device presents information from the policy along with a traditional proof of identity (e.g., password) for authentication. Sun et al. [25] suggest using images for authentication. In this scheme, face image hashing is combined with traditional graphical password to provide a two-factor authentication scheme. Information extracted from the user's image and user supplied graphic (not of the user's face) are used together to provide two-factor authentication.

13.4.1.1.3 Reauthentication Based on Mobility

An approach to handle the case in which a mobile user authenticates to gain access to network services and then loses the mobile device is presented in Reference [26]. In this system, the mobility patterns of the device are monitored over a period of time to develop a mobility profile using a hierarchical hidden Markov model. In the event the mobility pattern of the device strays from the established profile, the user is asked to reauthenticate before further access to network resources is granted. This solution assumes a person who captures the device will deviate from the normal patterns of mobility for the device, which is not always the case. However, this system works at the cell tower level; movement of even a city block during a period in which the mobility pattern is typically static can be detected. The main drawback of this system is if the mobility pattern of the user is extremely random, an accurate mobility profile cannot be generated.

13.4.1.1.4 Limitations of Authentication Solutions

Although all the approaches presented here increase the security of the authentication process, they do have some disadvantages. Many assume an infrastructure (ticket servers, voucher services, etc.) of some type exists before devices can authenticate. The overhead (both in cost and complexity) of assembling these infrastructures can be quite high. Moreover, they often will not scale well as the domain increases in size or across multiple domains. The user interface for authentication is not well articulated in many of the solutions, which suggests there could be practical issues with implementing the actual schemes. Further, the time it takes to authenticate users and distribute tickets increases the overall response time of the system.

13.4.1.2 Data/Location Privacy

13.4.1.2.1 Location Privacy

The most popular approach for mitigating location privacy risks uses k-Anonymity and third-party servers that process and return location-based queries. k-Anonymity was first introduced by Sweeney [27] as a way to protect the linking of anonymized datasets to external data (e.g., linking private medical information with voter registration information) to identify a user. Its core feature is that, for any dataset, at least k records in the set should have exactly the same values for a set of attributes in the set, thus adding noise to the data so identification of an individual is made difficult or impossible. For example, if the attributes date of birth, age, and race are used with a value of $k = 10$, at least 10 records in the dataset should all have the same value for those attributes.

The most famous implementation of k-Anonymity is CASPER as suggested by Mokbel et al. [28]. In CASPER, the mobile user registers with a third-party anonymizer server to whom the user provides a privacy profile. The profile consists of the tuple (k, A_{min}), with k being the minimum number of users from which the mobile user is not distinguishable and A_{min} being the acceptable spatial region for cloaking (physical size of cell region). k-Anonymity is used to cloak the location of the user by mapping the user's location to a larger area which is defined by A_{min}. The system also uses a location-based query server. On the basis of the area defined by A_{min}, query results consist of candidates from which the user may choose. For example, suppose A_{min} is 15 square miles, the query server would return results for that entire area when the user only needs results for 1 square mile. The user chooses which candidate result matches his needs. The CASPER system is unaware of that choice. Performance results suggest that the amount of time to cloak the mobile user and the amount of privacy afforded by CASPER is adequate to serve the needs of most mobile users. The main limitation of CASPER is that it uses collaborative anonymization in the form of a third-party server that anonymizes user locations. If the server is compromised, the user location is compromised as well. Other implementations of k-Anonymity with third-party location servers are discussed in References [29–36].

A novel solution to address the problem of a compromised anonymizing server is presented by Ghinita et al. [37]. In this solution, devices register with a central certificate server (that does not know device location) which returns a list of other devices that have been registered. When a mobile user executes a location-based query, the nodes that are registered with the certificate server cooperate to generate a spatial region large enough to anonymize the request based on user parameters (e.g., k-Anonymity). The location-based query is then sent to a pseudonym server that hides the IP of the requester and forwards the request to the location-based service. The limitation of this approach is that a significant infrastructure needs to exist to provide anonymous service; collaboration between nodes and a server is still necessary.

13.4.1.2.2 Data Privacy

Protecting private data used by mobile devices are typically accomplished through symmetric key cryptography or asymmetric key cryptography (commonly known as PKI cryptography). Using a key, personal data on the device are encrypted so it is unreadable to anyone without the key. For data privacy to be successful, two domains have to be secured: the file system of the device itself and the wireless medium on which data are transmitted.

13.4.1.2.2.1 Local File Privacy — Encryption for local files is usually accomplished with symmetric key cryptography that is more suited to mobile devices because most of the frequently used symmetric encryption algorithms (DES, triple DES, and AES) use appreciably less computational power than PKI. Encryption of local data is generally a straightforward process. A cipher program running on the local device uses a key to encrypt data and the user provides the key to decrypt it. Ciphers can be devised so data are encrypted one byte at a time or in blocks. Specifications for common ciphers such as AES, Triple-DES, and Skipjack can be found in Reference [38]. One novel solution for encryption data is to use a two-device authentication system in which part of the authentication data is on the device and part of it is stored on the network [39]. This scheme uses two pieces of identification for authentication, a textual password (of 8 to 32 characters) and a finger print. Before a user can access a local file, s/he must be authenticated. First the password is divided into two parts. The first four characters are processed by the local OS, the remaining characters are sent to a network-based authenticator. If the password credential is validated, the thumbprint image is divided in half. The first half is validated by the local OS and the rest of the scan is sent to a network-based authenticator. The weakness of this type of solution is that, without network connectivity, the user is unable to access local files because only half of the authentication is completed.

13.4.1.2.2.2 Network Privacy — Besides local file security on mobile devices, network access security is another concern. As mobile devices often have access to personal or corporate data across a network, it is necessary to protect that data during transmission. Again, this is often accomplished with some version of encryption. Although symmetric key encryption is popular for encrypting local files, PKI encryption is considerably more secure and more appropriate for network transmission. PKI is a well-studied area of security; therefore, it will not be presented here. The interested reader is directed to References [40–43]. PKI is usually implemented at the transport layer of the TCP/IP stack in the form of secure socket layer/transport layer security (SSL/TLS). In the past, implementing PKI encryption on mobile devices was difficult because of device resource constraints. However, with improvements in hardware in recent years, this problem is abating. For devices that are still unable to perform PKI calculations, a PKI proxy has been suggested [44,45]. In these schemes, a mobile device uses a secure symmetric key to connect to the PKI server and then sends all of its PKI computations to the PKI proxy for processing. The obvious limitation of this scheme is that the connection between the server and the mobile device has to be secured before PKI operations can occur. So the schemes depend on symmetric keys to establish a secure connection between the devices, so there is overhead to establish this connection.

13.4.2 Solutions for Network Vulnerabilities

13.4.2.1 HWN Security

This section describes a simple and effective approach discussed in Reference [7] to secure connectivity framework for integrated heterogeneous networks. The security protocol conforms to mobile IP security as well as *ad hoc* routing security. The proposed approach for securing global connectivity and *ad hoc* network has the following properties:

1. Mobile IP security follows the security strategy of *minimal public-based authentication protocol* with certain extensions. HA acts as the authentication server.
2. Only after being registered with FA and certificated by FA, a mobile node (MN) can be trusted by other *ad hoc* MNs so that the MN can participate in *ad hoc* routing protocols.
3. MN's home address, its *ad hoc* address, and its certificate are bound for identifying the MN. Therefore, the trustable relationships between *ad hoc* nodes are enhanced.
4. The approach uses the cryptographic certificates, which are also used in single-hop 802.11 and ARAN [22], that makes *ad hoc* secure.

The security protocol for integrated wired and *ad hoc* networks includes two parts: the global security of mobile IP and the security of the *ad hoc* network. In order to communicate with other nodes in the *ad hoc* network or the wired network, MN performs the following operations:

1. *FA advertisement and discovery*: FA advertises and MN finds a route to FA.
2. *MN registration with FA and HA*: MN follows minimal public-based authentication protocol to register with FA and HA.
3. *Binding MN mobile IP home address and its ad hoc identifier*: MN's home address, its *ad hoc* identifier, and its certificate are bound for tracing the history of the MN.
4. FA issues the certificate to the registered MNs from the *ad hoc* network.
5. MN creates a local data structure of certificates.

The sequence diagram in Figure 13.10 illustrates the basic process of security implementation in integrated wired and ad hoc networks. The example assumes it is the first time that MN_A is communicating in the *ad hoc* network. Other *ad hoc* hosts are already successfully registered with FA and certified by FA. To find a path from MN_A to FA, MN_A initiates the FA discovery by issuing a message *R_Request* with a signature. FA replies to the MN_A with *R_Reply*. Then MN_A sends a message *Solicitation* to FA for requesting an *Advertisement* with a COA from FA. The MN_A registers with HA via FA. If the registration is successful, then FA issues MN_A with a certificate and other MN certificates. FA also issues other MNs with MN_A's certificate. Therefore, MN_A can initiate a route discovery for searching another mobile host, such as MN_x or a correspondent node in a wired network. MN_x will respond to MN_A with a route. The detailed exchange of messages is found in Reference [7].

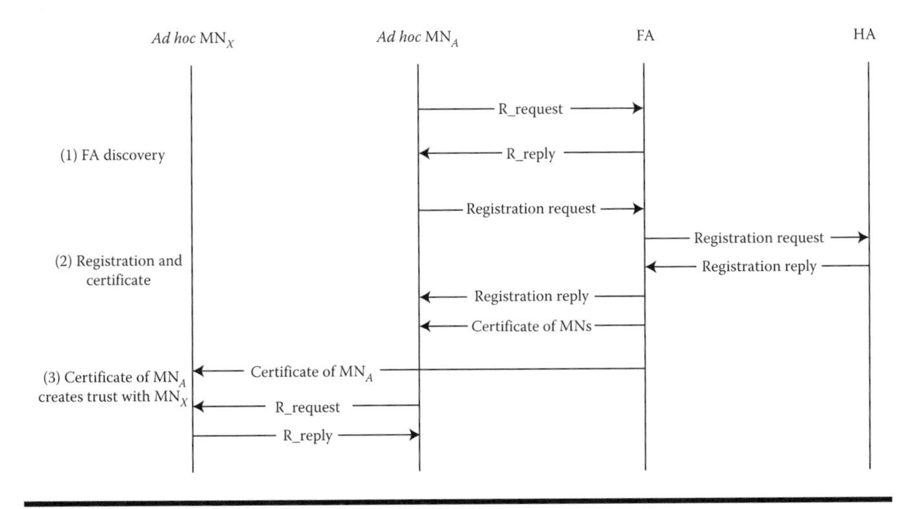

Figure 13.10 An example of routing security.

13.4.2.2 Security Analyses

13.4.2.2.1 Authentication and Access Control

HA authenticates MN and FA. FA issues certificates to each MN that is used for authentications of other *ad hoc* hosts. Before participating in *ad hoc* routing protocol, MN must register with FA to obtain the certificate from the FA. From the view of MN, once the MN receives a successful registration reply from the HA, it is assured that the FA is valid. Meanwhile from the view of the *ad hoc* network, if MN registers with FA successfully, the MN is a trustable one. Furthermore, FA issues a certificate to the MN. Therefore, registration with FA plays two main roles here:

a. MN's registration with FA for mobility binding at HA to determine whether the MN obtains its mobility from HA.
b. Access control in an *ad hoc* network to determine whether the MN can participate in an *ad hoc* routing protocol, which enhances the trust relationships in the *ad hoc* network.

Any node cannot generate a valid route discovery message by spoofing or inventing an IP address. In the process of route discovery, the control message created by a node must be signed and validated by a receiving node. Thus, the route discovery prevents antiauthenticating attacks, such as *creating routing loop fabrication* because no node can create and sign a packet in the name of a spoofed or invented node [7].

13.4.2.2.2 Identification

Without any centralized administration, one of the crucial problems of malicious nodes in *ad hoc* networks is that it is very easy for these nodes to change their identities in the *ad hoc* network. It is because of the lack of consistent identities for *ad hoc* hosts that the histories of *ad hoc* hosts could not be traced. On the other hand, when global connectivity is available, the proposed approach binds the *ad hoc* hosts' home addresses with their identities in the *ad hoc* network. The bindings are unique because of the uniqueness of the *ad hoc* host's home address. The binding is also associated with the *ad hoc* host's secret key and certificate. Therefore, it becomes difficult for any ad hoc host to masquerade itself by spoofing or creating a valid address.

13.4.2.2.3 Nonduplication

Nonce* and timestamp make a route request or reply containing unique data, to battle against the attack of forwarding a duplicated message from a malicious

* An abbreviation of "number used once"—a random or pseudorandom number issued in an authentication protocol to ensure that old communications cannot be reused in replay attacks.

node. The first route request message to a certain destination contains a nonce. A new nonce in the reply message indicates the next nonce in the next request. When an intermediate or target node receives a routing control message, the node compares certain fields of the received packet with corresponding data in the local table to avoid duplicate processing. If it has timestamp in the received packet, the received node also makes sure the timestamp is close enough to current estimated time.

13.4.2.2.4 Integrity

Packets in ad hoc networks are signed using a private key at each node. Then the receiver verifies the signature and certificate of the sender. It fights again attacks of anti-integrity, such as the attack of *modification*.

13.4.3 Solutions for Application Vulnerabilities

13.4.3.1 Access Control

Controlling access to data that reside on a mobile device is a significant security concern. Virtualization has garnered considerable attention in the recent literature as a mechanism for enforcing access control on mobile devices. There are three possible levels of virtualization: paravirtualization, full virtualization, and hardware-assisted virtualization [46]. The predominant approach for implementing virtualization on mobile devices is paravirtualization through an entity called the Virtual Machine Monitor (VMM) (see Figure 13.11). Unlike full virtualization, the system call interface of the guest OS has to be modified to work with the VMM which hampers developers' ability to use any OS as a guest on arbitrary hardware. Currently, Linux is the most popular OS for running VMMs because it is open source and the kernel can be modified to work with the VMM.

One implementation of VMM access control is presented by Lee et al. [47]. The proposed architecture includes a secure-on-chip ROM that stores a master key that keeps data in the flash memory confidential. At boot time, the boot loader verifies the authenticity of the VMM with a digital certificate that has been preloaded on to the device. Once the VMM has been authenticated, it authenticates the two

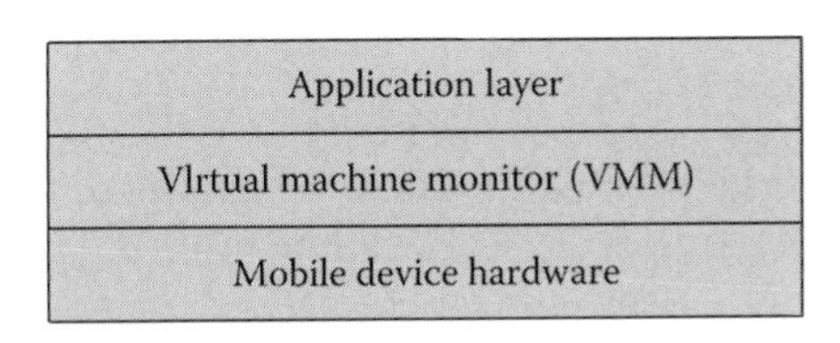

Figure 13.11 Mobile device architecture with VMM layer.

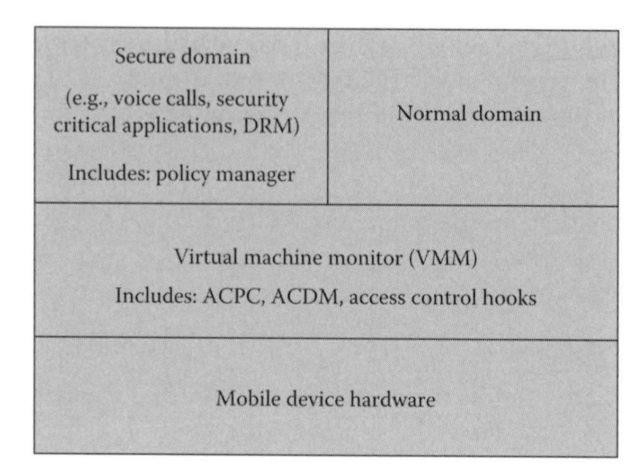

Figure 13.12 VMM architecture proposed by Lee et al. (Adapted from Lee, S.-M. et al., In *Proceedings of the 5th IEEE Consumer Communications and Networking Conference (CCNC 2008)*, 2008.)

domains that run on top of it (e.g., secure domain and normal domain). The secure domain hosts applications that require access control such as voice calls, online banking applications, shopping applications, and stock trading applications.

Actual access control functionality resides mostly into the VMM portion of the stack (see Figure 13.12) and is embodied in four modules:

1. *Access Control Policy Conductor*: loads the access control models that will be used by the device and decrypts the encrypted policies at boot time.
2. *Access Control Decision-Maker*: grants access to resources on the basis of the policies of the access control policy conductor.
3. *Access Control Hooks*: provide an interface between the access control decision-maker and the guest OS. These hooks are OS specific.
4. *Policy Manager*: provides an interface to the access control policy conductor to add additional security policies when necessary at the level of the secure domain.

Similar mechanisms have also been discussed by Winter [48] and Lee et al. [49]. In both of these solutions, the VMM provides access control services for peripherals and storage.

13.4.3.2 Application Code

Several strategies have been proposed to prevent running untrusted code or at least sandboxing it so malicious applications can cause limited damage on a device. The most common approach for application code security involves policy-based systems in which the security profile of a device and/or the security profile of an application are checked before execution of the application.

One method for implementing policy-based security is a security-by-contract model [50]. In this model, applications are managed and distributed to devices according to well-defined security policies. For example, the domain administrator for a corporate environment will certify what applications will run on which devices in the network and generate a policy that can be distributed to the mobile devices outlining the certifications. Once an application is distributed, it is installed, but execution is blocked until the application security profile and the device security profile have been reconciled. The model supports a flexible layer to allow various methods for verifying the security profiles. If the application is allowed to run, it is constantly monitored during runtime by the policy manager that will block any attempts to execute operations that are outside its defined policy. For example, if an application policy prohibits writing to flash memory on the device and it tries to do so, the policy manager will prevent the operation.

Similarly, Enck et al. [51] propose using a security service named Kirin. The Kirin service takes advantage of the security model in Android that restricts application permissions in two ways: user confirmation and developer application signatures. It adds to that architecture a checking service that reviews the security implications of installing a particular application against a collection of security rules. If the application fails any rule, Kirin prevents the application from being installed. Alternatively, the user is notified of the risks of installing that particular application and is given the opportunity to install it anyway.

13.4.3.3 Web Services Security

As SOAs are extended to support mobile devices, security techniques for Web services have to be re-examined in the context of the limitations of mobile devices. In this section, we present mechanisms for securing Web services in general and adaptations and strategies for securing mobile devices that use Web services.

13.4.3.3.1 XML Document Security

The foundation of Web services is the SOAP message which is the Extensible Markup Language (XML), a subset of the Standard Generalized Markup Language (SGML). XML allows programmers to define custom tags, attributes, and document types to organize and retrieve any type of structured data. Originally envisioned as a convenient method for formatting data in a human-readable and machine-parsable format for presentation over the Web, XML is also being used in many environments to transfer data from disparate systems in a structured manner. Because XML was not originally designed with security in mind, several mechanisms have been developed for implementing security. The following few sections briefly introduce these mechanisms (see Table 13.1). More comprehensive treatments of XML security extensions can be found in References [52–54].

Table 13.1 XML Security Provisions

Security Operation	*Standard*	*Governing Standards Body*
Integrity and signatures	XML digital signatures [55]	W3C
Confidentiality	XML encryption [56]	W3C
Key management	XML key management specification [57]	W3C
Authentication and authorization assertions	Security Assertion Markup Language [58]	OASIS
Authorization rules	Extensible Access Control Markup Language [59]	OASIS
Rights management	Extensible Rights Markup Language [60]	OASIS

13.4.3.3.2 XML Digital Signatures

As with traditional digital signatures, XML digital signatures (XMLDS) are designed so a document user can "sign" a document as proof of the document's integrity as well as confirmation of who created it. XMLDS differ from traditional signatures in two important ways:

1. It supports the ability to digitally sign on specific portions of a document, not the entire document tree.
2. It supports *canonicalizations*, transformations that ensure that the same XML document with small formatting differences (e.g., extra white space) will yield the same digital signature.

The XMLDS standard specifies three types of signatures: enveloped and enveloping signatures, in which the signatures are part of the files they sign, and detached signatures that are used to sign separate documents.

13.4.3.3.3 XML Encryption

Like XMLDS, XML encryption (XMLE) can be applied to specific portions of a document or the entire document tree, depending on application security requirements. XMLE supports both symmetric and asymmetric encryption and may be used together. For example, a common strategy is to encrypt a document with a symmetric key and then encrypt the key with asymmetric cryptography and send both the document and the encrypted key to the recipient. The advantage of this strategy for mobile devices in particular is clear. Encrypting a key with asymmetric cryptography rather than an entire document can yield much more efficiency and reduce processing time on the mobile device. Another significant advantage of XMLE is that encryption can be used for storage as well as transmission; documents can be stored in encrypted format that can provide increased security for a mobile device if it is lost or stolen.

13.4.3.3.4 XML Key Management Specification

To support PKI encryption, the XML key management specification (XKMS) standard specifies protocols between XKMS clients and servers. It defines message formats for requests and responses for key management such as registration, validation, key discovery, and key revocation. It was designed specifically to work in tandem with XMLDS and XMLE. One of the advantages of XKMS is that a server may be used as a proxy for encryption/decryption for mobile devices that do not have enough resources to perform PKI operations by themselves.

13.4.3.3.5 Security Assertion Markup Language

The Security Assertion Markup Language (SAML) is used to exchange authorization and authentication information between disparate systems. SAML is specifically designed to provide two features: single sign-on (SSO) and federated identity (FI). Prior to SAML, Web-based SSO typically used browser cookies to maintain user authentication state. Cookies, however, are not transmitted between DNS domains, so they are not available for use in multiple DNS domains. Products claiming to support SSO actually used a proprietary system for exchanging authentication state between domains. SAML provides a standard vendor-independent grammar and protocol for transferring information about a user from one Web server to another, regardless of DNS domains. Specifically, SAML supports assertions, which are statements specifying identity information for subjects (e.g., an end-user) and a subject's attributes such as group membership and other credentials, as well as a framework for sharing this information with other Web servers. For example, authentication information could be shared between eBay and PayPal during a transaction so that each service does not have to maintain a separate cookie for this information.

Although SSO allows users to use one sign-on for authentication, it still requires a user have a sign-on for each system on which it wants to authenticate. SAML further enumerates standards for FI, which allows online services that want to build a collaborative application environment for mutual users to use a single identity for that user that is shared between all parties involved. The advantage of this standard is that each entity does not have to maintain user accounts and provide identity management services; instead it uses a single identity management that all other services use as well. Using the previous example of eBay and

PayPal, FI would negate the need for separate user accounts in each domain. Instead, the user has the same identity for both, with the identity information stored in a single location.

SAML is not intended for access control directly, but provides information that can be used for access control. It assumes a trust relationship already exists between the applications and is therefore not concerned with confidentiality, integrity, or any other security controls.

13.4.3.3.6 Extensible Access Control Markup Language

Whereas SAML is concerned with exchanging authorization and authentication information, the Extensible Access Control Markup Language (XACML) is used in conjunction with SAML to implement access control of actual resources (e.g., XML documents in this case). XACML is a common language for expressing security policy. It defines how to encode and bundle XACML rules that contain four pieces of information:

1. *Subject*: includes user identification, groups, or role names.
2. *Target object*: the resource for which the rule is applicable and like XMLDS and XMLE can be applied to the document element level.
3. *Permitted action*: specifies read, write, create, or delete access.
4. *Provision*: includes any action that must be executed on the rule's activation (e.g., logging in, requesting further credentials, or alerting another user that a resource is being accessed).

These rules are defined at configuration time. At runtime, an application can send these rules along with text or binary data to indicate who may use it and under what conditions. For example, digital rights management can be implemented with XACML rules that accompany audio or video data; the application decodes the rules and uses them to implement security on the end-user device.

13.4.3.3.7 Extensible Rights Markup Language

Extensible Rights Markup Language (XrML) overlaps with XACML and provides many similar features. It is more general purpose than XACML and is easier to implement, but not nearly as flexible. XrML differs from XACML in the data model it uses and it refers to its rules as grants. A grant uses data components to define access to a resource:

1. *Principal*: person or object to whom the grant is issued.
2. *Right*: the right the grant specifies (e.g., read, write, or execute access).
3. *Resource*: the object to which the principle can be granted a right.
4. *Condition*: the circumstances under which the grant applies.

Like XACML, XrML can be used for digital rights management. For instance, an XrML document might include expiration dates or times for accessing a particular resource after which a user would have to be reauthenticated to continue accessing it.

13.4.3.3.8 Web Services Security

Although XML document security solves many security dilemmas, it is not the only strategy for securing Web services. In this section, we present several other solutions that may incorporate XML document security along with other mechanisms to secure Web services.

Web services security leverages many existing standards (e.g., XMLE, XMLDS) to articulate extensions to SOAP to implement message content integrity and confidentiality. It is independent of transport (e.g., SSL/TLS) and application layer protocols, which makes it extremely flexible. The primary goal is for the standard to define building blocks that can be used with Web services to ensure security; it does not describe explicit security protocols.

Web services security establishes the concept of security tokens, which is part of a message that is passed across trust domains. The security token verifies a claim, which is defined as: "a declaration made by an entity (e.g., name, identity, key, group, privilege, capability, etc.)" [61:11]. With this construct, an authority (e.g., VeriSign) can vouch for or endorse the claims made by an organization by using its key to sign the message. In the absence of an authority, the recipient of a security token can choose to accept the claim on the basis of other criteria (e.g., personal trust of the sender). Security tokens are implemented SOAP headers where they can be imbedded. This extension to SOAP is simple to accomplish and does not affect the data contained in the message itself.

13.4.3.3.8.1 Delegation — One of the challenging features of Web services is that services can be chained together dynamically at runtime. This complicates security because an application may be exposed to unsecured services in the chain that could have implications for both the client and the service. For example, suppose user X invokes a Web service S1, which in turn invokes Web service S2 (see Figure 13.13); further, assume S1 has privilege to access S2, but user X does not and user X has privilege to access S1. In this scenario, user X may be exposed to information from S2 to which it is not allowed because S1 used its privilege level to invoke the S2 on behalf of user X.

To handle this situation, She et al. [62] suggest a delegation model. In this scheme, an infrastructure for delegating privilege is established so Web services consumed in the invocation chain are invoked on the basis of the privilege of the user, not intermediate services. This prevents users from being exposed to confidential information while at the same time protects users from invoking services from untrusted domains. The disadvantage of this approach is that a trust

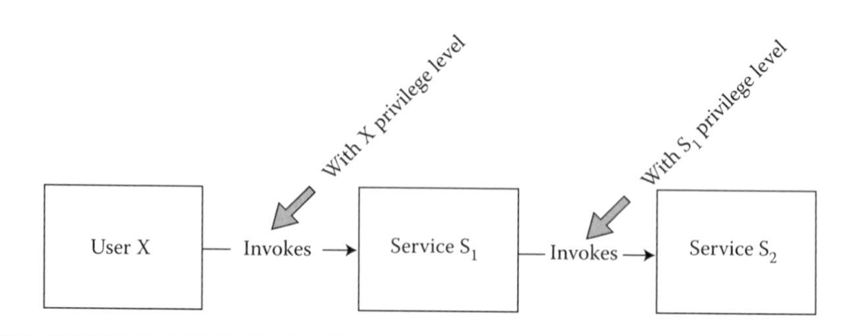

Figure 13.13 Web service invocation chain.

relationship has to be established between domains (e.g., departments, separate companies, anonymous services). The overhead for creating and maintaining the trust infrastructure will become unmanageable as the relationships scale and invocation chain gets larger. Still, it is an important first step in establishing a secure chain.

13.4.3.3.8.2 Proxy-Based Web Services — An attractive approach for securing Web services is using a Web service proxy server. The advantage of this solution is that the security services can be handled by a central authority for the mobile device that does not have adequate resources to fully implement them. Such a solution is suggested by Wu and Huang [63]. Although this solution is not targeted specifically for mobile devices, it can be adapted (see Figure 13.14). In this approach, devices forward all the requests to a proxy server that intercepts all HTTP traffic and strips out unsecured SOAP messages, applies relevant security processing (e.g., authentication, encryption, digital signature, etc.) and then forwards the secured messages to the destination Web service. The main advantage of this approach is that the mobile device does not have to expend processor time or battery power to implement security. The problem with this approach is that an infrastructure has to be built to support it before mobile devices can be used. Moreover, the original articulation of this approach assumes a secure connection from the client to the proxy server. This may not be possible in the case of mobile devices such as phones and netbooks that use cellular connections to access the proxy server. This can be solved by using symmetric key encryption to secure the messages as they are transferred to the proxy service where more stringent security can be applied before consuming a Web service.

13.4.3.4 Impact of Security on the Performance of Web Services

Security for any system is never a cost-free endeavor. In most instances, adding security increases the time it takes to process data that have to be secured, transferred, authenticated, and released to the receiver. For Web services, the latency of SOAP messages is often adversely affected by security. For example, in a study [64] of the

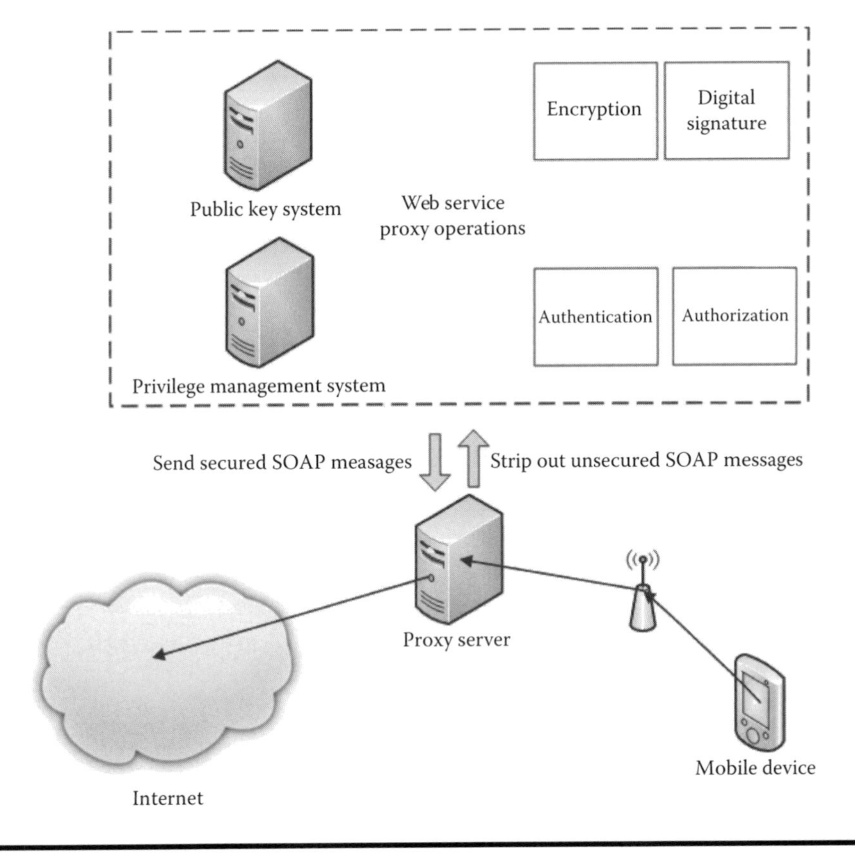

Figure 13.14 Securing Web services through a proxy.

impact of security on the performance of Web services, it is observed that for SOAP messages that have been encrypted, latency increases by 30–70%. Further, results from this study show that using digital signatures introduces latency of 125–190% and using both encryption and signatures increases latency by 160–260%. For mobile environments, this is a significant difference and shows that traditional methods for securing Web services may be infeasible for mobile devices. Nguyen et al. [65] further elucidate this issue. Measurements here indicate decryption time ranges anywhere from 450 to over 2000 ms for file sizes ranging from 50 to 300 KB, while the encryption process on the server side was far less (that data was not presented by the study). Unfortunately, this study does not present decryption times for desktop machines for comparison so that the actual effect on performance between these two classes of machines can be measured. Still, any operation on a modern machine that takes more than 2 s to complete is significant when most operations occur in the millisecond or nanosecond timescale.

13.5 Best Practices for Designing and Implementing Security for Mobile Services

Implementing security on mobile devices is a significant challenge because there are numerous vulnerable points in the service chain. Despite these challenges, designing secure mobile systems is possible; however, it requires a disciplined approach to design. From the research presented in this chapter, a set of best practices for designing and implementing security for the mobile domain has emerged. Here, we present those best practices with a brief discussion on each.

1. *Understand the security problems of each layer of the mobile services stack as well as the interactions between each layer*: Completely understanding the whole stack is no trivial task—but vital to successfully implementing security. The complex interactions between layers can have security implications that are not obvious at first glance. For instance, without understanding the complexity of encryption algorithms, a designer may not realize which candidate algorithms will work on mobile devices.
2. *Security should be considered as early in the design process as possible*: It is not unusual for security issues to be considered after a system has been designed or as a separate issue from the development process. Therefore, security solutions are often "bolted on" to systems afterwards. This approach does not usually lead to the most efficient and secure implementations. Security should be built into systems and devices in much the same way we build hardware or software interfaces. They should be an intrinsic part of the system and every design decision should be considered in the context of how it will affect security.
3. *Understand existing security standards and how to use them*: Viega and Epstein [66] underscore this issue and present three pitfalls for implementing security standards including:
   a. Security standards may not be secure
   b. Using the wrong standard
   c. Ignoring what the standard does not do

   Although this work [66] was targeted specifically to securing Web services, the hazards are applicable to any situation in which a security standard is implemented. The incorrect application or implementation of security standards is a common occurrence and may lead to a false sense of security for the mobile user. To assist with the analysis of existing security standards, Gutierrez et al. [67] propose a security process framework. Again, this framework is developed in the context of Web services, but may be adapted as a model for analyzing the mobile services stack. It is divided into three stages:
   a. *Gathering security requirements*: determine what the specifications for security are for a particular system.

   b. *Identifying appropriate security standards*: determine what standards will apply for the security issue being addressed.
   c. *Selecting the standards that will be implemented*: from the candidate standards, choose the standard(s) that will be implemented in the system to provide security.
4. *Generalize security solutions as much as possible*: The obvious advantage to generalizing security solutions is that they can be adapted to a wide variety of devices, applications, and services. Development costs can be reduced dramatically with this approach as the cost of developing the solution can be amortized over many devices or infrastructures. Generalized solutions are more likely candidates for becoming standards than specialized solutions. Once a solution becomes suitable for further investigation to implement as a standard, the research community scrutinizes it thoroughly to remove (or to understand) its deficiencies, which benefits the wider development and research communities. Moreover, attack vectors also change as adversaries determine new ways to compromise networks and the devices on them. Generalized security solutions may provide more protection, especially in the case of attacks that are similar to other attacks whereas solutions that are targeted to a specific realization of an attack may be more difficult to adapt.
5. *Give users the ability to choose the level of security necessary for a particular transaction*: There is no "one-size-fits-all" solution, for designing security for mobile devices. Not every application needs to be secured at the same level. Browsing the Internet for daily news is typically not as important to secure as online banking or other transactions in which credit cards are used. User convenience is improved because overbearing security solutions that affect every network transaction may discourage the user from using mobile services. Moreover, security typically increases the power consumption on mobile devices. Although most can be easily recharged and battery life is improving, reducing power consumption is still a primary concern for designers. Not using expensive security when it is not necessary helps to preserve battery life.

13.6 Conclusion

In this chapter, we present the security problems that affect mobile services and describe various solutions for handling them that have been presented in the literature. Specifically, we divide security into three areas: core, network, and application vulnerabilities.

Authentication attacks, data and location privacy attacks, and data integrity attacks are considered core vulnerabilities. These attacks are designed to compromise either the mobile device user's identity so an attacker may get access to a larger system or to compromise data stored on the (e.g., personal information) device. Approaches for mitigating core vulnerabilities include Kerberos and multifactor authentication, mobility-based authentication, as well as various forms of data encryption for network traffic and local files.

Network vulnerabilities include Internet denial service attacks, handoff attacks, and mobile IP station attacks. Solutions for these attacks are packet authentication and access control, encryption, prevention of duplicating packets, and binding home address identities to mobile devices on the network.

Application vulnerabilities engender access control, application code risks, and attacks against the numerous Web services standards such as SOAP and XML. Solutions include using virtual machine layers or policies that mediate between application and hardware to prevent applications from using network connections or accessing file systems without permission. In addition, numerous standards for securing Web services and XML documents (XrML, SAML, XACML, etc.) exist.

We further suggest a set of "best practices" that mobile service engineers should consider when designing mobile devices and the infrastructure/systems that will interact with them.

References

1. Beji, S. and N. El Kadhi, An overview of mobile applications architecture and the associated technologies, in *The Fourth International Conference on Wireless and Mobile Communications (ICWMC '08).* 2008.
2. Simple Object Access Protocol Specifications. Available from: http://www.w3.org/TR/soap/. Last accessed on November 14, 2011.
3. OASIS UDDI Specification. Available from: http://www.oasis-open.org. Last accessed on November 14, 2011.
4. Web Services Description Language (WSDL) 1.1. Available from: http://www.w3.org/TR/wsdl. Last accessed on November 14, 2011.
5. Sanchez-Nielsen, E., S. Martin-Ruiz, and J. Rodriguez-Pedrianes, An open and dynamical service oriented architecture for supporting mobile services, in *Proceedings of the 6th International Conference on Web Engineering.* ACM: Palo Alto, California, USA, 2006, pp. 121–128.
6. Papazoglou, M.P. and W.-J. Heuvel, Service oriented architectures: Approaches, technologies and research issues. *The VLDB Journal,* 2007, 16(3): 389–415.
7. Xie, B., On secure communication in integrated Internet and heterogeneous multi-hop wireless networks, *PhD Thesis,* University of Louisville, Kentucky, 2006, p. 203.
8. Cavalcanti, D., D. Agrawal, C. Cordiero, B. Xie, and A. Kumar, Issues in integrating cellular networks WLANs, AND MANETs: A futuristic heterogeneous wireless network. *IEEE Wireless Communications,* 2005, 12(3): 30–41.
9. Voelcker, J., Stalked by satellite—An alarming rise in GPS-enabled harassment. *IEEE Spectrum,* 2006, 43(7): 15–16.
10. Decker, M., Location privacy—An overview, in *Proceedings of 7th International Conference on Mobile Business (ICMB '08),* 2008.

11. Serror, J., H. Zang, and J. Bolot, Impact of paging channel overloads or attacks on a cellular network, in *Proceedings of the 5th ACM Workshop on Wireless Security*. ACM: Los Angeles, California, 2006, pp. 75–84.
12. Chakrabarti, A. and G. Manimaran, Internet infrastructure security: A taxonomy. *IEEE Network*, 2002, 16(6): 13–21.
13. Trojan: SymbOS/Viver.A. Available from: http://www.f-secure.com/v-descs/trojan_symbos_viver_a.shtml. Last accessed on November 15, 2011.
14. Pfleeger, C. and S.L. Plfeeger, *Security in Computing*, 4th ed. Upper Saddle River, NJ: Prentice Hall, 2007.
15. The Kerberos Network Authentication Service. Available from: http://tools.ietf.org/html/rfc4120. Last accessed on November 14, 2011.
16. Stallings, W., *Network Security Essentials Applications and Standards*, 3rd ed. Upper Saddle River, NJ: Pearson, 2007.
17. Wang, H., X. Huang, and G.R. Dodda, Ticket-based mobile commerce system and its implementation, in *Proceedings of the 2nd ACM International Workshop on Quality of Service: Security for Wireless and Mobile Networks*. ACM: Terromolinos, Spain, 2006, pp. 119–122.
18. Pirzada, A.A. and C. McDonald, Kerberos assisted authentication in mobile ad hoc networks, in *Proceedings of the 27th Australasian Conference on Computer Science*, Vol. 26. Australian Computer Society, Inc.: Dunedin, New Zealand, 2004, pp. 41–46.
19. Fox, A. and S.D. Gribble, Security on the move: Indirect authentication using Kerberos, in *Proceedings of the 2nd Annual International Conference on Mobile Computing and Networking*. ACM: Rye, New York, 1996, pp. 155–164.
20. Microsoft, Microsoft Developers Network. Available from: http://www.msdn.com. Last accessed on November 14, 2011.
21. Sun, H.-M. and H.-T. Yeh, Password-based authentication and key distribution protocols with perfect forward secrecy. *Journal of Computer and System Sciences*, 2006, 72(6): 1002–1011.
22. Chang, C.-C., S.-Y. Lin, and J.-H. Yang, Efficient user authentication and key establishment protocols with perfect forward secrecy for mobile devices, in *Proceedings of Ninth IEEE International Conference on Computer and Information Technology (CIT '09)*, 2009.
23. Soleymani, B. and M. Maheswaran, Social authentication protocol for mobile phones, in *Proceedings of International Conference on Computational Science and Engineering (CSE '09)*, 2009.
24. Jansen, W. and V. Korolev, A location-based mechanism for mobile device security, in *Proceedings of 2009 WRI World Congress on Computer Science and Information Engineering*, 2009.
25. Sun, Q., Z. Li, and X. Jiang, An interactive and secure user authentication scheme for mobile devices, in *Proceedings of IEEE International Symposium on Circuits and Systems (ISCAS 2008)*, 2008.
26. Yan, G., S. Eidenbenz, and B. Sun, Mobi-watchdog: You can steal, but you can't run!, in *Proceedings of the Second ACM Conference on Wireless Network Security*. ACM: Zurich, Switzerland, 2009, pp. 139–150.
27. Sweeney, L., k-Anonymity: A model for protecting privacy. *International Journal of Uncertainty, Fuzziness & Knowledge-Based Systems*, 2002, 10(5): 557.
28. Mokbel, M.F., C.-Y. Chow, and W.G. Aref, The new Casper: Query processing for location services without compromising privacy, in *Proceedings of the 32nd*

International Conference on Very Large Data Bases. VLDB Endowment: Seoul, Korea, 2006, pp. 763–774.
29. Bamba, B., L. Liu, P. Pesti, and T. Wang, Supporting anonymous location queries in mobile environments with privacy grid, in *Proceeding of the 17th International Conference on World Wide Web*. ACM: Beijing, China, 2008, pp. 237–246.
30. Beresford, A.R. and F. Stajano, Location privacy in pervasive computing. *IEEE Pervasive Computing*, 2003, 2(1): 46–55.
31. Gedik, B. and L. Ling, Protecting location privacy with personalized k-Anonymity: Architecture and algorithms. *IEEE Transactions on Mobile Computing*, 2008, 7(1): 1–18.
32. Gruteser, M. and D. Grunwald, Anonymous usage of location-based services through spatial and temporal cloaking, in *Proceedings of the 1st International Conference on Mobile Systems, Applications and Services*. ACM: San Francisco, California, 2003, pp. 31–42.
33. Hoh, B., M. Gruteser, R. Herring, J. Ban, D. Work, J. C. Herrera, A. Bayen, M. Annavaram, and Q. Jacobson, Virtual trip lines for distributed privacy-preserving traffic monitoring, in *Proceeding of the 6th International Conference on Mobile Systems, Applications, and Services*. ACM: Breckenridge, Colorado, 2008, pp. 15–28.
34. Lu, H., C.S. Jensen, and M.L. Yiu, PAD: Privacy-area aware, dummy-based location privacy in mobile services, in *Proceedings of the Seventh ACM International Workshop on Data Engineering for Wireless and Mobile Access*. ACM: Vancouver, Canada, 2008, pp. 16–23.
35. Man Lung, Y. et al., SpaceTwist: Managing the trade-offs among location privacy, query performance, and query accuracy in mobile services, in *Proceedings of IEEE 24th International Conference on Data Engineering (ICDE 2008)*, 2008.
36. Pan, X., J. Xu, and X. Meng, Protecting location privacy against location-dependent attack in mobile services, in *Proceedings of the 17th ACM Conference on Information and Knowledge Management*. ACM: Napa Valley, California, 2008, pp. 1475–1476.
37. Ghinita, G., P. Kalnis, and S. Skiadopoulos, PRIVE: Anonymous location-based queries in distributed mobile systems, in *Proceedings of the 16th International Conference on World Wide Web*. ACM: Banff, Alberta, Canada, 2007, pp. 371–380.
38. Computer Security Division, Computer Security Resource Center, National Institute of Standards and Technology. Available from: http://csrc.nist.gov. Last accessed on November 14, 2011.
39. Asghar, M.T., J. Ahmad, and S. Safdar, Security model for the protection of sensitive and confidential data on mobile devices, in *Proceedings of International Symposium on Biometrics and Security Technologies (ISBAST 2008)*, 2008.
40. Diffie, W. and M. Hellman, New directions in cryptography. *IEEE Transactions on Information Theory*, 1976, 22(6): 644–654.
41. Diffie, W. and M.E. Hellman, Privacy and authentication: An introduction to cryptography. *Proceedings of the IEEE*, 1979, 67(3): 397–427.
42. Elgamal, T., A public key cryptosystem and a signature scheme based on discrete logarithms. *IEEE Transactions on Information Theory*, 1985, 31(4): 469–472.
43. Rivest, R.L., A. Shamir, and L. Adleman, A method for obtaining digital signatures and public-key cryptosystems. *Communications of the ACM*, 1978, 21(2): 120–126.
44. Jalali-Sohi, M. and P. Ebinger, Towards efficient PKIs for restricted mobile devices, in *Proceedings of International Conference on Communication and Computer Networks*, 2002.

45. Park, K.-W., S.S. Lim, and K.H. Park, Computationally efficient PKI-based single sign-on protocol, PKASSO for mobile devices. *IEEE Transactions on Computers*, 2008, 57(6): 821–834.
46. Brakensiek, L., A. Droge, M. Botteck, H. Hartig, and A. Lackorzynski, Virtualization as an enabler for security in mobile devices, in *Proceedings of the 1st Workshop on Isolation and Integration in Embedded Systems*. ACM: Glasgow, Scotland, 2008, pp. 17–22.
47. Lee, S.-M. et al., A multi-layer mandatory access control mechanism for mobile devices based on virtualization, in *Proceedings of the 5th IEEE Consumer Communications and Networking Conference (CCNC 2008)*, 2008.
48. Winter, J., Trusted computing building blocks for embedded Linux-based ARM trust-zone platforms, in *Proceedings of the 3rd ACM Workshop on Scalable Trusted Computing*. ACM: Alexandria, Virginia, 2008, pp. 21–30.
49. Lee, S.-M., S. Suh, and C. Jong-Deok, Fine-grained I/O access control of the mobile devices based on the Xen architecture, in *Proceedings of the 15th Annual International Conference on Mobile Computing and Networking*. ACM: Beijing, China, 2009, pp. 273–284.
50. Desmet, L., W. Joosen, F. Massacci, K. Naliuka, P. Philippaerts, F. Piessens, and D. Vanoverberghe, A flexible security architecture to support third-party applications on mobile devices, in *Proceedings of the 2007 ACM Workshop on Computer Security Architecture*. ACM: Fairfax, Virginia, 2007, pp. 19–28.
51. Enck, W., M. Ongtang, and P. McDaniel, On lightweight mobile phone application certification, in *Proceedings of the 16th ACM Conference on Computer and Communications Security*. ACM: Chicago, Illinois, 2009, pp. 235–245.
52. Sun, L. and Y. Li, XML and web services security, in *Proceedings of 12th International Conference on Computer Supported Cooperative Work in Design (CSCWD 2008)*, 2008.
53. Naedele, M., Standards for XML and Web services security. *Computer*, 2003, 36(4): 96–98.
54. Nordbotten, N.A., XML and Web services security standards. *IEEE Communications Surveys & Tutorials*, 2009, 11(3): 4–21.
55. XML Signature Syntax and Processing. Available from: http://www.w3.org/TR/xmld-sig-core/. Last accessed on November 14, 2011.
56. XML Encryption Syntax and Processing. Available from: http://www.w3.org/TR/xmlenc-core/. Last accessed on November 14, 2011.
57. XML Key Management Specification (XKMS). Available from: http://www.w3.org/TR/xkms/. Last accessed on November 14, 2011.
58. Security Assertion Markup Language (SAML). Available from: http://xml.coverpages.org/saml.html. Last accessed on November 14, 2011.
59. Extensible Access Control Markup Language (XACML). Available from: http://xml.coverpages.org/xacml.html. Last accessed on November 14, 2011.
60. Extensible Rights Markup Language (XrML). Available from: http://xml.coverpages.org/xrml.html. Last accessed on November 14, 2011.
61. OASIS, Web Services Security: SOAP Message Security 1.1 (WS-Security 2004). 2006.
62. She, W., B. Thuraisingham, and I.L. Yen, Delegation-based security model for Web services, in *Proceedings of the 10th IEEE High Assurance Systems Engineering Symposium (HASE '07)*, 2007.

63. Wu, J. and Z. Huang, Proxy-based Web service security, in *Proceedings of IEEE Asia-Pacific Services Computing Conference (APSCC '08)*, 2008.
64. Chen, S., J. Zic, K. Tang, and D. Levy, Performance evaluation and modeling of Web services security, in *Proceedings of IEEE International Conference on Web Services (ICWS 2007)*, 2007.
65. Nguyen, T.T., I. Jorstad, and T. Do van, Security and performance of mobile XML Web services, in *Proceedings of the Fourth International Conference on Networking and Services (ICNS 2008)*, 2008.
66. Viega, J. and J. Epstein, Why applying standards to Web services is not enough. *IEEE Security & Privacy*, 2006, 4(4): 25–31.
67. Gutierrez, C., E. Fernandez-Medina, and M. Piattini, PWSSec: Process for Web services security, in *Proceedings of International Conference on Web Services (ICWS '06)*, 2006.

Chapter 14

GEO-PRIVACY

Enforcing Privacy Policies Considering User Location

Matthias Farwick, Berthold Agreiter, Basel Katt, Thomas Trojer, and Patrick C. K. Hung

Contents

14.1 Introduction

In recent years, many countries have enacted laws that regulate how personal data of citizens must be treated in the context of globally networked and fast-changing economic environment. Even Hong Kong, with basically no prior history of legal privacy protection, has enacted laws to protect the privacy of their citizens (Birnhack, 2008.) However, since the first laws came into force, implementation vendors struggled to realize the requirements of these to actually make them enforceable in running access control systems. This fact has been amplified by the fast pace in which technology and its use has changed in the last decade. One major example here is the exploding use of mobile devices. As mobile handheld devices gain increasingly more memory and processing power, they can be utilized in the health sector to aid medical staff inspecting health records (Tikkanen, 2006), or they can be used by businessmen in the field to work with company data. This kind of usage of personal data was not foreseen by the creators of privacy laws although it poses serious threats to its safety. One of many examples is the E.U. Data Privacy Directive that was created in 1992 when mobile computing was still in its infancy. However, aspects that are highly relevant to the use of private data in a mobile context can still be found in those legal texts. For example, the E.U. Data Privacy Directive regulates circumstances under which private data can be transferred to third countries outside of the European Union. Nonetheless, current privacy access control models are not built with the aspect of location in mind, which arises from the possibility to retrieve and manipulate data in a mobile context. Conversely, location-based access control systems are not built around the concept of privacy.

In this chapter, we identify the need for location information in privacy models by analyzing two current privacy laws. Namely, the E.U. Data Privacy Directive and the Health Insurance Portability and Accountability Act (HIPAA) in the United States. By looking at different access control models that provide the means for expressing location restrictions within their policies, we conclude that there is no access control model that simultaneously takes location and privacy into account. As a result we develop the formal location-based access control model GEO-PRIVACY which extends GEO-RBAC with obligations, purposes, data access conditions, and consent. We focus on creating a model that is actually implementable without making too many assumptions. For example, the privacy concept of retention is only solvable by means of usage control (Sandhu and Park, 2003) in the mobile context. However, this is a research topic on its own and will therefore not be covered in this chapter.

In the following sections, we give two motivating examples for our approach to integrate the location concept into privacy policies.

14.1.1 Motivating Examples

In this section, two motivating examples are introduced to exemplify possible scenarios in which the GEO-PRIVACY model is useful. These scenarios are both motivated by the actual need from a legal point of view (see Section 14.3), as well as experience from many documented privacy breaches in recent years. It is an interesting fact that location information in access control can be applied on a macroscopic and microscopic scale. For example, in the macroscopic case, the information whether a user wants to access data from a different country, can, and in some legal situations, has to be used to enforce access control restrictions. On the other hand, there are cases where very small change in the location of a user can make a difference in the access control decision. After having analyzed different privacy laws in Section 14.3, the following example policies are formulated as more precise natural language policies in Section 14.3.3.

14.1.1.1 Example 1 (Microscopic)

The IT infrastructure of a hospital provides mobile devices with access to diverse information such as patient room occupancy for the hospital's medical staff. Also, doctors and nurses are able to use these devices to access the health record of a patient according to their roles. As users of mobile devices can freely move around and can potentially carry the devices out of the hospital, there clearly arises a need for an access control model that takes the location of a user into account and also guarantees adequate privacy for patients. For example, a doctor could use her mobile device to access patient records while being outside of the hospital. This should, in most cases, be prohibited as patient data are much more prone to misuse when stored on mobile devices that are outside of the physical boundaries of the hospital. Also, a home-care nurse should only be allowed to access the e-health record of her patient when being in the home-care location, as suggested in the HIPAA Security Guidelines (Phillips, 2006) (see Section 14.3.2). This need has become evident through many recent data scandals. In one of these scandals, the laptop of a doctor was stolen which contained several thousand unencrypted patient records.* Furthermore, many important principles of privacy, for example, purposes and obligations, come into play in this example. A doctor, for instance, should only be allowed to access the health information of a patient for specific purposes such as making a diagnosis. Other purposes, such as research studies, should only be allowed if personal consent by the patient is obtained.

14.1.1.2 Example 2 (Macroscopic)

For this larger-scale example, assume that there exist laws that restrict the flow of private information between countries. One example for such a law is the European Data Privacy Directive 95/46/EC that regulates (see Section 14.3.1) the flow of private data between member states of the European Union and third countries. These restrictions often indicate that data export is only allowed for explicitly stated purposes. Furthermore, the receiving side often has to fulfill obligations such as

* http://news.bbc.co.uk/1/hi/wales/7849212.stm

providing reasonable security measures. Additionally, the transfer is only allowed with the explicit and unambiguous consent of the data owner. This example again shows the need for a location-based access control system with integrated privacy capabilities.

From the two examples above we derive the main claim of this chapter: *location is an important aspect in controlling access to private data and current access control models do not cater to this need.* This claim will be verified by looking at current access control models in Section 14.2 and analyzing different privacy laws in Section 14.3. We will now highlight the contributions of the work presented in this chapter.

14.1.2 Contributions

The contribution of the work presented in this chapter is threefold. It provides a discussion on several current international privacy laws, namely, the E.U. Data Privacy Directive 95/46/EC and HIPAA in the United States. As the first contribution, we *analyze location aspects in the requirements defined in these privacy laws.* From these aspects, we conclude that there is a need for including location information into privacy policies. Secondly, in Section 14.4, the *location-based access control model* GEO-RBAC *is extended with conceptual means for expressing privacy conditions in a novel access control framework*: GEO-PRIVACY. To the best of our knowledge, this is the first approach documented in the research literature that attempts to combine aspects of privacy and location in an access control model. As the final contribution, we *present a prototypical implementation of the proposed access control model* including a corresponding policy authoring tool in Section 14.5. The implementation is based on the eXtensible Access Control Markup Language (XACML) as the policy language, the Web Ontology Language (OWL) for representing domain knowledge, OpenGIS* standards for working with location information, and the Eclipse rich client policy (RCP) platform† as the basis for the policy authoring tool.

14.2 Background and Related Work

In this section, we first introduce several key technologies relevant to our study and then discuss related work.

14.2.1 GEO-RBAC

Here, we describe the location-based access control model GEO-RBAC (Damiani et al., 2007) because it forms the basis for our new model GEO-PRIVACY. The core idea behind this location-based access control model is the usage of the well-established

* http://www.opengeospatial.org
† http://www.eclipse.org/rcp

and open standard for geospatial information OpenGIS (Herring, 2006) as the means to express geospatial locations. This standard allows for the precise definition of geospatial information and also provides means to describe containment of areas. This makes the definition of different granularity of location-based access control policies feasible. In OpenGIS types of geographical features are called *feature types*. Instances of such types are called *features*. For example, the concept of a street is a feature type, where Route 66 in the United States is a feature. GEO-RBAC extends the RBAC96 models (Sandhu et al., 1996) and adopts its notation.

The central parts in GEO-RBAC are role schemas and role instances. Role schemas (R_S) are used to define the pattern, consisting of permissions and a feature type, in which a given role is enabled for a user. It therefore defines a common name for a set of permissions that can be reused for different instantiations of a role. A role instance (R_i) is the instantiation of a role schema with concrete features from the real world. For example, a role schema could denote that the role can only be enabled in a hospital, leaving out the specification to which specific hospital this applies. A role instance fills this gap by exactly defining which feature is referred to, for example, a specific hospital building. Both role instances and role schemas can be ordered in hierarchies that allow for inheritance of permissions.

Permissions can be assigned to R_S and R_i by the permission-to-spatial role schema/instance assignment relation. Permissions (PRMS) consist of operation (OPS) on objects (OBJ). Users (U) are assigned to role instances with the user-to-spatial role instance assignment relation. At runtime, when a user logs into the system, a new session from the set SES is created and assigned to the user by the *SessionUsers* relation. Roles that can potentially be enabled during this session are defined by the *SessionRoles* relation. Depending on the real position of the user, specific roles are enabled for that user denoted by the *EnabledSessionRoles* relation. The real position of the user is mapped to a logical position corresponding to an OpenGIS feature defined in a mapping function in the role schema. If this feature is contained in the so-called role extent* of the role schema, the role is enabled. Refer to the study by Damiani et al. (2007) for a complete introduction to this access control model.

14.2.2 Used Technologies

For the prototypical implementation of our work we used several technologies that are briefly introduced here.

14.2.2.1 eXtensible Access Control Markup Language

XACML is a language and architecture specification for the access control domain. It is widely used for the implementation of various access control models. The

* Defines the OpenGIS feature of a role schema in which the role is enabled.

policy language has XML syntax and allows for the extension with custom functions and data types. XACML policies can be grouped in policy sets by specifying policy-combining algorithms, which govern the result of the evaluation of each contained policy. For this work, we used XACML for the GEO-PRIVACY policy syntax and SUN's XACML* reference implementation as the foundation for the implementation of the access control engine.

14.2.2.2 Web Ontology Language

In the implementation of GEO-PRIVACY, we used OWL (Smith et al., 2004) to represent knowledge about an access control domain, for example, user accounts, roles, and relationships between stakeholders in the system. For example, we used OWL to describe the relation between a private data requester (e.g., a doctor requesting a health record) and a data subject (e.g., a patient). OWL is logic-based and can express constraints and inference rules on the domain to discover inconsistencies in the knowledge and to infer new knowledge from the knowledge base. We also used these properties in the implementation of GEO-PRIVACY to evaluate complex access control conditions on the variables in the OWL knowledge base.

14.2.2.3 User-Friendly Desktop-Integrated GIS

To implement a policy authoring tool for GEO-PRIVACY we needed a framework that allows for the handling of geospatial data. For this reason, we utilized the Eclipse RCP-based user-friendly desktop-integrated GIS (UDIG)† as the basis for the privacy policy authoring tool. UDIG can handle OpenGIS conformant geospatial data from various sources. In addition, we used the GeoTools application programming interface‡ to programmatically work with geospatial data.

14.2.3 Related Work

In the context of our study, there exists a plethora of previous literature in the fields of access control, location-based services, location-sensing technologies, and privacy. However, to the best of our knowledge, no previous work tries to combine a location-based access control model with privacy-preserving techniques. For a broad overview, the work of Langheinrich (2001) provides a high-level overview of privacy concerns in mobile computing. This section gives the reader an overview of the existing literature in the different fields. We begin by looking at literature that discusses the combination of existing access control models and privacy.

---

* http://sunxacml.sourceforge.net

† http://udig.refractions.net

‡ http://www.geotools.org

14.2.3.1 Extensions to Existing Access Control Models

There are several privacy extensions to existing access control models. Ni et al. (2007) present an extension to RBAC called P-RBAC that provides a formal model to express purposes, conditions, and obligations of access permission. Their work introduces a simple constraint language LC_0 to denote conditions for permission. Furthermore, P-RBAC is capable of expressing hierarchies for purposes, data, and roles, thereby simplifying the expression of policies. As opposed to our work, P-RBAC does not cater to the location aspect of privacy in access control.

14.2.3.2 Location-Based Access Control Models

Location-based access control has gained a lot of attention in research in recent years, as the realization of such systems has become feasible owing to technological advancements. The major contribution in this area is GEO-RBAC that has already been introduced in Section 14.2.1. Another contribution in this field is the work of Ardagna et al. (2006). They propose an access control infrastructure that foresees that the access control engine retrieves the location information from the user through a location service to be incorporated into the access control decision. This reliance on a third-party location service gives rise to numerous privacy concerns such as those addressed by Bettini et al. (2005). In their model, the authors take into account technical deficiencies such as imprecise measurements and timeouts. This approach does not define how a semantic location such as "office" can be derived and whether there is a hierarchy associated with such a location that can be used for abstraction and reasoning.

The work by Chandran and Joshi (2005) is closely related to the GEO-RBAC approach. The authors propose a location and time-based access control model on the foundation called LoT-RBAC. This model is, like GEO-RBAC, capable of expressing logical and also physical location-based access control policies, but unlike GEO-RBAC the model does not make use of the widely used OpenGIS standard. This makes the architecture less likely to be applied in a real-life scenario.

Hansen and Oleshchuk (2003) present another approach of extending RBAC for location-based access control. In their work, permissions are assigned to roles according to the location of the user who exercises the role. Therefore, roles have different authorizations in different locations. A major drawback of this approach is that it does not mention the possibilities of how the knowledge of logical location hierarchies can be used to infer further information about the location, as possible with GEO-RBAC.

14.2.3.3 Semantic Web-Based Access Control Policies

Several recent efforts have made use of semantic Web technologies to implement access control frameworks. The primary motivation has been the inference capabilities inherent in ontology languages such as OWL that enable the access decision mechanism and can be used to detect inconsistencies within sets of policies. Among the first contributions in this field is the work of Di et al. (2005), which presents an approach for specifying access control policies with OWL. One on the most recent advancements in this area is the work by Ferrini and Bertino [2009], who propose XACML as the policy language and enforcement mechanism and OWL as the knowledge base for attributes and constraints. Further work on using semantic Web technologies for access control can be found in the studies by Finin et al. (2008) and Heilili et al. (2006).

In this section, we briefly introduced the location-based access control model GEO-RBAC, discussed technologies that are used in the prototypical implementation of this work, and highlighted related research work in the field of access control.

14.3 The Location Concept in Privacy Laws

The content of our work is built around the claim that location is important for private data access control policies. To justify this claim, we analyze two privacy laws from the European Union and the United States, and look for location aspects in these laws. On the basis of this verified claim, we extend the GEO-RBAC model with means to protect the privacy of data owners in the following section and show an implementation of this model in Section 14.5. In the following, we introduce the privacy concept in the context of this work and then continue to analyze the different laws.

Privacy is a broad term and needs to be precisely defined in the context of our discussion in this chapter as there is no general definition. When the term privacy is used, it mainly refers to *physical privacy*, implying that anybody has the "right to be let alone," which has already been formulated in the nineteenth century (Samuel and Brandeis, 1890).

In the age of information technology, however, this physical privacy is still important but has been augmented with the need to protect the collection, storage, and dissemination of personal identifiable information, also called *information privacy*. This need arises from the ease by which personal data can be collected using automation techniques and the ease by which data can be brought in relation with each other, allowing one to connect datasets to reach conclusions about persons and their private behavior. This has led to a commercialization of personal information that results in creating markets for companies dealing with personal identifiable information. Money lenders, for instance, frequently buy sets of personal information about potential debtors to calculate credit-worthiness from variables such as the living neighborhood, the owned car, or previous paying schemes. This has the consequence that owing to released personal information one can, for instance, be granted or refused a loan.

Some countries and international organizations have recognized the need for measures to ensure privacy as a result of the above-mentioned facts and to control the flow of personal data across borders to foster economic growth. In the following discussion on the two laws, a special emphasis is laid on how location aspects play a role in the law in order to justify the main claim of our work that location is an important aspect in privacy policies.

14.3.1 The European Data Privacy Directive 95/46/EC

This directive was implemented in 1995 and had to be transposed to a national law of member states by the end of 1998. It aims at specifying the circumstances under which the processing, storage, and/or distribution of personal data is rightful. Within its 34 articles it specifies several key points that are recurring in privacy laws. These key points are highlighted in *italics* within this section.

The first key area is the *quality* (Article 6) of collected personal data. It has to be correct and also updated when changes occur. The data owner is also granted the *right to access* (Article 12) his/her data, and the right to alter and delete it free of charge. Furthermore, the data collector has to prove the *legitimacy* (Article 7) of collecting information by collecting only for certain purposes.

The data subject also has to unambiguously give his/her consent to the data collection. It is also forbidden to process data of certain *categories* (Article 8) such as the race, political opinions, and religious beliefs of the data subject.

Another important aspect is the obligation of the data collector to provide the data owner the *information* about which organization collected the data (Article 10).

Furthermore, the directive prescribes that high standards of *confidentiality* and *security of processing* have to be assured.

Additionally, each member country has to implement a government-independent *supervisory authority* that has to be notified by any data controller, in case personal data are collected. These authorities are also responsible for monitoring the implementation of the directive within the territory of the member state.

14.3.1.1 Location Aspects

Article 25 of the directive regulates the transfer of personal data to third countries. It states that transfer to non-E.U. countries is only allowed if an "adequate level of protection" is guaranteed by this third country. This adequateness can only be reached if the third country implements privacy protection laws similar to those of the European Union. Also, in cases where a member country detects that a third country does not provide the necessary level of protection it shall inform the other member countries. This article justifies the need to consider location in privacy policies.

As very few countries guarantee the "adequate level of protection" by law, special agreements have been made between the European Union and third countries. The most important of these agreements is the E.U.–U.S. Safe Harbor program* that came into place in 2000.

14.3.2 The Health Insurance Portability and Accountability Act

HIPAA (1996) is a U.S. law that was enacted by the U.S. Congress in 1996. It consists of two parts of which the first one regulates the health insurance of workers and their relatives. The second part is more significant with regard to our work. It attempts to establish U.S.-wide security, privacy, and accountability standards for healthcare transactions. It is therefore narrower in its scope compared to the E.U. Data Privacy Directive that covers private data in general. The second part of the Act requests the U.S. Department of Health and Human Services to create rules that regulate the treatment of electronic protected health information (ePHI).

The *privacy rule* regulates the common privacy principles a "covered entity" must follow. Covered entities include health insurers or medical service providers. The principles are *openness*, *purpose*, *notification*, *accessibility*, *accountability*, and *complaints*. Openness requires a covered entity to disclose the ePHI to a data subject on request within 30 days. Purposes define the specific intention for which data can be collected. A notification rule states that a data subject must be notified of the use of their ePHI. Also, data subjects must be able to request alternation of their data because of the accessibility rule and be able to file official complaints.

Additionally, the HIPAA prescribes several *administrative safeguards* that govern the administrative tasks that have to be fulfilled when ePHI is processed. For example, access to ePHI should only be allowed on a need-to-know basis; employees should receive special privacy safeguard training and privacy audits need to be conducted on a regular basis and as a response to privacy incidents.

Furthermore, a number of rules prescribe physical and technical safeguards such as encryption to protect ePHI. Finally, the *enforcement rule* specifies the punishment for violating rules of the HIPAA.

14.3.2.1 Location Aspects

HIPAA does not directly address location aspects such as data transfer to another country, as the E.U. Data Privacy Directive (see Section 14.3.1) does. However, supplementary documents (Phillips, 2006) from the U.S. Department of Health and Human Services clearly identify the risks involved in remote or offsite use of

* http://www.export.gov/safeharbor

ePHI with portable devices such as personal digital assistants or laptops. The document first states:

> In general, covered entities should be extremely cautious about allowing the off-site use of, or access to, EPHI (Phillips 2006, p. 1).

Here, it becomes evident that the problem of remote access and offsite use already has been recognized by the government. Furthermore, the Act identifies the risk of

> Employees access(ing) EPHI when not authorized to do so while working off-site (Phillips 2006, p. 4).

To tackle this risk, the document recommends

> Establish(ing) remote access roles specific to applications and business requirements. Different remote users may require different levels of access based on job function (Phillips 2006, p. 4).

As possible risk-management strategies, the Act recommends "proper clearance procedures" (Phillips 2006, p. 4) and establishing "remote access roles specific to applications and business requirements" (Phillips 2006, p. 4).

This identified risk is another indicator that there is a need for privacy policies that take into account location.

14.3.3 Example Location-Based Privacy Policies

In the preceding section two international privacy laws have been introduced and their requirements with regard to taking into account location information in the access control enforcement process have been identified. In this section, these requirements are condensed into two (natural language) policies that resemble requirements from the E.U. Data Privacy Directive (Policy 1 in Section 14.3.3.1) and requirements from HIPAA (Policy 2 in Section 14.3.3.2). Like the example from the introductory section, these policies are characterized as macroscopic and microscopic, respectively.

14.3.3.1 Policy 1 (Macroscopic)

Accredited research groups from the United States (*location*) can access (read) personal data such as "ShoppingBehavior," for the purpose of "Research" (*purpose*), from individuals of 18 years and above, but only if they get the consent from the data subject (*consent*). Also, each access must be logged (*accountability*).

14.3.3.2 Policy 2 (Microscopic)

A nurse can access (read, write) the health record of a person if s/he is located in his/her specific hospital or in specific areas where s/he is treating home-care patients (*location*). Furthermore, s/he can only access the health record if s/he is the assigned nurse of that patient (*need-to-know* principle) and for the purpose of diagnosis and medication (*purpose*). In addition, after access is granted, the action has to be logged (*accountability*). In order to use the health record (*data category*) of the patient for a research study a doctor needs to acquire the consent of the patient (*consent*).

In this section, we identified the location aspects in two international privacy laws. From the requirements of these laws, it can be concluded that location information can play a crucial role in the enforcement of privacy policies. This underlines the main claim of our study that location is an important aspect in privacy policies. However, current privacy access control models neither provide the facilities to create policies that describe these requirements, nor provide the technical frameworks to enforce these policies (see Section 14.2.3).

14.4 The Formal GEO-PRIVACY Model

The discussion on privacy laws in Section 14.3 confirms our claim that in realistic privacy policies the location aspect can play a crucial role. We have also introduced the important notions of privacy such as security, data categories, legitimacy, accountability, accessibility, dissemination, retention, and consent. On the basis of this set of privacy requirements, our work focuses on the enforcement of security (from an access control viewpoint), accountability, legitimacy, data categories, and consent. The enforcement of retention time and dissemination are left out, as they are concerned with client-side policy enforcement that is a complex research topic on its own. Refer to the studies by Pretschner et al. (2008) and Sandhu and Park (2003) for current advancements in this field. We develop a formal model that is capable of enforcing the above-mentioned requirements. It is achieved by incorporating new privacy entities into the GEO-RBAC model, forming the novel privacy access control model GEO-PRIVACY.

Figure 14.1 shows the extended conceptual model of GEO-PRIVACY. The core idea of the new model is that each permission not only is bound to a role in the model but also depends on a set of entities that govern the conditions under which private data can be retrieved and used. Furthermore, we restrict the model to work with access requests instead of sessions, to avoid running into the problem of having to enforce continuous location-based access control. This problem occurs when a user creates a session and gets assigned different roles on the basis of his/her location. When the user then moves around, the roles might not be valid anymore with regard to the user's location; however, the roles will be enabled as long as the session is valid. For possible mitigation solutions for this scenario, refer to the study by

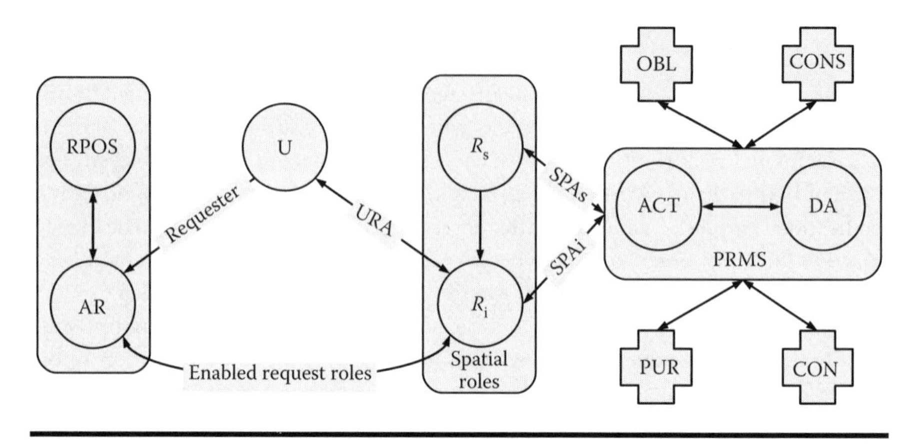

Figure 14.1 The formal model of geo-privacy extending GEO-RBAC with consent, obligations, purposes, and conditions.

Damiani and Silvestri (2008). The drawback of our approach is that, for each access request, the user needs to provide all credentials again.

In the following, the definitions of each of these new privacy entities are discussed. Refer to Figure 14.1 for an overview of how these entities relate to each other in the model. Optionally, some entities of the model can be ordered into hierarchies. Hierarchies add complexity to the formal model and verification attempts, but in turn add flexibility and increased compactness to the syntax of implementing languages. Table 14.1 gives an overview of the GEO-PRIVACY-specific notations. For a further introduction to GEO-RBAC, which is the foundation of GEO-PRIVACY, refer to Section 14.2.1 and the original body of work by Damiani et al. (2007).

In the following section, we will look at how access requests, users, and roles are defined in GEO-PRIVACY and how they differ from their definitions in GEO-RBAC. After that we introduce the main additions to GEO-RBAC that extend it with mechanisms to enforce privacy policies. We conclude this section by formally expressing the natural language privacy policies from Section 14.3.3, with the formal access control model introduced in this section.

14.4.1 Access Requests, Position, Users, and Roles

In this section, we look at the specifics of GEO-PRIVACY in comparison with GEO-RBAC, with regard to access requests, position, users, and roles. By doing so, we will describe the conceptual diagram in Figure 14.1 (from left to right).

14.4.1.1 Access Requests

As mentioned in the Introduction, we refrain from using the concept of sessions in this model as, in the context of movement, sessions can lead to unwanted role

Table 14.1 GEO-PRIVACY-Specific Notations

Notation	*Semantics*
PRM	Set of permissions restricted by role assignment and privacy entities
ACT	Actions that can be executed on a data object when permission is granted
DA	Data types/data categories
CONS	Boolean value that signifies whether consent has to be requested from the data owner/subject
OBL	Obligations applying to a permission
PUR	Set of purposes under which a permission is granted
CON	Set of conditions restricting the assignment of permissions
DSUB	Set owners or subjects of the requested data
URA	Assignment of roles to users
SUB	Assignment of data item to data subjects
SPA_s	Assignment of permissions to role schemas
SPA_i	Assignment of permissions to role instances

assignment over time. Therefore, an access request has the exact lifespan of any request/response pair.

Definition 1 (Access Request)

An access request, ar $\in$ AR, is defined as a tuple ar = (u,act,p,uri,loc), where $u \in U$ is the requesting user, act $\in$ ACT is the targeted action, $p \in$ PUR is the purpose of the data access, uri $\in$ URI is the identifier of the requested resource, and loc $\in$ RPOS is the real location of the user.

For further explanation of the sets introduced in the prior definition, refer to the subsequent sections.

14.4.1.2 Position

The position model in GEO-PRIVACY corresponds to the position model in GEO-RBAC. We will give a short reminder of these concepts here.

GEO-RBAC differentiates between real and logical positions. Real positions can be thought of as actual coordinates on the earth's surface. Logical positions correspond to features in the geometric system such as rooms, buildings, cities, or countries. Between features, certain functions are defined, for example, *contains*(f_1, f_2) and some others. To define role schemas in GEO-RBAC, so-called mapping functions are used. These abstract functions map an input real position to an output feature via the containment function. For example, the mapping function m_{city} returns the city in which a certain coordinate is located. For the full definition refer to the study by Damiani et al. (2007).

14.4.1.3 Users

Each access request, $ar \in AR$, is associated with a user, $u \in U$, via the *Requester* relationship. Additionally, each user is assigned with a set of role instances via the user role assignment relation URA. These are the potential roles a user can have enabled depending on his/her location.

14.4.1.4 Role Schemas and Role Instances

Role schemas and role instances are also defined analogously to GEO-RBAC. For example, GEO-PRIVACY requires the same spatial containment of roles that inherit from each other and role schema hierarchies are denoted by the $\preccurlyeq_{RS}$ symbol.

14.4.2 Privacy Entities in GEO-PRIVACY

In this section, we introduce the privacy features of GEO-PRIVACY that have been added to the GEO-RBAC model. These are *data types*, *purposes*, *obligations*, *consent*, *actions*, and *conditions*, including hierarchies of a subset of these entities.

14.4.2.1 Data Types

Data types are used to distinguish different types of (private) information from each other. Data types can be defined in a hierarchy in the model. Here, a data-type hierarchy is defined as a semantic data hierarchy and not the hierarchy in terms of structured data such as file systems or XML documents. For example, medication data are a special type of medical data. Also, data types can be used to distinguish certain *data categories* (see Section 14.3.1) that are defined by the E.U. Data Privacy Directive.

Definition 2 (Data Types and Data Identifiers)

Let DA be a set of all semantically distinguishable data types. Data types can be partially ordered in hierarchies defined by $DH = DA \times DA$ and denoted by the $\preccurlyeq_{DA}$ relation symbol. Each real data item is uniquely identifiable by IDs from the set URI. Formally: $\forall uri \in URI\ \exists dt \in DA: data\_type(uri) = dt$, where, $data\_type: URI \rightarrow DA$. Also, from each URI a concrete data subject can be resolved that identifies the owner of the data item. The set of data subjects is defined by DSUB. The abstract function $dataSubject: URI \rightarrow DSUB$ is used to resolve a data item to a data subject. This implies that $\forall uri \in URI\ \exists ds \in DSUB: dataSubject(uri) = ds$.

For an access request, the data should actually be identifiable by, for instance, a uniform resource identifier (URI). This is important to relate the requested data to an actual data subject to be able to request consent if needed, or to evaluate more complex conditions.

14.4.2.2 Actions

Actions are the operations that can potentially be executed on data items.

Definition 3 (Actions)

Let ACT be the set of all actions that can be executed on a data type DA. Actions can be ordered in action hierarchies AH that are defined by the partial order $AH = ACT \times ACT$ denoted by the $\preccurlyeq_{ACT}$ relation symbol.

14.4.2.3 Purposes

Purposes are used to restrict the purposes for which a permission can potentially be granted.

Definition 4 (Purposes)

Let PUR be the set of all purposes that can be assigned to permissions. Purposes can be ordered in hierarchies that are defined by $PH = PUR \times PUR$ denoted by the $\preccurlyeq_{PUR}$ relation symbol.

14.4.2.4 Obligations

Obligations are used to specify actions that have to be executed in order to grant permission. For example, obligations can be used to demand that each access to resource has to be logged.

Definition 5 (Obligations)

Let OBL be the set of all obligations that uniquely identify requirements that need to be fulfilled by a requester in order to be granted a permission p from the set of permissions PRMS. Obligations can be ordered in a partial order defined by $OH = OBL \times OBL$ that is denoted by the $\preccurlyeq_{OBL}$ relation symbol.

Here, the obligation hierarchy implies that, if a parent obligation needs to be executed, all subobligations also need to be fulfilled.

14.4.2.5 Consent

The consent entity is used to indicate whether the consent of a data owner has to be gathered in order to access a certain data item. Obviously, the consent could also be modeled as an obligation in each policy, but we chose to treat consent as a first-class object in this model as it is of high importance for privacy policies.

Definition 6 (Consent)

Let CONS be the set of boolean values {true, false}. If a true value is assigned to a permission instance, it means that the data subject's consent has to be gathered in order to enable this permission. The abstract function requestConsent:$DSUB \times A$ $CT \times PUR \rightarrow CONS$ is used to request the consent from the data subject for an action on a data item for a specific purpose.

14.4.2.6 Conditions

Conditions are important for preserving data privacy because they can be used to express complex requirements for the data usage and access control. Conditions should not be confused with constraints in RBAC. Constraints are used to express high-level constraints such as static and dynamic separation of duty, whereas conditions are used to define access control conditions that are based on attributes of the data requester, the data subject (in the case of privacy policies), and the context.

As each application domain for privacy access control (that is, each jurisdiction) has different complex conditional access control requirements, a simple condition expression language does not seem adequate. For example, Ni et al. (2007) present a privacy condition language that is based on binary comparison operators $\{=, \neq\}$. We argue that this form of simplicity is not adequate for the expression of complex privacy conditions in a realistic setting. Therefore, we introduce *n*-ary domain-specific predicates (DSPs) in our condition language. These predicates can be used to easily specify domain-specific conditions such as "only the doctor who is assigned responsible for a certain patient can read the patients' health record."

Definition 7 (Conditions)

Let $\neq$, $=$ denote in/equality; $v \in$ VAR be a variable from a set of variables VAR; DSP is a 2-ary predicate symbol; REQUESTER is the variable that denotes a data requester; DSUB is the variable that resolves to the data subject identifier of the requested data; $\wedge$ is the logical conjunction, and $c \in C$ is a constant from a set of constants C (i.e., the knowledge base). We then define a condition as follows:

1. $v_1 = c_1$ and $v_2 = c_2$ are conditions.
2. DSP_1(REQUESTER, DSUB_ID) is a condition.
3. DSP_2(REQUESTER, v_3) is a condition.
4. If con_1 is a condition, then $\neg con_1$ is a condition.
5. Let con_1 and con_2 be conditions, then $con_1 \wedge con_2$ is a condition.

Line 1 is an example of simple conditional checks on environment variables. In this case, the first condition is fulfilled if the variable v_1 is equal to the constant c_1, or v_2 is not equal to c_2, respectively. Line 2 is an example for a DSP, which is true if the requester has a certain relationship (defined by the predicate) with the data subject. Such a predicate could, for example, be *treatsPatient*, which checks whether the doctor who requests a health record is actually assigned to the patient who is the data subject of the request. Line 3 shows how domain-specific attributes of the requester can be checked by a DSP. Line 4 shows that conditional checks can be negated, and Line 5 shows that conditions can be unified by conjunction to a single condition. In the formal model we use the notation eval(con) to indicate the evaluation function for a condition. For an access to be granted, all conditions need to evaluate to *true*.

14.4.2.7 Permissions

Permissions are the collections of all actions that can be applied on a data object if purposes PUR, consent CONS, obligations OBL, and conditions CON are fulfilled.

Definition 8 (Permissions)

$PRMS = ACT \times DA \times 2^{(CONS \times OBL \times PUR \times CON)}$. Permissions are assigned to role schemas in the set SPAs and to role instances by the set SPAi.

In this section, the differences between GEO-PRIVACY and GEO-RBAC have been described and the major privacy-related additions have been explained. Owing to space limitations we have left out the description of the access control function that formally describes the evaluation of GEO-PRIVACY policies.

14.4.3 An Example GEO-PRIVACY Policy

In Section 14.4.2, we have introduced the formal notation of GEO-PRIVACY. We now proceed by giving a complete example on how it can be used to formally express a location-based privacy policy. For this example, we formalize the informal natural language policies presented in Section 14.3.3. Listing 14.1 shows the formal description of the mentioned policies.

First, the basic location objects are defined. The set FT contains the feature types *Countries*, *HospitalBuildings*, and *Coordinate*. The feature type *Countries* is needed for the macroscopic policy discussed in Section 14.3.3, whereas the feature type *Hospital* is used for the microscopic policy. The feature type *Coordinate* is used for the actual determination of the location of a data requester. After that, several instances of the feature types *Countries* and *HospitalBuildings* are defined.

In the next step, we define the sets and hierarchies for privacy entities. ACT specifies three actions, of which the *readHealthRecord* and *updateHealthRecord* are in a hierarchical relationship defined in AH. Next, data types and data type hierarchies are defined by DA and DH. Here, we define a fine-grained hierarchy of data types in order to be able to create precise permissions for different types of data. Also, we introduce several purposes in the set PUR and define a hierarchy for them in PH. For example, we define the *MedicationChangePurpose* as well as the *DiagnosisPurpose*. In addition, as both natural language policies demand that any access is logged, we define a *LoggingObligation* in the OBL set which is a superobligation of *NoObligation* defined in the OH set.

In the following, the policy-specific conditions are defined in the set CON. For the microscopic policy, we define a condition that yields *true* if and only if the requester is the responsible doctor of the data subject (patient), that is, con_1. In the second condition (con_2), the DSP *hasAge* is used to verify that the data subject is older than 18 years.

Now, we have created all the components to construct the permissions. We first define two permissions for the microscopic case: p_{micro} states that *HealthData* can be read for *DiagnosisPurpose* if the case consent by the data subject has been given. Each access needs to be logged and there needs to be a *treatsPatient* relationship between the data subject and the requester. p_{micro2} states a similar permission, but is defined for the update of medication data for the *MedicationChangePurpose*. The last permission, p_{macro}, states that it is allowed to *retrieveShoppingBehavior*, in the case that consent has been given by the data subject, the access is logged, and the use of the data is solely for a *ResearchPurpose*. Also, con_2 states that the data subject has to be at least 18 years old.

We now specify three role schemas in the set RS: rs_1 and rs_2 for the microscopic case and rs_3 for the macroscopic case. rs_1 defines the role schema with the name "Nurse" that can be enabled in the feature type *HospitalBuildings*, and the feature type that is collected in order to determine the location of the requester is *Coordinate*. Additionally, the function m_{coord} defines an abstract function that checks the

Location Objects:
FT = {Countries, HospitalBuildings, Coordinate} with:
{USA, Austria} ∈ countries
{University Clinic Innsbruck, KaiserJosefStr1} ∈ Hospital

Privacy Objects:
ACT = {readHealthRecord, updateHealthRecord, retrieveShoppingBehavior}
AH = {readHealthRecord $\leqslant_{ACT}$ updateHealthRecord}

DA = {PersonalData, ShoppingBehaviorData, HealthData, MedicationData,
HealthBillingData, HealthRecordData} with:
DH = {PersonalData $\leqslant_{DA}$ ShoppingBehaviorData,
PersonalData $\leqslant_{DA}$ HealthData,
HealthData $\leqslant_{DA}$ MedicationData,
HealthRecordData $\leqslant_{DA}$ HealthBillingData}

PUR = {AnyPurpose, MedicalPurpose, DiagnosisPurpose,
MedicationChangePurpose, ResearchPurpose} with:
PH = {AnyPurpose $\leqslant_{PUR}$ MedicalPurpose,
MedicalPurpose $\leqslant_{PUR}$ DiagnosisPurpose,
MedicalPurpose $\leqslant_{PUR}$ MedicationPurpose,
AnyPurpose $\leqslant_{PUR}$ ResearchPurpose}

OBL = {NoObligation, LoggingObligation}
OH = {NoObligation $\leqslant_{OBL}$ LoggingObligation}

Conditions:
CON = {con1, con2}
con_1 = "treatsPatient(REQUESTER, DATA SUBJECT)"
con_2 = "hasAge(DATA SUBJECT) >= 18"

Permissions:
PRMS = {p_{micro1}, p_{micro2}, p_{macro}} with:
p_{micro1} = <(readHealthRecord, HealthData), true, LoggingObligation,DiagnosisPurpose,con_1>
p_{micro2} = <(updateHealthRecord,MedicationData), true, LoggingObligation, MedicationChangePurpose, con_1>
p_{macro} = <(retrieveShoppingBehavior, ShoppingBehaviorData), true, LoggingObligation, ResearchPurpose, con_2>

Role schemas:
RS = {rs_1, rs_2, rs_3} with:
rs_1 = <"Nurse", HospitalBuildings, Coordinate, m_{coord}>
rs_2 = <"OffsiteNurse", HospitalBuildings, Coordinate, m_{coord}>
rs_3 = <"ResearchGroup", Countries, Coordinate, m_{coord}>
RH = {$rs_1 \leqslant_{RS} rs_2$}
SPA_S = {(rs_1, p_{micro1}), (rs_2, p_{micro2}), (rs_3, p_{macro})}

Instances:
RI = {ri_1(rs_1, Pediatric Clinic Innsbruck),
ri_2(rs_2,KaiserJosefStr1),
ri_3(rs_3,USA)}

Listing 14.1 The formal description of the policies described in natural language in Section 14.3.3.

```
User Assignment:
U = {NurseAlice, AAAResearch}
URA = {(NurseAlice, ri2), (AAAResearch, ri3)}

Access Requests:
AR = {
 ar1 <NurseAlice,
    updateHealthRecord,
    MedicationChangePurpose,
    http://exampleclinicX.com/records/patientX,
    coordinate(14.5063, 47.5259)>,

 ar2 <AAAResearch,
    retrieveShoppingBehavior,
    researchPurpose,
    http://exmamplestatisticsX.com/shopping-behavior/stats,
    coordinate(-96.3603, 42.5237)>
```

Listing 14.1 (continued)

containment of a coordinate in a hospital building. The second role schema, rs_2, follows the same pattern but has the role name *OffsiteNurse*. This role schema will later be assigned with the offsite feature, to allow remote access, while working at a specific patient's house. rs_3 is responsible for the macroscopic policy and defines the role schema *ResearchGroup*, which can be instantiated with instances of the *Countries* feature type. The definition of *Coordinate* and m_{coord} follows along the lines of the two previous role schemas. We then define rs_2 as a super-schema of rs_1 and therefore it inherits all permissions from rs_1. We now assign permissions to role schemas in the set SPA_s and then instantiate role instances of the previously defined role schemas in the set RI. ri_1 instantiates rs_1 with the feature *Pediatric_Clinic_innsbruck*. Similarly, the role schema of the offsite nurse (rs_2) is instantiated with the feature *KaiserJosefStr1* which represents an offsite nursing home. The third role schema (rs_3) is instantiated with the feature *USA* in order to allow access for research purposes as demanded by the macroscopic policy.

Additionally, we define two users, *NurseAlice* and *AAAResearch* and assign them roles according to the macroscopic and microscopic use case. Finally, we introduce two access requests ar_1 and ar_2. The first access request is issued by the user *NurseAlice*, with the action *updateHealthRecord*, the purpose *MedicationChangePurpose*, a request-uri, and a coordinate that actually corresponds to a building in the Innsbruck University Clinic. This means that Alice's assigned role ri_2 is actually enabled if the data subject's consent is successfully retrieved and the access logged. Analogously, ar_2 defines an access request where the user *AAAResearch* accesses the system from a coordinate that corresponds to a location in the United States. Therefore, in this case the roles assigned to *AAAResearch* could potentially be enabled, if the logging obligation is fulfilled and the data subject's consent is gathered. This example concludes the formal description of the natural language policies introduced in Section

14.3.3. We have seen that the requirements from these natural language policies have successfully been translated to the formal model. By this we have created a model that is able to enforce the privacy requirements of security (from an access control viewpoint), accountability, legitimacy, data categories, and consent, which is also able to take the location of a requester into account. Owing to space limitations we have left out the formal definition of the GEO-PRIVACY access control function.

14.5 Implementation of GEO-PRIVACY

As we have emphasized in previous sections, we laid special emphasis on the feasibility of the server-side implementation of our approach to location-based privacy enforcement. Many challenges of access control models, for a specific domain, are only discovered when implementing the model. Those challenges might lie not only in technical problems but also in methodologies and ergonomics of the implementation. For example, if a formal policy model is developed, how can one express those policies in a realistic scenario? What does a potential modeling tool look like? Should abstractions be applied in order to shield the policy engineer from low-level complexities? How are policies stored and retrieved?

In order to address these issues for GEO-PRIVACY, we created a prototypical implementation of both an access control engine and a policy specification tool for our model. In the implementation we have only focused on the server side of the access control engine. In a realistic scenario, a technical solution for establishing trust in the data sent by the client also needs to be considered. For our implementation, we assume that the location and purpose information from the client can be trusted.

14.5.1 Technological Considerations

Most access control frameworks consist of a basic set of components. These components are (i) the access control engine,* containing a policy enforcement point (PEP) and a policy decision point (PDP); (ii) a knowledge base, containing context information such as user data; (iii) a back-end data service, containing the actual data; and (iv) a policy authoring tool. In addition to these basic components, our solution also needs to cater to location processing via a geospatial data repository and the evaluation of conditions via a logic component. In Table 14.2, we show the technological decisions we made to implement GEO-PRIVACY. In the following subsections, we explain the rationale behind choosing each technology for a specific component.

* See ISO/IEC 10181-3, part Security Frameworks for Open Systems.

Table 14.2 Technological Decisions for the Implementation of GEO-PRIVACY

Component	*Technology Used*
Access control engine	Extended Sun XACML reference implementation
Knowledge base	OWL ontology built with Protégé; integration with policy authoring tool via generated Java binding
Policy authoring tool	RCP-based extension to UDIG with policy generator based on JET templates
Back-end private DB	Simple REST-based hierarchical data service
Geospatial data repository	Built with the GeoTools application programming interface
Logic component	Extended XACML functions on OWL knowledge base

14.5.1.1 Policy Authoring Tool

Common policy authoring tools often simply consist of modifiable tree data structures, tables, and drop-down menus to create roles and to enter user data, as well as to perform user role assignment. In addition to these relatively simple requirements for a user interface, GEO-PRIVACY needs a means to display, select, create, and modify diverse geographical features and feature types according to the OpenGIS standard. Therefore, the decision was made to extend the GIS tool UDIG (see Section 14.2) with the means to express the GEO-RBAC entities, as well as the GEO-PRIVACY-specific privacy entities.

14.5.1.2 Knowledge Base

In our model, we use conditions to reason over the knowledge base. The OWL is a perfect fit for representing domain knowledge in this context. It allows for reasoning over a knowledge base, for example, to discover inconsistencies such as violated static separation of duty constraints. Another big advantage of the OWL is that it can be used to express complex conditions via predicates. This feature is utilized by the implementation to create domain-specific condition languages. We used OWL instead of OWL 2 because tool support was still restricted for OWL 2 at the time of writing. Furthermore, the OWL sublanguage OWL-DL was used as it is a decidable logic opposed to OWL-Full and efficient reasoners for its exist. Additionally, OWL can be leveraged to create public purpose ontologies that can be reused to create a common understanding of specific purposes, thereby defining the exact meaning of specifying a purpose for an access request.

14.5.1.3 Access Control Engine

For the implementation of the access control engine, two different approaches were considered. Because of the decision to use an ontology language like OWL to represent knowledge in the implementation, it was possible to represent policies in OWL as well. Access control decisions can potentially be made by reasoning over existing policies as proposed by Finin et al. (2008). The other option was the approach by Ferrini and Bertino (2009) who proposed the use of XACML as the policy language, and OWL as the means to express knowledge and constraints. For the implementation of GEO-PRIVACY, we chose the latter approach for the following reasons. First, there exist a variety of well-studied implementations of XACML that can be reused and extended. No equivalent exists for an OWL-based access control mechanism. Additionally, XACML has important aspects, such as obligations, already built-in. Second, with the hybrid approach by Ferrini and Bertino (2009), one can utilize the best of both worlds, the proven implementation, architecture, and usability of XACML as well as the reasoning capabilities of an ontology language like OWL.

14.5.1.4 Geospatial Data Repository

The decision regarding the geospatial data repository was already made by deciding on the policy authoring tool. As UDIG is tightly integrated with the GeoTools API, it is used to create, store, and modify geospatial data as well as to check the containment of features for the access control mechanism.

14.5.1.5 Logic Component

The implementation of the logic component consists of two parts. The first part is the policy selection mechanism and the rule-combining algorithm that are already integrated in SUN's XACML reference implementation. The second part is the integration of conditions and facts from the OWL knowledge base. This integration is realized by creating new XACML functions that are able to retrieve truth values from the knowledge base. Here, the selection of the technologies was natural because it is based on prior decisions regarding the technologies for knowledge representation and the access control engine.

14.5.1.6 Back-End Private Data Database

As an exemplary back-end private data service that provides access to the protected private data, a simple hierarchical REST service was implemented. The reason for this choice is that with REST data can be hierarchically ordered in the hierarchy of URIs.

14.5.2 The Prototype of the GEO-PRIVACY XACML/OWL Access Control Engine

As we have seen in the previous section, each technical component for the implementation is chosen for specific reasons. We now describe how components fit together and explain how they interact with each other in order to put into reality the formal model described in Section 14.4. The following sections then describe the access control engine that can enforce this policy language and the policy specification tool that is used to model policies and generate the enforceable code.

14.5.2.1 High-Level Architecture

Figure 14.2 shows the actual architecture of the GEO-PRIVACY implementation.

We will now describe an example access request, and explain each component's responsibility at each step. In step 1, a policy administrator has to author policies with the policy administration point (PAP). In this case, it is the extended UDIG Eclipse RCP application that is described in more detail in Section 14.5.3. To create and edit policies, the PAP needs to access and write to the OWL knowledge base and the geospatial data source. In step 2, a user wants to access a private resource from the system through a mobile device. The access request is intercepted by the

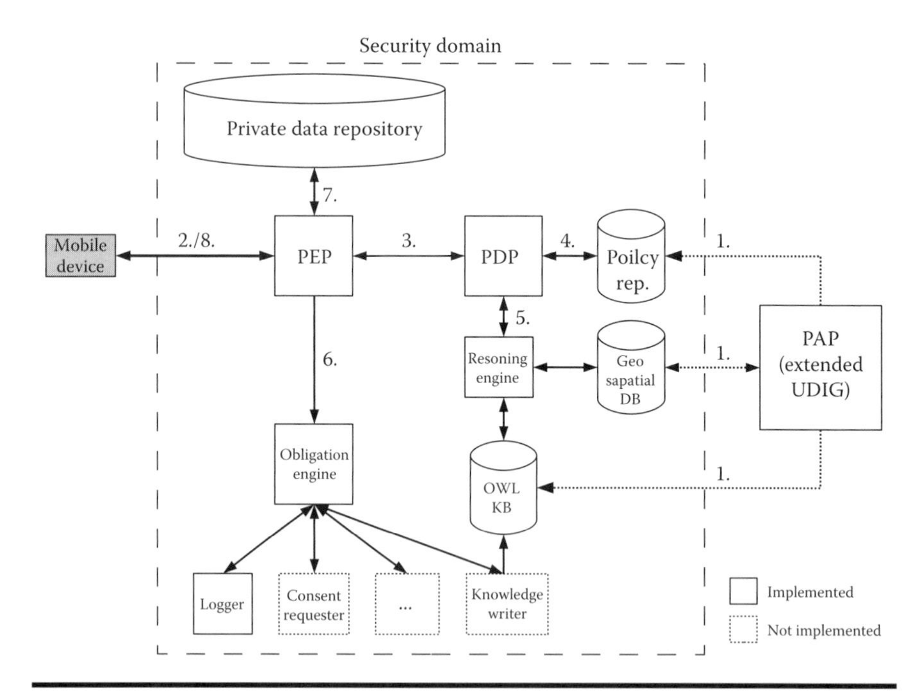

Figure 14.2 The architecture of the GEO-PRIVACY XACML/OWL access control engine.

PEP that is implemented as an extension to SUN's XACML reference implementation. The PEP is responsible for analyzing the request and filtering all access information from the request, such as the user's location, the purpose of the request, and the user's access credentials. In step 3, the PEP forwards this information in the standard format of an XACML access request to the PDP. According to the information provided by the XACML access request, the PDP consults the policy repository for applicable policies in step 4. For any found policy the PDP consults the reasoning engine, which first checks if the role of the user is activated for his/her current location, by mapping the user-provided coordinate to an actual feature, and then compares this to the required feature of the role schema (step 5).

The OWL knowledge base is consulted to check whether the requested action, the purpose, and the data-type hierarchies match the given policy via the reasoning engine. It is also responsible for evaluating the conditions of a policy with the truth values from the knowledge base. If an access-granting policy is found, the PDP signals this to the PEP including any necessary obligation identifiers that are found in the policy. The PEP forwards these obligation identifiers to the obligation engine. For this prototype, we only implemented the logging obligation. Future extensions could include a consent requester that asks data owners for their consent via an email or Web interface, or a knowledge writer that writes data about past access to resources to the knowledge base. Once all obligations are executed successfully, the request is finally forwarded to the private data repository to return the requested data. If a necessary obligation could not be executed, access to the resource is not granted.

Logically, the architecture is separated into three different parts. A policy enforcement engine containing a PEP and PDP that are implemented by extending SUN's XACML reference implementation, a knowledge base providing attribute information implemented as an OWL ontology with the Pellet reasoner,* and a policy authoring tool forming the PAP. These three main components are explained in the following sections.

14.5.2.2 XACML Access Control Implementation

The access control engine is the crucial part of any access control framework. It is the entity that decides on the access rights of a request and denies or grants access. As stated before, the access control engine of GEO-PRIVACY is implemented by extending SUN's XACML reference implementation that is written in Java.† Here, the current implementation from SUN's CVS repository was checked out. This version is, to a large percentage, XACML 2.0 compliant as stated on the referenced Web site.

The XACML specification recommends a standard architecture for an XACML access control engine. This architecture is, except for small differences, implemented by the GEO-PRIVACY engine. The main difference is that, unlike the

* http://www.mindswap.org/2003/pellet/index.shtml

† http://java.sun.com

recommended architecture, the GEO-PRIVACY engine does not retrieve all attributes from a context handler. Rather, attributes that are not retrieved from the access request set AR (see Section 14.4.1) are fetched via custom functions in the PDP from the OWL knowledge base. This follows along the lines of the approach by Ferrini and Bertino (2009). However, their approach is not feasible without rewriting the XACML policy engine and losing the conformity with the XACML specification. The reason is that they use custom functions with *<Apply/>* within the target element. This does not conform to any of the current XACML standards. In contrast, we implement the function in the target via the *Match* element.

GEO-PRIVACY follows the RBAC approach in which a role is assigned with permissions and also inherits all permissions from super roles in the role hierarchy. Therefore, for implementing an XACML policy repository and designing policy templates, one has to consider not allowing the work to become redundant, as permissions can be reused. This is important to make the permissions in the policy repository more manageable. For this reason, the XACML 2.0 RBAC specification was developed by OASIS.* This specification explains a standard way to cope with the issue of separating roles and permissions in XACML policies. GEO-PRIVACY role instances are defined as policy sets whereas permissions are imported into the policy set as policies in different repositories.

14.5.2.3 GEO-PRIVACY XACML Policies

As we have explained in the previous section, roles and permissions are separated from each other in the implementation of GEO-PRIVACY. Table 14.3 gives a general overview of how the entities of the formal model are translated into XACML. How these are modeled by the underlying OWL ontology is described later in this section. As one can see from the table, only role instances are actually translated into enforceable XACML policies, because role schemas are abstract and have been created in the GEO-RBAC model to simplify reuse. We will now explain a policy set that represents a role instance and then show how policies are used to craft permissions.

Listing 14.2 shows an XACML policy set that describes the role instance ri_1 from Section 14.4.3. The policy described here assumes that all information from an access request (ar) is contained in the XACML request context.

As said before, the *<target/>* of the policy set is used to determine whether a policy set is applicable to a certain request. That is the requester has the given role assigned to him/her. This is done in lines 8–12 in the given PolicySet. The SubjectMatch tag is used to retrieve the user's ID from the request context and check whether this user has the role Rs1 directly or via role inheritance assigned to him/her. This is achieved through the custom function *check_transitive_role* following Ferrini and Bertino (2009). How this transitive check is executed on the underlying OWL knowledge base is explained in Section 14.5.2.

* http://www.oasis-open.org/committees/tc_home.php? wg_abbrev = xacml

Table 14.3 GEO-PRIVACY Implementation Mappings

GEO-PRIVACY Model Element	*OWL Mapping*	*XACML Mapping*
Role schema: R_s	owl:Class RoleSchema	abstact—no mapping
Role instance: R_i	owl:Class RoleInstance	xacml:PolicySet
Action: ACT	owl:Class Action	XACML action attr.
Data: DA	owl:Class DataType	XACML resource attr.
Real position: RPOS	—	XACML subject attr.
User: *U*	owl:Class User	—
Purpose: PUR	owl:Class Purpose	XACML actionMatch
Obligation: OBL	owl:Class Obligation	xacml:Obligation
Condition: CON	owl:Class Condition	xacml:Condition
Permissions: PRMS	owl:Class Permission	xacml:Policy
Access request: AC	—	XACML request ctx.
User-role assignment: URA	owl:ObjectProperty	—

If the user has this role assigned to him/her, the policy checks whether the user is located in a certain geographical feature. This is achieved through a special policy in lines 24–43. The policy simply contains one *<Rule>* with a *<Condition>* that applies the function *eval-coordinate-contained-in-feature* function. This function has four parameters, namely, the latitude and longitude comprising the position of the requester from the access request, the expected feature ID, and the type of feature.

After this follows another policy set that contains all permissions of the roles in lines 45–55. The set starts with a policy that simply denies access. Subsequently, follow the permissions that are assigned to a role instance. As the policy set is initialized with the *permit-overrides* policy-combining algorithm, the policy set only evaluates to *PERMIT* if one of the imported policies permits access. Otherwise the initial *DENY* will be applied which will deny access. This role (policy set) imports a permission via the *<PolicyIdReference>* in line 54. Listing 14.3 shows this imported permission.

As permissions in RBAC can be assigned to more than one role, the *Subjects* tag in the permission policy is not set to match any subject. However, in GEO-PRIVACY permissions have to match with the data type, the action, and the purpose of an access request. Therefore, lines 10–17 describe the check whether the requested data is of the type *HealthRecordData*. This is achieved through the custom function *check-transitive-datatype*, which retrieves the data type from the URI of the requested resource and checks its transitive relation to the *HealthRecordData* in the

```
PolicySet xmlns= "urn:oasis:names:tc:xacml:1.0:policy" xmlns:xsi="..."
PolicySetId="GeneratedPolicy1" PolicyCombiningAlgId="...deny-overrides">
 <Description> PolicySet Generated by GEO-Privacy code generator.</Description>

 <Target>
 <Subjects>
 <Subject><!-- Ri -->
   <SubjectMatch MatchId="check-transitive-role">
   <AttributeValue DataType="...string">Rs1</AttributeValue>
   <SubjectAttributeDesignator
   DataType="...string" AttributeId="userID"/>
   </SubjectMatch>
 </Subject>
 </Subjects>
 <Resources>
 <AnyResource/>
 </Resources>
 <Actions>
 <AnyAction/>
 </Actions>
 </Target>

<!-- Location Policy -->
 <Policy xmlns="...policy" PolicyId="RoleLocationPolicy"
  RuleCombiningAlgId="deny-overrides">
  <Target/>

  <Rule RuleId="LocationRule" Effect="Permit"><!-- RPOS -->
  <Condition FunctionId="...boolean-equal">
   <Apply FunctionId="...eval-coordinate-contained-in-feature">
    <SubjectAttributeDesignator
     DataType="...double" AttributeId="latitude"/>
    <SubjectAttributeDesignator
     DataType="...double" AttributeId="longitude"/>
    <AttributeValue
     DataType="...string">Pediatric_Clinic_Innsbruck</AttributeValue>
    <AttributeValue
     DataType="...string">HospitalBuildings</AttributeValue>
   </Apply>
   <AttributeValue DataType="...boolean">true</AttributeValue>
  </Condition>
  </Rule>
 </Policy>
 <!-- Permission PolicySet-->
 <PolicySet xmlns="...policy" PolicySetId="..."
        PolicyCombiningAlgId="...permit-overrides">
 <Target/>
 <Policy xmlns="...policy" PolicyId="..." RuleCombiningAlgId="...
deny-overrides">
   <Target/>
   <Rule RuleId="DenyRule" Effect="Deny"/>
 </Policy>

  <!-- Permissions -->
  <PolicyIdReference>pmicro2</PolicyIdReference>
 </PolicySet> </PolicySet>
```

Listing 14.2 XACML policy set defining a role instance.

```
<Policy xmlns="...policy" PolicyId="pmicro2"     RuleCombiningAlgId="...
ordered-permit-overrides">
 <Target>
  <Subjects> <AnySubject/>  </Subjects>

  <Resources>
  <!-- Data Type:HealthRecordData -->
   <Resource>
    <ResourceMatch MatchId="check-transitive-datatype">
     <AttributeValue DataType="&xsd;#string">HealthRecordData</AttributeValue>
      <ResourceAttributeDesignator DataType="&xsd;#string"
AttributeId="resource-id"/>
    </ResourceMatch>
   </Resource>
  </Resources>

  <Actions>
  <!-- Action:readHealthRecord, Purpose:DiagnosisPurpose -->
   <Action>
    <ActionMatch MatchId="check-transitive-action">
     <AttributeValue DataType="&xsd;#string">readHealthRecord</AttributeValue>
      <ActionAttributeDesignator DataType="&xsd;#string"
AttributeId="urn:oasis:names:tc:xacml:1.0:action:action-id"/>
    </ActionMatch>
    <ActionMatch MatchId="check-transitive-purpose">
     <AttributeValue DataType="&xsd;#string">DiagnosisPurpose</AttributeValue>
     <ActionAttributeDesignator DataType="&xsd;#string" AttributeId="purpose"/>
    </ActionMatch>
   </Action>
  </Actions>
 </Target>

 <Rule RuleId="DenyAllOthers" Effect="Deny"/>

 <!-- Condition:con1 -->
 <Rule RuleId="con1" Effect="Permit">
  <Target/>
  <Condition FunctionId="urn:oasis:names:tc:xacml:1.0:function:boolean-equal">
   <Apply FunctionId="get-object-property">
    <AttributeValue DataType="&xsd;#boolean">treatsPatient</AttributeValue>
    <Apply FunctionId="get-user">
     <SubjectAttributeDesignator DataType="...string" AttributeId="userID"/>
    </Apply>
    <Apply FunctionId="get-data-owner">
     <ResourceAttributeDesignator DataType="&xsd;#string"
AttributeId="resource-id"/>
    </Apply>
   </Apply>
  </Condition>
 </Rule>

 <!-- Obligtion:LoggingObligation -->
 <Obligations>
    <Obligation ObligationId="LoggingObligation" FulfillOn="Permit"/>
    <Obligation ObligationId="Consent" FulfillOn="Permit"/>
  </Obligations>
</Policy>
```

Listing 14.3 XACML policy defining a permission.

knowledge base. The same is done for the action and the purpose of the request in lines 19–33. In line 36, a rule is defined that denies all access as the standard behavior. If the rule in lines 39–54 returns a permit, this former deny rule is overwritten. This complex rule makes use of the DSP *treatsPatient* to find out whether the data requester is the doctor responsible for the treatment of the data owner. This is achieved through the *get-object-property* custom function that has three parameters. The first parameter (line 43) is the identifier of the DSP. The second parameter retrieves the URI that identifies the user in the knowledge base. The last parameter is the ID of the data owner that is requested via the *get-data-owner* function. From this, the inference engine can evaluate whether the *treatsPatient* relation exists between the data requester and the data owner. Additional conditions can be added here by adding them within the conjunction function *urn:oasis:names:tc:xacml:1.0:function:and*. Finally, lines 57–60 specify the logging obligation that needs to be executed in case the result of this policy is PERMIT. In addition, the consent requirement is encoded as an obligation here.

14.5.2.4 OWL Knowledge Base

In the implementation of GEO-PRIVACY, all knowledge, including the access control policies themselves, is described and made persistent using OWL. OWL is based on the formal groundings of computational logic, and has therefore many properties that make it interesting for the knowledge representation language in such highly complex and nonfailure-tolerant environments such as access control frameworks. In such environments, potentially thousands of users, user attributes, policies, and environment variables can be stored and are brought in relation with each other. This can lead to inconsistencies, redundancy, and eventually unwanted behavior, for example, by conflicting policies. With an ontology language like OWL, rules can be defined that help to detect such inconsistencies, thus making the system safer. Another important advantage of OWL for an access control system is the ability to use an inference engine to query the knowledge base for truth values. In our study, we use this capability to find the transitive hierarchical relationships of roles, purposes, actions, data types, and obligations. Also, and equally importantly, we use the inference mechanism to implement conditions on environment attributes in our framework. Additionally, OWL axioms can be used to implement complex constraints such as static separation of duty as described by Ferrini and Bertino (2009). Finally, OWL as the way of storing structured semantic information is extensible, which allows for the adaption of the underlying knowledge base according to the requirements of a (privacy) access control domain. This fact is utilized for GEO-PRIVACY conditions that require DSPs in the condition language. The aim of this domain specificity is to simplify condition formulation for policy authors as a general-purpose language can become too complex.

For the above-mentioned reasons, we implement the knowledge base of the access control framework with OWL. In the following, we give the reader an overview of the underlying base ontology that stays static for each access control domain. In addition, we give an example of a domain-specific ontology extension that allows for DSPs, which can be used in conditions.

Figure 14.3 shows a graphical representation of the OWL class inheritance tree that comprises the GEO-PRIVACY OWL base ontology. For additional reference, Table 14.3 shows how the most important classes map to their representation in the formal model.

According to the OWL notation, the arrows in Figure 14.3 signify an "is–a" relationship similar to the generalization notation in UML. To make this ontology more readable and to make it easier to distinguish the concepts of the original work of GEO-RBAC from the new contribution of our work, we have divided the ontology into three main classes of which the sets of their children are intersecting. This is possible as OWL allows for multiple inheritance. These three base classes are *GeospatialObject*, *GEORBACObject*, and *GEOPrivacyObject*. We will now briefly expand on these subclasses and their corresponding purpose.

As it is the aim of the OWL knowledge base to capture as much knowledge as possible, we included the knowledge about feature types and their corresponding features into the knowledge base. This information is automatically loaded from the geospatial database, as described in Section 14.5.3. All the knowledge about containment and the relation between features and feature types still resides in the geospatial database. As both feature types and features are part of the GEO-RBAC model they also inherit from *GEORBACObject*.

The other classes inheriting from *GEORBACObject* are the original objects from the GEO-RBAC model, such as role schemas, role instances, users, permissions, and actions. The latter two classes also inherit from the *GEOPrivacyObject* class as they have been extended in the new model.

The largest set of classes inherits from the *GEOPrivacyObject* class. Most of those classes are self-explanatory on the basis of their name, except for the *ExecutionHistory* class. This class is a placeholder in case the implementation later supports history-based access control.

To realize hierarchy for roles, data types, obligations, actions, and purposes, instances of these classes can have transitive super and subrelations between instances of the same type. How this is achieved for the super data-type relation is shown in Listing 14.4.

In this OWL snippet, lines 2–8 define a transitive object property between two data types with the name *hasSuperDataType*. Line 3 defines this property as the inverse property of the property *hasSubDataType*. Lines 4 and 5 state that the range and the domain of the property both have to be instances of the class *DataType*. Finally, lines 6 and 7 state that the property is an object property and that it is the subproperty of a property with the ID *DataProperty*.

The base ontology covers the basic objects that are likely to occur in any access control domain. Listing 14.5 shows how this base ontology can be extended with domain-specific properties that can be used in conditions.

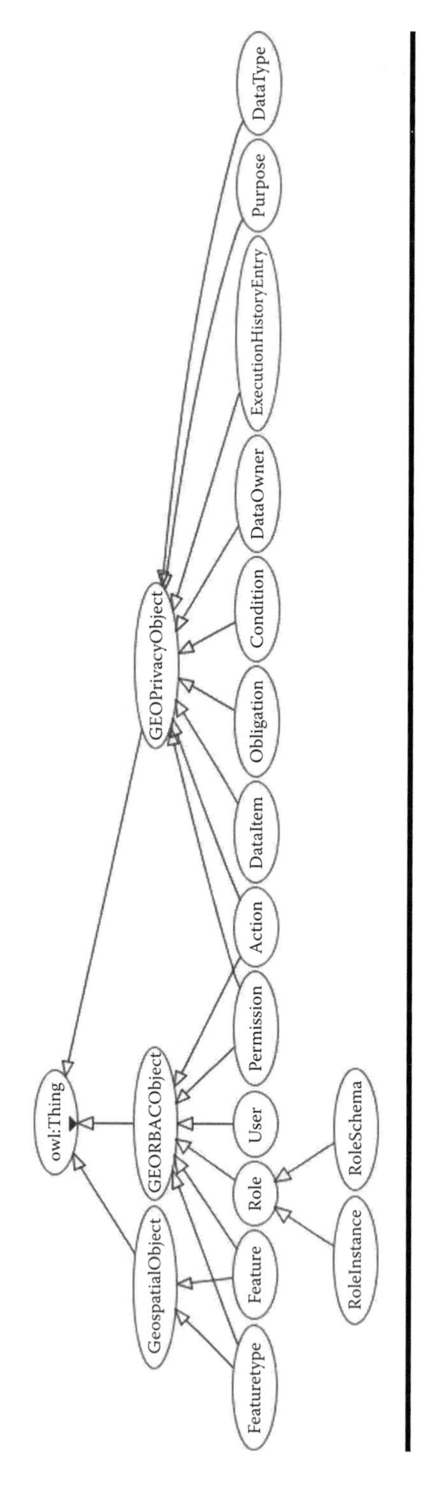

Figure 14.3 The underlying OWL base ontology of GEO-PRIVACY.

```
...
  <owl:TransitiveProperty rdf:about = "#hasSuperDataType">
    <owl:inverseOf rdf:resource = "#hasSubDataType"/>
    <rdfs:range rdf:resource = "#DataType"/>
    <rdfs:domain rdf:resource = "#DataType"/>
    <rdf:type rdf:resource="http://www.w3.org/2002/07/
owl#ObjectProperty"/>
    <rdfs:subPropertyOf rdf:resource = "#DataProperty"/>
  </owl:TransitiveProperty>
...
```

Listing 14.4 An extraction from the OWL Base Ontology showing how the datatype hierarchy is implemented in the OWL ontology.

The example from Listing 14.5 introduces two new properties that can be applied to data owners or users requesting access, respectively. The first property has the ID *treatsPatient* and the second one has the ID *hasAge*. Both are used to show how the information needed by the conditions con_1 andcon_2, from the formal policy of Section 14.4.3, can be included into the ontology end and ultimately be used as DSPs in conditions.

Lines 4–6 indicate that the ontology imports the GEO-PRIVACY base ontology. This way, the extension can make use of all elements of the base ontology and can also extend its concepts and properties. To create a new property that inherits the domain from the *UserProperty*, a new object property with the ID *treatsPatient* is declared in lines 8–13. The range of this new property is set to be *DataOwner*. In lines 15–20, the *hasAge* data type property is declared. Its range is set to integer in line 16 and lines 17–19 set it as a subproperty of *DataOwnerProperty*. In lines 22–30, the ontology makes use of the important restriction mechanism. This restriction is applied to the *DataOwner* class and enforces that each data owner may only have one *hasAge* property assigned to him/her.

This concludes this very basic extension of the base ontology. For any new domain such an extension needs to be created in order to be able to write domain-specific conditions.

14.5.3 Policy Authoring Tool

In previous sections, we have shown how the XACML access control engine is implemented and how the underlying OWL ontology for knowledge representation and retrieval is structured. As we have argued before, it is not realistic in a real-world scenario for a policy administrator to author XACML policies directly for complex policies as the GEO-PRIVACY policies. Therefore, we created an Eclipse-based GEO-PRIVACY policy authoring tool that enables policy administrators to create policies without knowledge about XACML, OWL, or geospatial containment functions. All that a policy author needs to know are the requirements with regard to the privacy of personal data and the locations from which the data should be available. The tool

```
<rdf:RDF
...
 xmlns:geo_priv="&geo_priv;">
 <owl:Ontology rdf:about="&ext;">
  <owl:imports rdf:resource="&geo_priv;"/>
 </owl:Ontology>

  <owl:ObjectProperty rdf:about="&ext;#treatsPatient">
    <rdfs:subPropertyOf>
      <rdf:Property rdf:about="&geo_priv;#UserProperty"/>
    </rdfs:subPropertyOf>
    <rdfs:range rdf:resource="&geo_priv;#DataOwner"/>
  </owl:ObjectProperty>

  <owl:DatatypeProperty rdf:about="&ext;#hasAge">
    <rdfs:range rdf:resource="&xsd;#int"/>
    <rdfs:subPropertyOf>
      <owl:DatatypeProperty rdf:about="&ext;#DataOwnerProperty"/>
    </rdfs:subPropertyOf>
  </owl:DatatypeProperty>

  <rdfs:Class rdf:about = "&geo_priv;#DataOwner">
    <rdfs:subClassOf><owl:Restriction>
        <owl:onProperty>
          <owl:DatatypeProperty rdf:about="&ext;#hasAge"/>
        </owl:onProperty>
        <owl:cardinality rdf:datatype="&xsd;#int"
        >1</owl:cardinality></owl:Restriction>
    </rdfs:subClassOf>
  </rdfs:Class>

</rdf:RDF>
```

Listing 14.5 An OWL ontology that extends the base ontology with domain specific predicates.

uses a mixture of tree structures, drop-down menus, and text fields, as well as the view of geographical features and feature types from UDIG, in order to create GEO-PRIVACY policies. The following list summarizes the core features of the tool:

- Creation of GEO-PRIVACY purposes, obligations, data types, and actions in hierarchical order.
- Assignment of data types to resource URIs.
- Creation of permissions.
- Creation of role schemas from a list of feature types in a feature type repository.
- Creation of feature types (UDIG functionality).
- Creation of role instances from role schemas and features.
- Persistence of knowledge.

- User management and user role assignment.
- Generation of XACML policies.

For a visual impression of the tool, see Figure 14.4, specifying GEO-PRIVACY policies. The layout of the tool follows the conventions of Eclipse RCP applications and is fully customizable via views and perspectives. The central piece of this view is the map (1) that lets the user choose features from the currently selected feature type in the catalog. In this case, it is a world map of the feature type *Country* with all countries as features. Here, the feature *USA* is selected for a given role instance. The right part of the image (2) is reserved for specifying hierarchies. Here, we can see that the purpose hierarchy is currently edited and the data type hierarchy tab is also located left of the purpose hierarchy tab. The left side of the screen (3) is reserved to select the different layers of existing features. For example, in this case, the *countries* layer is checked and selected whereas the *UniversityClinicBuilding* layer is neither selected nor checked. This means that only the country features are presented on the map and can be selected as well as edited. The topmost part of the application is reserved for editing tools that can be used to create and modify features and feature types. These tools represent the core functionality of UDIG and have been reused without change. The bottom part of the editor can also be used to edit hierarchies. Here, action and

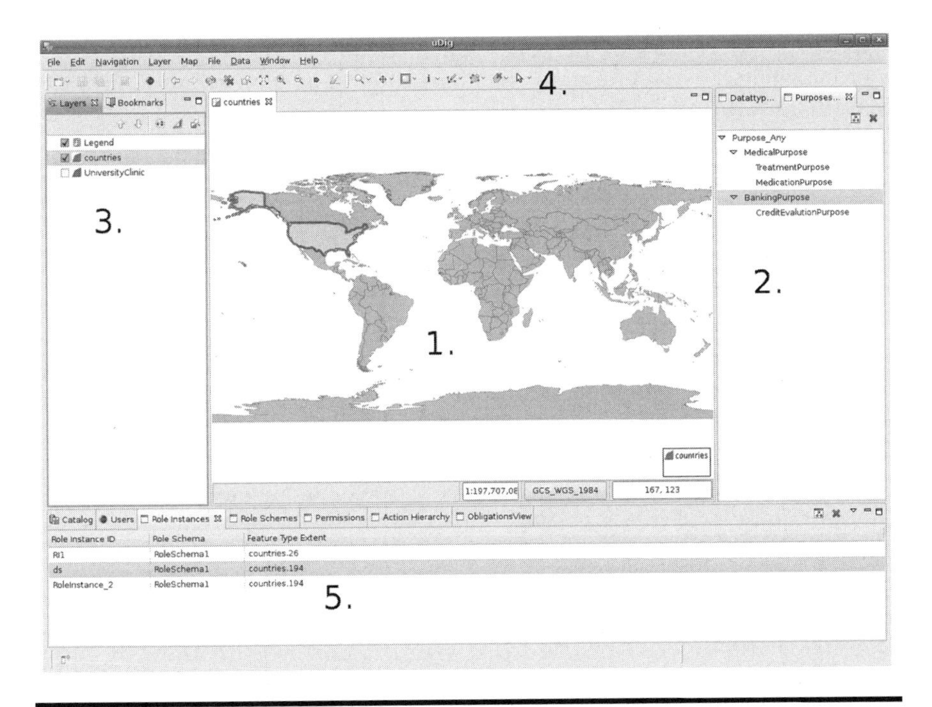

Figure 14.4 Screenshot of the policy specification tool showing a layer with country features and the editing functionality.

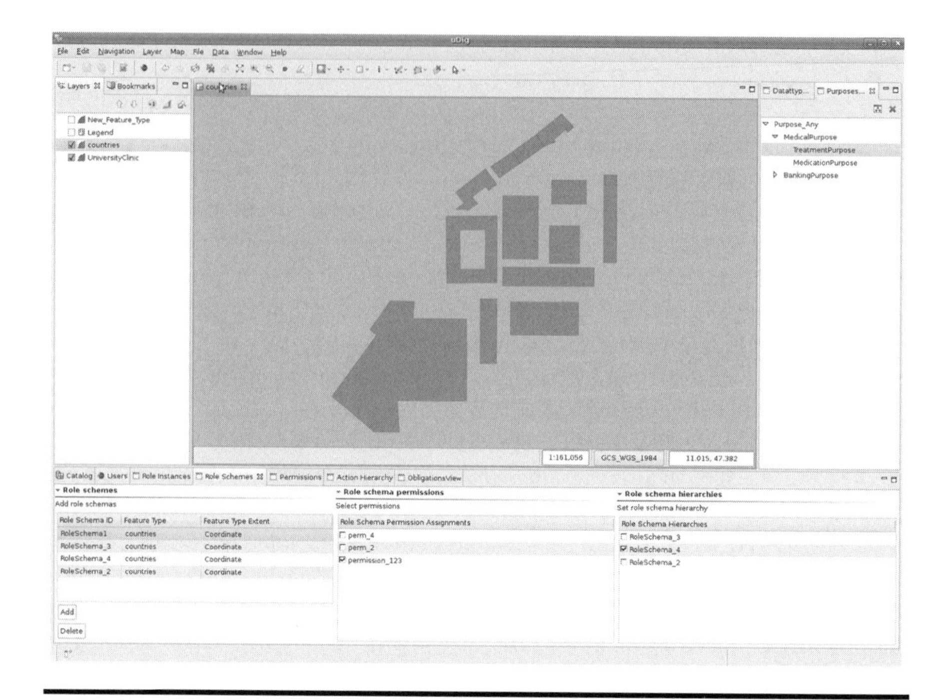

Figure 14.5 Screenshot of the policy specification tool showing a hand-drawn layer modeling the exact location and dimensions of buildings of the University Clinic in Innsbruck, Austria.

obligation hierarchies can be edited. More importantly, this area is reserved for editing user information, such as creating new users or changes to the user role assignment, as well as creating and editing role schemas and role instances. Additionally, the tab area contains the catalog tab that lets the user administrate the geographic data sources, which are locally hosted or gathered from Web feature services.

Figure 14.4 shows a prebuilt map that is provided with UDIG. This map denoting country borders is also useful for creating macroscopic policies as in the case that was introduced in Section 14.3.3.1. However, it cannot be used to describe microscopic policies. Therefore, the tool allows for the creation of self-defined feature types corresponding to a specific coordinate system. Figure 14.5 shows a self-defined feature-type *UniversityClinic* that denotes the exact boundaries of the buildings of the University Clinic of Innsbruck in Austria.

14.5.3.1 High-Level Architecture and Implementation

Figure 14.6 conceptually shows how the features of the UDIG are extended by interfaces to the geospatial database and the OWL knowledge base. Changes to the knowledge base are committed to the OWL ontology, and changes to the

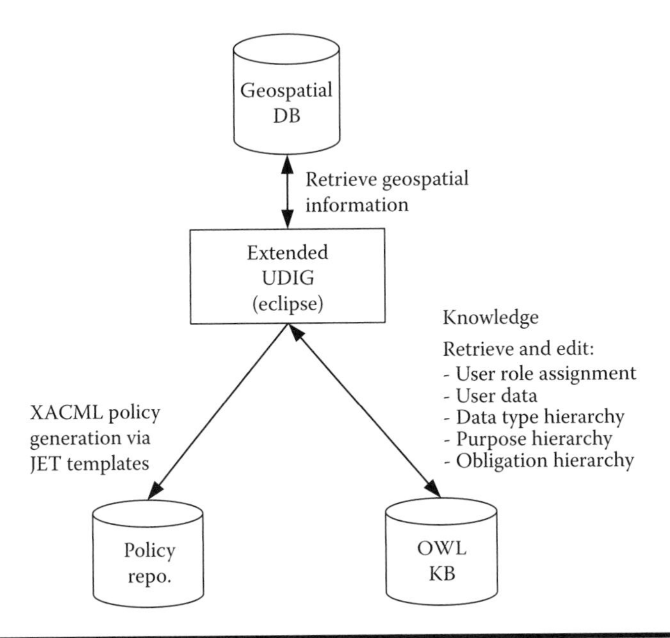

Figure 14.6 Policy specification mechanism.

geospatial information, for example, the creation of a new feature type, are committed to the geospatial database. After the policy definition process, the XACML representation of the policies can be generated from the tool through JET templates. Permissions are generated as XACML policies whereas role instances are generated as XACML policy sets, as explained in Section 14.5.2.

In this section, we have introduced the reader to the GEO-PRIVACY access control engine, the underlying knowledge base implemented in OWL, and have outlined the GEO-PRIVACY policy authoring tool. The architecture consists of an XACML access control engine that was described first and a knowledge base implemented with OWL. After the description of the knowledge base, we introduced the GEO-PRIVACY policy authoring tool that can be used to generate the XACML policies, to enforce access control, and to administer the knowledge, for example, by editing user data.

14.6 Challenges and Future Research

For the implementation of GEO-PRIVACY, we faced the challenge of having to integrate several highly complex technologies such as XACML, OWL, OpenGIS programming interfaces, and the Eclipse RCP platform. However, despite the complexities of implementation, many open issues still have to be solved before production use of our model.

First of all, we assume in our implementation that the information sent by mobile devices to the access control engine can be trusted, that is, the location information sent to the engine is correct and the retrieved data are only used for the specified purpose. This is certainly not always the case. A possible solution for this issue could be the use of remote attestation techniques (Haldar et al., 2004).

Another missing piece in our implementation that is related to the client side is the enforcement of retention time that is required by some privacy laws. Client-side usage control is an active research area and not yet fully explored. Theoretical approaches to this problem can be found in the studies by Sandhu and Park (2004) and by Pretschner et al. (2008).

Another issue that was not tackled by our work is the problem of movement of mobile devices. This problem has been avoided by removing the session concept from the model and enforcing access control on the basis of each request. However, there are cases in which sessions are required. Damiani and Silvestri (2008) describe an approach to mitigate this problem.

In future work we will also investigate how to automatically generate textual editors for the domain-specific condition languages of GEO-PRIVACY.

14.7 Conclusion

In this work we started out with two usage scenarios and claimed that they underline the requirement for location detection in privacy policies. After introducing relevant background technologies and literature, we analyzed several international privacy laws to verify this claim. We discovered that these laws actually require the checking of data requesters' physical location to decide on his/her access rights on private data. We then captured the requirements of these laws in two natural language privacy policies at the end of Section 14.3.

Finding the location requirements in the privacy laws led us to the development of the formal access control model GEO-PRIVACY, described in Section 14.4. GEO-PRIVACY is a privacy extension to the formal location-based access control model GEO-RBAC by Damiani et al. (2007). In this new model, we introduce four different privacy concepts in GEO-RBAC. These are the concepts of purpose, consent, obligations, and privacy conditions. All of these concepts stem from the requirements of the above-mentioned privacy laws. After we introduced the formal model, we gave an example policy that uses GEO-PRIVACY to model the requirements that were introduced by the natural language policies in Section 14.3.

To prove the feasibility of enforcing the GEO-PRIVACY policies on the server side, we developed a prototype of an access control framework for GEO-PRIVACY. This prototype is based on XACML as the policy definition language, leveraging SUN's XACML reference implementation and the ontology language OWL as the means for describing domain knowledge such as role hierarchies. This prototype is described in Section 14.5 alongside an Eclipse RCP authoring and domain knowledge editing tool.

In this work, we have shown that by extending the location-based access control model GEO-RBAC with privacy entities we can express a large number of privacy requirements in this location-based model. These requirements are location detection, consent, accountability, obligations, purpose, and data categories. To the best of our knowledge, no approach exists in the literature that combines location aspects with the privacy aspects mentioned above.

QUESTIONS

Q1: What is a motivating scenario for location-based access control with privacy features?

Q2: Which location aspects does the EU data privacy directive contain?

Q3: Which location aspects does the U.S. law HIPAA contain?

Q4: What are common privacy requirements?

Q5: What is the difference between a microscopic and a macroscopic location-based access control policy?

Q6: Why is the current implementation not secure if the mobile client implementation cannot be fully trusted? Are there approaches to solve this problem?

Q7: Which advantages can a formal basis for an access control language bring?

Q8: Which feature of the XACML does the implementation use to access the knowledge base?

References

Ardagna, C. A., Cremonini, M., Damiani, E., De Capitani di Vimercati, S., and Samarati, P. Supporting location-based conditions in access control policies. In *Asiaccs '06: Proceedings of the 2006 ACM Symposium on Information, Computer and Communications Security* (pp. 212–222). New York, NY, USA: ACM, 2006.

Bettini, C., Wang, X. S., and Jajodia, S. Protecting privacy against location-based personal identification. In *Proceedings of the 2nd VLDB Workshop on Secure Data Management* (pp. 185–199). Berlin: Springer-Verlag, 2005.

Birnhack, M. D. The EU data protection directive: An engine of a global regime. *Computer Law & Security Report*, 24(6), 508–520, 2008.

Chandran, S. M., and Joshi, J. B. D. Lot-RBAC: A location and time-based RBAC model. In A. H. H. Ngu, M. Kitsuregawa, E. J. Neuhold, J.-Y. Chung, and Q. Z. Sheng (eds), *Web Information Systems Engineering—WISE* (*LNCS* Vol. 3806, pp. 361–375). Berlin: Springer-Verlag, 2005.

Damiani, M. L., Bertino, E., Catania, B., and Perlasca, P. GEO-RBAC: A spatially aware RBAC. *ACM Transactions on Information and System Security*, 10(1), 2007.

Damiani, M. L., and Silvestri, C. Towards movement-aware access control. In *Proceedings of SPRINGL '08: Proceedings of the SIGSPATIAL ACM GIS 2008 International Workshop on Security and Privacy in GIS and LBS* (pp. 39–45). New York, NY: ACM, 2008.

Di, W., Jian, L., Yabo, D., and Miaoliang, Z. Using semantic web technologies to specify constraints of RBAC. In *Pdcat '05: Proceedings of the Sixth International Conference on*

Parallel and Distributed Computing Applications and Technologies (pp. 543–545). Washington, DC: IEEE Computer Society, 2005.

Ferrini, R., and Bertino, E. Supporting RBAC with XACML + OWL. In *SACMAT 2009: Proceedings of the 14th ACM Symposium on Access Control Models and Technologies* (pp. 145–154). New York, NY: ACM, 2009.

Finin, T., Joshi, A., Kagal, L., Niu, J., Sandhu, R., and Winsborough, W. H.)ROWLBAC—Representing role based access control in OWL. In *Proceedings of the 13th Symposium on Access Control Models and Technologies* (pp. 73–82). New York, NY: ACM, 2008.

Haldar, V., Chandra, D., and Franz, M. Semantic remote attestation—A virtual machine directed approach to trusted computing. In *Proceedings of USENIX Virtual Machine Research and Technology Symposium* (pp. 29–41). Berkeley, CA: USENIX Association, 2004.

Hansen, F., and Oleshchuk, V. Spatial role-based access control model for wireless networks. In *Proceedings of the 58th IEEE Vehicular Technology Conference (VTC '03)* (Vol. 3, pp. 2093–2097). Washington, DC: IEEE Computer Society, 2003.

Health Insurance Portability and Accountability Act of 1996, Pub. L. No. 104–191, 110 Stat. 1936, 1996.

Heilili, N., Chen, Y., Zhao, C., Luo, Z., and Lin, Z. OWL-based approach for RBAC with negative authorization. In *Knowledge Science, Engineering and Management* (*LNCS* Vol. 4092, pp. 164–175). Berlin: Springer-Verlag, 2006.

Herring, J. R. OpenGIS implementation specification for geographic information—simple feature access—Part 1: Common architecture. *Technical Report*, OpenGIS Consortium, 2006.

Langheinrich, M. Privacy by design—Principles of privacy-aware ubiquitous systems. In *UbiComp '01: Proceedings of the 3rd International Conference on Ubiquitous Computing* (pp. 273–291). London, UK: Springer-Verlag, 2001.

Ni, Q., Trombetta, A., Bertino, E., and Lobo, J. Privacy-aware role based access control. In *SACMAT '07: Proceedings of the 12th ACM Symposium on Access Control Models and Technologies* (pp. 41–50). New York, NY: ACM, 2007.

Phillips, M. (eds). *HIPAA Security Guidlines* (pp. 1–7), Department of Health & Human Services, USA, 2006.

Pretschner, A., Hilty, M., Schütz, F., Schaefer, C., and Walter, T. Usage control enforcement: Present and future. *IEEE Security and Privacy*, 6(4), 44–53, 2008.

Samuel, W., and Brandeis, L. The right to privacy. *Harvard Law Review*, *IV*(5), 192–220, 1890.

Sandhu, R., Coyne, E. J., Feinstein, H. L., and Youman, C. E. Role-based access control models. *Computer*, 29(2), 38–47, 1996.

Sandhu, R. and Park, J. Usage control: A vision for next generation access control. In V. Gorodetsky, L. Popyack, and V. Skormin (eds), *Computer Network Security* (*LNCS* Vol. 2776, pp. 17–31). Berlin: Springer-Verlag, 2003.

Smith, M. K., Welty, C., and McGuinness, D. L. OWL Web ontology language guide. *Technical Report*, W3C [online] (retreived 28/09/09: http://www.w3.org/TR/owl-guide/).

Tikkanen, J. (2006). BlackBerry and Health Insurance Portability and Accountability Act (HIPAA) guidelines. *Technical Report*, JJT Consulting Group, 2004.

Chapter 15

Service-Based Connectivity for Wireless Systems

Abraham George

Contents

15.1 Introduction

Future generations of wireless networks are expected to migrate to service-based architectures from connection-oriented architectures. The key difference between these two architectures is that in the latter the user negotiates for all his service needs with a single network attachment point, whereas in the former the user may negotiate for each required service with network attachment points that can provide the required service. It is envisioned that the next generation of wireless systems would provide convergence of various wireless network technologies so as to have architectures with global connectivity [1–6]. In an integrated network, it is essential to have mechanisms and architectures to cater to needs of each user as requirements may vary widely from high data rates to low latency [3,7–10]. In traditional cellular networks, it is assumed that all service requests will have the same characteristics and, therefore, service-based architectures may not be relevant in traditional networks, but with a multitude of data and voice services, the problem of quality of service (QoS) and radio resource provisioning becomes more interesting and complex. Vertical handover can be an effective way of load balancing and providing QoS [2,11]. In integrated networks, multiple services or sessions pertaining to a user have to be serviced on the basis of the requirements of each session. This selective handling of services or sessions by the network is termed as service-oriented architecture. Service-oriented network architecture requires loose coupling of services with the underlying network technology [3,8,12,13]. It segregates functions into distinct units or services, which can be made accessible over a network in order to allow users to combine and reuse them in the production of applications [14–18]. In other words, in service-oriented architecture, the user need not be aware of the underlying technology as the application is loosely coupled with the underlying architecture. Furthermore, with an all-IP network, internetwork handover will be imminent [19]. In an all-IP network, it is imperative for network service providers to make best possible use of the combined resources of all available networks to serve the users. The key issues addressed by service-oriented architectures are QoS-based provisioning and decongestion of spectrum.

This chapter addresses the network availability constraint (*insufficient radio resources*) to serve all the mobile services originating from a single-user terminal. The key thrust here is the idea that a mobile user terminal may obtain service-based connectivity (SBC) by attaching itself to multiple attachment points, that is, base station (BS)/access points (APs). The term "service" refers to an application on a mobile device such as video streaming, gaming session, or a voice session. The user terminal may be equipped with a single network interface or multiple network interfaces. In this chapter, we will consider the design aspects in two cases: (1) a user terminal with a single-interface is simultaneously attached to multiple BSs or attachment points; (2) a user terminal is equipped with multiple network interfaces and each interface may serve one or more user services. In the first case, the main challenges are with respect to location management, call establishment, and radio resource management. In the second case, the challenges are with respect to interface management, interface selection, and handoff management for a service originating from a mobile station (MS). In conventional cellular network architectures, a single-interface user terminal is attached to a single BS and all user applications are serviced by the serving BS. The disadvantage of this approach is that the serving BS/AP may not be able to support all service application requests from one user because of resource constraints. As a result, the serving BS may decline a service request from a user if it is not able to support the user application request. This chapter addresses the weakly connected operation for mobile services by providing a one-to-many connectivity at the user terminal. Next-generation networks such as 3GPP long-term evolution (LTE) [20] and WiMax [21] support users with high data rate, high mobility, and multiple services. Also, these networks require the deployment to have overlapping cell coverage areas. Therefore, this chapter is relevant where we exploit the overlapping cell areas to support multiple user services and maximize network capacity. The key contribution of this chapter is (1) proposing SBC methodology in wireless networks for both single-interface and multi-interface devices; (2) enumerating and analyzing the challenges in designing SBC in wireless systems; and (3) discussing possible solutions to certain challenges. Although this chapter discusses separately the issues and challenges in single-interface and multi-interface service-based architectures, both architectures can coexist. The discussions in this chapter are mainly with respect to the BSs, although few references are made about design issues at MSs. The detailed design discussions of the MS terminal are outside the scope of this chapter.

The rest of this chapter is organized as follows. Section 15.2 explains the motivation behind this proposal and need for SBC architectures. Section 15.3 illustrates the service-based architecture for single-interface and multi-interface devices. Section 15.4 describes network discovery and attachment challenges in SBC architectures. Section 15.5 describes connection methodology challenges in SBC architectures. In Section 15.6, we describe the handoff management challenges in SBC. In Section 15.7, we summarize the challenges in SBC. Sections 15.4 through 15.7 separately describe the issues and challenges in single-interface SBC and multi-interface SBC. Finally, concluding remarks are given in Section 15.8.

15.2 Motivation and Need for Novel Network Architectures for Mobile Services

In existing network systems such as GSM and CDMA, all services running on an MS will communicate with a single BS, or in other words a mobile device is tightly coupled with a network element. If the current BS is not able to accept a new service request from a user who is already running one or more user services, the BS may hand over the user to a neighboring BS along with all the service flows. If the neighboring

BS is not able to accept the user because of resource constraints, then the service request of the user may not be accepted. In this chapter, we propose that one MS may be simultaneously connected to more than one BS, where each service flow from the user may be serviced by different BSs. Note that service flow refers to a user service. Multiple connections from an MS are established only if none of the BSs in the region are able to handle all the service requests from an MS. We will explain the need for novel architectures with an example. Consider a BS, namely BS1, and a user, namely user 1, has established connectivity. Here, establishment of connectivity implies that user 1 is synchronized with BS1 and the user is registered at BS1. When user 1 first initiates the connection, the intended service is a voice call that we will name as "service 1." Therefore, service parameters such as QoS negotiation occurs for a voice service with a set value for permissible packet latency and guaranteed bit rate. Later while service 1 is in session, the user wishes to initiate another session, named "service 2," which is a streaming video service. Consider that user 1 is at the outer boundary of the cell, where the interference from neighboring BSs are higher and as a result the user is assigned lower modulation indices. In most present-day systems [20,21], an adaptive modulation scheme is used where the modulation set consists of BPSK, QPSK, QAM, 16QAM, and so on. In this case, the modulation index is based on the user's capability to transmit and receive at the respective modulation indeces. Take the case of the 3PPP LTE system [20] where modulation indices are QPSK, 16QAM, and 64QAM. A user having very low signal-to-noise ratio (SINR) close to a lower threshold value may use lower modulation schemes such as QPSK, whereas a user with high SINR will use 64QAM. In the case of user 1 in the above scenario, let us assume that user 1 is using a modulation scheme in the mid-range, that is, 16QAM. This modulation scheme will suffice the user for the service 1 application. When the user sends a request to initiate service 2, the call admission control scheme at BS1 decides whether service 2 can be supported by BS1. BS1 also negotiates for the required service type with the gateway entity. In this case, BS1 is not able to provide service 2 to the user as BS1 does not have sufficient radio resources to serve the user at the specified modulation index. In this scenario, user 1 can do the following: (1) re-request for service 2 after a timeout period; as network traffic and channel conditions are highly dynamic, BS1 can try to accommodate service 2 after a timeout period. (2) BS1 may obtain the measurement report of user 1 and see if any of the neighboring BSs can accommodate both service 1 and service 2. If a neighboring BS, named BS2, can accommodate both services of user 1, then user 1 may be handed over to BS2.

In case 1 above, the delay incurred in setting service 2 can be very high and may not be acceptable to user 1. In case 2 above, handover is possible only if BS2 is able to accommodate both the services of user 1. In the event of BS2 having insufficient radio resources to accommodate both the services of user 1, BS2 may either reject the user completely or grant admission to services of the user for which it has resources. In both the above cases there is a very high probability of not meeting the user expectation, as service 2 requires high bandwidth and has high delay constraints.

In this chapter, we propose that the connection be established for a service on the user device with an available network attachment point. This proposition allows a user to be connected to several network attachment points on the basis of service needs. With respect to the above example, service 1 can be associated with BS1 and service 2 can be associated with BS2 using the SBC methodology described in this chapter. SBC architecture is based on establishing and maintaining connectivity of a user application or service with a network element rather than establishing connectivity of a user device with a network element. The benefits of SBC include network capacity increase, higher resource utilization, and higher user service quality. The SBC concept is supported by dynamic service composition methodology and ambient networks [9,22].

15.3 Proposed Network Architecture

In this section, we describe both single-interface SBC and multi-interface SBC network architectures.

15.3.1 Single-Interface SBC

Network architecture as illustrated in Figure 15.1 allows a single-interface MS to establish connection with multiple BSs/APs and perform data transmission through all these multiple connections. The MS may be connected to multiple BSs where each BS serves a set of application requests from an MS. We will explain this concept

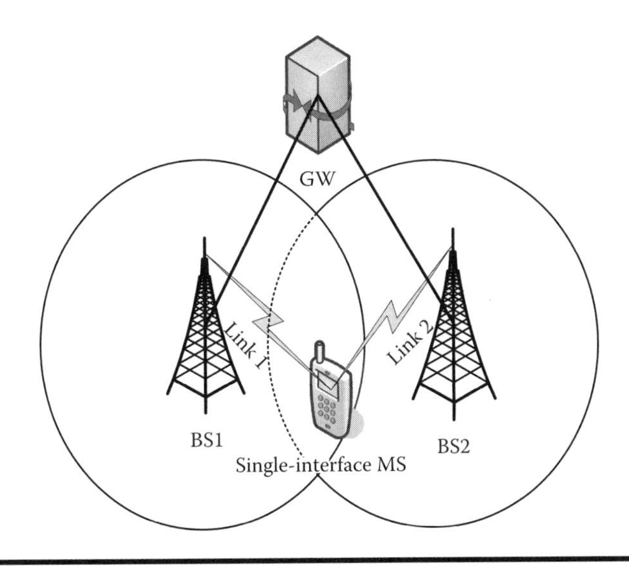

Figure 15.1 Multiple base station connectivity with single-interface devices.

with an example. Consider a scenario with one MS (single-interface) and two BSs (BS1 and BS2). Here, the MS is connected with BS1, or in other words a radio resource connection is established between BS1 and the MS. A communication session is ongoing in this connection. We will name this communication session as "session 1." During this time, the user wants to initiate another service namely "session 2," which requires a high data rate. In existing systems, if BS1 is unable to service this requirement for session 2, BS1 rejects the user's request for session 2 during the admission control process. Using our methodology, the user may get connected to BS2 (connection process initiated by BS1) and obtain service only for session 2 if BS2 permits. Note that the primary connection is established between BS1 and MS and also another parallel connection may be established between BS2 and MS. We will call this second and all subsequent connection as secondary connections. The primary advantage of this dual connectivity is that, individually, each of the BSs may not be able to satisfy both the bandwidth requirements of session 1 and session 2, but each BS can accept the service request it can support.

The underlying principle of this methodology is that applications can be multiplexed in the time domain. The resource scheduling of a session will be done at periodic time intervals of x milliseconds. The number of channel resources allocated during a transmission opportunity is at the control of the serving BS, depending on the application type and amount of data to send. In the case of voice application, the session is scheduled periodically at 4–6 timer intervals; for multimedia applications, this interval may be slightly higher. What can be exploited here is the scheduling time interval of a session. A single radio interface communicates with multiple BSs at different instances in the time domain. We will explain this statement using the above example. Consider that session 1 is scheduled with a periodicity of 4 ms and session 2 is scheduled with a periodicity of 8 ms. If session 1 starts at the first millisecond, then session 1 periodically repeats at 1, 5, 9, 11, 15, 19, . . .; similarly, if session 2 is configured to start at the second millisecond, then session 2 periodically repeats at 2, 10, 18, 26, ... Here, the MS will listen to the corresponding BS on the basis of the periodicity of the session and the starting time. Similarly, two sessions can also be multiplexed on the frequency domain. Theoretically, if a session is associated with a BS, a user terminal can be associated with n BSs to obtain n sessions. The key challenge here is to avoid overlap of two sessions in the time domain. The resource allocation manager and call admission control procedure is to be entrusted with this responsibility. A detailed discussion on these procedures is available in Sections 15.4.1 and 15.5.1.

15.3.2 Multi-Interface SBC

The network architecture illustrated in Figure 15.2 enables a multi-interface MS to simultaneously get connected to multiple BSs/APs to obtain services for multiple sessions. We will explain this architecture with an example. Consider a scenario with one MS where the MS has dual interfaces, one cellular interface, and an

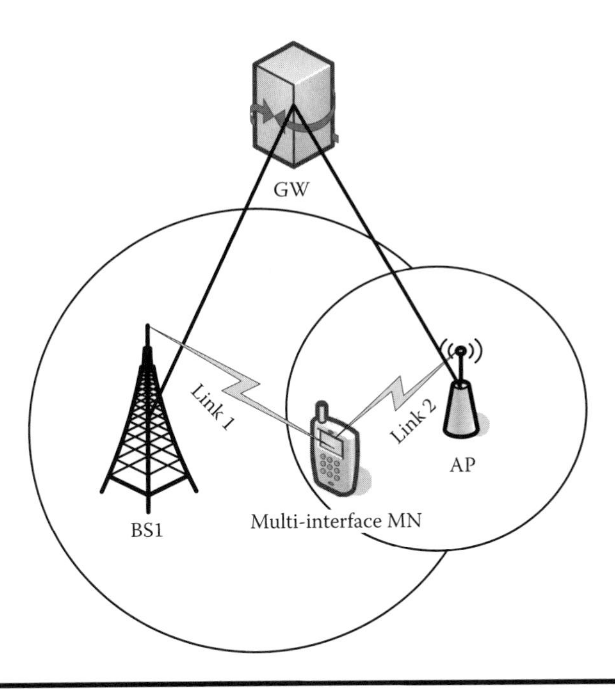

Figure 15.2 Multiple base station connectivity with multi-interface device.

802.11 interface. The MS has established a connection with a cellular BS, namely BS1, and we call this connection as "session 1." As session 1 is ongoing, the MS wants to initiate another service requiring a high data rate. We will name the second service as "session 2." If BS1 does not have sufficient radio resources to support this new service requested by the user, BS1 will reject the service request. Using the proposed architecture, the MS can initiate session 2 with any neighboring attachment point that can offer the service. In this case, the MS can request the neighboring 802.11 AP for session 2. The session 2 connection with the AP can be initiated by BS1. In other words, if BS1 does not have sufficient resources for session 2, it can forward the connection request to a neighboring BS or AP. The AP may accept the request if it has sufficient resources and complete the connection procedure with the MS. In this proposed methodology, session 1 may be established with BS and session 2 will be established with AP. For the user, each of the device interfaces is used to establish a service. The paradigm shift in the above-proposed architecture is that a set of network attachment points can be used to serve a user.

15.3.3 Design Issues in SBC

The SBC paradigm poses several challenges in mobility management and network communication both for single-interface and for multi-interface architectures as stated below.

1. How does an MS discover and associate with multiple attachment points?
2. How does an MS register with multiple attachment points of the same technology BS?
3. How does an MS hand over a session from one BS to another?
4. How does a BS allocate radio resources to minimize cochannel interference in case of an MS connected to multiple BSs of the same technology?
5. How is a service identified in the network, or in other words how to locate a session for incoming calls?
6. In the case of a multi-interface MS connected to multiple BSs, how to provide continuous service to all sessions as coverage of networks may be different?

The issues can be classified into single-interface SBC and multi-interface SBC. Issues 1, 3, and 5 are common to both single-interface and multi-interface SBC, issues 2 and 4 are relevant to single-interface SBC, and issue 6 is pertinent to multi-interface SBC. A detailed study of mobility management issues in heterogeneous networks can be found in Reference [6]. In subsequent sections of this chapter, we will explicate the design challenges of this proposed architecture for both single-interface and multi-interface SBC. The challenges with respect to SBC can be categorized into three types: (1) service-based discovery, (2) one-to-many connection methodology, and (3) handoff management. Service-based discovery is associated with the attachment of a service with a suitable network element and location management. Issues 1, 2, and 5 listed above fall under this category. One-to-many connection methodology is associated with the establishment of multiple radio connections with the BS. Issue 4 listed above fits into this category. Handoff management is associated with maintaining the connection as the MS switches between attachment points. Issues 3 and 6 listed above pertain to this category.

15.4 Service-Based Network Discovery and Attachment

In this section we will describe the concept of service-based discovery for single-interface SBC and multi-interface SBC. This section focuses on how to associate a mobile service with an attachment point and a network interface. Here, we will identify the metrics required to make this association. An inappropriate selection of BS may result in poor QoS or wastage of resources at the BS.

15.4.1 Single-Interface SBC

Here, the key challenges are mapping of a service with a suitable BS and the connection initiation of a session with a BS. We will explain the issues with pertinent examples and provide probable solutions to these issues. An MS may not be able to determine whether a single BS can satisfy all subsequent session request's of the user and which neighboring BSs can grant the user request because this will need large amounts of signaling information. Also, a question arises regarding which BS an MS will place a service request with when there are multiple neighboring BSs. One solution to this issue is to associate the MS with a parent BS or primary BS, and this parent BS will have to be entrusted to provide connectivity to a user's subsequent request. An MS will be associated with a primary BS and all subsequent attachment points of the MS can be termed as secondary BSs. The primary BS will evaluate all user requests and it will forward the connection request to neighboring BSs if it cannot provide connectivity. For this purpose, the primary BS may request the measurement report from the MS. Figure 15.3 illustrates the above methodology. The measurement response message will contain the list of neighboring BSs of the MS.

A BS that offers initial connectivity to an MS is termed as the primary attachment point for the MS and all subsequent attachment points are labeled as secondary attachment points. When a primary BS receives a new session request from the MS, it performs a session admission control to check whether it can support the session. If the BS cannot support the session request, the BS may request for a measurement report from the MS for all neighboring BSs. On receiving the measurement report from the MS, the primary BS may forward the session request to other BSs in the measurement report. If a secondary BS accepts the session request, the primary BS registers these details in its registry. At the time of forwarding a session request, the primary BS also sends the current periodicity of all current sessions of the MS. The secondary BS will use this information to avoid overlap with ongoing sessions. This issue is further discussed in Section 15.5.1. The secondary BSs may reject a session request if it cannot avoid time overlap with other sessions of the MS. The secondary BS may reject the request if it cannot meet the QoS requirements of the session. The secondary BS may also deny the request if the session periodicity at the secondary BS cannot satisfy the session throughput requirement.

Another key challenge here is to distinguish the services of the same user at the location registry. The network gateway has to forward an incoming call to the

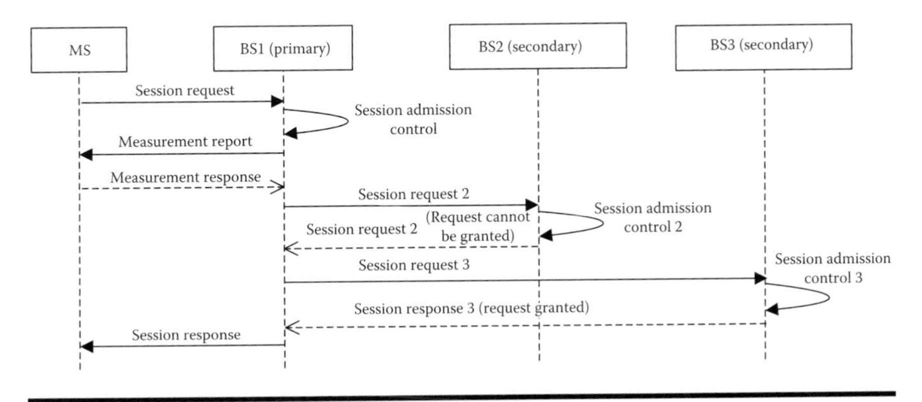

Figure 15.3 Illustration of a single-interface service-based connection.

corresponding BS. The network gateway may forward all incoming sessions to the primary BS, and then the primary BS will forward these to corresponding BSs. A session can be classified on the basis of the QoS tag assigned by the network. For example, session 1, which may be a voice service, will have a QoS tag of 1 and may be attached to BS1. Session 2, which may be a video service, will have a QoS tag of 2 and may be attached to BS2. At the location registry, a combination of user identifier and QoS tag may form a unique identifier. For an incoming call, the location registry at the primary BS should be able to forward the incoming request to the respective BSs or establish a tunnel between the primary and secondary BS. For an outgoing call, the MS should be able to map each session with a BS clearly. Each of the sessions can be directly registered with the location registry entity so that the call can be directly forwarded to the session. In the above example, the MS may register with the location registry entity both for session 1 and for session 2. The assumption here is that the gateway entity will be able to classify an incoming packet as session 1 or session 2 on the basis of the IP header. If the mobility management entity does not have provision to register an individual session, the call may be first forwarded to the primary BS and then in turn forwarded to the respective BSs. This forwarding procedure will involve additional latency. The open research areas for single-interface SBC include devising a location registry scheme and connection selection based on type of service requested.

15.4.2 Multi-Interface SBC

In a multi-interface device, the device should be able to exploit multiple networks to obtain multiple services. If a BS cannot support subsequent session requests of a user, the multi-interface MS may request other networks to offer service. Here, the MS is responsible for selecting the appropriate attachment point for a session. An MS device will require a connection manager module that can map different sessions onto different interfaces. This may be a logical module residing at the MS, whose key function is to associate a session with a network interface [8]. As shown in Figure 15.4, the MS may request for a service with a BS and if the session request is rejected it may request another available network. The concept of SBC or always-best connectivity (ABC) [23] allows an MS to attach to a network that best suits the application needs. In Figure 15.4, the connection manager module at the MS stores the association of the MS with BS1 and BS2 with respect to the individual sessions. The key question that promulgates research interest is: how does an MS discover an attachment point that best suits its application requirements? In other words, if there are multiple network options available, how does an MS identify the network that will meet its session requirement?

We will explain this problem with respect to Figure 15.4. If BS2/AP1 rejects a session request, the MS may request a session with BS1 or BS2.The simplistic solution will to be place a session request at each attachment point one after the other until the session is accepted. In the above example, the MS may first place a session

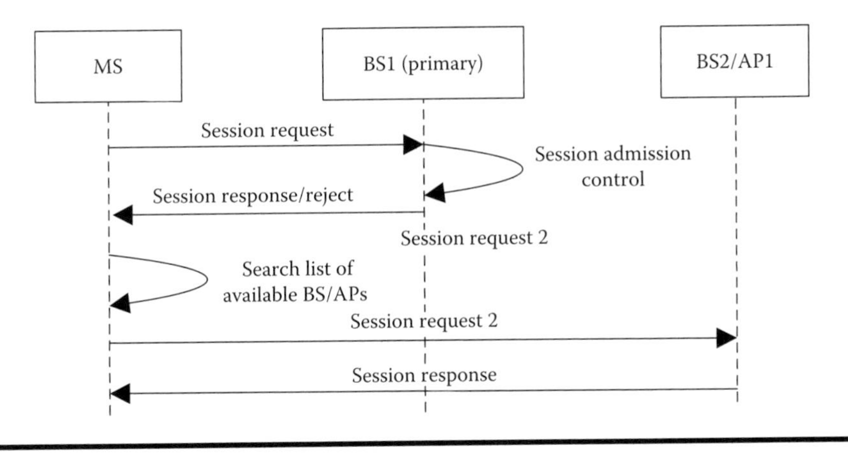

Figure 15.4 Illustration of a multi-interface service-based connection.

request with BS1 and then with BS2 if BS1 does not accept the session request. The order of placing the request could be on the basis of signal strength. The drawback of this approach is that the delay incurred in establishing the session can be exceedingly high if the initial attachment points cannot meet the session requirements. An enhancement to this approach could be that each attachment point sends out its load factor and the maximum service class it can support through broadcast or advertisement messages it sends out. A set of input parameters can be collated from all the available network attachment points at the BS and an MS can use a network selection algorithm to determine the best network attachment point. The above network selection model requires all network attachment points to periodically broadcast all the network selection parameters. The QoS class identifiers can vary from network to network; for example, 3GPP specifies eight QoS [20] class identifiers whereas WiMax [21] specifies only five service class identifiers. Therefore, the network selection model at the MS will need a scheme to normalize input parameters received from different networks, which may resemble a QoS translation layer [10]. Moreover, broadcasting network information in the case of multihop networks is not straightforward [6] as the attachment point may be another mobile node and it may not have access to its network parameters. The above scheme can be labeled as a mobile-initiated connection scheme as the MS discovers the appropriate attachment point. Purely network-initiated schemes may not be suited in this case as the network may not be aware of all the network attachment points in the vicinity of the MS.

An alternative approach could be a blend of mobile-initiated and network-initiated approaches. We will label this scheme as a hybrid connection initiation scheme. This hybrid scheme is illustrated in Figure 15.5. We will explain the scheme with respect to Figure 15.5. An MS wanting to initiate a new service will place a session request with the primary BS in the case that the MS is already associated with the BS. If the BS cannot accept the session request, the BS may send the request to a list

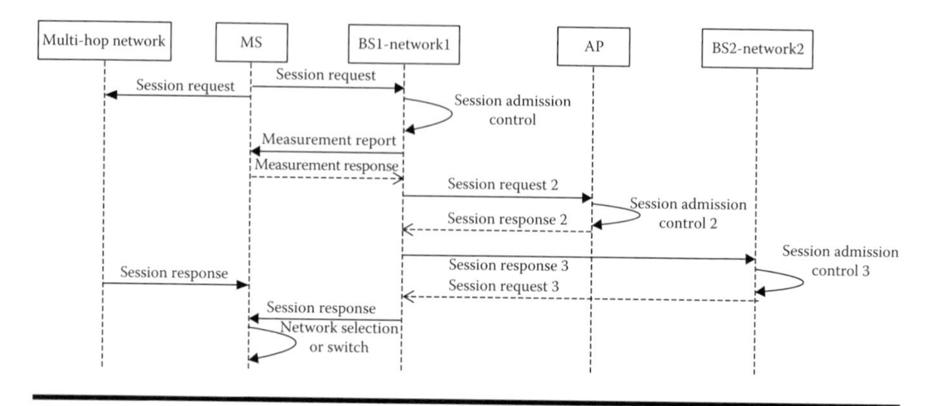

Figure 15.5 Illustration of a hybrid service-based connection initiation scheme.

of alternate network interfaces within its reach. The MS may respond with signal strength measurements of each of these network interfaces. BS1 will send a session request to each of the network interfaces on behalf of the MS. BS1 may use an algorithm to place the list of network interfaces in order of probability of session request acceptance for the requested session from the MS. This will minimize the session establishment delay by first requesting network attachment points that have higher probability to offer connectivity for a particular session from the MS. In the above example, BS1 will first send *a session request* to the AP; the AP will perform an admission control process and send the *session response* to BS1. If the AP rejects the request, the session request is again sent to BS2 and BS2 sends the response after the admission control process. In this case, let us say BS2 accepts the session request and BS1 notifies the MS of the session request from BS2. The MS may also request its multihop network [1,6] attachment points for session establishment. The MS will directly communicate with network attachment points that the BS1 may not be aware of. For example, in Figure 15.5, BS1 may not be aware of the multihop network interfaces of the MS. The advantage of this hybrid connection scheme is that the amount of signaling traffic required to communicate the network selection parameters to the MS can be minimized and connection establishment delay will decrease as the selection procedure is divided between the MS and the BS. Note that in this proposition an MS is assigned a primary BS as in the case of a single-interface SBC described in Section 15.4.1.

There are multiple issues to address in the above proposition. The first issue to address is the discovery of neighboring attachment points. Each BS should be able to communicate with adjacent network BSs, or in other words each BS should be equipped with multiple network interfaces. For example, a CDMA BS with a WiMAX interface can directly place a session request with a neighboring WiMAX BS on behalf of the MS. A BS attached with multiple interfaces should be able to discover adjacent BSs. For example, a CDMA BS with a WiMAX interface and a UMTS interface should be able to automatically discover all adjacent WiMaX and

UMTS BSs. The next question that arises is: how will a BS equipped with multiple interfaces communicate with neighboring BSs? The amount of signaling information generated will be huge if each interface of a BS were to communicate with adjacent BSs. One solution could be to develop a common interface for communication across BSs that is not specific to any technology. 3GPP suggests that its LTE BS communicates via the X2 interface with its neighboring BSs [20], but this is specific to LTE technology. Another issue to address in a unified network will be to specify a common QoS identifier scheme that is understood across networks. An alternative approach could be to specify an intermediate layer that translates the QoS identifier of one network to that of another network and vice versa.

15.5 A One-to-Many Connection Methodology for Mobile User Terminal

In this section, we present our one-to-many connection paradigm for single-interface SBC and multi-interface SBC. Here, we will analyze the internal working of this paradigm. Some of the areas analyzed in this section will be radio resource allocation, radio admission control, and the medium access control (MAC) layer operation at the MS.

15.5.1 Single-Interface SBC

An MS attached to a BS will receive periodic control messages stating on what frequency channels it can expect or receive data signals in subsequent frames [20,21]. In the case of single-interface SBC, an MS can receive/transmit data from/to multiple BSs when the communication stream is multiplexed. The question that arises here is: how will an MS attached to multiple BSs ensure that signals from independent BSs do not interfere with each other? We will explain this issue using an example. Consider a scenario where a single-interface MS is connected to two BSs, namely BS1 and BS2, and each BS has 50 channels to operate in a system based on time division access. Let us say BS1 communicates with the MS on channels 1–10. If BS2 were to also communicate with MS on the same frequency channels labeled 1–10 in the same time slot, the receiver would experience high interference from these two sources. The same is true in the case of control messages and data messages sent to the MS. This interference problem is more predominant in downlink transmission as the two BSs are independent of each other, whereas in uplink communication the BSs are separate from each other. Uplink communication presents an interesting problem when multiple BSs assign the same frequency channel to an MS in same time slot. As in the above example, let us say BS1 has assigned channel numbers 1–10 in the first time slot for uplink transmission of MS. Unaware of this assignment, BS2 that is also serving the same MS assigns channel numbers 1–10 in the same time slot. In the case of uplink, the MS can only communicate with a single BS at a particular time slot on a channel as there is only one radio interface. A solution to this downlink and uplink transmission problem is to multiplex each session in the time domain. The primary BS should cooperate with secondary BSs to operate in different time slots. The mechanism for cooperation between the primary and secondary BSs is an open research area. The research question will be: how many sessions can be multiplexed at maximum? Another issue is the high interference between the same technology BSs when they are in proximity. Consider the case where BS1 is serving a session of MS1, BS2 is serving a session of MS2, and BS1, BS2 are in proximity. Here, MS1 and MS2 may receive high interference if the same channel is allotted to both MSs in the same time slot. A possible solution to this problem is the cooperation among the adjacent BSs in assigning channels to an MS. In the above example, BS2 may avoid allotting channels allotted by BS1 in the same time slot or vice versa. This can be accomplished by cooperation among the BSs in assigning channels to an MS. In LTE, adjacent BSs cooperate with each other to minimize interference [20].

Another open research area is the protocol structure at the MS. As a single-interface MS may communicate with multiple BSs, the packet flow for each session or service needs to be distinct at all layers including the MAC and physical layers. An example is the case where an MS is connected to a primary BS and one secondary BS. In uplink, the service flows for both sessions need to be separated. At the MAC layer, packet formation is done separately for the respective session flows. Also, at the physical layer, the packets are formed separately. Therefore, the research issue is the efficient design of protocol stack at the MS with minimal computational and memory overhead. SBC systems will require dynamic reconfiguration capabilities capable of adapting their protocol stacks during runtime [24] to meet the ever-changing service demands. A detailed discussion on user service demands in beyond 3G networks can be found in References [12,22]. A generic object-oriented information model for describing the feasible combinations of protocol layers into protocol stacks is specified in Reference [24]. The model supports the derivation of an associated ontology specified in standard W3C semantic languages.

15.5.2 Multi-Interface SBC

In the case of multi-interface SBC, each session is mapped to only one network interface but the reverse is not true as each interface can handle one or more sessions. An interface at an MS may be equipped with a protocol stack with or without cross-layer interaction. Autili et al. [3] envision a multinetwork system with cross-layer interaction between vertical layers. Consider a case where two applications originate from the higher layers of an MS with two network interfaces. In the simplest case, each application may be mapped to a single network interface where there will be no horizontal communication between layers. A more complex case will be where a single application will be split across both the network interfaces. A detailed study of this case can be found in Reference [25]. In this chapter, we do not discuss cross-layer interaction. The open challenge in this area is the handling of multiple interfaces on a user terminal. This can be realized by using the software-defined radio [26] where the radio functionality is implemented in software. The advantage of the software-defined radio is that multiple technologies can be realized with a single software reconfigurable interface instead of having multiple network interfaces.

15.6 Handoff Management in the One-to-Many Connection

An important issue in this proposed network is the handoff procedure as a user is connected to multiple attachment points. We will describe the handover methodology in this section for both single-interface SBC and multi-interface SBC.

15.6.1 Single-Interface SBC

In single-interface SBC, the MS may be connected to multiple BSs. The difficulty during handover is to ensure synchronization of the MS with the new BS and to avoid usage of channels used by the primary BS if the secondary connection is being handed over to a new BS. One approach to this problem is, on disconnection from a network, the MS will request the primary BS to offer connectivity or help it to get attached to another BS. Consider the case where an MS is connected to two BSs, BS A and BS B where BS A is assigned as the primary BS. Three scenarios may arise with mobility: (1) the MS may go out of coverage of secondary BS B; (2) the MS may go out of coverage of primary BS A; and (3) the MS may go out of coverage of BS A and BS B. In case 1, where the MS goes out of the reach of the secondary BS, the key issue is how the session on interface B can be switched to another BS with minimal delay. As the MS is already associated with primary BS A, the MS may request BS A to accept this session from BS B. If BS A cannot accept the session request from the MS, it may find a suitable secondary BS and inform to the MS, as described in Section 15.4.1. In case 2, where the MS goes out of coverage of the primary BS, the MS may immediately assign primary status to the secondary BS and the MS may request the primary BS to accept the session on interface A. The primary challenge in both the cases is to minimize the delay in handover. Another open research area is the assignment of the primary BS. The primary BS is the anchor point for all secondary connections; hence, it has to be the most stable connection. One criterion for selection of the primary attachment point is to identify the network that has the maximum coverage area. For example, if an MS is connected with a BS closer to the MS and also connected with a distant BS, then the closer BS should be the primary BS, because the primary BS is required for all resource allocations and new service request procedures. Nonavailability of the primary BS will result in poor service of all secondary connections, as there will be no coordinator for service discovery and resource distribution procedures explained in Sections 15.4.1 and 15.5.1, respectively.

15.6.2 Multi-Interface SBC

In a multi-interface SBC network, a user may use all the network interfaces of the device to obtain multiple services. Each of these network attachment points will have different coverage areas, some less and some large. The key challenge is to provide uninterrupted connection and minimize handover latency. The coverage area of the primary and secondary BSs can be completely different. We will illustrate this problem with an example. Consider a scenario where the MS is connected to a primary CDMA-BS and a secondary 802.11-AP. Here, the coverage region of the CDMA-BS may be a few kilometers whereas the coverage area of the AP may be a few hundred meters. If the MS is moving at a fast pace, frequent handovers will be required for the secondary connection. The connection procedure is described in Section 15.4.2. The question here is: will the session on AP get interrupted if AP coverage is not available? The delay incurred in performing a handover is significantly large and therefore seamless connectivity is far from possible when frequent handovers are performed. If a new secondary BS is not available then the AP session may be shifted to the primary BS or any other available secondary interfaces, as described in Section 15.4.2. The key issues here are: (1) minimizing the number of handovers at any interface, and (2) transferring the application session from one attachment point to another with minimal delay if the current BS fails. The handoff process should be initiated early enough to avoid data disruption; packets will be lost if the MS starts the connection generation process after the network interface is down. During internetwork handover, the encapsulated packets must be translated into the data link format of each network and the addresses must be translated across various networks, which contribute to an increase in the overall handoff delay. One of the main issues with the internetwork handoff process is the latency incurred in re-establishing the connection at a new location, during which packets in transit may be lost. Other challenges for mobility support in multi-interface SBC include authentication between wireless systems, that is, authentication mechanism when a service is switched between interfaces [6].

One possible approach for handoff management in multi-interface SBC networks will be to have a session management layer beneath the application layer. The role of this layer will be to transmit the packets over the most appropriate network interface. We will explain the role of this session management layer with an example. Consider an MS with three interfaces, A, B, and C, attached to three network attachment points (BS-A), BS-1, BS-2, and BS-3, respectively. Three different application sessions are running on each of these interfaces. The session layer will be placed between the application layer and network interfaces, A, B, and C. When an MS is moving in the outer coverage region of BS-3 and no neighboring BS is available with this interface, the session manager may detect the nonavailability of the BS with interface C and may initiate the process to switch the session on interface C to either interface A or interface B. In other words, this layer functions as a handover decision module. A similar approach for a service-based handover using a transparent layer has been proposed in Reference [18]. The procedure to perform the connection initiation and handover from one interface to another with minimal delay is an open area. The design of the session management layer is also an open research area.

15.7 Open Research Areas

In this section, we summarize the key issues and provide definite directions on specific problems that have to be addressed to provide SBC in wireless networks. Section 15.7.1 will highlight the open issues for single-interface devices and Section 15.7.2 will highlight the open issues for multi-interface devices.

15.7.1 Single-Interface User Terminal

One of the key features of a single-interface SBC is to enable service adaptability on the basis of network parameters such as network traffic, service availability, and user requirements. Several questions arise in this connection methodology and several issues remain to be addressed. A single-interface MS multiplexes multiple applications on a single radio device. An ideal scenario for the single-interface SBC would be that the MS with low mobility in an urban area requires multiple applications such as voice, browsing, and email. In this scenario, multiple BSs can be used to service the user's needs as the characteristics of the three applications are dissimilar in terms of data rate, latency, and error rates. The periodicity of scheduling browsing application is higher when compared with voice application and the periodicity of scheduling email applications is even higher when compared with browsing application. Hence, the overlap of sessions in time domain can be avoided as characteristics of each session vary from each other. Therefore, application or session characteristic is the primary input to decide whether session multiplexing is possible. Session management becomes more difficult if the MS is moving briskly in the above scenario. In this case, the difficulty arises as the coverage regions of all the BSs may be nonoverlapping. Therefore, several considerations have to be made before initiation of a secondary session. The primary BS will have to make all these considerations before forwarding a session to a secondary BS. The primary BS may also determine the feasibility of connectivity with a secondary BS on the basis of user movement and application type. The open research areas with respect to single-interface SBCs are multiplexing of sessions, call handling, and handoff management. The key research areas and open issues of location management for single-interface SBCs are listed in Table 15.1.

Table 15.1 Summary of Location Management Challenges for Single-Interface Service-Based Connections

Category	*Challenge*	*Key Issue*
Network discovery and attachment	Discovery of base stations	How does a mobile station determine a base station that can support a user service?
Network discovery and attachment	Location registration	How to distinguish the services of the same user at the location registry?
Connection methodology	Mobile station protocol structure	How will a mobile station, attached to multiple base stations, ensure that signals from independent BSs do not interfere with each other?
Connection methodology	Protocol structure	How will the protocol stack handle multiple service flows?
Handoff management	Reconnection	How to reattach with a base station after the connection is lost?
Handoff management	Avoid disruption of session	How to minimize the delay in handover?

15.7.2 Multi-Interface User Terminal

This architecture is based on the key idea of using multi-interfaces to service multiple user sessions or requests. This can also be used for load balancing and increasing the capacity in single-hop networks. Here MSs are equipped with multiple network interfaces, so that they can selectively connect to a network on the basis of network conditions and the QoS required by the application. Some of the specific challenges in multi-interface SBCs are determining where to register the location information of an MS, as the MS moves between various wireless network domains, providing the location of an MS to a correspondent node in the fixed network, and connection reconfiguration. The key research areas and open issues of location management for multi-interface SBCs are listed in Table 15.2.

15.8 Conclusion

In this chapter, we have provided a comprehensive study of SBCs in the light of single-interface and multi-interface devices. In conventional architectures, a user device is mapped to the network, whereas in SBC and other heterogeneous architectures a service originating from the user device is mapped to the network.

TABLE 15.2 Summary of Location Management Challenges for Multi-Interface Service-Based Connections

Category	*Challenge*	*Key Issue*
Network discovery and attachment	Network selection	How does a mobile station discover an attachment point that best suits its application requirements?
Network discovery and attachment	Inter-base station communication	How will a base station equipped with multiple interfaces communicate with neighboring base stations?
Connection methodology	Handling multiple interfaces	How will multiple interfaces be handled at the mobile station?
Handoff management	Latency in handover	How to avoid connection disruption as coverage of networks will be different?
Handoff management	Internetwork handover	How to perform internetwork handover?

This allows increase in the network capacity and resolves the network availability constraint for multiple connections. We have divided challenges into three areas, namely, service-based network discovery and attachment, one-to-many connection methodology, and handoff management. Realizing multi-interface SBC is more complex, as the solution has to deal with varying underlying infrastructure support, mobility patterns, and issues at the MAC and physical layers.

QUESTIONS

1. Explain the term service-based connectivity? How different is service-based architecture from traditional cellular architecture?
2. Describe the need for service-based architectures.
3. What are two types of service-based architectures described in this chapter? Explain them briefly with relevant examples.
4. How does a single radio interface communicate with multiple base stations in a single-interface service-based connection?
5. Describe the location management challenges in a single-interface service-based connection?
6. Briefly explain the process of session establishment with secondary base stations for a single-interface service-based connection?
7. Explain how connection feasibility can be checked before establishing a connection with a secondary base station in a single-interface service-based connection. What parameters can be used to determine feasibility?

8. Explain the concept of always-best connectivity (ABC). How can be ABC be achieved in multi-interface service-based connections?
9. Describe the handover management challenges in multi-interface service-based connections.
10. Describe the role of session management layer in multi-interface service-based connections. Explain with relevant references.

References

1. S.C. Spinella, G. Araniti, and A. Iera, A. Molinaro, Integration of ad hoc networks with infrastructured systems for multicast services provisioning, In *The Proceedings of ICUMT*, St. Petersburg, Russia, pp. 1–6, October 13, 2009.
2. G. Medina, J. Ruben, P. Rico, Ulises, and S. Enrique, VIKOR method for vertical handoff decision in beyond 3G wireless networks, In *Proceedings of the 6th International Conference on Computing Science and Automatic Control (CCE 2009)*, Toluca, Mexico, pp. 1–5, November 10, 2009.
3. M. Autili, M. Caporuscio, and V. Issarny, Architecting service oriented middleware for pervasive networking, *ICSE Workshop on Principles of Engineering Service Oriented Systems (PESOS 2009)*, pp. 58–61, 2009.
4. L. Sarakis, G. Kormentzas, and F. Guirao, Seamless service provision for multi heterogeneous access, *IEEE Wireless Communications Magazine*, 16(5), 2009, 32–40.
5. F. Belqasmi, R. Glitho, and R. Dssouli, Ambient network composition, *IEEE Network*, 22(4), 2008, 6–12.
6. A. George, A. Kumar, D. Cavalcanti, and D.P. Agrawal, Protocols for mobility management in heterogeneous multi-hop wireless networks, *Elsevier Pervasive and Mobile Computing Archive*, 4(1), 2008, 92–116.
7. H.S.R. Babu, G. Shankar, and P.S. Satyanarayana, Call admission control approaches in beyond 3G networks using multi criteria decision making, In *Proceedings of the First International Conference on Computational Intelligence, Communication Systems and Networks (CICSYN 2009)*, Indore, India, pp. 492–496, 2009.
8. A. Bertolino et al., PLASTIC: Providing lightweight & adaptable service technology for pervasive information & communication, In *Proceedings of the 23rd IEEE/ACM International Conference on Automated Software Engineering (ASE Workshops 2008)*, L'Aquila, Italy, pp. 65–70, 2008.
9. M.N. Masikos, D.G. Nikitopoulos, D.I. Axiotis, and M.E. Theologou, Designing a service platform for B3G environments, In *Proceedings of the 16th Mobile and Wireless Communications Summit (MWC 2007)*, pp. 1–5, 2007.
10. Q. Lv, F. Yang, and Q. Cao, A general QoS-aware service composition model for ubiquitous computing, In *International Workshop on Intelligent Systems and Applications (IWISA 2009)*, Qingdao, China, pp. 1–4, November 21–22, 2009.
11. L. SuKyoung, K. Sriram, K. Kyungsoo, K. Hyuk, and N. Golmie, Vertical handoff decision algorithms for providing optimized performance in heterogeneous wireless networks, *IEEE Transactions on Vehicular Technology*, 58 (2), 2009, 865–881.
12. C. Fortuna and M. Mohorcic, Dynamic composition of services for end-to-end information transport, *IEEE Wireless Communications*, 16( )4, 2009, 56–62.

13. T. Gu et al., A service-oriented middleware for building context-aware services, *Journal of Network and Computer Applications*, 28(1), 2005, 1–18.
14. L. Auer, N. Kryvinska, and C. Strauss, Service-oriented mobility architecture provides highly-configurable wireless services, In *Proceedings of the IEEE Wireless Telecommunications Symposium (WTS 2009)*, Prague, Czech Republic, pp. 1–1, April 22–24, 2009.
15. R. Popescu-Zeletin, S. Abranowskib, I. Fikourasc, G. Gasbarroned, M. Geblere, S. Henninge, H. van Kranenburgf, H. Portschye, E. Postmanne, and K. Raatikaineng Service architectures for the wireless world, *Computer Communications*, 26(1), 2003, 19–25.
16. D. Wu, S. Ci, H. Luo, H. Wang, and A. Katsaggelos, A quality-driven decision engine for live video transmission under service-oriented architecture, *IEEE Wireless Communications*, 16 (4), 2009, 48–54.
17. Y. Fang, Service-oriented design for broadband wireless networks [Message from the Editor-in-Chief], *IEEE Wireless Communications*, 16(4), 2009,2–3.
18. S. Kashihara, K. Tsukamoto, and Y. Oie, Service-oriented mobility management architecture for seamless handover in ubiquitous networks, *IEEE Wireless Communications*, 14(2), 2007, 28–34.
19. P. Newman, In search of the all-IP mobile network, *IEEE Communications Magazine*, 42(12), 2004, S3–S8.
20. 3GPP LTE, 3GPP TS 36.300 V9.0.0 (2009–06), http://www.3gpp.org. Last accessed on March 2010.
21. IEEE 802.16e Task Group (Mobile Wireless MAN), http://www.ieee802.org/16/tge/. Last Accessed on February 2010.
22. C. Kappler, P. Poyhonen, M. Johnsson, S. Schmid, and Siemens Networks, Berlin, Dynamic network composition for beyond 3G networks: A 3GPP viewpoint, *IEEE Network*, 21(1), 2007, 47–52.
23. E. Gustafsson and A. Jonsson, Always best connected, *IEEE Wireless Communications Magazine*, 10(1), 2003, 49–55.
24. V. Gazis, N. Alonistioti, and L. Merakos, Discovering feasible protocol stack combinations in beyond 3G systems: Information model, search algorithms and performance, In *Proceedings of the 18th International Symposium on Personal, Indoor and Mobile Radio Communications (PIMRC 2007)*, Athens, Greece, pp. 1–6, September 3–6, 2007.
25. B. Xie, A. Kumar, and D.P. Agrawal, Enabling multiservice on 3G and beyond: Challenges and future directions, *IEEE Communications Magazine*, 41(12), 2008, 66–72.
26. M. Dillinger, K. Madani, and N. Alonistioti, *Software Defined Radio: Architectures, Systems, and Function*, John Wiley & Sons Ltd, England, 2003, ISBN 0-470-85164-3.

Chapter 16

Mobile Access to Printed Texts for People Who Are Blind or Visually Impaired

A. S. Shaik and M. Yeasin

Contents

16.1 Introduction

In 2006, the World Health Organization (WHO) estimated that approximately 37 million people are blind and 124 million people are visually impaired in the world. This number is growing and may reach 2 million per year in the near future. An ever-increasing number of people are at risk of visual impairment as the population grows and the demographic shift moves toward the predominance of older age groups. Despite their many abilities, visual impairment causes a host of challenges in daily life. Hence, cost-effective, functional, easy-to-use, reliable, and ergonomically designed assistive technology solutions are vital to help overcome these daily challenges.

One of the problems faced by people with visual impairment or blindness is their inability to access printed text that exists everywhere in the world. Ray Kurzweil, the designer of the Kurzweil machine [1] in 1976, the first device for the blind to access text, reported a blind person explaining to him that the only real handicap for blind people is their complete lack of access to print (text). The American Foundation for the Blind [2] found that information provided in Braille is accessed by not more than 10% of the legally blind people in the United States. The scenario may be similar all around the world. Almost 70% [3] of people who are blind or visually impaired are unemployed and isolated from the information world in everyday life because of this inability, thus never realizing their potential. Family or friends may not assist them every time or everywhere.

A number of assistive technology solutions for persons with disabilities are available to access print but most of these devices and software often require custom modification or are prohibitively expensive. Many persons with disabilities do not have access to custom modification of the available devices and other benefits of current technology. Moreover, when available, personnel costs for engineering and support make the cost of custom modifications beyond the reach of the persons who need them. Thus, existing solutions to access printed texts have fallen short of user expectations, and are expensive and not suitable for mobile use. This necessitates a mobile device to access text that satisfies requirements such as cost, convenience, usability, portability and accessibility, expandability, flexibility, compatibility with other systems, learnability and the learning curve of device, ergonomics, utility, and reliability. A solution can be the use of mobile handheld devices that satisfy the above requirements. Mobile phones are widely used and are constantly evolving with new features being added all the time. Third-generation mobile phones can process computing-intensive programs in real time.

Apple has developed some applications on its iPhone for people who are blind or visually impaired, such as the Braille reader, screen reader, and so on. However, development of applications on this platform is bit complex compared to a recent advancement in handsets—Google's mobile Android phone with the first free open-source Android [4] operating system. Many new functions can be easily programmed and customized in this software with no extra hardware. The development of applications on other platforms such as Windows CE, Pocket PC, Palm OS, and so on, is complicated, and, moreover, these platforms are not available for free use by others. Android mobile phones are also provided with a text-to-speech (TTS) [5] engine to make the applications speak for people who are blind or visually impaired. Hence, providing a reading service on this platform of mobile phone is easier.

Optical character recognition/recognizer (OCR) [6] helps bring information from our analog lives into an increasing digital world. An OCR embedded in a mobile device will allow people who are blind or visually impaired to see the print world untethered. These features of Android and OCR lead to the idea of developing an integrated system on the Android mobile phone with a simpler user interface, using the camera image capturing routines, OCR, and TTS engines to provide read-out-loud services. The objectives of this system include being lightweight, affordable, and untethered while performing a large number of tasks. Most devices developed with these same objectives have been too cumbersome with unusable user interfaces, no voice output, unaffordable, and/or not readily available to be practical and truly portable. Our system overcomes all these issues and is readily accessible in real time. The application is implemented on the device, with no connection to an external server. The user interface was designed to accommodate users with little to no vision. It can recognize text from simple to colored backgrounds, outdoor and indoor locations, simple curved and glass surfaces, and so on, to a large extent. Now, for the first time, many people who are blind or visually impaired can afford the technology that would allow them to read most printed material through independent means. The results analyzed and evaluated prove the technology to be promising.

16.2 Literature Review

This section deals with current technologies for people who are blind or visually impaired and reviews the literature on OCR, TTS technologies, and related projects. For people who are blind or visually impaired to access text, current technologies are large print, speech, Braille, and scanned material. Each technology is described below. However, more research needs to be done to develop a technology to satisfy adequately all the requirements and needs of people who are blind or visually impaired.

Large print technologies are ZoomText [7], Lunar [8], and closed-circuit television (CCTV). ZoomText is a family of products (ZoomText, ZoomText Plus, and ZoomText Xtra) that magnifies text and graphics. It can magnify the full screen or a portion of the screen or a single line of the computer from 2 to 16 times. This is useful only for the visually impaired, not the blind, and does not work well on Web browsers. This has a screen reader attached to it and though costly it is widely used. Lunar, the screen magnifier for computers, helps manage the enlarged screen more effectively. It can magnify from 2 to 32 times with different viewing modes and smoothes images to view clear text. Again, it is a program for the visually impaired, widely used across the world, but is not useful for the blind and is expensive. CCTVs enable the visually impaired to read newspapers, medicine bottles, telephone books, handwritten letters, and so on. The CCTV system uses a handheld video camera to project a magnified image onto a video monitor or a television (TV) screen. A CCTV has features such as a magnification range of 2 to 60 times, switchable polarity (black text on white background or white text on black background), and a variety of monitor sizes (5 in. to more than 20 in.). This setup is quite expensive and bulky. The usability is quite unsatisfactory and difficult to operate for blind users. As all these technologies are for the visually impaired, they are not suitable for people who are blind. The conditions of mobility and portability are not satisfied either. These systems are limited to being used at home and on computers only.

Speech technologies include speech synthesizers with hardware (DecTalk [9], Keynote GOLD [10], Artic SynPhonix) and software (Microsoft SDK [11], AT&T Natural Voices [12]) versions and screen readers (HAL [13], JAWS [14], etc.). The hardware versions have either internal cards or external serial devices that allow the software programs to integrate speech output. On the basis of the type of software programs used to read the screen, these synthesizers can work in the DOS and Windows environment. DecTalk and Keynote GOLD are widely used, but are useful only on computers and are thus immobile. Software versions work in the Windows environment with a compatible sound card. These too are widely used but are not cost-effective and their usability is out of reach for people who are blind or visually impaired. HAL is a screen reader that works by recognizing and reading aloud information on the computer screen through the computer's soundcard. The information is also displayed in Braille for users of many supported Braille devices that are available in the market. It needs the instructor to guide the usage and is useful for navigating through Web applications. It runs only on computers, not on any other mobile devices. JAWS stand for Job Access with Speech. JAWS for Windows (JFW) offers all the features that made JAWS for DOS popular. JAWS uses the screen and determines the information to be spoken such that any unfamiliar application can be used. Simple applications can be run on the computer, but training is required to use their features. These technologies are helpful for people who are blind or visually impaired working on a computer. These too are costly but are widely used. Nowadays, screen readers are also provided on smart phones making the applications and device much easier to use.

Braille technology provides Braille editors, translators, displays, and embossers. The editing tool Braille2000 [15] edits all kinds of Braille documents with Braille tasks to aid the production of literary, textbook, and music Braille. Manipulating these tasks is not easy for every blind user. The translation software Duxbury [16] is a Braille editing and translation software. It is available in versions for DOS, Windows, and Macintosh computers. Duxbury is easy to use and is compatible with speech and Braille output. It also supports a number of languages. It is widely used but is expensive and is not accessible to all. The displays (Braille Wave [17], Braillex [18]) use Braille terminals to navigate through Windows, DOS, and UNIX. These Braille terminals can be used with a desktop or a laptop that utilizes refreshable Braille cells to allow the computer screen to read line by line in Braille. This is costly and not portable. The embossers (Braille Blazer [19], ViewPlus [20], Juliet) provide Braille output from a printer. They can be used with any computer using a Braille translation software program. These are bulky but widely used. These technologies are especially for people who are blind. They are expensive but very useful and accessible for Braille users.

Scanned material technology gives access to scanned documents from a scanner on the computer to use devices such as Cicero [13], Kurzweil 1000 [1], Open Book [21], and so on. In Cicero, the documents with text are placed on the scanner and then translated into Braille or speech or simply held as a text document to be modified later. The extracted text is not the exact input text, but efficient to some extent. Kurzweil 1000 is an advanced version of the reading machine invented by Ray Kurzweil in 1976 that reads out loud the printed text fed into the scanner. It can read only black-and-white documents. Open Book uses a scanner to take a picture of the page, which it sends it to a computer, translates the picture into understandable text, and then speaks aloud the text or outputs to Braille. It can scan and read a page in less than a minute and also it is available in many languages other than English. The text output is read loud, which sometimes reads unwanted characters. When the output is Braille, sometimes it was found to be inappropriate. These technologies can be used by people who are blind or visually impaired to read printed text, and mostly blind users are benefited by this. All these devices almost run in a similar way where text is extracted from the print. These devices are popular and widely used, but are costly and limited to home-use only.

OCR engines are used to digitize the text information that can be edited, searched, and reused for further processing with ease. Therefore, OCR transforms textual information from the analog world to the digital world. Reference [22] provides a list of open-source software and projects related to OCRs. In addition, open-source communities offer systems such as GOCR [23], OCRAD [24], and OCRopus [25]. Tesseract [26] is also one such open-source OCR engine developed by HP Labs between 1985 and 1995 with Google acquiring it in 2006. It is probably the first OCR to handle white-on-black text easily. Commercially available OCRs include the ABBYY FineReader, the ABBY Mobile OCR engine, the Microsoft Office Document Imaging, and OmniPage. Commercial OCRs are efficient but costly and open-source OCRs are also proved to be efficient. TTS [27] engines are widely introduced in almost all mobile platforms to help people who are blind or visually impaired in using mobile phone applications with ease. The voices used in these applications are close to human voices but lack the liveliness found in natural speech. For capturing an image on a mobile phone, there is no proper literature for people who are blind or visually impaired trying to access the mobile phone camera.

Many projects and commercial products attempt to use a mobile phone in building applications for people who are blind or visually impaired. A camera-based document retriever [28] is designed with TTS technology to obtain the electronic versions of the documents stored in a database. Therefore, an article of a document is read out over a phone speaker if the content of the image captured matches a document in the database. This poses a limitation of reaching documents in the database only with no access to print existing in the external world. A currency reader [29] has been effectively designed, but its recognition ability is limited to U.S. currency only. Haritaoglu [30] and Nakajima et al. [31] have developed a mobile system using client–server architecture with a personal digital assistant (PDA) for dynamic translations of the text, but it is unsuitable, does not address the needs of people who are blind or visually impaired, and is not practical. The size and number of buttons and the PDA's touch screen make their use almost impossible. Some handheld assistants [32,33] designed using the same PDA technology have unusable user interfaces, no voice output, and are costly. A reading aid is provided by Optacon [34], an electromechanical device that converts the characters to vibrating tactile stimuli. This is introduced in replacement of Braille to read text but needs a lot of training. It is also costly, bulky, and for home use only. Applications, such as bar code readers [35] and business card readers [36], have been developed on mobile phones, but these fail to provide access to other printed text and have an unusable user interface. Products such as the KNFB Reader [37] and AdvantEdge Reader [38] have also been introduced into the market but are very expensive and use two or more linked machines to recognize text in mobile conditions. The Android mobile accessibility solution has the potential to be inexpensive and more sustainable than current accessibility solutions.

16.3 Application Usage

The reconfigured mobile Android phone (R-MAP) was built using an on-device OCR on the Android platform to provide reading services for people who are blind or visually impaired. It also takes advantage of the high camera resolution, the Android autofocus mechanism, and a built-in TTS engine. Using R-MAP, an image of the desired text is taken using a mobile Android phone and is then sent to the on-device OCR for text recognition. Finally, the result is passed to the TTS engine for a voice output. This can be applied to situations where the user needs to read text in different environments such as indoor and outdoor locations, complex backgrounds, different surfaces, and so on. This service is effective in terms of the user interface as the user cannot see the screen.

R-MAP can be used to help people who are blind or visually impaired in everyday activities such as reading menus in restaurants, helping them organize and arrange food items in the refrigerator/pantry by reading labels on them, identifying objects, reading signs on doors, making a distinction between men's and women's restrooms, exploring new indoor or outdoor locations by their names (for example, Engineering Building, Rose Theater, etc.), reading envelopes, CDs, books, magazines, newspapers, medicine bottles, expiration dates, warnings, packages, instructions, and so on. It will provide more independence to people who are blind or visually impaired in terms of usability in everyday activities.

Let us consider the scenario of Jack, a businessman who is visually impaired, taking a trip to the grocery store. Jack and Buddy, his Seeing Eye dog, walk to the store two blocks away, pulling a small cart behind. As Jack has been a patron at this store for years, he knows exactly how many steps to take to get to a particular aisle for the items he regularly needs. However, today he needs a speciality item. Here, he pulls out his Android mobile phone to take a picture of the aisle signs. Once the phone has stated what is on each aisle, he goes down the one that fits his needs. He continues to take pictures of items on the shelves and listens to the phone before he reaches for the item he needs. Just like any other customer, Jack wants to get his money's worth, so he takes a picture of the item to check its expiration date. After Jack has finished shopping, he starts back home. However, on the way back, a road is obstructed with an automobile wreck, so Jack must use his R-MAP to read the various street signs on the new route home. When he finally arrives home, he needs to carefully arrange his pantry to help his efficiency in the kitchen later. He uses his R-MAP to read the labels (Figure 16.1) of each item, so he knows exactly where to

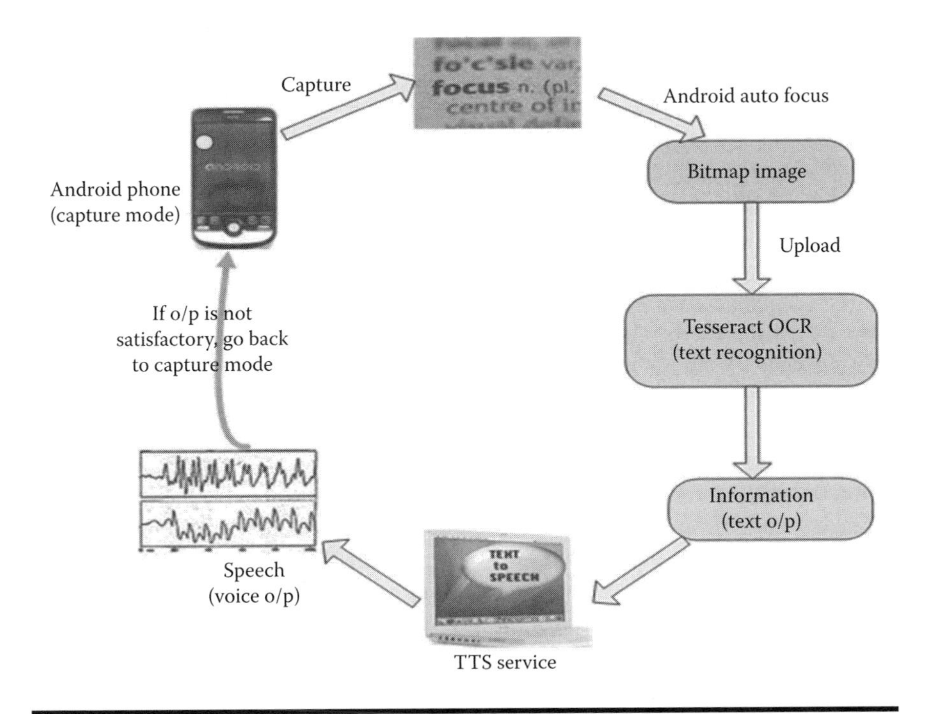

Figure 16.1 Architecture diagram.

put them for later use. A trip errand that before would have required help from someone was able to be performed independently by Jack through our application.

16.4 Architecture and User Interface

Mobile phones have lower processing power than desktops or notebooks. However, to provide users real-time response while interacting with a mobile phone, R-MAP is designed to minimize the processing time and user operation. Figure 16.1 shows the architectural diagram of R-MAP, where an Android mobile phone is used and all operations are done on-device and in real time. The application is developed on the Android developer HTC G1 phone, but it can be extended to any other platform with minimal effort.

When the application starts, the phone enters into capture mode. Android autofocus is used to focus the camera and take a clear image of the particular text. The image is sent to the Tesseract OCR engine in the form of a bitmap for text segmentation and recognition from where the information (text) is obtained. The text is then passed through the TTS engine, available in the mobile phone, to obtain the voice output. If the voice output tends to be unsatisfactory (i.e., the text recognition is not good) then the application can be started again from capture mode. The user interface has been designed for easy access to the application by minimizing user operation.

It can be a challenge for people who are blind or visually impaired to use the services provided on a mobile phone. A study [39] examined how people who are blind or visually impaired select, adapt, and use mobile devices. The simple user interface described below can be easily adapted and used in all day-to-day activities.

To operate the device, one needs to concentrate on the top right and bottom left positions of the touch screen, as shown in Figure 16.2. Icons are placed at these positions where it has to be clicked alternately. Assume that the left-hand thumb is used for icons on the bottom left and the right hand thumb for icons on the top right of the mobile phone. The home screen of the mobile phone is cleared and the icon of the application is placed on the top right where it has to be clicked by the right-hand thumb to start the application, as shown in Figure 16.2. Once the application starts to enter the capture mode, a click on the bottom left position (is made by the left-hand thumb. Another click, using the right-hand thumb on the top right position captures the image. The results are processed once the image is captured. It takes 5–20 s, depending on the amount of text and the lighting conditions. A final click on the bottom left position, using the left-hand thumb, gives the voice output. Hence, the entire user operation is designed to run the application on the basis of these two positions of the mobile phone which need to be clicked alternately.

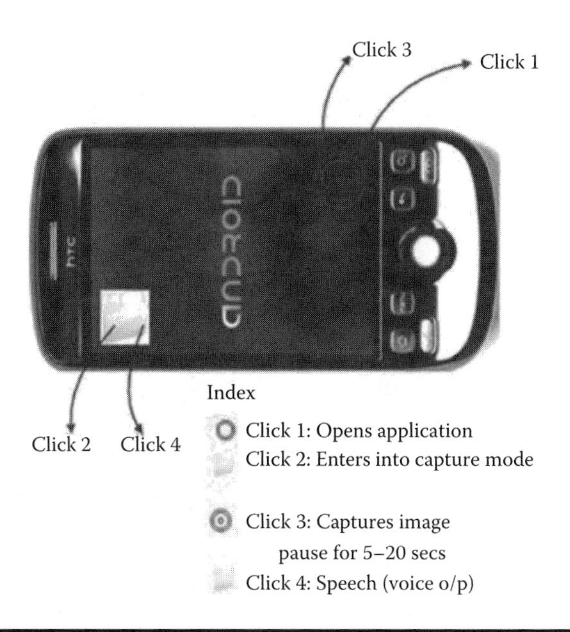

Figure 16.2 User interface.

16.5 Implementation

Taking a picture from a mobile phone camera may lead to various artifacts in the images (i.e., skew, blur, curved base lines, etc.), thereby causing the best available OCR to fail. In addition, as the system is enabled on-device, real-time response issues need to be considered critically. We developed R-MAP using Google's Android 1.6 version [40] of the platform in the Windows environment. The applications are written in the Java programming language. Android NDK [41] allows developers to implement parts of their applications using native code languages such as C and C++. It is used in conjunction with Android SDK and can provide benefits to applications by reusing an existing code and, in some cases, increasing the speed. We used Android NDK version 1.5. Various modules are available for the practical implementation of a completely integrated system.

Once the user initiates the application on the Android mobile phone, it asks the user to enter the capture mode. In this mode, the camera has a resolution of 3.2 mega pixels that satisfies the 300 dpi resolution requirement of the OCR engine and is provided with an autofocus mechanism. If the camera is not focused on the text properly, then it vibrates and the image is not captured. This way, we can overcome the issue of blur or focusing issues. As soon as the focus is acceptable, the image is captured and sent in compressed form to the OCR engine.

We have enabled an open-source Tesseract [42] OCR engine, version 2.03 from Google, on the Android mobile phone. Android NDK uses the existing C++ code of the Tesseract OCR engine with the help of a tutorial [43]. The text recognition experiments performed by UNLV [44] on Tesseract show over 95% of recognition accuracy, indicating it is the most efficient OCR engine. Currently, it can read only .tiff and .bmp images. The captured image is uploaded in the form of a bitmap to the on-device OCR engine. The OCR can provide skew correction for (−10, 10) degrees of rotation, thus saving the loss of image quality. It also handles curved baseline fitting and has characteristics such as noise reduction, color-based text detection, word spacing, chopping joined characters, and associating broken characters to cope with recognition of damaged images. Undergoing these processes, the OCR engine performs text segmentation and character recognition. The processing time from image uploading until obtaining the information in the form of text takes 5–20 s. The text is extracted and sent to the TTS engine for further processing.

TTS [45] needs to be enabled for the application to process the text extracted from the OCR engine to obtain the voice output. This engine is available in the Android mobile phone designed especially for people who are blind or visually impaired to access the applications. It can spell out words, read punctuation marks, and so on, with global prosodic parameters. It also has an adjustable speaking rate. In case the voice output is not satisfactory (i.e., the text is not recognized properly), the application can be started again from the capture mode.

Therefore, R-MAP serves some accessibility needs of people who are blind or visually impaired and is more effectively implemented than the current accessibility solutions.

16.6 Experiments and Evaluation

A number of experiments were conducted in an effort to evaluate the performance of the on-device OCR engine and the fully integrated R-MAP. The overall accuracy and speed of the on-device OCR engine were evaluated under various practical situations. These situations are very commonly encountered where text needs to be read out loud, such as indoor or outdoor locations, complex backgrounds, different surfaces, and images in various conditions (tilt, skew, lighting differences, etc.). We illustrate the text input to discuss performance metrics and, on the basis of the results obtained, performance analysis (Figures 16.3 through 16.7).

16.7 Test Corpora

R-MAP was applied to two test image corpora: a control corpus of four diverse black-and-white images, each one under four different conditions (indoor and outdoor lighting, skew, and tilt) and an experimental corpus of 50 color scene images to explain various situations (outdoor and indoor locations, different surfaces, and complex backgrounds). The test images were taken using the Google Android HTC G1.

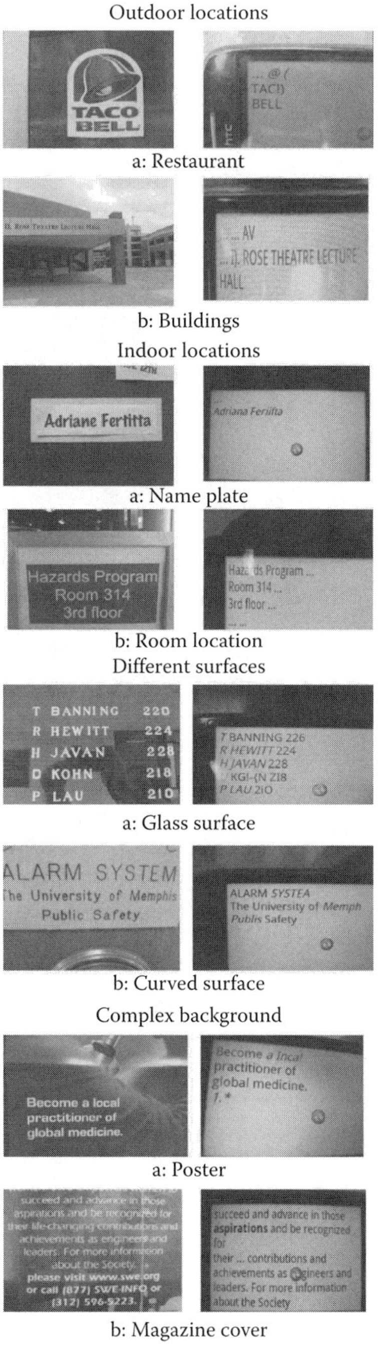

Outdoor locations

a: Restaurant

b: Buildings

Indoor locations

a: Name plate

b: Room location

Different surfaces

a: Glass surface

b: Curved surface

Complex background

a: Poster

b: Magazine cover

Figure 16.3 Samples of experimental corpus.

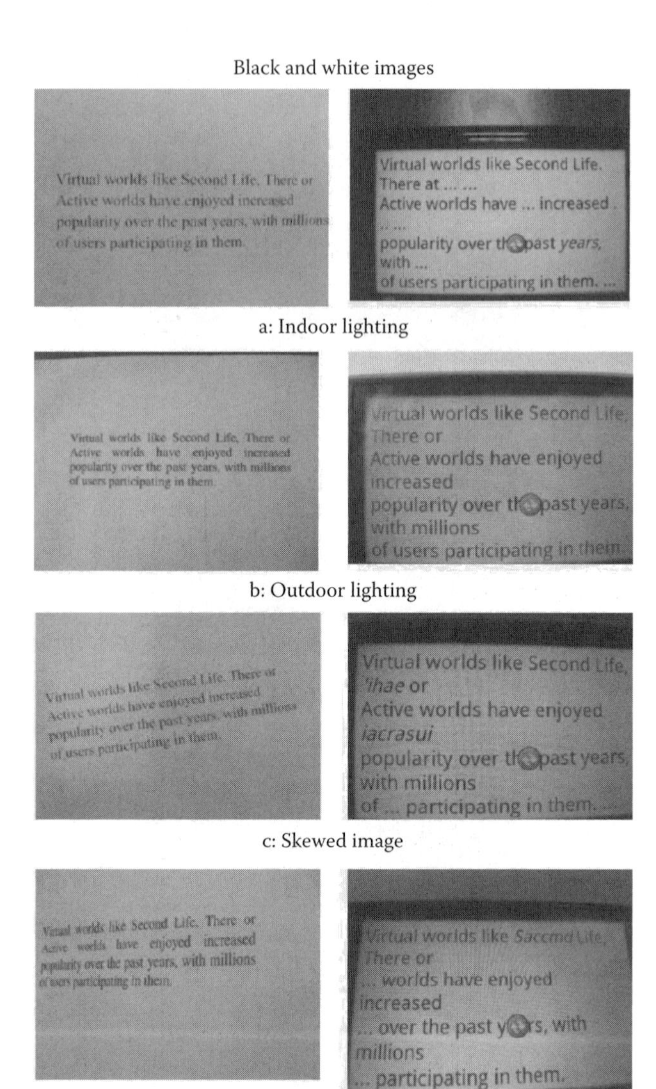

Figure 16.4 A sample of control corpus.

16.8 Performance Metrics

As the measure of accuracy of the OCR in various situations and conditions is our point of interest, we adopted two metrics proposed by the Information Science Research Institute (ISRI) at UNLV for the Fifth Annual Test of OCR Accuracy [46]. Those metrics are character accuracy and word accuracy.

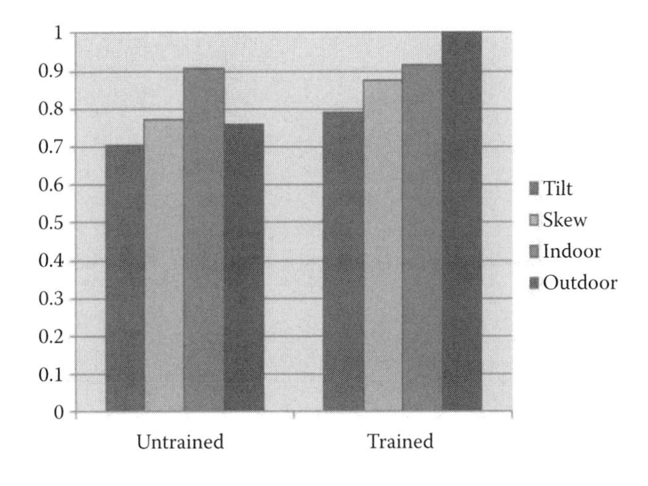

Figure 16.5 Word accuracy for untrained and trained optical character recognition.

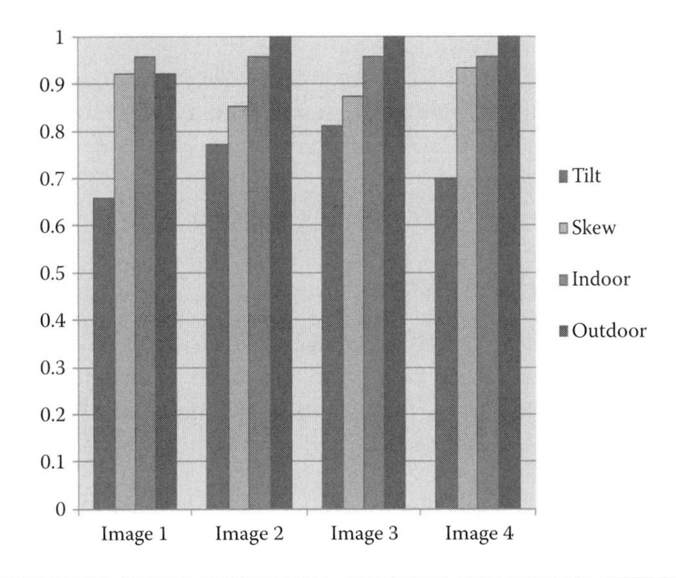

Figure 16.6 Word accuracy for control corpus.

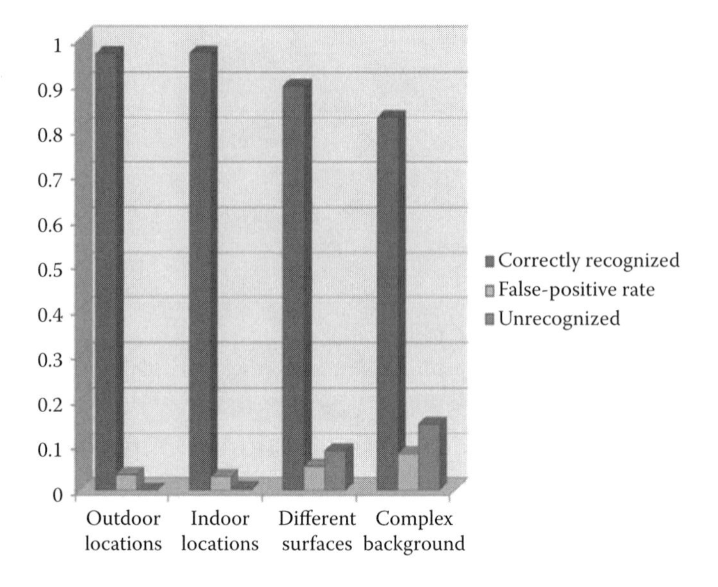

Figure 16.7 Character accuracy for experimental corpus.

16.8.1 Character Accuracy

This metric is defined as the ratio $r = (n - m)/n$, where n is the number of characters in the image and m is the number of errors resulting from the OCR engine.

16.8.2 Word Accuracy

A word is a sequence of one or more letters. The correct recognition of a word is more important than numbers or punctuation in text extraction. The word accuracy is given by the ratio m/n, where n is the total number of words in an image and m is the number of words recognized by the OCR engine.

16.9 Results and Analysis

In this section, we evaluate the OCR accuracy in the control corpus for different conditions and the experimental corpus for various situations on the basis of the performance metrics, and thereby the analysis of the experimental results to evaluate R-MAP.

16.9.1 Control Corpus

Let us consider four sets of diverse black-and-white images with each set taken in four different conditions. A sample of images in different conditions is shown in

Figure 16.4. These conditions normally arise when an image needs to be captured to read the text by people who are blind or visually impaired. As these images have sufficient amounts of text, we considered word accuracy over character accuracy. The word accuracy for these four images under each condition is shown in Figure 16.6.

An embedded camera in a mobile phone has far less lighting than a scanner. Binarization, a process to classify the image pixels, is undertaken by the OCR to solve this lighting issue. The experiments performed in dull lighting conditions gave poor results. Therefore, the experiments were performed under good indoor and outdoor lighting conditions. Word accuracy for images in good indoor lighting conditions was found to be 96% which was constant in all images and for the same set of images it had improved to 100% in good outdoor lighting conditions. This indicates that good lighting conditions improve OCR accuracy.

When the image is taken by a mobile phone camera, text lines may get skewed from their orientation. This results in poor OCR performance. The line finding algorithm in the OCR can recognize text without a need to deskew the image up to (−10, 10) degrees of rotation. Word accuracy for skewed images ranges from 85% to 90% under normal lighting conditions. Therefore, the image needs to be captured lowering the skew angle as much as possible to obtain good results.

Tilt, also known as the perception distortion, results when the text plane is not parallel to the image-capturing plane. It usually happens while capturing an image using a mobile phone camera. The effect is that the characters faraway look smaller and distorted. This degrades OCR performance. The simple solution can be the use of orientation sensors embedded in a mobile phone instead of applying image-processing techniques. Word accuracy for experiments conducted without orientation sensors range from 66% to 81% which can be improved by using an orientation sensor in the mobile phone.

Text recognition is possible on black-and-white images to a greater extent unless there are some issues with font, dull lighting conditions, and so on. The false positive rate (nontext recognized as text) was less than 2% in the case of indoor and outdoor lighting conditions, whereas, it was 11% for tilted and 5% for skewed images (percentage is based on the number of words). The unrecognized words for indoors was less than 3%, for tiled images it was 12%, for skewed it was 6%, but for outdoors it was less than 1%. These results prove R-MAP to be a promising application with little improvements to be made.

16.9.2 Training the OCR

We observed the text recognition results from Tesseract OCR, which produces, in some cases, bad text translations owing to previous text recognition failures. A possible solution could be to retrain the OCR engine as it benefits from the use of an adaptive classifier by capturing the images from where it gave good results, and then capture the image from where the results were bad to get good results. This is probably because of the OCR's use of polygonal approximation as input to the classifier instead of raw outlines.

To illustrate the training of the OCR, we performed an experiment (shown in Figure 16.5). The same experiment is performed twice under the same conditions (tilt, skew, indoor and outdoor lighting conditions) and its word accuracy is calculated. The results showed a 10% increase in word accuracy for tilt and skew conditions, with improvements in outdoor lighting conditions to 24%. In indoor lighting conditions, there was only a 2% increase, indicating the OCR is better trained for good indoor lighting conditions. This improvement in word accuracy indicates the OCR is trained every time with repeated capturing of the same image where partial results are obtained.

16.9.3 Experimental Corpus

The motivation behind these experiments was provided by the fact that OCR engines available fail miserably on anything but uniform text. Therefore, we captured the images of text located in indoor and outdoor locations, complex backgrounds such as on magazine covers and posters, different surfaces such as glass, curved surfaces, LCD (liquid crystal display) screens, and so on, to evaluate performance. The experiments were performed in each situation and character accuracy was calculated. A sample of images taken under these situations is shown in Figure 16.3. As the numbers of words in images are less, we targeted character accuracy rather than word accuracy, as shown in Figure 16.7.

In outdoor locations, text information is available in the form of names of buildings, restaurants, streets, and so on, where a user can capture an image and read the text. The experiments for a set of 10 images performed on the text available outdoors show an average character accuracy of 96.5%. The text available outdoors can pose problems such as reflection that gives no OCR output. However, good outdoor lighting conditions always help good performance of the OCR.

In indoor locations, text most commonly available includes that on name plates, male and female restrooms, room location, notice boards, and so on. In these cases, a set of 20 images was captured under good lighting conditions and their average character accuracy was measured. It was found to be equal to the average character accuracy of the text in outdoor locations, indicating that there was not much difference between the text indoor or outdoor unless good lighting conditions were satisfied.

The need to take images on curved surfaces, such as medicine bottles, glass surfaces such as doors, LCD screens, and so on, arises for people who are blind or visually impaired. To evaluate the performance of the R-MAP on these different surfaces, we took 10 images whose average character accuracy was found to be 89%. The experiments were also performed on a set of 10 images of complex backgrounds such as magazines, posters, and so on. The average character accuracy was found to be 83%. These experiments reveal that R-MAP is capable of reading text in these situations efficiently and is not just limited to uniform text.

In all these situations, there is a possibility of conditions such as tilt, skew, and so on, to arise but they are taken care of by the OCR as discussed in the case of the control corpus. The only difference with colored images is that it takes a longer processing time. The false-positive rate is more in complex backgrounds (8%) and different surfaces (5%) because of the consideration of nontext areas as text (percentage is based on the number of characters). The unrecognized text is also more in complex backgrounds (14%) and different surfaces (9%), indicating further analysis over these issues.

From the moment the application starts to the image captured, the runtime on-device corresponds to the completion of the whole process within 5–20 s, depending on the amount of text and lighting conditions. Therefore, the entire process meets the requirements of real-time processing.

16.10 Ergonomics

R-MAP has its own positive and negative attributes. Whereas the positive attributes include its ability to recognize text in various situations and under different conditions, the negative attributes are the limitations of a mobile phone and processing techniques required to overcome these limitations. Apart from these, care has been taken to satisfy the variety of reading needs of people who are blind or visually impaired. However, detailed user studies need to be carried out to resolve possible accessibility and usability issues. The presented challenges of hardware in mobile phone cameras, compared to expensive high-end cameras, are low resolution, motion blur, and lighting conditions arising. In some cases, print materials are quite long and cannot be captured with a single click because of the limited screen size of mobile phones. Also, OCR is not developed for camera use but for ideal pictures taken in scanners. These issues, in addition to finding text in a scene, the limitations to recognize text from handwritings, small or poor-quality print, and currencies, need to be investigated.

A good capture of the image is better than long processing to detect text, further improvement is needed for R-MAP to process more types of images (very complex backgrounds, dim lighting, highly curved, etc.) in order to be more robust. There is no literature for people who are blind or visually impaired using cameras. Therefore, issues related to capturing images need to be considered. People who are blind or visually impaired will be recruited to estimate the cognitive load using the R-MAP. The lessons learned from these studies can be extended to implement a more universal system design.

16.11 Summary

In this chapter, a fully integrated end-to-end application called R-MAP, developed for people who are blind or visually impaired, was discussed. R-MAP is able to provide mobile access to printed texts from different environments such as indoor, outdoor, complex background, different surfaces, and different orientations. A number of factors such as cost, learnability, portability, usability, and scalability were considered. An easy-to-use, intuitive interface is used to integrate R-MAP and is implemented using the Android platform that enables the streaming of captured images to the on-device OCR system and subsequently feeds that to the TTS to generate a voice output. OCR and TTS parameters were fine-tuned to make the application robust against various real and natural environments. R-MAP is a standalone application, built using the G1 Android phone, and requires no special hardware or Internet connection to external servers to provide this service. This application is easy to use and the interface is designed to minimize user operation. It is available in the English language but can be extended to other languages with minimum effort.

The Android mobile phone, with the on-device running application, can be reconfigured to develop more applications such as finding a reference in an open space, following a route map, and accessing current location with very high accuracy. Integration of such services will make the system more effective for people who are blind and visually impaired. This has the potential to ensure more opportunities in terms of education and employment. A detailed usability study to evaluate the system is currently in progress.

Acknowledgments

We would like to thank Chandra S. Kolli for helping with the experiments, and Trinity D. Owens and Iftekhar Anam for their help in editing. This research was partially funded by National Science Foundation USA Early Career Award (NSF-IIS-0746790). The opinions expressed in this research are the authors' view and has no bearing with the funding agency.

References

1. http://www.kurzweiledu.com/.
2. American Foundation for the Blind, Estimated number of adult Braille readers in the United States. *Journal of Visual Impairment and Blindness*, 90(3):287, 1996.
3. E. A. Taub. The blind lead the sighted: Technology for people with disabilities finds a broader market, *The New York Times*, October 1999.
4. Android, http://www.android.com/.
5. Android TTS, http://android-developers.blogspot.com/2009/09/introduction-to-text-to-speech-in.html.
6. Optical character recognition (OCR), http://en.wikipedia.org/wiki/optical character recognition.
7. ZoomText, http://www.compuaccess.com/.
8. Lunar, http://www.axistive.com/.

9. DecTalk, http://www.fonixspeech.com/.
10. Keynote GOLD, http://assistivetech.net/.
11. Microsoft SDK, http://msdn.microsoft.com/.
12. At&t Natural Voices, http://www.naturalvoices.att.com/.
13. HAL Screen Reader and Cicero, http://www.dolphinuk.co.uk/index dca.htm. http://www.synapseadaptive.com/dolphin/Default.htm.
14. JAWS, http://www.freedomscientific.com/.
15. Braille2000, http://www.braille2000.com/.
16. Duxbury, http://www.duxburysystems.com/.
17. Braille Wave, http://www.handytech.de/.
18. Braillex, http://www.indexbrailleaccessibility.com/products/papenmeier/elba.htm.
19. Braille Blazer, http://www.nanopac.com/.
20. ViewPlus, http://www.viewplus.com/.
21. OpenBook, http://www.openbookmn.org/.
22. Open source OCR resources, http://www.ocrgrid.org/ocrdev.html.
23. GOCR—A free optical character recognition program, http://jocr.sourceforge.net/.
24. OCRAD—The GNU OCR, http://www.gnu.org/software/ocrad/.
25. OCRopus—Open source document analysis and OCR system, http://code.google.com/p/ocropus/.
26. R. Smith. An overview of the Tesseract OCR engine. In *ICDAR '07: Proceedings of the Ninth International Conference on Document Analysis and Recognition*, Washington, DC, USA, IEEE Computer Society, pp. 629–633, 2007.
27. T. Portele and J. Kramer. Adapting a TTS system to a reading machine for the blind. In *Proceedings of the Fourth International Conference on Spoken Language (ICSLP '96)*, Vol. 1, pp. 184–187, October 1996.
28. X. Liu and D. Doermann. Mobile retriever—Finding document with a snapshot. In *Proceedings of CBDAR*, pp. 29–34, September 2007.
29. X. Liu and D. Doermann. A camera phone based currency reader for the visually impaired. In *Proceedings of the Tenth International ACM SIGACCESS Conference on Computers and Accessibility*, pp. 305–306, October 2008.
30. I. Haritaoglu. Infoscope: Link from real world to digital information space. In *UbiComp '01: Proceedings of the 3rd International Conference on Ubiquitous Computing*, pp. 247–255, Springer-Verlag, London, UK, 2001.
31. H. Nakajima, Y. Matsuo, M. Nagata, and K. Saito. Portable translator capable of recognizing characters on signboard and menu captured by built-in camera. In *ACL '05: Proceedings of the ACL 2005 on Interactive Poster and Demonstration Sessions*, pp. 61–64, Association for Computational Linguistics, Morristown, New Jersey, 2005.
32. Braillnote, http://www.pulsedata.com.
33. PacMate, http://www.freedomscientific.com/.
34. N. Efron. Optacon—A replacement for Braille? *Australian Journal of Optometry*, 60 (4): 118–129, 1977.
35. E. Ohbuchi, H. Hanaizumi, and L. A. Hock. Barcode readers using the camera device in mobile phones. In *Proceedings of the 2004 International Conference on Cyberworlds*, Washington, DC, USA, pp. 260–265, 2004.
36. X.-P. Luo, J. Li, and L.-X. Zhen. Design and implementation of a card reader based on build-in camera. In *Proceedings of the Pattern Recognition, 17th International Conference on (ICPR'04)*, Vol. 1, pp. 417–420, 2004.

37. KNFB Reader, http://www.knfbreader.com/.
38. AdvantEdge Reader, http://www.atechcenter.net/.
39. S. K. Kane, C. Jayant, J. O. Wobbrock, and R. E. Ladner. Freedom to roam: A study of mobile device adoption and accessibility for people with visual and motor disabilities. In *Assets '09: Proceedings of the 11th international ACM SIGACCESS Conference on Computers and Accessibility*, pp. 115–122, ACM, New York, NY, USA, 2009.
40. Android SDK, http://developer.android.com/.
41. http://android-developers.blogspot.com/.
42. http://code.google.com/p/tesseract-ocr/.
43. IT wizard, http://www.itwizard.ro/.
44. F. R. Jenkins, S. V. Rice and T. A. Nartker. The Fourth Annual Test of OCR Accuracy. *Technical Report*, Information Science Research Institute, University of Nevada, Las Vegas, April 1995.
45. TTS Stub, http://code.google.com/p/eyes-free/.
46. S. V. Rice, F. R. Jenkins, and T. A. Nartker. The Fifth Annual Test of OCR Accuracy. *Technical report*, Information Science Research Institute, University of Nevada, Las Vegas, 1996.

Index

A

J

K

L

M

N

O

P

S

T

X

Y

Z